Lecture Notes in Artificial Intelligence 4434

Edited by J. G. Carbonell and J. Siekmann

Subseries of Lecture Notes in Computer Science

Gerhard Lakemeyer Elizabeth Sklar
Domenico G. Sorrenti Tomoichi Takahashi (Eds.)

RoboCup 2006: Robot Soccer World Cup X

Springer

Series Editors

Jaime G. Carbonell, Carnegie Mellon University, Pittsburgh, PA, USA
Jörg Siekmann, University of Saarland, Saarbrücken, Germany

Volume Editors

Gerhard Lakemeyer
RWTH Aachen, Germany
E-mail: gerhard@cs.rwth-aachen.de

Elizabeth Sklar
City University of New York, USA
E-mail: sklar@sci.brooklyn.cuny.edu

Domenico G. Sorrenti
Università di Milano-Bicocca, Milan, Italy
E-mail: sorrenti@disco.unimib.it

Tomoichi Takahashi
Meijo University, Nagoya, Japan
E-mail: ttaka@ccmfs.meijo-u.ac.jp

Library of Congress Control Number: 2007931901

CR Subject Classification (1998): I.2, C.2.4, D.2.7, H.5, I.5.4, J.4

LNCS Sublibrary: SL 7 – Artificial Intelligence

ISSN 0302-9743
ISBN-10 3-540-74023-6 Springer Berlin Heidelberg New York
ISBN-13 978-3-540-74023-0 Springer Berlin Heidelberg New York

Springer is a part of Springer Science+Business Media

springer.com

Printed in Germany

Typesetting: Camera-ready by author, data conversion by Scientific Publishing Services, Chennai, India
Printed on acid-free paper SPIN: 12102426 06/3180 5 4 3 2 1 0

Preface

The 10th RoboCup International Symposium was held during June 19–20, 2006 at the Fair & Convention Center in Bremen, Germany, immediately after the 2006 Soccer, Rescue and Junior Competitions. RoboCup is increasingly seen by the robotics community as a significant approach to the evaluation of the effectiveness of the proposed solutions to the many difficult robotics problems.

The RoboCup International symposium hosted scientific contributions in all the areas relevant to RoboCup Competitions. The number of submissions to the Symposium increased again and totalled 143. Each paper was reviewed by at least three Program Committee members. The Program Committee included researchers involved in RoboCup and other scientists from outside the RoboCup community. Papers that received dissenting recommendations were discussed among the reviewers, moderated by the Co-chairs. The final decisions were made by the Co-chairs, who selected 22 submissions as full papers and 36 submissions as posters. This means an acceptance rate of less than 16% for full papers and less than 41% considering posters.

The symposium was run in single-track to allow coverage of all robotic-related topics by all attendees. We had five sessions for oral presentations and two poster sessions. We were also delighted to have two outstanding invited speakers. Hod Lipson (Cornell University, USA) spoke about his work on biologically inspired robotics in his talk "Biologically Inspired Robotics: From Evolving to Self-Reproducing Machines." Sebastian Thrun (Stanford University, USA) described how his team won the DARPA Grand Challenge in his talk "Winning the DARPA Grand Challenge."

The Symposium Co-chairs selected a few papers as nominees for the Best Paper Award and for the Best Student Paper Award. The RoboCup trustees made the final decision and selected "A 3D Simulator of Multiple Legged Robots Based on USARSim" by M. Zaratti, M. Fratarcangeli, L. Iocchi as Best Paper and "Half Field Offense in RoboCup Soccer: A Multiagent Reinforcement Learning Case Study" by S. Kalyanakrishnan, Y. Liu, P. Stone as Best Student Paper.

As the quality of the symposium depends heavily on the quality of the generous Program Committee members, we wish to particularly thank them for their work, which was very hard and concentrated in a very short time. We would also like to thank the Local Organizing Committee, headed by Ubbo Visser, for turning Robocup 2006 into such a memorable and enjoyable event. Last but not least we thank Stefan Schiffer for his help in preparing these proceedings.

Congratulations to the RoboCup Competitions and the RoboCup Symposium, which celebrated their tenth anniversary this year in Bremen!

December 2006

Gerhard Lakemeyer
Elizabeth Sklar
Domenico G. Sorrenti
Tomoichi Takahashi

Organization

Symposium Co-chairs

Gerhard Lakemeyer	RWTH Aachen, Germany
Elizabeth Sklar	City University of New York, USA
Domenico G. Sorrenti	Università di Milano - Bicocca, Italy
Tomoichi Takahashi	Meijo University, Japan

International Symposium Program Committee

Ahmed Tawfik, Canada
Alan Schultz, USA
Amy Eguchi, USA
Andrea Bonarini, Italy
Andreas Birk, Germany
Andreas Zell, Germany
Ansgar Bredenfeld, Germany
Bernhard Nebel, Germany
Brahim Chaib-draa, Canada
Brett Browning, USA
Carlos Cardeira, Portugal
Claude Sammut, Australia
Daniel Polani, UK
Daniele Nardi, Italy
David Jahshan, Australia
Dieter Fox, USA
Emanuele Menegatti, Italy
Emanuele Frontoni, Italy
Fernando Ribeiro, Portugal
Fiora Pirri, Italy
Francesco Amigoni, Italy
Franz Wotawa, Austria
Gal A. Kaminka, Israel
Gerd Mayer, Germany
Gerhard Kraetschmar, Germany
Giovanni Indiveri, Italy
Giuseppina Gini, Italy
Gordon Wyeth, Australia
H. Levent Akin, Turkey
Hod Lipson, USA
Holly Yanco, USA
Igor Verner, Israel
Ituki Noda, Japan
Jacky Baltes, Canada
Jeffrey Johnson, UK
Kamal Karlapalem, India
Kaspar Althoefer, UK
Luca Iocchi, Italy
Luis Almeida, Portugal
Luis Paulo Reis, Portugal
M. Bernardine Dias, USA
Maja Mataric, USA
Manuela Veloso, USA
Marcello Restelli, Italy
Maria Hybinette, USA
Marie desJardins, USA
Martijn Schut, The Netherlands
Martin Riedmiller, Germany
Mary-Anne Williams, Australia
Masayuki Ohta, Japan
Matteo Matteucci, Italy
Michael Beetz, Germany
Michael Bowling, Canada
Michele Folgheraiter, Italy
Mikhail Prokopenko, Australia
Milind Tambe, USA
Mohan Sridharan, USA
Nabil Ouerhani, Switzerland
Nobuhiro Ito, Japan
Pedro Lima, Portugal
Peta Wyeth, Australia
Peter Stone, USA

Pieter Jonker, The Netherlands
Raul Rojas, Germany
Riccardo Cassinis, Italy
Saeed Shiry, Iran
Sergio A. Velastin, UK
Sheila Tejada, USA
Simon Levy, USA
Simon Parsons, USA
Stefano Carpin, Germany
Steffen Gutmann, Japan
Tadashi Naruse, Japan
Tairo Nomura, Japan
Testuya Kimura, Japan
Tomoharu Nakashima, Japan
Tucker Balch, USA
Vincenzo Caglioti, Italy
Yashutake Takahashi, Japan
Yoshitaka Kuwata, Japan

Additional Reviewers

Brenna Argall
Alejandra Barrera
Sonia Chernova
Antonio D'angelo
Tomas de Boer
Enrique Munoz de Cote
Ugo Di Profio
Alessandro Farinelli
Alexander Ferrein
Michele Folgheraiter
Daniel Goehring
Giorgio Grisetti
Yang Gu
Manfred Hild
Matthias Juengel
Kemal Kaplan
Hatice Kose-Bagci
Alessandro Lazaric
Luca Marchetti
Colin McMillen
Cetin Mericli
Utku Tatlidede
Vittorio Amos Ziparo

Table of Contents

Full Papers

Simulation and Control

Learning

Learning, MAS, and Education

Vision

Vision and Localization

Posters

Session 1

Session 2

Session 3

Bridging the Gap Between Simulation and Reality in Urban Search and Rescue

Stefano Carpin[1], Mike Lewis[2], Jijun Wang[2], Steve Balakirsky[3], and Chris Scrapper[3]

[1] School of Engineering and Science
International University Bremen – Germany
[2] Department of Information Sciences and Telecommunications
University of Pittsburgh – USA
[3] Intelligent Systems Division
National Institute of Standards and Technology – USA

Abstract. Research efforts in urban search and rescue grew tremendously in recent years. In this paper we illustrate a simulation software that aims to be the meeting point between the communities of researchers involved in robotics and multi-agent systems. The proposed system allows the realistic modeling of robots, sensors and actuators, as well as complex unstructured dynamic environments. Multiple heterogeneous agents can be concurrently spawned inside the environment. We explain how different sensors and actuators have been added to the system and show how a seamless migration of code between real and simulated robots is possible. Quantitative results supporting the validation of simulation accuracy are also presented.

1 Introduction

Urban search and rescue (USAR) can be depicted as the research field that experienced the most vigorous development in recent years within the robotics community. It offers a unique combination of engineering and scientific challenges in a socially relevant application domain [5]. The broad spectrum of relevant topics attracts the attention of a wide group of researchers, with expertise as diverse as advanced locomotion systems, sensor fusion, cooperative multiagent planning, human-robot interfaces and more. In this framework, the contest schema adopted by the RoboCup Rescue community, with the distinction between the real robots competition and the simulation competition, captures the two extremes of this growing community. Looking back at the past RoboCup events, tremendous progresses in short time characterized both communities. In 2002 the real rescue robots competition was described as a competition where teleoperated robots were mainly used because of the complexity of the problem [3]. In the simulation competition, emphasis was instead on the inter-agent communication models adopted [9]. The huge gap between these two extremes is evident. Only two years later [6], the real robot competition saw the advent of teams with three dimensional mapping software, intelligent perception, and the

G. Lakemeyer et al. (Eds.): RoboCup 2006, LNAI 4434, pp. 1–12, 2007.

first team with a fully autonomous multi-robot system. Within the simulation competition, teams exhibited cooperative behaviors, special agent programming languages and learning components. With these premises, it is evident that soon a mutual migration of relevant techniques will materialize. Nevertheless, certain logistic obstacles still prevent a seamless and profitable percolation of ideas and knowledge. Having set the scene, in this paper we present the latest developments of a simulation environment, called USARsim, that naturally plays the role of an in-between research tool where multi-agent and multi-robot systems can be studied in a artificial environment offering experimental conditions comparable to reality. After a demo stage during Robocup 2005 in Osaka, USARsim has been selected as the software infrastructure underlying the Virtual Robots competition, that was approved as the third competition within the RoboCup rescue simulation framework. In addition, we also offer an overview of the MOAST API, a component based software framework that can be used to quickly prototype control software, both in reality and on top of USARsim. Finally, we provide results supporting a quantitative evaluation of the simulator fidelity.

2 Software Structure

The current version of USARsim is based on the UnrealEngine2 game engine released by Epic Games with Unreal Tournament 2004. The simulation is written as a combination of levels, describing the 3-D layout of the arenas and modifications, and scripts redefining the simulations behavior. The engine to run the simulation can be inexpensively obtained by buying the game. The Unreal Engine provides a sophisticated graphical development environment and a variety of specialized tools. The engine includes modules handling input, output (3D rendering, 2D drawing, sound), networking and physics and dynamics. The games level defines a 3-D environment in much the same way as VRML (virtual reality markup language) and may use many of the same tools. The game code handles most of the basic mechanics of simulation including simple physics. Multiplayer games use a client-server architecture in which the server maintains the reference state of the simulation while clients perform the complex graphics computations needed to display their individual views. USARsim uses this feature to provide controllable camera views and the ability to control multiple robots. Unreal Tournament has two types of entities, human players who run individual copies of the game and connect to the server (typically running on the first players machine), and "bots" (short for robots), simulated players running simple reactive programs. Gamebots, a modification to the Unreal Tournament game that allows bots to be controlled through a normal TCP/IP socket [1], provides a protocol for interacting with Unreal Tournament. Because the full range of bot commands and Unreal scripts can be accessed over this connection GameBots provides a more powerful and flexible entry into the simulation than the player interface. The GameBot interface is ideal for simulating USAR robots because it can both access bot commands such as Trace to simulate sensors and exert complicated forms of control such as adjusting motor torques to control a simulated robot. One of the client options, the spectate mode, allows the clients viewpoint (camera location and orientation from which the simulation is viewed) to be attached to any other player including "bots". By combining a bot controlled by GameBots with a spectator client we can simulate a robot with access to both simulated sensor data through the bot and a simulated video feed through the spectating client. By controlling the simulated robot indirectly through GameBots rather than as a normal client we gain the additional advantage of being able to simulate an autonomous robot (controlled by a program) a teleoperated robot (controlled by user input) or any level of automation in between.

3 Robot Interfaces

An intelligent system must translate a mission command into actuator voltages. While this may be done in a simple monolithic module, USARsim/MOAST implements a hierarchical control structure that compartmentalizes the control system responsibility and domain knowledge necessary to create each controller. The knowledge and control requirements of a typical robotic platform may be decomposed into the two broad areas of sensing and behavior generation. In turn, behavior generation may be decomposed into mobility behaviors and mission package behaviors. In this decomposition, mobility refers to the control aspects of the vehicle that relate only to the vehicle's motion (e.g. drive wheel velocities), sensing refers to systems that acquire information from the world (e.g. cameras), and mission packages are controllable items on the platform that are not related to mobility (e.g. camera pan/tilt or robotic arm). It is the authors' belief that decomposing a system in this way allows for the creation of a generic internal representation and control interface that is able to fully control most aspects of robotic platforms. USARsim is designed to implement this decomposition and provides developers with a modular interface into the low-level simulated hardware of the robotic platform. It provides for component discovery, and independent control of mobility, sensors, and mission packages. Coupling USARsim to the Mobility Open Architecture Simulation and Tools (MOAST) framework adds modularity in time by providing a set of hierarchical interfaces into these components. Two different physical control interfaces exist into the system. The first allows low-level control into USARsim and is based on sending ASCII text over a TCP/IP socket. Higher-level commands and status utilize the Neutral Message Language (NML) [8] that permits a physical interface of various types of sockets as well as serial lines.

3.1 USARsim Socket API

During the development of the interface to USARsim many factors were taken into account to ensure that the interface was both well-defined and standardized. Scientific standards and conventions for units, coordinate systems, and interfaces were used whenever possible. USARsim decouples the units of measurement used inside Unreal by ensuring that all units meet the International System of Units

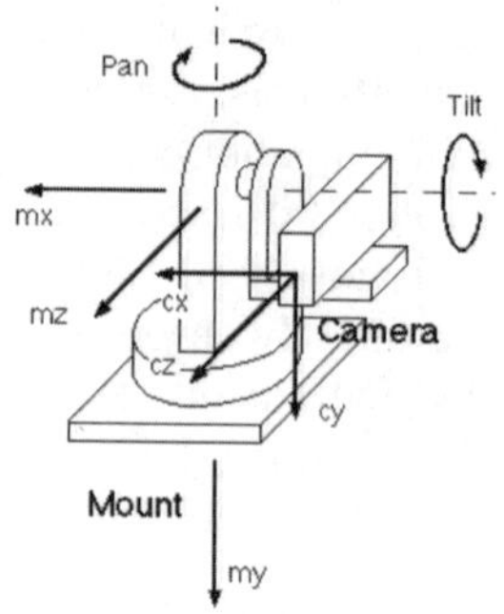

Fig. 1. Depiction of the Mission Package and the Sensor and their corresponding coordinate systems

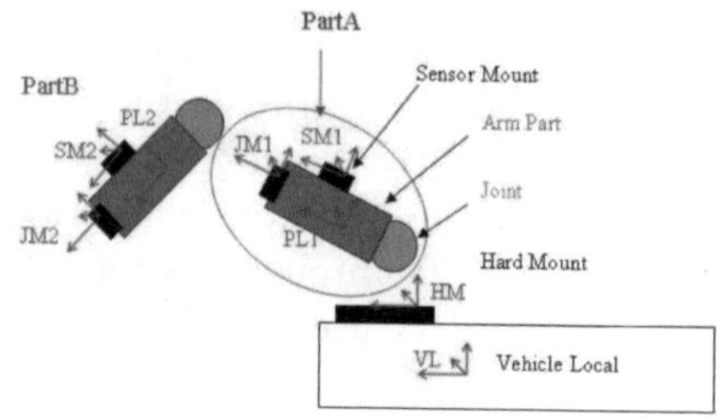

Fig. 2. Internal Representation of an Robotic Arm

(SI) standard conventions. SI Units are a National Institute of Standards and Technology (NIST) developed convention that is built on the modern metric system, and is recognized internationally. The coordinate systems for various components must be consistent, standardized, and anchored in the global coordinate system, as illustrated in Figure 1. USARsim leverages the previous efforts of the Society of Automotive Engineers, who published a set of standards for vehicle dynamics called SAE J670: Vehicle Dynamic Terminology. This set of standards is recognized as the American National Standard for vehicle dynamics and contains a comprehensive set of standards that describes vehicle dynamics through illustrated pictures of coordinate systems, definitions, and formal mathematical representations of the dynamics. Finally, the messaging protocol, including the primitives, syntax, and the semantics must be defined for the interface. Messaging protocols are used in USARsim to insure that infrequent and vital messages are received. The primitives, syntax, and semantics define the means in which a system may effectively communicate with USARsim, namely to speak USARsim's language. There are three basic components that exist currently in USARsim: robots, sensors, and mission package. For each class of objects there are defined class-conditional messages that enable a user to query the component's geography and configuration, send commands to, and receive status back. This enables the embodied agent controlling the virtual robot to be self-aware and maintain a closed-loop controller on actuators and sensors. The formulation of these messages are based on an underlying representation of the object, includes their coordinate system, composition of parts, and capabilities. This highlights a critical aspect underlying the entire interface; the representation of the components and how to control those components. For example, take a robotic arm, whose internal representation of an arm is visualized in Figure 2. In order for there to be a complete and closed representation of this robotic arm, the following aspects are defined as individual class conditional messages that are sent over the USARsim socket.

Configuration: How to represent the components and the assembly with respect to each other.

Geography: How to represent the pose of the sensor mounts and joints mount with respect to the part, and the pose of the part with respect its parent part.

Commands: How to represent the movements of each of the joints, either in terms of position and orientation or velocity vectors.

Status: How to represent the current state of the robotic arm.

3.2 Simulation Interface Middleware (*SIMware*)

Residing between USARSim and MOAST is the SIMware layer. This layer provides a modular environment and allows for a gradient of configurations from the purely virtual world to the real world. SIMware is designed to enable MOAST to conect to interfaces or APIs for real or virtual vehicles. It seemlessly connect to platforms with different messaging protocols, semantics, or different levels of abstraction. SIMware is made up of three basic components: a core, knowledge repository, and skins. The core of SIMware is essentially a set of state tables and interfaces that enables SIMware to administer the transference of data between two different interfaces. This transference is enabled through the use of knowledge repositories that provide insite into the target platform's capibilities and abstraction. The skins are an interface specific parsing utility that utilize the knowlege repository in order to enable the core to translate incoming and outgoing message traffic to meet the appropriate level of abstraction for the target interface.

3.3 MOAST API

The MOAST framework connects into USARsim via SIMware and provides additional capabilities for the system. These capabilities are encapsulated in components that are designed based on the hierarchical 4-D/RCS Reference Model Architecture [2]. The 4-D/RCS hierarchy is designed so that as one moves up the hierarchy, the scope of responsibility and knowledge increases, and the resolution of this knowledge and responsibility decreases. Each echelon (or level) of the 4-D/RCS architecture performs the same general type of functions: sensory processing (SP), world modeling (WM), value judgment (VJ), and behavior generation (BG). Sensory processing is responsible for populating the world model with relevant facts. These facts are based on both raw sensor data, and the results of previous SP (in the form of partial results or predictions of future results). The world model must store this information, information about the system self, and general world knowledge and rules. Furthermore, it must provide a means of interpreting and accessing this data. Behavior generation computes possible courses of action to take based on the knowledge in the WM, the systems goals, and the results of plan simulations. Value judgment aids in the BG process by providing a cost/benefit ratio for possible actions and world states.

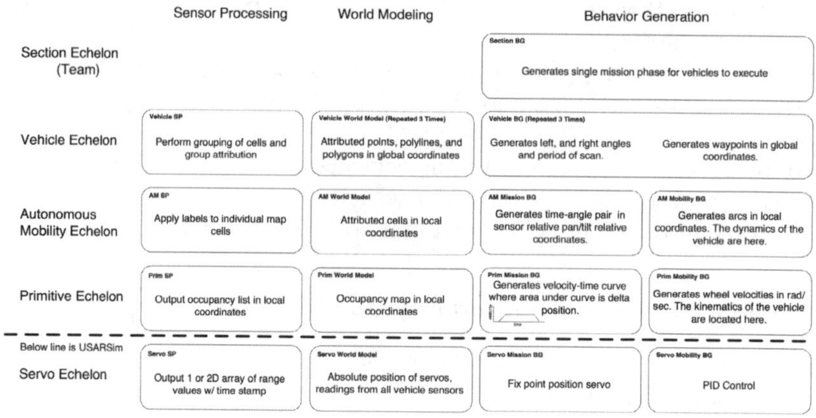

Fig. 3. Modular Decomposition of MOAST framework that provides modularity in broad task scope and time

The regularity of the architectural structure in 4-D/RCS enables scaling to any arbitrary size or level of complexity. Each echelon within 4-D/RCS has a characteristic range and resolution in space and time. Each echelon has characteristic tasks and plans, knowledge requirements, values, and rules for decision-making. Every component in each echelon has a limited span of control, a limited number of tasks to perform, a limited number of resources to manage, a limited number of skills to master, a limited planning horizon, and a limited amount of detail with which to cope.

This decomposition is depicted in Figure 3. Under this decomposition, the USARsim API may be seen as fulfilling the role of the servo echelon, where both the mobility and mission control components fall under BG. The sensors are able to output arrays of values, world model information about the vehicle self is delivered, and mission package and mobility control are possible. The remainder of this section will concentrate on the functioning and interfaces of the remaining echelons of the hierarchy.

Primitive Echelon. The primitive echelon behavior generation is in charge of translating constant curvature arcs or position constraints for vehicle systems into velocity profiles for individual component actuators based on vehicle kinematics. For example, the AM Mobility BG will send a dynamically correct constant curvature arc for the vehicle to traverse. This trajectory will contain both position and velocity information for the vehicle as a whole. For a skid steered vehicle, the Primitive Echelon BG plans individual wheel velocities based on the vehicle's kinematics that will cause to vehicle to follow the commanded trajectory. During the trajectory execution, BG will read vehicle state information from the Servo Echelon WM to assure that the trajectory is being maintained and will take corrective action if it is not. Failure to maintain the trajectory within the commanded tolerance will cause BG to send an error status to the AM Mobility BG.

The Primitive Echelon SP is in charge of converting sensor reports from sensor local coordinates to vehicle local coordinates. This information is read by the world model process which performs spatial-temporal averaging to create an occupancy map of the environment in vehicle local coordinates. This map is of fixed size and is centered on the current vehicle location. As the vehicle moves, distant objects fall off of the map. Future enhancements will allow for the population of newly added map area with any information that may be stored in the larger extents AM WM.

Autonomous Mobility Echelon. The Autonomous Mobility Echelon behavior generation is in charge of translating commanded way-points for vehicle systems into dynamically feasible trajectories. For example, the Vehicle Echelon mission controller may command a pan/tilt platform to scan between two absolute coordinate angles (e.g. due north and due east) with a given period. BG must take into account the vehicle motion and feasible pan/tilt acceleration/deceleration curves in order to generate velocity profiles for the unit to meet the commanded objectives. BG modules at this level may take advantage of all of the world model services provided o the Primitive Echelon in addition to the occupancy maps that have are maintained by the Primitive Echelon WM.

SP at this level extracts environmental attributes and in conjunction with WM labels the previously generated occupancy map with these attributes. Examples of attributes include terrain slope, and vegetation density.

Vehicle Echelon. The Vehicle Echelon behavior generation is in charge of accepting a mission for an individual vehicle to accomplish and decomposing this mission into commands for the vehicle subsystems. Coordinated way-points in global coordinates are then created for the vehicle systems to follow. This level must balance possibly conflicting objectives in order to determine these way-points. For example, the Section Echelon mobility BG may command the vehicle to arrive safely at a particular location by a certain time while searching for victims of an earthquake. The Vehicle Echelon mobility BG must plan a path that maximizes the chances of meeting the time schedule while minimizing the chance of an accident; and the Vehicle Echelon mission BG must plan a camera pan/tilt schedule that maximizes obstacle detection and victim detection. Both of these planning missions may present conflicting objectives.

SP at this level works on grouping cells from the AM WM into attributed points, lines, and polygons. These features are stored in a WM knowledge-base that supports SQL based spatial queries.

Section (Team) Echelon and Above. The highest level that has currently been implemented under the MOAST framework is the Section or Team Echelon. This level of BG has the responsibility of taking high-level tasks and decomposing them into tasks for multiple vehicles. For example, the Section Echelon mobility may plan cooperative routes for two vehicle to take in order to explore a building. This level must take into account individual vehicle competencies in order to create effective team arrangements. Higher echelon responsibilities would include such items as planning for groups of vehicles. An example of this would be commanding Section 1 to explore the first floor of a building and Section 2 to explore the second floor. Based on the individual teams performance, responsibilities may have to be adjusted or reassigned.

4 Validation

The usefulness of a simulation such as USARsim as a research tool is strongly dependent on the degree to which it has been validated and the availability of validation data for use in choosing models and assessing the generalizability of results. The provision of common and standard tools allows researchers to compare results, share software and advances, and collaborate in ways that would be impossible otherwise. While many of these benefits accrue simply from standardization, others require a closer correspondence between simulation and reality. While a human-robot interaction (HRI) experiment may not demand full realism in the behavior of a PID controller, replicating constraints such as a narrow field of view and invisibility of obstacles obstructing wheels may be essential to achieving results relevant to the operation of actual robots. Researchers wishing to port code developed in simulation to a real robot by contrast may need the highest fidelity model of the control system attainable to get useful results. In validating USARsim we are attempting to measure correspondences as precisely as possible so they also may serve for lower fidelity uses and where this is not possible identify those areas in which only low fidelity results are available.

A comparison of feature extraction for the Orange Arena using a laser range finder (Hokuyo PB9-11) on an experimental robot and its simulation in USARsim was reported already in [4]. The mapped areas along with their Hough transforms were practically identical and adjustable parameters tuned using the simulation did not require change when moved to the real robot. We have since conducted validation studies investigating HRI for the Personal Explorer Rover (PER) [7] and the Pioneer, P2/P3-DX. Some of these results for the PER were reported in [10]. This HRI validation testing was conducted at Carnegie Mellons replica of the NIST Orange Arena using both point-to-point and teleoperation control modes for the PER and teleoperation only for the pioneer P2-AT (simulation)/P3-AT (robot). In this study driving performance was observed for different surfaces and simple and complex courses using point-to-point or teleoperation control modes. Participants controlled the real and simulated robots from a personal computer located in the Usability laboratory at the School of Information Sciences, University of Pittsburgh. For simulation trials the simulation of the Orange Arena ran on an adjacent PC. For the real robotic control trials the participants controlled robots over the Internet in a replica of the Orange Arena in the basement of Newell Simon Hall at Carnegie Mellon University (see figure 4). Measures such as the distance from the stopping point to the target cone were collected for both the physical arena and the simulation. A standard interface developed for RoboCup USAR competition [7] was used under all conditions. Participants in the direct control mode controlled the robots

Fig. 4. On the left side the orange arena at CMU. On the right side the simulated arena within USARsim. The yellow cone to be reached can be observed in both images.

using a joystick. Both robots were skid steered so forward backward movements of the joystick led to movement while right/left movements produced changes in yaw. In the waypoint control mode participants selected waypoints by clicking on locations on the video display. This input was interpreted by the control software as specifying a direction and duration of travel. Manual adjustments in the point-to-point condition were made using the cursor keys.

Procedure. In Stage 1 testing of the PER and Pioneer (direct control mode) we established times, distances, and errors associated with movements over a wood floor, paper, and lava rocks. These data were used to adjust the speed of the simulated PER and Pioneer and alter the performance of the simulated PER when moving over scattered papers. In Stage 2 testing, PER robots were repeatedly run along a narrow corridor with varying types of debris (wood floor, scattered papers, lava rocks) while the sequence, timing and magnitude of commands were recorded. Participants were assigned to maneuver the robot with either direct teleoperation or waypoint (specified distance) modes of control. There were five participants in each of the PER groups (real-direct, real-waypoint, simulation-direct, simulation-waypoint) and four in the Pioneer (real-direct, simulation-direct) groups. In the initial three exposures to each environment, participants had to drive approximately three-meters, along an unobstructed path to an orange traffic cone. In later trials, obstacles were added to the environments, forcing the driver to negotiate at least three turns to reach the destination. The distances from stopping position to the goal and task times were recorded for both simulated and real trials. A time-stamped log of control actions and durations were collected for both real and simulated robots.

Terrain effects. The paper surface had little effect on either robots operation. The rocky surface by contrast had a considerable impact, including a loss of trac-

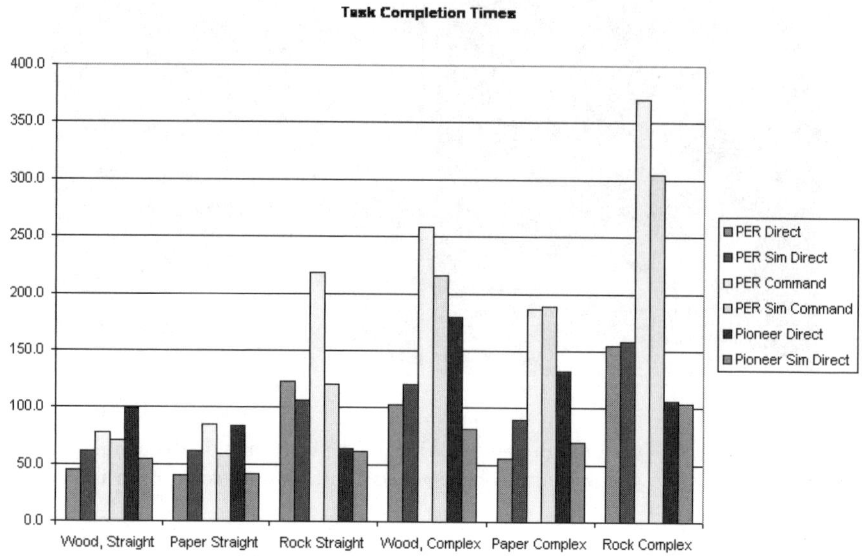

Fig. 5. Distribution of the times to complete the mission

tion and deflection of the robot. This was reflected by increases in the odometry and number of turn commands issued by the operators even for the straight course. A parallel spike in these metrics is recorded in the simulator data. As expected the complex course also led to more turning even on the wood floor.

Figure 5 shows these data for the simulated and actual PER and Pioneer.

Proximity. One metric on which the PER simulation and the physical robot consistently differed was the proximity to the cone acquired by the operator. Participants were given the instruction to get as close to the cone as possible without touching it. Operators using the physical robot reliably moved the robot to within 35cm from the cone, while the USARsim operators were usually closer to 80cm from the cone. It is unlikely that the simulation would have elicited more caution from the operators, so this result suggests that there could be a systematic distortion in depth perception, situation awareness, or strategy. In both cases the cone filled the cameras view at the end of the task. Alternatively, the actual PER was equipped with a safeguard to prevent running into objects while the simulated PER was not. Although this feature was not included in the instructions participants may have discovered it in controlling the robot and adopted a strategy of simply driving until the robot stopped. Figure 6 shows the distribution of these stopping distances. Another issue addressable from these data is the extent to which similarities in performance are a function of the platforms being simulated or differences between the simulation and control of real robots. Figure 5 suggests that both influences are present. As with our other data there are clear differences associated with platform and control mode. Note for instance the consistently shorter completion times shown in figure 5 for both

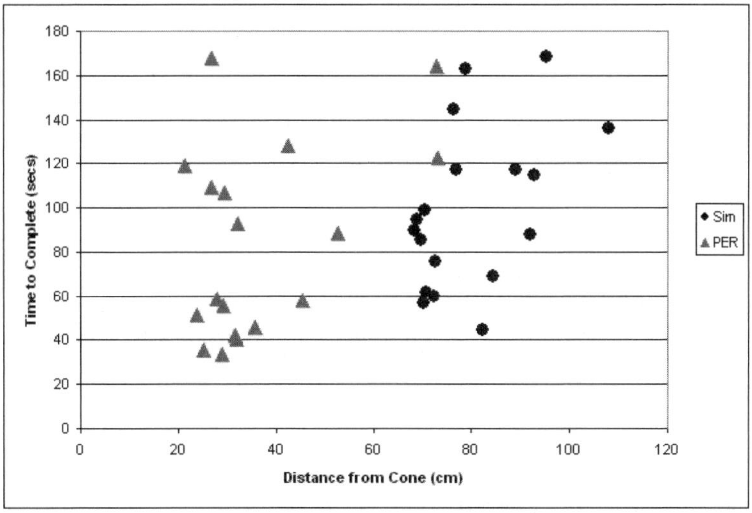

Fig. 6. Distribution of the stopping distances of the PER robot from the cone

actual and simulated Pioneers. Idle times, however, were much closer between the simulated PER and Pioneer than between the simulations and the simulated platforms. These substantially longer pauses between actions in controlling the real robot occurred despite matching frame rates although slight differences in response lag may have played a factor. Despite the difference in length of pauses completion times remain very close between the robot and the simulation. The average number of commands were also very similar between the simulation and the PER for control mode and environment except for straight travel over rocks in command mode where PER participants issued more than twice as many commands as those in the simulation or direct operation modes. A similar pattern occurs for forward distance traveled with close performance between simulation and PER for all conditions but straight travel over rocks, only now it is the teleoperated simulation that is higher.

5 Conclusions

In this paper we have presented the latest developments concerning the USARsim simulation environment, a natural candidate to meet the demands of researchers involved both in simulation and with real robots. Initial validation results show an appealing correspondence between experiences gained with USARsim and the corresponding real robots. Further benefits from USARsim can be obtained using MOAST, a framework that aids the development of autonomous robots. USARsim software and MOAST can be obtained for free from sourceforge.net/projects/usarsim and souceforge.net/projects/moast, respectively. As our library of models and validation data expands we hope to begin incorporating more rugged and realistic robots, tasks and environments. Accurate modeling tracked robots which will be made possible by the release of UnrealEngine3 would be a major step in this direction. The open source model adopted for the development of these software foster the active involvement of multiple developers and already gained quite some popularity.

References

1. Kaminka, G., Veloso, M., Schaffer, S., Sollitto, C., Adobbati, R., Marshall, A., Scholer, A., Tejada, S.: GameBots: A flexible test bed for multiagent team research. Comm. of the Association for Computing Machinery 45(1), 43–45 (2002)
2. Albus, J.: 4-D/RCS Reference Model Architecture for Unmanned Ground Vehicles. In: Proc. IEEE Int. Conf. on Robotics and Automation, pp. 3260–3265 (2000)
3. Asada, M., Kaminka, G.A.: An overview of RoboCup 2002 Fukuoka/Busan. In: Kaminka, G.A., Lima, P.U., Rojas, R. (eds.) RoboCup 2002. LNCS (LNAI), vol. 2752, pp. 1–7. Springer, Heidelberg (2003)
4. Carpin, S., Wang, J., Lewis, M., Birk, A., Jacoff, A.: High fidelity tools for rescue robotics: Results and perspectives. In: Robocup 2005 Symposium (2005)
5. Kitano, H., Tadokoro, S.: Robocup rescue: a grand challenge for multiagent and intelligent systems. AI Magazine (1), 39–52 (2001)
6. Lima, P., Custódio, L.: RoboCup 2004 Overview. In: Nardi, D., Riedmiller, M., Sammut, C., Santos-Victor, J. (eds.) RoboCup 2004. LNCS (LNAI), vol. 3276, pp. 1–17. Springer, Heidelberg (2005)
7. Nourbakhsh, I., Sycara, K., Koes, M., Young, M., Lewis, M., Burion, S.: Human-Robot Teaming for Search and Rescue, IEEE Pervasive Computing, pp. 72–78 (2005)
8. Shackleford, W.P., Proctor, F.M., Michaloski, J.L.: The Neutral Message Language: A model and Method for Message Passing in Heterogeneous Environments. In: Proceedings of the 2000 World Automation Conference (2000)
9. Tomoiki, T.: RoboCupRescue Simulation league. In: Kaminka, G.A., Lima, P.U., Rojas, R. (eds.) RoboCup 2002. LNCS (LNAI), vol. 2752, pp. 477–481. Springer, Heidelberg (2003)
10. Wang, J., Lewis, M., Hughes, S., Koes, M., Carpin, S.: Validating USARsim for use in HRI Research. In: Proceedings of the Human Factors and Ergonomics Society 49th Annual Meeting, pp. 457–461 (2005)

A 3D Simulator of Multiple Legged Robots Based on USARSim

Marco Zaratti, Marco Fratarcangeli, and Luca Iocchi

Dipartimento di Informatica e Sistemistica
Università "La Sapienza", Rome, Italy
Via Salaria 113 00198 Rome Italy
lastname@dis.uniroma1.it

Abstract. This paper presents a flexible 3D simulator able to reproduce the appearance and the dynamics of generic legged robots and objects in the environment at full frame rate (30 frames per second). Such a simulator extends and improves USARSim (Urban Search and Rescue Simulator), a robot simulator in turn based on the game platform Unreal Engine. This latter provides facilities for good quality rendering, physics simulation, networking, highly versatile scripting language and a powerful visual editor. Our simulator extends USARSim features by allowing for the simulation and control of legged robots and it introduces a multi-view functionality for multi-robot support. We successfully tested the simulator capabilities by mimicking a virtual environment with up to five network-controlled legged robots, like AIBO ERS-7 and QRIO.

1 Introduction

Robotic simulation is very important in developing robotics applications, both for rapid prototyping of applications, behaviors, scenarios, and for debugging purposes of many high-level tasks. Robot simulators have been always used in developing complex applications, and the choice of a simulator depends on the specific tasks we are interested in simulating. Moreover, simulators are also very important for robotic education: in fact, they are powerful teaching tools, allowing students to develop and experiment typical robotic tasks at home, without requiring them to use a real robot.

2D simulators are widely used to evaluate the behaviors of robotic applications, they are very effective for many kinds of robots and applications, and are easy to use and to customize. However, there are cases in which a 2D simulator is not sufficient. For example, for mobile robots with higher mobility than wheels (e.g., legged or snake-like robots) and in 3D environments, a 2D simulator may be too simplistic to correctly model some behaviors.

A 3D simulator for mobile robots must also correctly simulate the dynamics of the robots and of the objects in the environment, thus allowing for a correct evaluation of robot behaviors in the environment. The required accuracy of dynamics simulation depends on the particular behavior we are interested in evaluating. Moreover, real-time simulation is important in order to correctly model interactions among the robots and between the robots and the environment.

G. Lakemeyer et al. (Eds.): RoboCup 2006, LNAI 4434, pp. 13–24, 2007.

Since simulation accuracy is computationally demanding, it is often necessary an approximation to obtain real-time performance.

Another important feature of a robotic simulator is easy integration of different robotic platforms, different scenarios, different objects in the scene, as well as support for multi-robot applications.

Finally, visual realism is not fundamental for robotic simulation, since may be not adeguate to experiment and evaluate low-level sensor processes, such as image processing. However, visual realism usually has a minimum impact on the performance of the simulator (since most of the computation can be demanded to graphics adapters of the PCs), while a simulator with visual realism can be more attractive.

3D Robotic Simulators. There exist already several simulators that handle the issues discussed above. As our aim is to address 3D physics simulation, let us see how this feature is integrated in the simulators that have been used within the RoboCup Four Legged league and in general for 3D modeling of complex robots. There are several factors that make realistic robotics simulation hard to achieve. In order to represent a valid tool for the robotics researcher, the simulator must fulfill a number of requirements:

Flexibility: the simulator must allow for the simulation of different robots, not even know *a priori*, as well sensors and actuators. The generic virtual environment where the robots are placed, should be easy to model as well;
Physics Realism: to obtain plausible results, interaction among robots and between robots and the virtual environment must be carefully modeled through the physical laws of rigid body dynamics;
Visual Realism: the appearance of the whole system must be as accurate as possible to guarantee consistent sensor readings (e.g., images, audio);
Efficiency: simulation must be carried out in the most efficient way, hopefully in real-time, with a visualization frame rate of 30 frames per second;
Modularity: it must be easy to add and modify the features of the environment and of the robots, including the sensors input/output;
Effective Control: the simulator should be flexible enough to be easily interfaced to the same programming code that is used on the real robots.

The Asura Team[1] provides a development kit, namely the ASURA RoboCup Software, aiming at reproducing the Four-legged League environment. Such a development kit permits to develop strategies and sensor acquisition and processing. However, it lacks in flexibility since it can simulate only the AIBO ERS-210 robot in the RoboCup framework and it does not permit dynamics simulation, leading to a poor representation of the virtual system.

Zagal and Ruiz-del-Solar introduced UCHILSIM [7] in 2004. Such a simulator reproduces with high fidelity the dynamics of AIBO motions and its interactions with the objects in the game field. Physical simulation is carried out through

[1] www.asura.ac.

the open-source physical engine Open Dynamics Engine[2] (ODE) and objects (e.g., robots) are defined in the VRML standard. The goal of this project was to became a standard framework for learning complex AIBO behaviors. This simulator was quite promising but it seems to be no longer developed.

Gazebo [3] is a multi-robot 3D simulator with graphical interface and dynamics simulation (through ODE). It is able to simulate a wide range of sensors and it comes with models of existing robots even if the simulator does not allow to define complex objects (e.g., dummies for rescue arenas, moving people in the scene, a ball in the soccer field). The robots and sensors can be controlled by the Player [2] server or controllers can be written using a library provided with the simulator. Simulated environment are described in XML and new robot/sensor models can be created as plug-ins. Simulation of legged robots is supported but not extensively used in the current release.

SimRobot [4] by Laue *et al.* simulates arbitrary user-defined robots in three-dimensional space. To allow an extensive flexibility in building accurate models, a variety of different generic bodies, sensors and actuators has been implemented and specified in XML. The robot controllers are directly linked with the simulator library to produce an executable file. Furthermore, the simulator follows a user-oriented approach by including several mechanisms for visualization, direct actuator manipulation, and interaction with the simulated world. Dynamics is simulated through the ODE engine.

Webots[3] is a commercial general purpose mobile robotics simulation software. It uses ODE to simulate dynamics and it has an extensive library of actuators, sensors and robots like Aibo, Lego Mindstorms, Khepera, Koala and Hemisson. While the mechanical features of the robots are well defined, the main limitation of this simulator resides in the poor quality of the 3D graphical representation of the virtual environments, including robots and in the lack of adequate modeling tools.

USARSim (Urban Search and Rescue Simulator) [6,1] is a robot simulator based on the industrial game engine Unreal Engine[4]. It simulates the reference test arenas developed by National Institute of Standards and Technology (NIST) and robots intended for the Urban Search & Rescue (USAR) tasks. Since Unreal Engine has been deployed for the development of networked multi-player 3D games, it solves many of the issues related to modeling, animation and rendering of the virtual environment. It is a complete game development framework targeted at today's mainstream PC, it provides tools to rapidly develop objects and environment (Unreal Editor) and it is possible to define the behavior of the objects through an ad-hoc script language (Unreal Script). Dynamics of rigid bodies is transparently handled by the Karma physical engine [5]. As in Gazebo, robot controllers use a TCP/IP interface to control the robots and so may be programmed in any language that supports networking. The Client/Server architecture can be advantageously used to carry out complex computations on dedicated machines decoupling the simulation from intelligence processing.

---

[2] Open Dynamics Engine www.ode.org.

[3] www.cyberbotics.com/products/webots/.

[4] www.unrealtechnology.com.

However, the current version of USARSim is not able to simulate legged robots, like AIBOs, and it has several limitations like the total number of joints allowed for each robot, a limited support for multi-robot scenarios and an approximative collision handling.

Discussion. From the analysis of the existing simulators it appears that a general 3D simulator for legged robots with good dynamics simulation, multi-robot support, realistic appearance, and easy-to-use editing tools is not currently available. USARSim is the most promising, since it already implements many required features and it can be easily extended. Unreal Engine has a significant industrial support and a simulator based on it will benefit from new releases of this engine as soon as they are available, with minimum effort. Unreal Engine uses a different physical engine (Karma) than other simulators (using ODE). To our knowledge there are no comparative studies between these two engines and we believe the choice of ODE is given only by its open-source code. Unfortunately, the current version of Unreal Engine (and USARSim) makes a partial (and sometimes incorrect) use of the Karma engine. Finally, USARSim has been chosen as the standard simulator for a new RoboCup Rescue simulation league.

In this paper we present a 3D simulator based on USARSim[5] that allows for modeling complex legged robots (such as quadrupeds and humanoid ones) and for simulating their interaction. The performance of the system allows for simulating in real-time up to five of these robots in the environment. The simulator presented in this paper extends USARSim by introducing some important features: 1) it allows for the simulation and control of legged robots (four-legged, humanoids, etc.) 2) it introduces a multi-view functionality for multi-robot support. Moreover, we fixed a few problems in the use of the physical engine Karma, that was not fully and correctly integrated in the Unreal Engine.

We successfully tested the simulator by implementing AIBO ERS-7 and QRIO robots, controlling five of them at full frame rate (30 fps). The simulator is actually in use for the development of our team competing in the RoboCup Four-Legged League.

2 3D Multi-robot Simulator Architecture

In order to use a simulator for multi-robot applications, it is important to provide an effective interface to multiple robot control programs. A typical choice (e.g., Gazebo [3]) is to run the simulator as a server allowing robot control programs to act as clients. The robot control programs receive from the simulator data emulating sensor readings and send commands for the actuators. The simulator server manages interaction among robots and between robots and objects in the environment, maintaining an up to date representation of the world. For a realistic simulation, it is thus important that the simulation is asynchronous with respect to robot control programs. Moreover, the simulator must process data in real-time, otherwise it will feed robot control programs with unrealistic data.

[5] The simulator and demo videos are available from `www.dis.uniroma1.it/~spqr`.

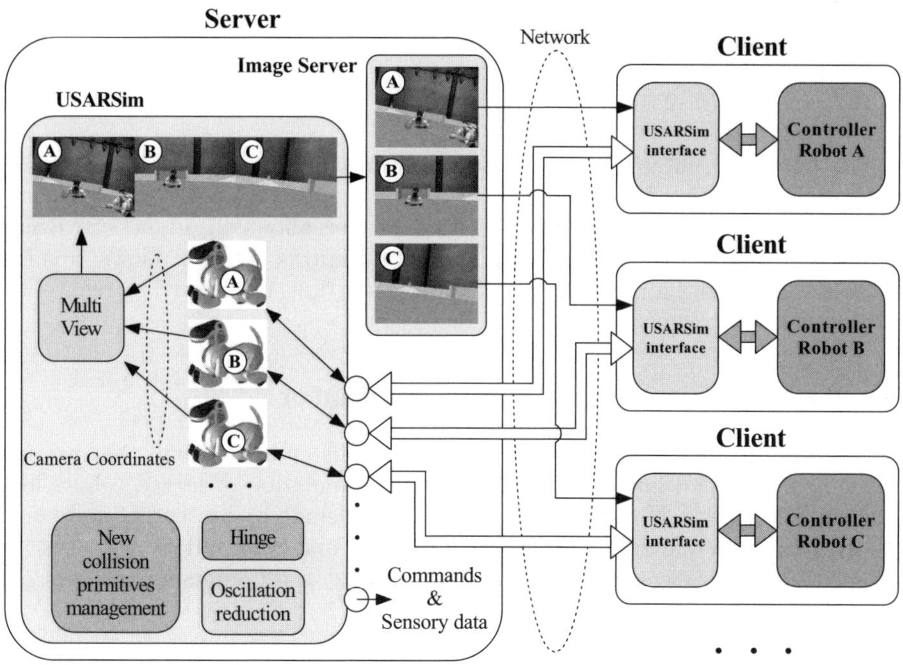

Fig. 1. Multi-Robot architecture

When using a 3D simulator with realistic modeling of appearance and dynamics of the environment, computational time is a main requirement and therefore it is usually necessary to run this process on a different machine with respect to robot control programs. Therefore, a networked client/server architecture should be used for 3D multi-robot simulation. The server machine runs the 3D simulator as a multi-client server. Other machines connected through TCP/IP act as robot control programs.

The choice of USARSim as the basis for the simulator described in this paper has required the implementation of another module, to overcome the problems due to the fact that the underlying game engine is not designed to act as a multi user server on a single machine. In fact, it is possible to use USARSim for multi-robot applications as long as only one robot is provided with a video camera. To simulate a multi-robot system with one camera per robot, it is necessary to run one USARSim process (actually an Unreal client) for each robot. Moreover, it is possible to run only one client per machine. This solution requires too many resources for a single simulation: for example, if robot control programs are separated from simulation we may need up to $2\,n$ machines for an n-robot simulation. Otherwise it is possible to use a single machine switching among robot cameras every t seconds, where t depends on the simulation load on the machine. However, also this solution is not advisable because it offers extremely low image acquisition rates, and it is not possible to access multiple cameras at the same time. The solution proposed in this paper allows for using a single machine for the simulation in presence of multiple robots with multiple cameras without the aforementioned drawbacks.

The architecture of the multi-robot simulator we have implemented is depicted in Figure 1.

The USARSim interface module on the clients manages all the communications with USARSim. It was created to translate or modify both outgoing commands and incoming sensory data. This interface allows to adapt USARSim to already existing controllers with minimal changes. It also allows to pre-process sensory data if necessary (e.g., apply distortions introduced by real robot cameras).

3 3D Multiple Legged Robot Simulator

In this section, we describe the main modifications and extensions that we implemented in USARSim in order to allow the simulation of legged robots like quadrupeds, bipeds, hexapods and so on. We conclude by presenting an example with two different robots, Sony AIBO ERS-7 and Sony QRIO, operated by independent controllers, coexisting and interacting in the same environment.

Oscillations of Rigid Parts. The first problem we faced was due to the oscillation of all the rigid parts of the robots. The amplitude of the oscillations was $\pm 2mm$ on X, Y and Z axis and it has been observed only when the simulation has been run in networking mode, when server and clients were running on separate machines. The cause was in the replication mechanism, used by Unreal Engine to synchronize server and clients. We fixed this issue in a straightforward way by using in the USARSim code a different data type not subject to network optimizations.

Collision Handling. A collision primitive is an invisible volume with a simple shape (e.g. a box, a cylinder), embedding a 3D mesh and it is used to simplify the collision detection process with other meshes. It is also useful to define dynamic properties like center of mass and inertia tensor. A well-defined collision handling is crucial for a plausible physical simulation and so the correct definition of the collision primitives.

In the original USARSim, each single part of a robot is defined by the class KDpart where, beside other information, it is specified the shape of the robot part and a default collision primitive. When the simulator loads a robot in the virtual scene, the original USARSim assembles all its parts with the corresponding collision primitives on the fly. The process however is prone to a sneaky Unreal Engine bug causing the loading of the same default collision primitive for each mesh. To bypass this bug, our approach has been to write a script, derived from KDpart, for each part of the modeled robots. In this way, meshes and collision primitives have not to be defined at run-time (with a new static-mesh properties) but it is sufficient to specify them in the definition script of the robot.

Because this major change, our implementation lost compatibility with the robots deployed with the original USARSim. To be able to continue to use these robots, we defined new scripts for each one of them, including the fine-tuning of physical parameters like mass, center of mass, inertia tensor and friction.

Hinge Joint. Original USARSim defines one motorized joint to connect the different parts of the robot, the CarWheelJoint (Fig. 2). Such a joint is provided by Karma physics engine integrated in the Unreal Engine and allows two or three DOFs, depending on its configuration. We did not use this joint because for our objectives, only one DOF was needed. USARSim allows for locking the suspension DOF, leading to a one-DOF joint, however the resulting joint is not stable enough for rigid parts, and it leads to instability of the simulation. Furthermore, with the CarWheelJoint, it is rather difficult to obtain the relative rotation angle among joined parts. Such angles cannot even be set precisely, since there is at least an error of ± 0.5 degrees, and this error drift away (i.e. increase) over time.

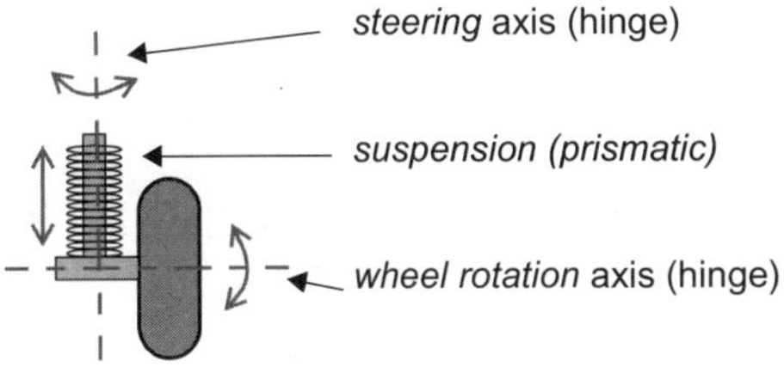

Fig. 2. CarWheelJoint [5]

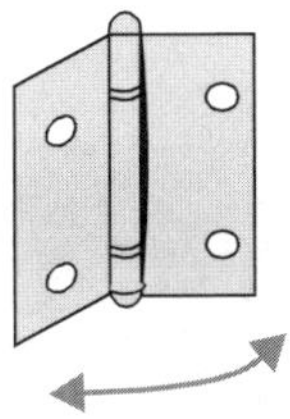

Fig. 3. Hinge Joint [5]

For these reasons, we modified USARSim in order to use another joint model, the Hinge (Fig. 3), also provided by Karma engine and not suffering from the aforementioned issues. Hinge allows one DOF, may be controlled in angle, angular velocity and torque, the maximum allowed precision is 0.0055 degrees, and it implements a feedback mechanism providing a stable control of the angle among joined parts.

Original USARSim allows up to 16 joints for each robot. This bound have been increased by modifying the size of some internal structures in the simulator code.

Multiple Views Support. Unreal Engine allows only one robot camera to be accessed at each frame. If more robots are placed in the virtual environment, only one robot controller may be correctly feed with image data. To allow the simulation of multiple robots, we attempted to interleave images obtained by accessing a different camera at every frame, but this solution is not feasible since simulation and controller modules run asynchronously. The solution provided by the original USARSim is to use the Unreal Engine multi-player support. Each robot runs on a different computer with its graphical client and a central server handles the whole simulation. This solution is obviously not suitable for multi-robot development since it requires too much hardware.

To solve this problem, we introduced a special kind of robot, namely the MultiView. MultiView collects camera locations and orientations from each camera

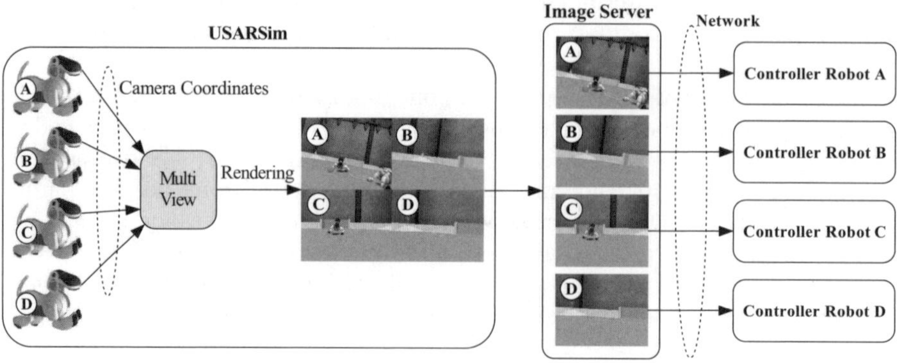

Fig. 4. Each robot camera is rendered by MultiView. Those images are then collected and splitted by ImageServer which sends them to the controllers.

Fig. 5. Multi-Robot example

on the robots and renders each robot view in a different subview. This image mosaic is then grabbed by ImageServer, a thread running on the same machine where the USARSim server resides, by direct access to the Direct3D frame buffer. The views from each robot are extracted from the mosaic and sent across the network to the corresponding controllers (Fig. 4). This solution allowed us to simulate and control multiple robots using cameras by running the simulator on a single machine. The only limitation is in the reduced resolution of the images. Observe that this is not a problem, since simulation can not anyway be used to validate image processing. Moreover, for robots with low-resolution cameras (such as AIBOs), actual resolution can be obtained.

AIBO Sensors. To simulate the AIBO ERS-7 we also added 3 new sensors: a simplistic instant acceleration sensor, a contact sensor and a more flexible IR distant sensor than the one already defined in the original USARSim. In particular, our IR distant sensor permits to set the maximal bound of the error magnitude in function of the measured distance.

MultiView and the client/server architecture allow to mix easily different robots in the same simulation. Figure 5 shows an example of two different robots, an AIBO and a QRIO humanoid robot, in the same map and handled remotely by different controllers.

4 Run-Time Environment Management

The flexibility of Unreal Engine and Unreal Script allowed us to define the behavior of the virtual environment in real-time.

We define the complex behavior of the objects in the virtual environment through events and triggers. An event is casted when an object, like a robot, comes in contact with a trigger, that is an invisible volume that can be placed anywhere in the map. Each event corresponds to an action like, for example, an affine transformation applied to an object (e.g., open and close a door or a passage), can turn on and off lights or motors. An event can activate users own script routines, permitting endless possibilities. Events can be chained, scheduled, dispatched to many objects or randomly generated.

To manually control objects and robots in the simulation environment, we defined new Unreal Tournament client console commands able, for example, to reset the simulation, to change lighting conditions, to transfer objects and robots from one place to another and so on.

Console commands can also be embedded into the script code defining the behavior of the robots. For example, it would be possible for a robot to modify the simulation laws in order to let it fly from one place to another.

5 Results

This section provides an investigation on physical behavior and the overall performance of our simulator. The testing machine has an AMD Athlon XP Burton 3000+ CPU with 1Gb DDR400 RAM and the nVidia FX 5900XT graphic adapter.

Physical Simulation. A plausible physical behavior is a primary concern if we want to test and simulate algorithms operating on real robots. USARSim uses Karma physics engine which is designed for video-games and not for robotics simulation. This means that the realism of the physical simulation is always sacrificed in order to achieve a smooth rendering frame rate (at least 30 fps). However, the accuracy of the simulation is not severely compromised because the approximations introduced by the Karma engine are comparable to the measurement errors due to the real robot sensor and actuator noise. Thus, it is not crucial to precisely quantify the approximation error, but it is important to qualitatively estimate the behavior of the robots involved in the simulation. Figures 6 and 7 show the results of the two principal interactions that may happen on the RoboCup field: ball kicking and AIBO collisions. We experienced a

Fig. 6. AIBO kicking the ball

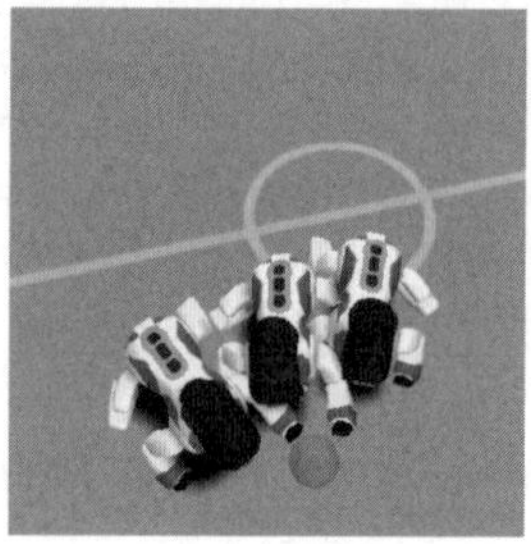

Fig. 7. Three AIBOs collision

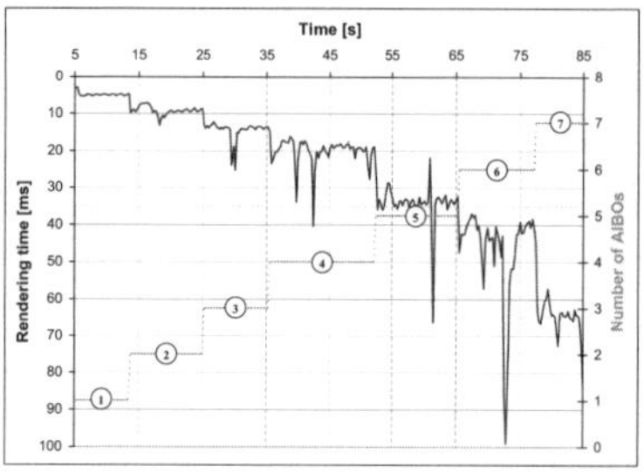

Fig. 8. AIBO creation stress test

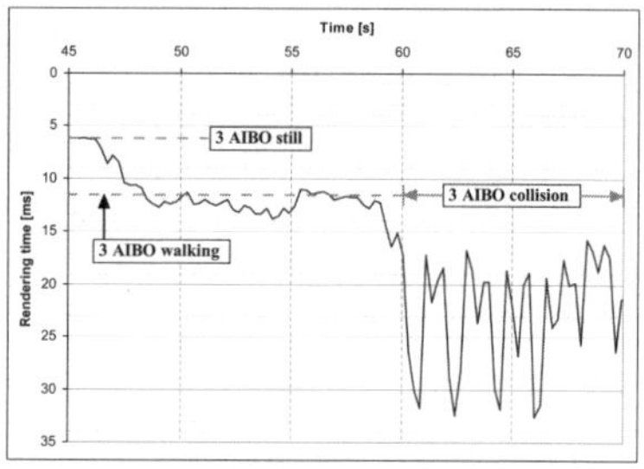

Fig. 9. 3 AIBOs collision

visually convincing behavior of the physical simulation and we can conclude that it is adequate both for a RoboCup simulation and for more general robot-robot and robot-environment interactions.

Performance. The first experiment, made to assess the simulator performance, determines the maximum number of supported AIBOs at the same time in the map using only the test bed PC. Each AIBO is created and set in walking state to make full use of the physics engine. The graph in Fig. 8 shows that the performance is still acceptable with five AIBOs, since the visualization frame rate is still greater than 25 fps.

The second test stresses directly the physics engine making AIBOs to collide with each other. When collisions occur, more than 75% of CPU time is devoted to collision detection and response. The graph in Fig. 9 corresponds to the collision of 3 AIBOs. It is very jagged and each performance drop corresponds to a contact between two or more robots. This result confirms that collision handling is the computation bottleneck.

In conclusion, the simulator can sustain at full frame rate five complex robots, three of which can collide at the same time.

6 Discussion

In this paper we have described the implementation of an extremely flexible simulator for multiple legged-robots. It supports rigid-body dynamics, realistic 3D environments, client/server architecture and real-time rendering. The simulator handles up to five legged robots at 30 frames per second on a middle-class hardware. Furthermore, it is coded in Unreal Script, the scripting language of Unreal Engine. This means that, although Unreal Engine is not open source, our extension is open and will be shared through the community, therefore everyone can access its script code, change its behavior, add new robots, sensors and any other functionality that may be required employing little effort.

As stated before, the choice of using Unreal Engine solves many of the main practical problems faced during the implementation of a robot simulator. However, such an engine has been devised primarily for games, not for robotic simulation. Thus, designers chose to sacrifice physical realism to obtain smoother animation and they bounded the physical time step to the visualization frame rate (i.e., the time step is equal to the frame rate), whereas the simulation and visualization could be clearly separated.

However, the advantages of using an industrial product are in the great development support they provide and in the availability of many effective tools to create contents such as scenarios and objects. For example, the Unreal Editor allows for easy creation of generic environments (Fig. 10), even with scripted objects reacting to changes in the system.

Moreover, improvements to the engine will directly reflect to improvements in the simulator with little effort, while obtaining significant advantages. For example, we intend to upgrade the simulator to use Unreal Engine 3 as soon as it will be available. This will dramatically enhance rendering quality, physics simulation, script and net code.

Applications. The primary use of this simulator is to evaluate the behavior of legged robots in a dynamic environment, such as RoboCup soccer. The advantages of using this simulator are evident in multi-robot contexts. As described above, the simulator is able to simulate in real-time up to 5 robots. We believe that with increasing CPU power it will soon possible to simulate a 4 vs. 4 game.

We have used the simulator to evaluate different situations, such as an attacker robot against three defenders and a goalie, two attacker robots against two defenders and a goalie. A first important process is debugging and refining plans, i.e., evaluating if the robots take the correct decisions according to the current state of the game. Note that a 2D simulator would have some limitations in this process: for example, partial occlusions of the ball, contacts between robots cannot be realistic modeled in 2D. A second process is to evaluate coordination strategies: e.g., position of the robots that are not in possess of the ball, position of the defenders, and decisions about when and how to pass.

In order to make development more effective we have implemented a USARSim interface (as described in Section 2), thus the same control code can be run connected to the simulator or on the real robot. This allows for a fast and effective development of many tasks by the students of our group.

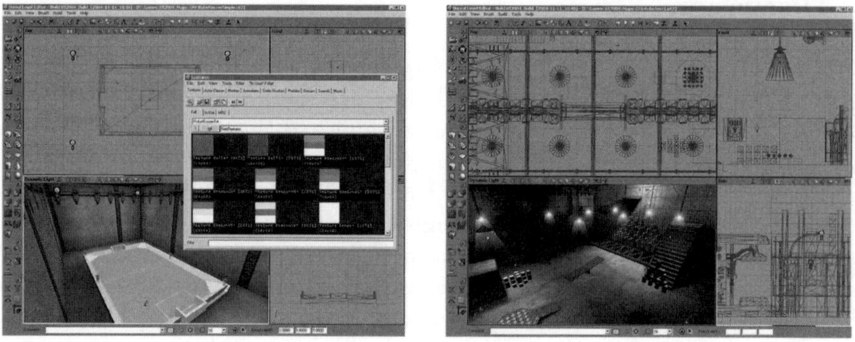

Fig. 10. Unreal Editor used to create RoboCup soccer field and a test arena

Future works. The extensions described in this paper will be integrated in the USARSim simulator. Moreover, we are planning to make some further improvements like enhancing sensor data message handling and making simpler and more rapid the creation of new robots. A further major task will be the modification of the Unreal Tournament deathmatch code in such a way to provide tools to interact with the environment during the simulation run; for example, we intend to implement tools like a game controller interface and a virtual referee placing the robots for the RoboCup soccer setting.

References

1. Carpin, S., Wang, J., Lewis, M., Birk, A., Jacoff, A.: High fidelity tools for rescue robotics: results and perspectives. In: Proc. of RoboCup Symposium (2005)
2. Gerkey, B., Vaughan, R.T., Howard, A.: The Player/Stage Project: Tools for Multi-Robot and Distributed Sensor Systems. In: Proceedings of the 11th International Conference on Advanced Robotics (July 2003)
3. Koenig, N., Howard, A.: Gazebo - 3D Multiple Robot Simulator with Dynamics (2006), `http://playerstage.sourceforge.net/gazebo/gazebo.html`
4. Laue, T., Spiess, K., Röfer, T.: SimRobot - A General Physical Robot Simulator and its Application in RoboCup. In: Proc. of RoboCup Symposium. Universität Bremen (2005)
5. Mathengine: MathEngine Karma User Guide (March 2002)
6. Wang, J., Lewis, M., Gennari, J.: A game engine based simulation of the NIST Urban Search & Rescue arenas. In: Proceedings of the 2003 Winter Simulation Conference (2003)
7. Zagal, J.C., Ruiz-del, J.: Solar. UCHILSIM: A Dynamically and Visually Realistic Simulator for the RoboCup Four Legged League. In: Nardi, D., Riedmiller, M., Sammut, C., Santos-Victor, J. (eds.) RoboCup 2004. LNCS (LNAI), vol. 3276, Springer, Heidelberg (2005)

3D2Real: Simulation League Finals in Real Robots

Norbert Michael Mayer[1,2], Joschka Boedecker[1], Rodrigo da Silva Guerra[1], Oliver Obst[3], and Minoru Asada[1,2]

[1]Dept. of Adaptive Machine Systems,
Graduate School of Engineering, Osaka University, Osaka, Japan
[2]Asada S.I. Project, ERATO JST, Osaka, Japan
[3]Center for Computing Technologies, University of Bremen, Bremen, Germany
{joschka,guerra}@er.ams.eng.osaka-u.ac.jp
{norbert,asada}@ams.eng.osaka-u.ac.jp
fruit@tzi.de

Abstract. We present a road map for a joint project of the simulation league and the humanoid league that we call 3D2Real. This project is concerned with the integration of these two leagues which is becoming increasingly important as the research fields are converging. Currently, a lot of work is duplicated across the leagues, collaboration is sparse, and knowhow is not transfered effectively. This binds resources to solve the same problems over and over again. To address this, we discuss the current situation of both leagues with respect to these points and focus on open issues that have to be fixed. In addition, we describe existing open standards and contributions from the RoboCup community that we plan to use for the project. As a milestone, we propose to conduct the finals of the 3D simulation tournament on real robots by the year 2008. Finally, we propose a database of simulated parts and algorithms in which each league can benefit and contribute with their expertise. These contributions facilitate synergies to be used across individual leagues for the benefit of the RoboCup project and the year 2050 goal.

1 Introduction

Looking at the stated goal of RoboCup to present a team of humanoid robots able to win against the human soccer champion in 2050 [1,2] it is apparent that the different leagues we see in RoboCup today will have to move closer to one another and eventually be merged. There are certain unique features in every league that make it attractive as an environment for researchers to focus on a set of specific problems on the way to the final goal. In the end, however, it will be humanoid robots taking on the challenge in 2050.

To ensure steady progress, the competitions held at RoboCup are made more complex and challenging every year. Through this evolutionary process, we already start to see some boundaries getting blurred across the leagues. The Small Size League, for instance, is expanding the field size and is coming closer to the

G. Lakemeyer et al. (Eds.): RoboCup 2006, LNAI 4434, pp. 25–34, 2007.

Middle Size League. In the humanoid league, we saw the first real games in 2005. More players will be introduced, giving rise to the need of tactics in addition to the low-level control methods which have been the traditional focus of this league. The simulation league, on the other hand, in which researchers had concentrated on those high-level strategies is starting to use more realistic models for their agents, targeting simulation of 11 vs. 11 humanoid robots within the next few years.

One problem is that a lot of work is being repeated in the different leagues while solutions for the same (or at least similar) issues exist in another league. Nearly every team uses more or less advanced simulators for their robots as part of their development tools, for instance. Designing and implementing a good robot simulator is a difficult and time consuming task, so it makes sense to reuse the existing work. It is obvious that the knowhow of the different leagues has to be integrated in order to achieve synergy effects and free resources for other challenging tasks.

First steps in direction of a league-independent soccer theory were outlined in [3]. Some documented examples of collaborations between researchers from different leagues can be found. In [4], the authors describe the revision of a software framework for behavior development for a humanoid robot according to a design which had been successfully used in simulation league before. At the same time, it was planned to integrate a model of the humanoid robot into the simulation league 3D simulator.

Keeping the pace towards the ultimate goal, both hardware and software complexity tend to grow fast. This tendency makes it difficult for the current structural division of the leagues to keep developing their independent architectures in an isolated way. Particularly, problems like this can already be observed both in humanoid and simulation league teams. Development ends up covering technical issues not directly related to the interests of a particular league.

This paper is focused on the aforementioned problems. It is very clear for the authors of this work that in the long term there would be gradually fewer platforms of very high complexity. This makes apparent that the current league-oriented division of architectures would not be a feasible endeavor for current teams. We provide a well-grounded road-map and suggest tools for helping the gradual long-term shift from the current league-based division of architectures into a new cooperative and modular labor division. It is our hope that efforts in this sense will help the coordination of the work of different leagues, complementing and completing each other towards the 2050 challenge.

2 Current State of the Leagues

The humanoid league (HL) underwent a profound development since it was introduced in the RoboCup of 2002 in Fukuoka. The rules maturated in many points and gained focus on the issues that are essential from a technical point of view. Thus, the center of mass of all robots has to be on a certain height in relation to the size of the feet. The competitions and challenges have changed in various ways. In the RoboCup 2005 regular 2-2 games have been conducted for the first time. Like in other leagues, the organizers see a maturation process also in the design of the robots. The typical robot of the current competition is a small robot that uses servo motors as actuators and a simple but robust control structure. One aim of the technical committee is to lead the development towards important research problems. Dynamic walking and stability are currently the most important issues, which are enforced by the technical challenge and the rules about the shape of the robot. As a consequence, we see a significant progress within this relatively new league. The HL also grew in the number of participants. Between 2002 and 2004 around 10 teams participated in the HL. In 2005, there were around 20 participants already. For the RoboCup 2006 we received 23 pre-registration for the KidSize League and the 12 pre-registrations for the TeenSize League.

One of the first leagues of RoboCup was the two-dimensional soccer simulation league. The actual hardware of the simulated robots, the actuators and also the perception are simulated on a relatively high level as opposed to the robots in the current hardware leagues. The motivation for the high level of abstraction was the desire to create a league where participants can concentrate mainly on coordination and cooperation of robot teams. The rationale was that in the (quite far) future, many "lower level" problems of the hardware leagues would be solved, leaving cooperation among agents in a team as main challenge. In fact, two-dimensional soccer simulation league helped to address many different open problems of creating cooperative multiagent systems.

Because of the simplified model of 2D simulation league, a three-dimensional physical simulation was created. The three-dimensional physical simulator used in Soccer Simulation League addresses additional classes of problems:

- Articulated agents create the problem of coordinating several actions of the same agent among each other, as well as with global team behavior.
- Decision making procedures have to deal with a much higher complexity of the decision space, compared to 2D Soccer Simulation League.

At least the latter of these applies already to the current 3D simulation, where agents are very much simplified. Methods to create soccer playing agents for a team have to deal with a higher complexity of the environment, and hopefully can be transferred to humanoid robotics more easily. The current development of 3D Soccer Simulation League leads to simple two-legged agents used in technical challenges already this year (see also Fig. 2).

One of the problems of making Soccer Simulation League closer to humanoid robotics is that solely researching high level coordination and cooperation becomes intractable, when lower level controllers have also to be implemented by everybody. One of the advantages of the 2D simulator however was the possibility to research cooperation in a team quite easily. In order to keep the advantages of the 2D simulator while adding new possibilities for the additional research problems listed above, two different levels of interfaces should be provided for users of a Simulation League Simulator: one high-level interface granting the possibility of researching high-level coordination only. This way, existing approaches can be

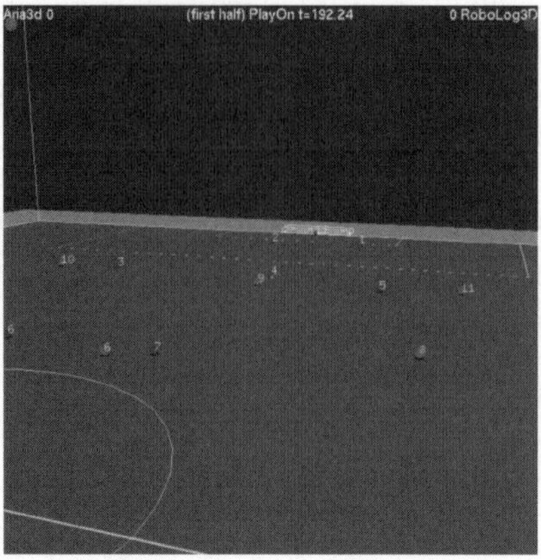

Fig. 1. The current version of the 3D simulation using spheres as agent models. Every agent has an omni-vision camera which delivers noisy data about the environment, a kick-effector to shoot the ball, and an actuator simulating omni-drive to move on the field.

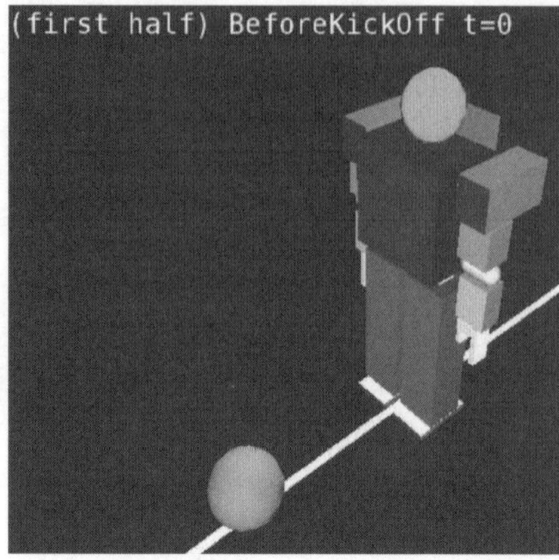

Fig. 2. An exemplary robot model of Fujitsu's HOAP-2 that could be used in the RoboCup-2007 simulation league competition.

transferred to the domain of robotic soccer easily. The lower level interface has to provide full control over all features of the simulated robots, so that developers can research and take care of dependencies between lower-level and higher level control.

Currently, the development in Soccer Simulation League leads towards humanoid robots, which already can be controlled by a lower level interface. However, controllers for these robots have to be developed in order to provide an easy-to-use interface.

As a result, humanoid and simulation league have more common qualities. This way, joint competitions of Soccer Simulation League and Humanoid League become possible, which promotes a faster progress in both leagues.

3 Road Map

In both the HL and the SL significant changes are underway. We suggest a time frame for the development of the joint events and propose for both leagues a number of synchronized steps in the following subsections.

3.1 RoboCup 2007: 3D2Real Competition: Technical Challenge in Simulation League with a Real Robot Model

We propose for 2007 an additional tournament called *3D2Real* competition in the SL. The competition consists of an obstacle run with a humanoid robot.

The layout of the competition is going to be identical to a technical challenge in the HL of the same year. It is planned to simulate a real existing humanoid robot. The type of the robot is decided within the next year by the SL technical committee in collaboration with the HL technical committee.

The SL 3D2Real competition is done in simulation first by applying the same criteria as in the corresponding HL competition. The three best participants of the simulation round qualify for a second round in which real robots are used. The programs of the virtual agents therefore have to be able to run on the real robot.

The simulation environment is going to be derived from the existing simulation environment of the SL 3D league and the RoSiML [5] modeling language. RoSiML is an XML-based modelling language successfully used for a simulator in Sony legged league.

The robot model, as well as the physical and control parameters are planned to be as close as possible to the real robot. A standardized interface for the controller commands will be provided.

The first step is intended to get an overview of the problems that arise from porting a simulated behavior into the real world. In particular we are interested in the following questions:

- What differences exist between the real world robot and the simulated environment. How similar are they?
- What kind of tools are necessary?
- How reliable are the control parameters, and what kind of noise model is appropriate to simulate real world fluctuations and randomness?

Results from the 3D2Real competition are intended to be integrated into the 3D SL simulator for the following year. It is also intended to automate the upload of a behavior program to the robot.

3.2 2008: The 3D Simulation League Final Is Played on Real Robots

In this year it is intended that the 3D SL players are simulated versions of a real existing robot type. It is intended that the round robin is done in computer simulation, whereas the finals are done in real world robots. It might be appropriate, however, to reduce the number of players from 11 to five robots.

The robots and the playground are provided by the organizers of the RoboCup competition 2008.

In the HL, a description of each of the participating robots in the RoSiML language is going to become part of the qualification process. The intention is that beginning from the year 2008 the RoSiML files of all robots participating in the HL are published and integrated into a online repository that is available for research and development. More details on this repository are given in section 5.

	Simulation League	Humanoid League
until RC 2007	simlation environment simulates a real robot-type	
RC 2007	3D SL TC 2nd round in real robots	
2007-2008	Development of CPR	
RC 2008	3D finals in real robots (one type)	RoSiML models become part of the HL qualification
RC 2009	3D SL finals with sev- eral types of robots	HL team commit to CPR

Fig. 3. 3D2Real project: Overview of the roadmap towards simulation league finals in real robots

3.3 RoboCup 2009: Games with Several Types of Robots

Based on this repository, the organizers of the 2009 SL competition select several types of robots that can be used as models for the 2009 SL 3D competition.

4 Requirements on Humanoid Robot Systems Eligible for the 3D2Real

A real existing humanoid robot type eligible for the 3D2Real competition should fulfill a set of requirements and should come with a certain software environment (see also Fig. 4). We suggest the following necessary requirements:

- The robot has to be compliant with the rules of the RoboCup humanoid league. The architecture should include a small IBM PC(386-architecture).
- The software environment should be published in source code, the programming language is C/C++, with a preference to C++. The vision processing comes with the robot.
- The robots mechanical design has to be described in the RoSiML language.
- The robot comes with a compatibility layer for ODE that consists of two parts: The first part covers the sensor processing. Generic classes for camera, touch sensors, attitude sensors, actuator states are to be provided by the SL organizers. A detailed description of these sensors and their noise levels have has to be worked in. The output of the vision processing is a list of recognized

objects, i.e. ball, posts, goals, line crossings, and their position in a list. The second part consists of compatibility layer for the actuator processing. ODE type of motors are assumed at the simulation layer. An encapsulation of the real actuators has to be provided, this may include high level motion primitives e.g. *walk*, *turn move camera.*

The aim is that a control program coming from SL participant results in the same robot behavior in the simulation as in the real robot, as far as this is possible. The organizing/technical committee chooses the first robot type for the competitions in the years 2007 and 2008 in an open and fair selection process.

5 Central Parts Repository

Traditionally, simulation league has focused on *game strategy* while the humanoid league has a major focus on *robot design and control.* Simply making more realistic simulations, or simply forcing more strategy on games of real-robots would not be effective ways of helping the future cross-development towards a common goal. This *strategy* vs.*design and control* division is not just a casual one – it is deeply rooted in the researchers of both leagues, reflecting their particular backgrounds and interests, and this should be respected. The *Central Parts Repository* (CPR) is here proposed as a common framework for allowing professionals in multidisciplinary fields to help each other within their different spheres of interest and backgrounds.

The CPR is conceptualized as a database of parts and algorithms in which each league should contribute with their expertise and at the same time enjoy out of the box solutions for the problems that are out of their sphere of interests (expertise of others). The database would cover a diversity of items ranging from single robotic parts, such as servos, to entire assembled robots, including controllers and algorithms. Special care would be needed for assuring realistic constraints, especially in regard to physical behavior of fundamental parts and their controllers.

The software architecture of the current 3D Soccer Simulator is a rather complex piece of engineering, result of years of development by experienced experts in computer-related fields. Developed with very powerful plug-in mechanism, the 3D Simulator brings great flexibility for development of independent modules in a decentralized way. Moreover, the current implementation of the 3D Soccer Simulator allows the use of a modular and convenient script language for the geometric and functional assemble of simulated entities. The strong plug-in architecture along with the support for RSG files provides already all the necessary tools for the development of a modular CPR with little interference into the current course of development.

6 Discussion

In our paper, we have argued for shared competitions between humanoid soccer league and Soccer Simulation League, and presented a joint road map for both

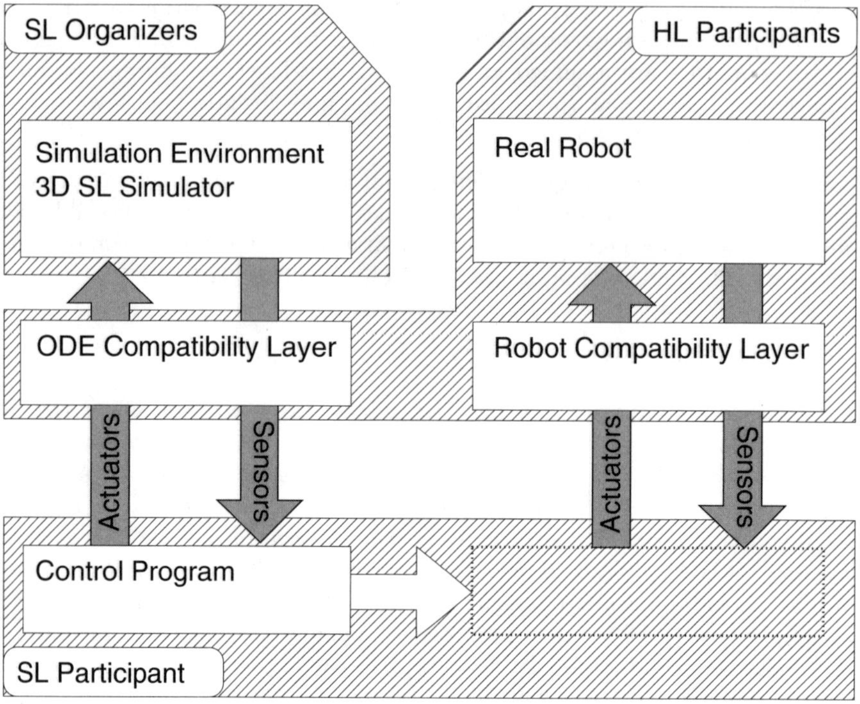

Fig. 4. 3D2Real project: Layout of the control architecture. The hatched boxes show how the different leagues contribute to the complete system architecture of the 3D2Real project. The control program for simulation system and real robot system are identical.

leagues. We suggest to establish the 3D2Real project. A part of the project is to conduct the finals of the simulation league in real robots. It is further suggested to establish a central part repository in which the parts of real existing robots are described in RoSiML. The RoboCup project can benefit in several ways from this project. In the following we outline a subset of the possible benefits:

Compare simulation and real robot. The performance of a behavior program in the simulation can be compared with the performance of a behavior program in a real robot. Differences may result from unrealistic assumptions about the statistics of the sensory input.

Sensory input in real robots is very noisy, biased input. The difference between the simulated sensory input and the sensory input that comes from a real world humanoid robot system can be directly recorded and compared. In this way we can get accurate statistics and can integrate the results into the simulated sensory environment.

In this way a stepwise improvement of the 3D SL simulator is possible. In this way it is possible to establish feedback from the reality to the simulation league. In particular, it can be seen how applicable are the strategies that have been developed in the SL in real robots.

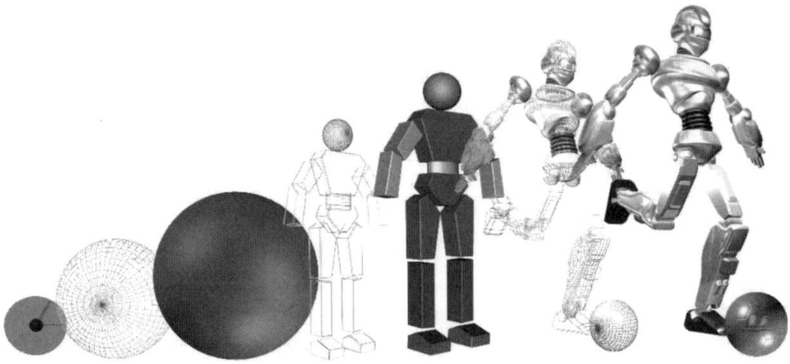

Fig. 5. Simulation league past and future: the 3D2Real project can help to program realistic robots in the SL. (Rendering by Heni Ben Amor)

Real world humanoid robots in the SL. The simulation league is aiming to become a more and more realistic environment with realistic robots as players (Fig. 5). The development of humanoid robots, however is dynamic. In the HL every year different types of robots are going to appear. The 3D2Real project gives a natural link between the two leagues. It keeps the SL automatically on track with the most recent developments in humanoid robotics.

Standard simulation environment. In the HL the improved 3D SL simulation environment can be standard tool to simulate their robots. Many teams participating in the RoboCup soccer competition develop at some point of their work a simulation environment in order to be able to test their behaviors. The aim of the authors is to establish the 3D SL simulator as an easy to use standard tool for the HL teams.

Central Parts Repository. The proposed central repository can help in several ways to establish a fruitful interaction among the HL and between the SL and HL. It can help the HL participants to create rapidly RoSiML files describing their humanoid robot. In addition, it may be in later stages be used for SL participants to construct hypothetical, but realistic robots that might show improvements. These robots can be shown in a demonstration and give hints to the HL.

Merging of two leagues. We propose an example how separate leagues can contribute to a joint project. In present day RoboCup it is a challenge to make the knowledge of one league available for the other leagues in the RoboCup project. The simulation league started around 10 years ago with the proposition to be 10 years ahead the real robots. The present work describes how the knowhow of can be made available for the current real world teams and thus, 10 years of work and development available.

Similar project in the Rescue League. Finally, we would like to note that a similar project is underway in the rescue simulation league, which has recently shown remarkable progress with the introduction of their simulator USARsim[6].

Similar as in our proposal the aim of the USARsim simulator is to give a physically correct description of the enviornment (here soccer; there a desaster area) and the robots (here biped robots; there ususally wheeled or tracked robots). Although environment and robots are different, we see on the long time scale some potential to benefit from synergies in the two simulators.

Acknowledgements

The authors would like to thank Markus Rollmann, Heni Ben Amor, and Tim Laue for their support. N. M. M. thanks Matthew Browne for his help. Furthermore, thanks go to the anonymous reviewers for the useful comments on this paper. The work was supported by a JSPS fellowship for young researchers, the Handai FRC, and several KAKEN Wakate projects.

References

1. Kitano, H., Asada, M., Kuniyoshi, Y., Noda, I., Osawa, E., Matsubara, H.: RoboCup: A Challenge AI Problem. AI Magazine (1997)
2. Kitano, H., Asada, M.: The Robocup humanoid challenge as the millennium challenge for advanced robotics. Advanced Robotics 13(8), 723–736 (2000)
3. Dylla, F., Ferrein, A., Lakemeyer, G., Murray, J., Obst, O., Röfer, T., Stolzenburg, F., Visser, U., Wagner, T.: Towards a League-Independent Qualitative Soccer Theory for RoboCup. In: Nardi, D., Riedmiller, M., Sammut, C., Santos-Victor, J. (eds.) RoboCup 2004. LNCS (LNAI), vol. 3276, pp. 611–618. Springer, Heidelberg (2005)
4. Boedecker, J., Mayer, N.M., Ogino, M., da Silva Guerra, R., Kikuchi, M., Asada, M.: Getting closer: How simulation and humanoid league can benefit from each other. In: Murase, K., Sekiyama, K., Kubota, N., Naniwa, T., Sitte, J. (eds.) AMiRE, pp. 93–98. Springer, Heidelberg (2005)
5. Laue, T., Spiess, K., Röefer, T.: Simrobot - a general physical robot simulator and its application in robocup. In: Bredenfeld, A., Jacoff, A., Noda, I., Takahashi, Y. (eds.) RoboCup 2005. LNCS (LNAI), vol. 4020, Springer, Heidelberg (2006)
6. Wang, J., Lewis, M., Hughes, S., Koes, M., Carpin, S.: Validating USARsim for use in HRI research. In: Proceedings of the Human Factors and Ergonomics Society 49th Annual Meeting, pp. 457–461 (2005)

Motion Control of Swedish Wheeled Mobile Robots in the Presence of Actuator Saturation

Giovanni Indiveri[1], Jan Paulus[2], and Paul G. Plöger[2]

[1] DII Dipartimento Ingegneria Innovazione, University of Lecce, Lecce, Italy
giovanni.indiveri@unile.it
http://persone.dii.unile.it/indiveri/
[2] Univ. Apl. Science Bonn-Rhein-Sieg and Fraunhofer AIS
Sankt Augustin, Germany
http://www.inf.fh-bonn-rhein-sieg.de
http://www.ais.fraunhofer.de

Abstract. Swedish wheeled mobile robots have remarkable mobility properties allowing them to rotate and translate at the same time. Being holonomic systems, their kinematics model results in the possibility of designing separate and independent position and heading trajectory tracking control laws. Nevertheless, if these control laws should be implemented in the presence of unaccounted actuator saturation, the resulting saturated linear and angular velocity commands could interfere with each other thus dramatically affecting the overall expected performance. Based on Lyapunov's direct method, a position and heading trajectory tracking control law for Swedish wheeled robots is developed. It explicitly accounts for actuator saturation by using ideas from a prioritized task based control framework.

1 Introduction

In the last few years Swedish wheeled omnidirectional mobile robots have had a large attention among the mobile robotics research community. A Swedish wheel differs from a common wheel in the fact that rollers are mounted on its perimeter. If all the rollers are parallel to each other and misaligned with respect to the wheel hub axis, they will provide an extra degree of mobility with respect to a traditional perfectly rolling wheel. As reported in [1], the Swedish (or mecanum) wheel was invented in 1973 by Bengt Ilon, an engineer working for the Swedish company Mecanum AB. The interest in such kind of wheels is related to the possibility of developing omnidirectional robots in the sense of [2], i.e. robots that "have a full mobility in the plane which means that they can move at each instant in any direction without any reorientation" [2]. Notice that several references make a misleading use of the term omnidirectional, as they refer to vehicles equipped with fully steering traditional wheels. Such systems can eventually move in any direction, as the unicycle model, but only after reorienting their wheels appropriately and not at any given instant of time. The need to reorient the wheels or not prior to implementing any desired linear velocity is related to the presence or not of nonholonomic constraints. As

G. Lakemeyer et al. (Eds.): RoboCup 2006, LNAI 4434, pp. 35–46, 2007.

opposed to traditional wheel car-like or differentially driven mobile robots, the translational velocity vector of a Swedish wheeled vehicle can point in an arbitrary direction at any time without re-orienting the wheels. Otherwise stated, Swedish wheeled vehicles are not affected by nonholonomic constraints: as far as the structural properties of the kinematics model of a Swedish wheeled robot is concerned, angular and linear velocities are independent. As a consequence one can design separate and independent trajectory tracking guidance control laws for position and heading. Yet if these control laws are implemented in the presence of unaccounted actuator saturation, the resulting saturated linear and angular velocity commands could interfere with each other and thus affect the overall performance of the motion control schema. A novel trajectory tracking control law is presented in this paper that explicitly accounts for actuator saturation within a prioritized task approach. Heading and position tracking are treated as independent control objectives (tasks) having different priorities: by allocating control effort to the different tasks based upon their assigned priorities it is possible to guarantee the independence of the heading and position control actions in spite of actuator saturation. Overall convergence of tracking errors to zero is theoretically guaranteed using Lyapunov methods. In a RoboCup scenario priorities are an immediate consequence of the currently active behaviour roles taken by the robot. In the defend mode, by example, we want to block a ball as fast as possible thus maximum linear speed is called for, while during dribbling angular velocity takes the highest priority.

After deriving and discussing the vehicle's kinematics model in Section 2, a kinematics (guidance control) tracking control law accounting for actuator saturation is designed in Section 3 based on Lyapunov techniques. Experimental validation results are reported in Section 4. Final remarks and conclusions are discussed in Section 5.

2 Robot Kinematics Modeling

With reference to Fig. 1, a three wheel omnidrive mobile robot is considered. All wheel main axis, i.e. hub axis, are assumed to always lie parallel to a fixed ground plane $\mathcal{P}$ having unit vector $\mathbf{k} \perp \mathcal{P}$. An orthonormal body fixed frame $< B >= \{\mathbf{i}_B, \mathbf{j}_B, \mathbf{k}_B\}$ is chosen such that $\mathbf{i}_B \times \mathbf{j}_B = \mathbf{k}_B = \mathbf{k}$. Let $\mathbf{b}_h \ \forall\ h = \{1, 2, 3\}$ denote the position of the $h-$ths wheel hub in the body fixed frame and $\mathbf{n}_h$ the unit vector of each wheel hub axis, i.e. $\mathbf{n}_h := \mathbf{b}_h / \|\mathbf{b}_h\|$. At last for each wheel we define the unit vector $\mathbf{u}_h := \mathbf{n}_h \times \mathbf{k}$. Calling $\mathbf{v}_c$ the linear velocity of the robots center and $\omega\mathbf{k}$ its angular velocity vector, the velocity vector $\mathbf{v}_h$ of the center of each omnidirectional wheel hub will be given by:

$$\mathbf{v}_h = \mathbf{v}_c + \omega\,\mathbf{k} \times \mathbf{b}_h \quad \forall\ h = \{1, 2, 3\} \tag{1}$$

implying:

$$\mathbf{v}_c = \frac{1}{3}\left(\sum_{h=1}^{3} \mathbf{v}_h \ - \ \omega\,\mathbf{k} \times \sum_{h=1}^{3} \mathbf{b}_h\right) . \tag{2}$$

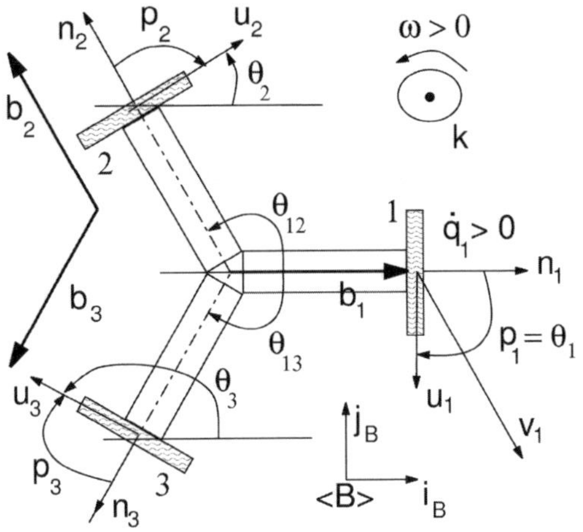

Fig. 1. Three wheel omnidrive robot: geometrical model

Based on this equation and on the non skidding hypothesis

$$\mathbf{v}_h^T \mathbf{u}_h = \rho\, \dot{q}_h \quad \forall \;\; h \tag{3}$$

where ρ is the wheel's radius (all wheels are assumed to have equal radius ρ) and $\dot{q}_h\, \mathbf{n}_h$ its angular velocity component along the hub axis (rolling angular velocity), the vehicle kinematics model can be expressed through the linear and angular velocity jacobian matrices as:

$$^B\mathbf{v}_c = J_{lv}\, \dot{\mathbf{q}} \;\; : \;\; J_{lv} \in R^{2\times 3} \tag{4}$$

$$\omega = J_\omega\, \dot{\mathbf{q}} \;\; : \;\; J_\omega \in R^{1\times 3}. \tag{5}$$

where the superscript B in $^B\mathbf{v}_c$ indicates that the components of vector $J_{lv}\, \dot{\mathbf{q}}$ are given in the body fixed frame $< B >$. If the three wheels should be mounted symmetrically at 120° degrees from each other at a same distance $b = \|\mathbf{b}_h\| \;\; \forall \; h = \{1, 2, 3\}$ from the robot's center, and assuming the body fixed frame $< B >$ to have its x axis $\mathbf{i}_B$ aligned with $\mathbf{n}_1$ ($\mathbf{k}_B$ is normal to the plane $\mathcal{P}$ and $\mathbf{j}_B$ such that $\mathbf{k}_B = \mathbf{i}_B \times \mathbf{j}_B$) as depicted in figure (1) it can be shown after lengthy, but straightforward kinematics calculations, that:

$$J_\omega = -\frac{\rho}{3\,b}\,(1 \;\;\; 1 \;\;\; 1) \tag{6}$$

$$J_{lv} = \frac{\rho}{3} \begin{pmatrix} 0 & \sqrt{3} & -\sqrt{3} \\ -2 & 1 & 1 \end{pmatrix}. \tag{7}$$

Similar kinematics derivations for 3 and 4 Swedish wheeled robots are discussed, by example, in [3] -[9]. It is important to notice that both J_ω and J_{lv} given in equations (6) and (7) are full rank and that

$$J_{lv}\, J_\omega^T = \mathbf{0}. \tag{8}$$

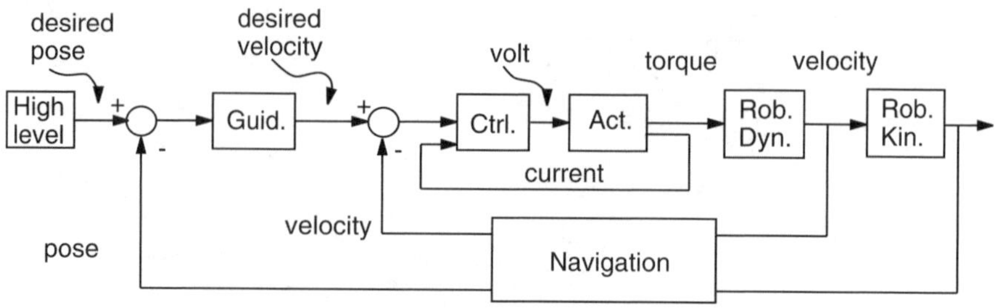

Fig. 2. Schematic view of an NGC architecture. Refer to text for details.

As shown in the sequel, this last equation allows to design separate kinematics control laws for linear and angular velocities. Interestingly the derivation of equations (6) and (7), here not reported for the sake of shortness, shows that property (8) is a consequence of having assumed the wheels to be symmetrically located, namely that $\sum_i \mathbf{b}_i = \mathbf{0}$.

3 Trajectory Tracking Control Law Design

Following a standard and most common approach for autonomous robots, the overall control architecture is organized on three level: navigation, guidance and control (NGC). Navigation takes care of the vehicles motion state estimation based upon available proprioceptive and exteroceptive information. Guidance is a closed loop control system, fed by the navigation subsystem, designed on the kinematics model of the system that generates desired angular and linear velocity reference signals. Within the described NGC framework, a trajectory tracking guidance law is derived: given an inertial (global) frame $< G >= (\mathbf{i}, \mathbf{j}, \mathbf{k})$ with $\mathbf{k} := (\mathbf{i} \times \mathbf{j}) \perp \mathcal{P}$ being $\mathcal{P}$ the floor plane, a reference (planar) trajectory is a differentiable curve in $\mathcal{P}$

$$\mathbf{r}_d(t) = \mathbf{i}\,\left(\mathbf{r}_d^T(t)\mathbf{i}\right) + \mathbf{j}\,\left(\mathbf{r}_d^T(t)\mathbf{j}\right) \tag{9}$$

with curvilinear abscissa

$$s(t) := \int_{t_0}^{t} \left\| \frac{d\,\mathbf{r}_d(\tau)}{d\tau} \right\| d\tau \tag{10}$$

and unit tangent vector

$$\mathbf{t}_d = \frac{d\,\mathbf{r}_d}{ds}. \tag{11}$$

The kinematics trajectory tracking problem consists in finding a control law for the systems input $\dot{\mathbf{q}}$ such that the position and heading tracking errors

$$\mathbf{e}_r(t) := \mathbf{r}_d(t) - \mathbf{r}_c(t) \tag{12}$$

$$e_\varphi(t) := \varphi_d(t) - \varphi(t) \tag{13}$$

converge to zero, namely such that:

$$\lim_{t\to\infty} \mathbf{e}_r(t) = \lim_{t\to\infty} (\mathbf{r}_d(t) - \mathbf{r}_c(t)) = 0 \tag{14}$$

$$\lim_{t\to\infty} e_\varphi(t) = \lim_{t\to\infty} (\varphi_d(t) - \varphi(t)) = 0 \tag{15}$$

being $\mathbf{r}_c(t)$ the position in $< G >$ of a reference point (e.g. the geometrical center or the center of mass) of the robot, $\varphi(t)$ its heading, $\varphi_d(t)$ the desired reference heading, $\mathbf{e}_r(t) = (\mathbf{r}_d(t) - \mathbf{r}_c(t))$ the position tracking error and $e_\varphi(t) = (\varphi_d(t) - \varphi(t))$ the heading error. Notice that for nonholonomic vehicles having a unicycle or car-like kinematics model, the reference heading $\varphi_d(t)$ is *not* arbitrary, but needs to coincide with the heading of the trajectories unit tangent vector $\mathbf{t}_d$. To the contrary given any position reference trajectory $\mathbf{r}_d(t)$, a Swedish wheeled vehicle will be free to track any arbitrary heading $\varphi_d(t)$ that does not need to coincide with the heading of $\mathbf{t}_d$.

3.1 Trajectory Tracking Controller Design

In accordance with the notation previously introduced, consider equations (4-5) being $\mathbf{v}_c = \dot{\mathbf{r}}_c(t)$ and $\omega = \dot{\varphi}(t)$ the time derivatives of the robots position $\mathbf{r}_c(t)$ and heading $\varphi(t)$. To solve the above stated trajectory tracking problem, consider the Lyapunov candidate function

$$V = \frac{1}{2}\mathbf{e}_r^T\, K_r\, \mathbf{e}_r + \frac{1}{2}e_\varphi^T\, K_\varphi\, e_\varphi \tag{16}$$

being $K_r \in R^{2\times 2}$ a symmetric positive definite ($K_r > 0$) matrix and K_φ a positive constant. The time derivative of V results in

$$\dot{V} = \mathbf{e}_r^T\, K_r\, (\dot{\mathbf{r}}_d(t) - J_{lv}\dot{\mathbf{q}}) + e_\varphi^T\, K_\varphi\, (\dot{\varphi}_d(t) - J_\omega \dot{\mathbf{q}})\,. \tag{17}$$

Denoting with $J_{lv}^\dagger$ and $J_\omega^\dagger$ the right pseudo-inverse matrices of full rank J_{lv} and J_ω respectively (J_{lv} and J_ω are full rank by hypothesis),

$$J_{lv}^\dagger = J_{lv}^T\, \left(J_{lv}\, J_{lv}^T\right)^{-1} \quad \text{and} \quad J_\omega^\dagger = J_\omega^T\, \left(J_\omega\, J_\omega^T\right)^{-1} \tag{18}$$

a possible value for $\dot{\mathbf{q}}$ making $\dot{V}$ in equation (17) negative definite is:

$$\dot{\mathbf{q}}_d(t) = \dot{\mathbf{q}}_{lvd}(t) \;+\; \dot{\mathbf{q}}_{\varphi d}(t) \tag{19}$$

$$\dot{\mathbf{q}}_{lvd}(t) = J_{lv}^\dagger\, (\, \dot{\mathbf{r}}_d(t) + K_r\, \mathbf{e}_r(t)\,) \tag{20}$$

$$\dot{\mathbf{q}}_{\varphi d}(t) = J_\omega^\dagger\, (\, \dot{\varphi}_d(t) + K_\varphi\, e_\varphi(t)\,) \tag{21}$$

implying in closed loop

$$\dot{V} = -\mathbf{e}_r^T\, K_r\, K_r\, \mathbf{e}_r - (K_\varphi\, e_\varphi)^2 < 0. \tag{22}$$

As for standard tracking controllers, the solution in equation (19) is a combination of feedforward terms proportional to the reference linear and angular velocities and a feedback term. The proposed solution guarantees global exponential stability of equilibrium $\mathbf{e}_r = \mathbf{0}, e_\varphi = 0$ of the error dynamics, thus (robustly) solving the trajectory tracking problem. Control law (19) is the sum of two contributions: the first (20) relative to position tracking and the second (21) to heading tracking. In the light of property (8), it should be noticed that the two contributions do not interfere with each other, namely the contribution of $\dot{\mathbf{q}}_{lvd}$ to the robots angular velocity and the contribution of $\dot{\mathbf{q}}_{\varphi d}$ to the robots linear velocity are both null, i.e.

$$J_\omega \dot{\mathbf{q}}_{lvd} = J_\omega \left(J_{lv}^T \left(J_{lv} J_{lv}^T \right)^{-1} \right) \left(\dot{\mathbf{r}}_d(t) + K_r \, \mathbf{e}_r(t) \right) = 0 \tag{23}$$

$$J_{lv} \dot{\mathbf{q}}_{\varphi d}(t) = J_{lv} \left(J_\omega^T \left(J_\omega J_\omega^T \right)^{-1} \right) \left(\dot{\varphi}_d(t) + K_\varphi \, e_\varphi(t) \right) = \mathbf{0} \tag{24}$$

due to the fact that they are proportional to $J_\omega J_{lv}^T$ and $J_{lv} J_\omega^T$ respectively. As discussed above, when designing vehicle kinematics guidance laws it must be assumed that the lower level (actuator) dynamics should be much faster than the kinematics. This requirement is reflected on design choices such as actuator power and desired reference trajectories: the former needs to be sufficiently large for the given inertial properties of the vehicle so that maximum vehicle accelerations can be much larger than the maximum reference accelerations $\ddot{\varphi}_d(t)$ and $\ddot{\mathbf{r}}_d(t)$. As far as the ratio of maximum vehicle acceleration over maximum reference acceleration is sufficiently large the dynamic behaviour of the kinematics guidance law will be fine. Thus, as for any other kinematics designed guidance solution, the proposed control law should be implemented on Swedish wheeled vehicles with sufficiently powerful actuators with respect to the maximum reference accelerations $\ddot{\varphi}_d(t)$ and $\ddot{\mathbf{r}}_d(t)$. As for actuator saturation, the situation is slightly more complex. Given the proportional nature of the control law (19), the tracking error (either in position or heading) or the desired reference velocities can always happen to be large enough for the actuators to saturate, namely calling $\dot{q}_{\max} > 0$ the maximum absolute angular velocity that the vehicles actuators are able to generate, whatever the gains K_r and K_φ should be, depending on $\dot{\varphi}_d(t)$, $\dot{\mathbf{r}}_d(t)$, $\mathbf{e}_r(t)$ or $e_\varphi(t)$ the saturation condition[1]

$$\|\dot{\mathbf{q}}_d\|_\infty \leq \dot{q}_{\max} \tag{25}$$

may always be violated. Notice that while the feedforward signals $\dot{\varphi}_d(t)$ and $\dot{\mathbf{r}}_d(t)$ can eventually always be bounded, the tracking error's initial conditions are not design parameters. Hence a commanded $\dot{\mathbf{q}}_d$ with exceeding infinity-norm due to odd initial conditions cannot be *a priori* excluded.

3.2 Actuator Saturation

The presence of actuator saturation has a severe impact on performance: in particular given the additive structure of equation (19), saturation can affect the decoupling between commanded angular and linear velocities. In order to

[1] Given $\mathbf{q} \in R^{N \times 1}$, $\|\mathbf{q}\|_\infty = \max\{|q_1|, |q_2|, \ldots, |q_N|\}$ is its infinity norm.

cope with actuator saturation and to guarantee a prioritized execution of the position and heading tracking tasks, the following modification of the proposed control law is suggested: the sum in equation (19) should be weighted with error and reference dependent weights such that (*i*) the resulting $\dot{\mathbf{q}}_d$ command has norm within the actuator limits, (*ii*) the tasks (position and heading in the present case) are executed with a priority based time order (higher priority tasks earlier) and (*iii*) the tracking error converges to zero.

To reach these goals, consider the saturation function

$$\sigma : R \times [0, \infty) \longrightarrow R \quad \text{such that} \quad \sigma(x, c) = \begin{cases} 0 & \text{if} \quad x = 0 \\ 1 & \text{if} \quad 0 < |x| < c \\ c/|x| & \text{otherwise.} \end{cases} \tag{26}$$

In the sequel the non negative second argument c of $\sigma(x, c)$ will be called the capacity of x. Notice that by definition $\sigma(x, c)$ is simply a nonnegative scalar scaling factor such that $x\,\sigma(x, c)$ is "clipped" to $c\,\mathrm{sign}(x)$ whenever $|x|$ should exceed the capacity c and is equal to x otherwise, i.e. $x\,\sigma(x, c)$ is simply the saturated version of x in the range $[-c, c]$. Also notice that by its very definition

$$\sigma(x, 0) = 0 \ \ \forall \ \ x, \tag{27}$$

namely if x should be assigned zero capacity, then $x\,\sigma(x, 0) = 0$ for any value of x. Assume that

$$\dot{\mathbf{q}}_d = \sum_{l=1}^{n} \dot{\mathbf{q}}_l \tag{28}$$

is the actuator input being $\dot{\mathbf{q}}_1, \dot{\mathbf{q}}_2, \ldots, \dot{\mathbf{q}}_n$ n independent task inputs ordered by decreasing priority with increasing index ($\dot{\mathbf{q}}_1$ has highest priority). Each term on the right hand side of equation (28) and $\dot{\mathbf{q}}_d$ itself should be bounded by $\dot{q}_{\max}$. Considering that each task should be executed in a prioritized fashion, the sum in (28) may be replaced by a weighted sum as follows:

$$\dot{\mathbf{q}}_d = \dot{\mathbf{q}}_1\,\sigma\left(\|\dot{\mathbf{q}}_1\|_\infty, c_1\right) \,+\, \dot{\mathbf{q}}_2\,\sigma\left(\|\dot{\mathbf{q}}_2\|_\infty, c_2\right) \,+\, \ldots \,+\, \dot{\mathbf{q}}_n\,\sigma\left(\|\dot{\mathbf{q}}_n\|_\infty, c_n\right) \tag{29}$$

where each task capacity is recursively and dynamically computed as:

$$\begin{aligned} c_1(t) &\leq \dot{q}_{\max} \quad (\text{constant, i.e. } \dot{c}_1(t) = 0) \\ c_2(t) &= c_1 - \|\dot{\mathbf{q}}_1\|_\infty\, \sigma\left(\|\dot{\mathbf{q}}_1\|_\infty, c_1\right) \\ c_3(t) &= c_2(t) - \|\dot{\mathbf{q}}_2\|_\infty\, \sigma\left(\|\dot{\mathbf{q}}_2\|_\infty, c_2(t)\right) \\ \vdots &= \vdots \\ c_n(t) &= c_{n-1}(t) - \|\dot{\mathbf{q}}_{n-1}\|_\infty\, \sigma\left(\|\dot{\mathbf{q}}_{n-1}\|_\infty, c_{n-1}(t)\right). \end{aligned} \tag{30}$$

$c_1, \ldots, c_n$ can be chosen according to the state of the environment, i.e. in RoboCup they can depend on global external game states. Notice that by construction all the above task capacities are non negative, i.e.

$$c_j \geq 0 \ \ \forall \ \ j \in [1, n],$$

and that

$$c_j \le c_{j-1} \quad \forall \quad j \in [2, n]$$
$$c_i = 0 \implies c_j = 0 \quad \forall \quad j > i$$

namely if a given task is assigned zero capacity, all the lower priority tasks will also automatically have zero capacity and all their weights in the sum (29) will be zero. The capacity of task i can be viewed as the residual capacity after the higher priority task $i-1$ has been commanded; thus, by example, c_2 will be zero (and also $c_j : j > 2$) if the task 1 input $\dot{\mathbf{q}}_1$ is saturating all its capacity c_1. In words, each task will be commanded with a non null weight only if the higher priority task have not saturated. The fact that c_1 needs not to exceed $\dot{q}_{\max}$ is due to the fact that task 1 alone should not saturate the actuator capacity $\dot{q}_{\max}$; moreover given that $c_{j+1} \le c_j \; \forall \; j \in [1, n-1]$ the condition $c_1 \le \dot{q}_{\max}$ guarantees that *each* term in the sum (29) will have infinity norm smaller or equal to the threshold $\dot{q}_{\max}$. Indeed the proposed dynamic update rule for the task's capacities also guarantees that the total control signal (29) has infinity norm smaller or equal than $\dot{q}_{\max}$. In order to implement the above described schema in the present trajectory tracking case, assume that the reference feedforward linear and angular velocities are sufficiently small, namely that

$$\left\| J_{lv}^{\dagger} \, \dot{\mathbf{r}}_d(t) \right\|_{\infty} < \frac{1}{2} \; \dot{q}_{\max} \quad \forall \, t \tag{31}$$
$$\left\| J_{\omega}^{\dagger} \, \dot{\varphi}_d(t) \right\|_{\infty} < \frac{1}{2} \; \dot{q}_{\max} \quad \forall \, t. \tag{32}$$

These conditions are necessary to guarantee that the tracking task is asymptotically feasible, namely that when the position and heading tracking errors are null the control effort of the control law (19) is compatible with the actuator saturation limit, i.e.

$$\mathbf{e}_r = \mathbf{0}, e_{\varphi} = 0 \quad \implies$$
$$\|\dot{\mathbf{q}}_d(t)\|_{\infty} = \left\| J_{lv}^{\dagger} \, \dot{\mathbf{r}}_d(t) + J_{\omega}^{\dagger} \, \dot{\varphi}_d(t) \right\|_{\infty} \le \left\| J_{lv}^{\dagger} \, \dot{\mathbf{r}}_d(t) \right\|_{\infty} + \left\| J_{\omega}^{\dagger} \, \dot{\varphi}_d(t) \right\|_{\infty} < \dot{q}_{\max}.$$

As a first example, assume that position tracking is assigned highest priority with respect to heading tracking. Then define:

$$\dot{\mathbf{q}}_1 := J_{lv}^{\dagger} \, \dot{\mathbf{r}}_d(t) \tag{33}$$
$$\dot{\mathbf{q}}_2 := J_{lv}^{\dagger} \, K_r \, \mathbf{e}_r(t) \tag{34}$$
$$\dot{\mathbf{q}}_3 := J_{\omega}^{\dagger} \, \dot{\varphi}_d(t) \tag{35}$$
$$\dot{\mathbf{q}}_4 := J_{\omega}^{\dagger} \, K_{\varphi} \, e_{\varphi}(t). \tag{36}$$

With these definitions consider the control law (29-30) with

$$c_1(t) = \dot{q}_{\max} > 0 \;\; \forall \;\; t$$

that together with the feasibility condition (31) implies

$$0 < \frac{1}{2} \dot{q}_{\max} \;\; \le \;\; c_2 \;\; \le \;\; \dot{q}_{\max},$$

i.e. the tasks 1 and 2 have always non null capacity. Moreover as by hypothesis $\|\dot{\mathbf{q}}_1\|_\infty < 0.5\,\dot{q}_{\max}$ (equation (31)) and $c_1 = \dot{q}_{\max}$, it follows that

$$\dot{\mathbf{q}}_1\,\sigma(\|\dot{\mathbf{q}}_1\|_\infty, c_1) = \dot{\mathbf{q}}_1 \quad \forall \quad t.$$

Consequently

$$\begin{aligned} V_1 &= \frac{1}{2}\mathbf{e}_r^T\,K_r\,\mathbf{e}_r \quad \Longrightarrow \\ \dot{V}_1 &= \mathbf{e}_r^T\,K_r\,\left(\dot{\mathbf{r}}_d(t) - J_{lv}\left[J_{lv}^\dagger\,\dot{\mathbf{r}}_d(t) + J_{lv}^\dagger\,K_r\,\mathbf{e}_r(t)\,\sigma\left(\left\|J_{lv}^\dagger\,K_r\,\mathbf{e}_r(t)\right\|_\infty, c_2\right)\right]\right) = \\ &= -\mathbf{e}_r^T\,K_r\,K_r\,\mathbf{e}_r(t)\,\sigma\left(\left\|J_{lv}^\dagger\,K_r\,\mathbf{e}_r(t)\right\|_\infty, c_2\right) < 0 \end{aligned} \tag{37}$$

i.e. $\dot{V}_1$ is negative definite that proves asymptotic global Lyapunov stability of $\mathbf{e}_r = \mathbf{0}$. Notice that $\dot{\mathbf{q}}_3$ and $\dot{\mathbf{q}}_4$ do not contribute to $\dot{V}_1$ as they belong to the null space of J_{lv} (equation 8). As far as the secondary (heading) task is concerned, convergence can also be proven through a Lyapunov argument. The global asymptotic stability of $\mathbf{e}_r = \mathbf{0}$ guarantees that

$$\lim_{t\to\infty} \dot{\mathbf{q}}_2(t) = \mathbf{0} \quad \Longrightarrow \quad \lim_{t\to\infty} c_3 = c_2 \;\geq\; \frac{1}{2}\dot{q}_{\max}.$$

Given the feasibility condition (32), this means that

$$\exists\ t^* \quad : \quad \dot{\mathbf{q}}_3\,\sigma(\|\dot{\mathbf{q}}_3\|_\infty, c_3) = \dot{\mathbf{q}}_3 \ \text{ and } \ c_4 > 0 \ \forall\ t \geq t^*.$$

It follows that

$$\begin{aligned} V_2 &= \frac{1}{2}e_\varphi^T\,K_\varphi\,e_\varphi \quad \Longrightarrow \\ \dot{V}_2(t)\Big|_{t\geq t^*} &= e_\varphi^T\,K_\varphi\,(\dot{\varphi}_d(t) - J_\omega \dot{\mathbf{q}}_d) = \\ &= e_\varphi^T\,K_\varphi\,\left[\dot{\varphi}_d(t) - J_\omega(J_\omega^\dagger\,\dot{\varphi}_d(t) + J_\omega^\dagger\,K_\varphi\,e_\varphi(t)\,\sigma\left(\|J_\omega^\dagger\,K_\varphi\,e_\varphi(t)\|_\infty, c_4\right))\right] = \\ &= -e_\varphi^T\,K_\varphi^2\,e_\varphi(t)\,\sigma\left(\|J_\omega^\dagger\,K_\varphi\,e_\varphi(t)\|_\infty, c_4\right) < 0 \end{aligned} \tag{38}$$

namely there exists a finite time t^* after which the time derivative of V_2 is always negative, thus proving convergence to zero of the heading error $e_\varphi(t)$. Prior to t^* the heading error $e_\varphi(t)$ is not guaranteed to be decreasing. Notice that $\dot{\mathbf{q}}_1$ and $\dot{\mathbf{q}}_2$ do not contribute to $\dot{V}_2$ as they belong to the null space of J_ω (equation 8).

As a second example, heading can be selected to be the highest priority task, it is then sufficient to select $\dot{\mathbf{q}}_1, \ldots, \dot{\mathbf{q}}_4$ as

$$\dot{\mathbf{q}}_1 := J_\omega^\dagger\,\dot{\varphi}_d(t) \tag{39}$$

$$\dot{\mathbf{q}}_2 := J_\omega^\dagger\,K_\varphi\,e_\varphi(t) \tag{40}$$

$$\dot{\mathbf{q}}_3 := J_{lv}^\dagger\,\dot{\mathbf{r}}_d(t) \tag{41}$$

$$\dot{\mathbf{q}}_4 := J_{lv}^\dagger\,K_r\,\mathbf{e}_r(t) \tag{42}$$

in equations (29-30); Lyapunov stability of the heading error and asymptotic convergence of the position error could be proven accordingly.

4 Experimental Results

The proposed control law has been experimentally tested on the Volksbot platform (*www.volksbot.de*) [10] developed at the Fraunhofer AiS - Autonomous intelligent Systems Institute of Sankt Augustin, Germany. The robot is about 8Kg in weight and is actuated by three 90 Watts, 24 volt DC motors with a 1 : 8 gear ratio. Low level wheel speed control is achieved through a 3 channel PID motor driver (the AiS TMC200 board) interfaced to an onboard laptop via a regular serial RS232 line. The presented kinematics trajectory tracking control law, i.e. the guidance loop in figure (2), is implemented on the onboard laptop. Motor power is supplied through NiMH batteries with 3, 5 Ah capacity. The three omnidirectional wheels have a 5cm radius, are made of lightweight plastic and are mounted at 120° from each other. The robot is equipped with an omnivision system made by a 30Hz, 640 × 480 pixels YUV color FireWire camera pointing towards a 70mm diameter hyperbolic mirror. Such systems are used for map-based Monte Carlo self-localization [11] [13]. Details can be found in [12].

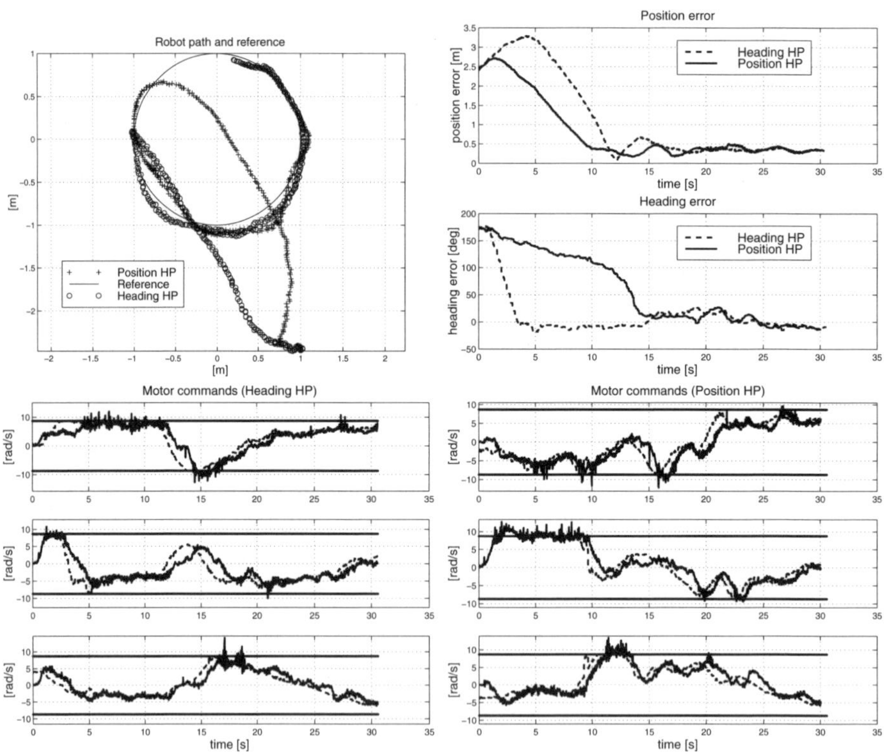

Fig. 3. Experimental results. HP stands for High Priority. Refer to text for details.

In order to evaluate the performance of the proposed control solution the position and heading of the robot must be measured reliably and compared with the desired reference values. To this extent an experimental setup has been designed where the position of the robot was measured by a fixed laser range finder pointing on the robot and its heading was measured by the robot itself using its omnivision system. All collected data was suitably synchronized with the desired references. Extensive experimental trials with several different references have shown the effectiveness of the proposed solution: the case of a circular reference trajectory with constant (with respect to a fixed frame) heading is reported in figure (3). The position and heading error plots with respect to time clearly confirm the effectiveness of the priority assignment policy. The growth of the position errors in the first few seconds of the experiment (top right plot) are due to the robots dynamics that was neglected in the control law design. As expected, as long as the actuators guarantee large enough accelerations with respect to the reference accelerations, the kinematics designed control law exhibits good dynamic performance, i.e. there is only a small lag with respect to the ideal purely kinematics case.

In the motor command plots, the commanded ($\dot{q}_d$, dashed lines) and encoder measured wheel speeds (solid lines) are reported with respect to time. Notice that for the sake of performance measurement accuracy, the saturation threshold was artificially set to the value of ± 8.7rad/s (thick solid lines) via software in order to achieve saturation at acceptable linear speeds.

5 Conclusions

A trajectory tracking control law for Swedish wheeled robots has been derived that takes explicitly into account motor saturations. Motor saturation is always present and may have a sever impact on motion control performances of mobile robots. This is particularly relevant for omnidirectional mobile robots equipped with Swedish wheels: these offer a lower grip with the floor with respect to traditional wheels resulting in a higher probability of exhibiting skidding and/or sliding when high velocity commands are issued. As a consequence the possibility of commanding motor speeds always compatible with the saturation limits is extremely important for omnidirectional mobile robots. Moreover the introduction of a task based prioritization of heading and position tracking may have a relevant impact on the behaviour control level. The selection of heading or position tracking tasks as higher priority ones will generally depend on the (dynamic) role assignment: by using the described lower level control solution the highest priority tasks errors are guaranteed to converge faster to zero without ever commanding motor speeds exceeding the maximum HW allowed values. Future work directions should include studies on how the behaviour system should take advantage of a guaranteed prioritized convergence of the tracking errors.

References

1. Diegel, O., Badve, A., Bright, G., Potgieter, J., Tlale, S.: Improved Mecanum Wheel Design for Omni-directional Robots. In: Proc. 2002 Australian Conference on Robotics and Automation, Auckland, pp. 27–29 (November 2002)
2. Campion, G., Bastin, G., D'Andréa-Novel, B.: Structural Properties and Classification of Kinematic and Dynamic Models of Wheeled Mobile Robots. IEEE Transactions on Robotics and Automation 12(1), 47–62 (1996)
3. Agulló, J., Cardona, S., Vivancos, J.: Kinematics of Vehicles with Directional Sliding Wheels. Mech. Mach. Theory 22(4), 295–301 (1987)
4. Muir, P.F., Neuman, C.P.: Kinematic Modeling For Feedback Control Of An Omnidirectional Wheeled Mobile Robot. In: Cox, J., Wilfong, G.T. (eds.) Autonomous robot vehicles, pp. 25–31. Springer, New York (1990)
5. Pin, F.G., Killough, S.M.: A New Family of Omnidirectional and Holonomic Wheeled Platforms for Mobile Robots. IEEE Transactions on Robotics and Automation 10(4), 480–489 (1994)
6. Asama, H., Sato, M., Bogoni, L., Kaetsu, H., Matsumoto, A., Endo, I.: Development of an Omni-Directional Mobile Robot with 3 DOF Decoupling Drive Mechanism. In: Proc. of the 1995 IEEE International Conference on Robotics and Automation (ICRA 95), Nagoya, Aichi, Japan, pp. 1925–1930 (1995)
7. Loh, W.K., Low, K.H., Leow, Y.P.: Mechatronics Design and Kinematic Modelling of a Singularityless Omni-Directional Wheeled Mobile Robot: In: Proc. of the, IEEE International Conference on Robotics and Automation (ICRA 03), Tapei, Taiwan, September 14 - 19, 2003, pp. 3237–3242 (2003)
8. Kim, W.K., Yi, B.-J., Lim, D.J.: Kinematic Modeling of Mobile Robots by Transfer Method of Augmented Generalized Coordinates. Journal of Robotic Systems 21(6), 275–300 (2004)
9. Song, J.-B., Byun, K.-S.: Design and Control of a Four-Wheeled Omnidirectional Mobile Robot with Steerable Omnidirectional Wheels. Journal of Robotic Systems 21(4), 193–208 (2004)
10. Wisspeintner, T., Nowak, W., Bredenfeld, A.: Volksbot: A flexible component based mobile Robot system. In: ROBOCUP Symposium 2005, Springer LNAI, http://www.ais.fraunhofer.de/BE/2005/Wisspeintner_2005_RoboCup_VolksBot.pdf (to appear)
11. Menegatti, E., Pretto, A., Pagello, E.: A new omnidirectional vision sensor for monte-carlo localization. In: Int. RoboCup Symposium CD-ROM, Lisbon, Portugal (2004)
12. Olufs, S.: Vision-Based Probabilistic State Estimation using Omnidirectional Cameras. Master Thesis FH-BRS, St. Augustin (2005)
13. Schulz, D., Burgard, W., Fox, D., Cremers, A.: People tracking with a mobile robot using sample-based joint probabilistic data association filters. International Journal of Robotics Research (IJRR). Springer (2003)

Imitative Reinforcement Learning for Soccer Playing Robots

Tobias Latzke, Sven Behnke, and Maren Bennewitz

University of Freiburg, Computer Science Institute, D-79110 Freiburg, Germany
{latzke,behnke,maren}@informatik.uni-freiburg.de
http://www.NimbRo.net

Abstract. In this paper, we apply Reinforcement Learning (RL) to a real-world task. While complex problems have been solved by RL in simulated worlds, the costs of obtaining enough training examples often prohibits the use of plain RL in real-world scenarios. We propose three approaches to reduce training expenses for real-world RL. Firstly, we replace the random exploration of the huge search space, which plain RL uses, by guided exploration that imitates a teacher. Secondly, we use experiences not only once but store and reuse them later on when their value is easier to assess. Finally, we utilize function approximators in order to represent the experience in a way that balances between generalization and discrimination. We evaluate the performance of the combined extensions of plain RL using a humanoid robot in the RoboCup soccer domain. As we show in simulation and real-world experiments, our approach enables the robot to quickly learn fundamental soccer skills.

1 Introduction

Reinforcement learning (RL) is an established machine learning technique. In a trial and error based procedure, an agent acquires knowledge about the consequences of its actions and strategies to attain a certain goal [1]. While RL methods have been successfully applied to complex problems in simulated environments [2,3], they have rarely been used for real-world scenarios. The high costs of obtaining enough training examples for real systems often prohibits the acquisition of successful behavior by means of plain trial and error. The central intention of our work is the reduction of training expenses for RL methods so that they are applicable to real-world scenarios. In this paper, we propose three approaches to achieve this goal:

1. Speeding-up the exploration through imitation of a teacher
2. Repeated reevaluation of past experiences
3. Application of function approximators for better generalization

Imitation allows for knowledge transfer by observation between sufficiently similar agents. Imitation learning is a well established concept in robotics [4,5,6,7,8,9]. The idea is that the learning agent observes the actions of an experienced agent as well as the corresponding consequences. These observations give the learner clues about successful strategies to reach the goal. Through the imitation of the teacher, the exploration of the huge search space is guided to regions that are promising. The learning agent no longer

G. Lakemeyer et al. (Eds.): RoboCup 2006, LNAI 4434, pp. 47–58, 2007.

depends on random exploration but rather on meaningful indications which actions to choose.

In addition to the guidance by other agent's experiences, the extensive exploitation of own experiences is crucial for learning in real-world systems. Many learning algorithms rely on the online processing of experiences at execution time and discard them immediately afterwards. Often however, the full merit of an experience can be assessed only later in the learning process when more information is available. We reduce the amount of training data that has to be collected by storing and reusing experiences. The extensive use of both, own and observed experiences, provides the necessary knowledge to choose promising actions that lead to a successful performance.

In complex domains, however, it is highly unlikely that exactly the same situation is encountered twice. Generalization to similar situations is therefore essential. On the other hand, generalization should not prevent discrimination between different situations. The function approximation technique presented in this paper combines the advantages of quick generalization and accurate long-term discrimination.

We present extensive experiments to evaluate the performance of a combination of the proposed extensions to plain RL. In simulation as well as in real-world experiments with a humanoid robot in the RoboCup soccer domain, we demonstrate how the robot is able to quickly learn fundamental soccer skills.

This paper is organized as follows. In the following section, we briefly introduce Q-learning, which is a popular RL algorithm. In Section 3, we describe function approximation and its application to Q-learning. In Section 4, we present our approach to function approximation that allows for quick generalization as well as for sufficient discrimination. Section 5 explains our use of imitation and memory to guide the exploration. Section 6 describes the robot hardware as well as the application of RL to a specific task and Section 7 presents the experimental results obtained in simulation and with a real robot. Finally, in Section 8 we discuss related work.

2 Q-Learning

The framework underlying RL is that of Markov decision processes (MDPs), which describe the effects of actions in a stochastic environment and the possible rewards at the different environmental states. The goal of the agent is to maximize the expected (discounted) future reward, without knowing the MDP or the reward function in advance.

The action selection according to the current state is called the agent's policy. RL methods use an estimate of the expected cumulated future reward, the utility function, to derive a policy in order to maximize the long-term reward. In our work we use an ϵ-greedy policy, i.e. the agent chooses a random action with probability ϵ and the action with the highest utility expectation otherwise.

Temporal-difference (TD) methods [10] like Q-learning perform an update of the utility estimate after each transition (s, a, s'), where s is the state in which action a was executed and s' is the resulting state. As the cumulated future reward is not known in advance, TD methods use an estimate of the utility of s' to update the utility function. In Q-learning, the utility function is called Q and maintains utility values for each state-action pair. The update rule after a transition is given by

$$Q_{t+1}(s,a) = Q_t(s,a) + \alpha \cdot \left(r_{t+1} + \gamma \cdot \max_{a' \in \mathcal{A}} Q_t(s',a') - Q_t(s,a) \right) . \tag{1}$$

Here, r is the immediate reward, $0 < \alpha \leq 1$ is the learning rate and $0 < \gamma \leq 1$ is the discount factor.

3 Function Approximation

A tabular representation of the Q-function that stores one estimation for each state-action pair is only useful for small problems. In practice however, the number of possible states is often very large or even infinite, making it impossible to maintain a table of all Q-values. Furthermore, the completely isolated evaluation of each state-action pair is not appropriate since it does not happen often that exactly the same situation is encountered twice. In practice, many situations appear similar and are discriminable only in detail. Experiences should therefore be generalized. Function approximation is a method that is often used for generalization.

The general notation of function approximation is that given we have input-output pairs (x, y), we want to compute a function f that is an approximation of the unknown function f^* that produced (x, y). First, a prototype for f has to be defined, i.e. a class of functions, that allows to express f^* with a suitable parameterization. The approximation f is represented by a parameter vector $\boldsymbol{\theta}$, which can be seen as instantiation of the function prototype.

3.1 Function Approximation Using Gradient Descent Methods

The idea of gradient descent methods is to modify the parameter vector $\boldsymbol{\theta}$ toward the direction that yields the greatest error reduction for the example under consideration. This is done by calculating the gradient of the local error term with respect to $\boldsymbol{\theta}$

$$\begin{aligned} \boldsymbol{\theta}_{t+1} &= \boldsymbol{\theta}_t + \frac{1}{2}\eta \nabla_{\boldsymbol{\theta}_t} (y - f_t(x))^2 \\ &= \boldsymbol{\theta}_t + \eta (y - f_t(x)) \nabla_{\boldsymbol{\theta}_t} f_t(x) . \end{aligned} \tag{2}$$

Here, η is called the step size parameter and $\boldsymbol{\theta}_t$ and f_t indicate the parameter vector as well as the approximation of f^* at time t.

In this paper, we consider the special case of linear gradient descent. We extract a set of features from the input x and represent them as the feature vector $\boldsymbol{\phi}_x$. The number of parameters n is equal to the number of features and the function prototype is given by the weighted sum of the features

$$f(x) = \sum_{i=1}^{n} \boldsymbol{\theta}(i) \boldsymbol{\phi}_x(i) . \tag{3}$$

The components of $\boldsymbol{\theta}$ are called the weights of the corresponding features. The advantage of this prototype is that the gradient $\nabla_{\boldsymbol{\theta}} f(x)$ corresponds to the feature vector $\boldsymbol{\phi}_x$. However, care has to be taken that suitable features are extracted from x to provide linear correlation between $\boldsymbol{\phi}_x$ and y in the case that the correlation between x and y is non-linear.

3.2 *Q*-Learning with Function Approximation

A training example (x, y) in RL consists of the previous state s and the executed action a as input and the target for Q-learning $r_{t+1} + \gamma \max_{a' \in \mathcal{A}} Q_t(s', a')$ as output. The update rule for Q-learning with linear gradient descent is then given by:

$$\boldsymbol{\theta}_{t+1} = \boldsymbol{\theta}_t + \eta(r_{t+1} + \gamma \max_{a' \in \mathcal{A}} Q_t(s', a') - Q_t(s, a))\, \boldsymbol{\phi}_{s,a} \,. \tag{4}$$

The update rule contains the feature vector $\boldsymbol{\phi}_{s,a}$, which is the function's gradient. This implements responsibility assignment: Those features, that have been most active in the last decision, are most responsible for the current local error. Thus the corresponding weights have to be adjusted most.

4 Feature Construction

A typical approach for feature construction is the use of input features that indicate whether or not the input is inside a certain region of the state space. Such a region is called the receptive field of the corresponding feature. We construct the features following the principles of coarse coding [11]. The essential idea is to use multiple large receptive fields, which may overlap, so that an input can activate multiple features. Two inputs are similar in the features that are present for both or absent for both and they differ in the features that are present for only one of the inputs. The simultaneous consideration of similarity and difference is one important instrument to deal with the conflict between generalization and discrimination, which typically occurs in learning systems. In this work, we use a variant with continuous features that have an activation value between 0 (i.e. the input is outside the receptive field) and 1 (i.e. the input is in the center of the receptive field). The activation function Φ is called the feature's shape.

In the one-dimensional case the receptive field is an interval $[b_0, b_1]$. Let $c = \frac{b_0+b_1}{2}$ be the center of the receptive field. Then the activation function is:

$$\Phi(x) = \begin{cases} 1 - \frac{2|x-c|}{b_1-b_0} & \text{if} \quad x \in [b_0, b_1] \\ 0 & \text{else} \end{cases} \,. \tag{5}$$

For the transfer to n dimensions we propose the product of the activities on each dimension respectively:

$$\Phi(x^{(1)}, \ldots, x^{(n)}) = \begin{cases} \prod_{i=1}^{n} \left(1 - \frac{2|x^{(i)}-c^{(i)}|}{b_1^{(i)}-b_0^{(i)}}\right) & \text{if} \quad \forall i \leq n \colon x^{(i)} \in [b_0^{(i)}, b_1^{(i)}] \\ 0 & \text{else} \end{cases} \,, \tag{6}$$

where $x^{(i)}$ and $b_{0/1}^{(i)}$ denote the input and the borders of the receptive field in the dimension i.

The whole state space is covered completely by a group of features. We call this group a generic feature. The dimensions are partitioned into uniform intervals with the borders $a_0, \ldots, a_e$, where e is the total number of intervals on this dimension (the resolution). In one dimension each receptive field covers two adjacent intervals with its

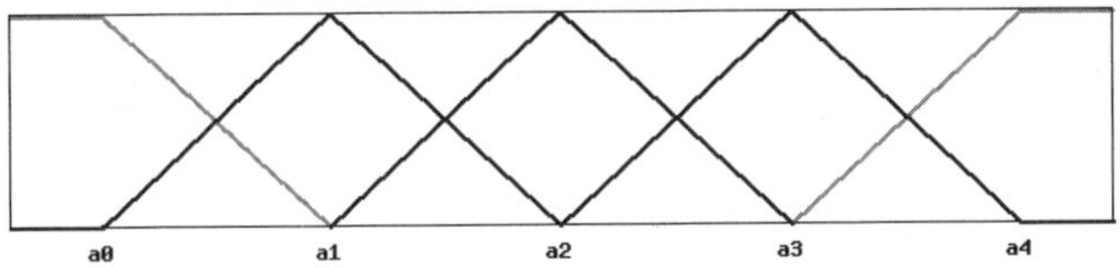

Fig. 1. 1-dimensional generic feature

peak exactly on their border. For each pair of intervals there is one feature plus two extra features at the border of the domain. Fig. 1 shows the generated features for a resolution of $e = 4$. For each input value there are two active features with a total activation of 1. Each feature contributes to the approximation according to its activation. This can be interpreted as the feature's responsibility for a certain input. The systematic overlapping within a generic feature corresponds to a smooth shift of responsibility between two adjacent features. The resulting approximation is a linear interpolation of the adjacent weights.

5 Imitation and Memory

Imitation is an important mechanism for the transfer of knowledge and skills between similar agents. An imitating agent can observe the behavior of an experienced agent and thus draw conclusions about the consequences of certain actions.

In this paper, the imitating agent has full access to experiences of a teacher. These experiences are provided as sequences of states and actions in the learning agent's representation along with the corresponding rewards. Our imitation approach relies on the evaluation of these sequences according to own criteria. One possibility to do so, is the application of the Q-learning algorithm with the stored sequence of actions. The imitation effect can be increased by processing the sequences in temporally inversed order. This is possible because the whole sequence of transitions and rewards is known in advance.

One advantage of storing and evaluating such sequences is the fact that the same algorithm can be applied to processing own experiences. This can be seen as a form of episodic memory, which allows to "revive" own experiences when their utility is easier to access. If, e.g., the agent chooses a very good action by chance, but does not see immediately that it was a good action or why it was good, then it can be helpful to revive that experience later in the learning process when increased knowledge allows for a better assessment of this observation.

The evaluation of observations in temporally inversed order is related to another technique for faster information propagation in temporal-difference methods. Eligibility traces [10,12] use the *last* local error to update the value not only for the current state but for the recently visited states as well. So there is no individual local error used to update the previous states of the stored sequence. In contrast to that, we calculate a *new* local error for *each* transition in the stored sequence. Thus, we treat each step of the sequence as if it was the actual observation. Convergence proofs for table-based Q-learning (e.g. [13]) require to visit all states and to choose all actions infinitely often, but they do not make any assumptions about the ordering of these observations. Thus, the guarantees on convergence remain unchanged under the appropriate conditions.

6 Task and Implementation

The proposed concepts of function approximation, imitation, and memory are evaluated for a humanoid toy robot called RoboSapien, which has been augmented with a camera and a Pocket PC as to allow for autonomous behavior as proposed in [14]. The task of the agent is to dribble the ball from different positions into the empty goal. Note that dribbling is achieved by walking against the ball. There is no explicit movement to kick the ball. The exact task setup is taken from the scoring test defined in [14]. In this test, the robot stands on the most distant point of the center circle, facing the empty goal. The ball is placed at ten different positions on the other half of the center circle in the robot's field of view. One advantage of the adoption of this existing scoring test is the possibility to compare the performance to an existing hard coded behavior.

According to the given task, suitable state variables have to be chosen that contain all relevant information for learning a good policy. Obviously, the positions of the ball and the goal are necessary to perform the task. While the ball position can be expressed by two variables (e.g. angle and distance), this is not sufficient for the goal position. Since the goal has a significant width, an additional variable is required (e.g. left post angle and right post angle instead of one angle). These five variables define the unambiguous position of all relevant objects and their orientation to each other. There are many possibilities to represent this information.

The previous paragraph implicitly assumes an egocentric representation of the information, i.e. the relative positions of the ball and the goal from the agent's point of view. Another possibility is the transformation to an allocentric representation, i.e. the absolute positions of ball and robot on the field (including the robot's orientation). Further possibilities are e.g. egocentric Cartesian coordinates (x and y) instead of polar coordinates (angle and distance).

In this work, we use a combination of these representations and additional variables like the square roots of the distances to the ball and to the goal. In total we use 27 variables to represent the state space. This redundant representation allows the simultaneous consideration of different aspects of the situation. Each single representation is well-suited for the detection of some properties and similarities of situations whereas others can hardly be distinguished. The combination of several representations allows to benefit from the advantages of each representation. The additional state variables do not cause significant additional costs of computation since they just provide another perspective on the same data. As our function approximation method relies on many partitionings of the state space anyway, there is no difference (concerning the number of features) in using two partitionings of the same representation and using one partitioning of two different representations.

The action space of our robot consists of four possible actions: *walk_forward*, *walk_backward*, *turn_right*, and *turn_left*, that are executed for 3.2s each and separated by a short break of 0.4s. This pause is required to ensure that the actions' effects do not depend on previous actions but only on the current state.

As explained above, we use Q-learning with linear gradient descent to learn the policy. The function approximator consists of multiple generic features (see Section 4) that consider only a small subset of the available state variables each. First, a minimal representation is chosen from the redundant state representation, i.e., five independent variables that unambiguously describe the positions of all relevant objects and their orientation to each other. Then a generic feature is created for all possible combinations of two and of three dimensions out of these five. This is done for several minimal representations resulting in a total number of 140 generic features, that only consider the state space. These features are important to estimate the value of the current situation. However, they do not distinguish between the different possible next actions. Hence, another 140 features are added that additionally consider the action space.

To reduce the complexity of the search space, we exploit symmetry. This technique is widely used in search or optimization problems [15,16]. Here, we use mirroring along the horizontal axis of the field. The value of turning left is equivalent to the value of turning right in the mirrored situation. Thus, it is sufficient to learn the value of turning right. Similarly, for walking forward or backward it is sufficient to learn the value for only one half of the state space.

7 Experimental Results

We evaluate the performance of the proposed concepts as follows. The task is described as an episodic RL task. Each action introduces an immediate reward of -1 per time step. The slight punishment of each action is sometimes called the costs of the actions. Each episode has three possible outcomes with the rewards:

- *Success:* The Ball is inside the opponent goal, reward: +1000
- *Failure:* The ball is outside the field, reward: -400
- *Time-out:* Abortion after 300sec, no additional reward

In addition to the ten situations defined for the scoring test, we define a set of training situations. This is to ensure that the agent regularly encounters the different regions of the state space. The agent alternately starts in a situation from the scoring test and in a training situation. A pair of training episode and scoring test episode is called a trial. The performance is evaluated separately for the scoring test and for the training situations, since we are interested in the performance on the given task, i.e. the scoring test. The presented experimental results refer to the performance in the scoring test.

Our main performance criterion is the success rate, i.e. the number of goals divided by the number of episodes. The success rate is averaged over the last 50 episodes. Since we initialized the success history with 50 failed episodes, the success rate is 0 in the beginning and only after the 50 episodes, the average is based on actual episodes. This is indicated by a vertical line in the plots. With the real system, fewer episodes can be run due to time constraints. In this case, the performance is averaged over 20 episodes. Our second criterion is the duration of successful episodes. It allows to distinguish the performance of policies that have a high success rate. This value is averaged over 50 successful episodes.

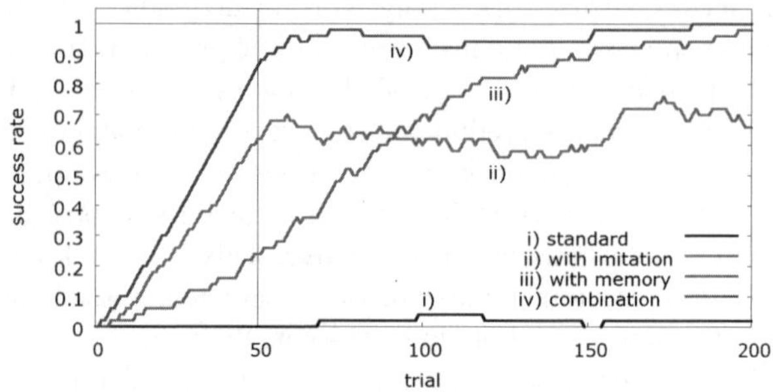

Fig. 2. The effects of imitation and memory. As can be seen, by applying our proposed approach using imitation and memory (Setup iv), the robot needs only few trials in order to come up with a good policy.

The influence of the different parameters can be evaluated faster and more systematical in a simulated environment, compared to using the real system. We first present the results obtained in a simulator and show the performance on the real robot afterwards.

7.1 Accelerating Learning by Using Imitation and Memory

First, we show that imitation and the reevaluation of own experiences seriously accelerate learning. Both approaches are evaluated separately and in combination using the following setups:

i) Standard Q-learning with $\eta = 0.2$ (constant); $\gamma = 0.98$; $\epsilon = 0$;[1]
ii) Same as i), additionally after each episode: evaluation of 36 successful episodes of an experienced agent
iii) Same as i), additionally after each episode: evaluation of the last 36 own episodes
iv) Same as iii), additionally: memory initialized with 36 successful episodes of an experienced agent (that will be replaced by own experiences after each trial)

The experienced agent is a human controlled RoboSapien with a success rate of 100%. As Fig. 2 shows, classical Q-learning does not lead to noticeable success within a reasonable time. Isolated imitation as employed in Setup ii) results in a behavior with a success rate of about 70% already after few trials. However, the robot does not improve further. It seems that the extensive use of stored experiences leads to a biased transition model that prohibits further progress in learning. When the robot uses a memory of own experiences only, learning starts more slowly but leads after 200 trials to a success rate near 100%. The combination of the two concepts imitation and memory (Setup iv) yields an almost immediate success rate of about 90%. As can be seen, after 70 trials the agent has learned a good policy with a stable success rate near 100%.

7.2 Influence of the Discount Factor

The parameter γ is used to discount rewards that will be gained in the future. One interesting effect of this future discount is that the robot prefers solutions with shorter

[1] Experiments with $\epsilon > 0$ ended up with similar results.

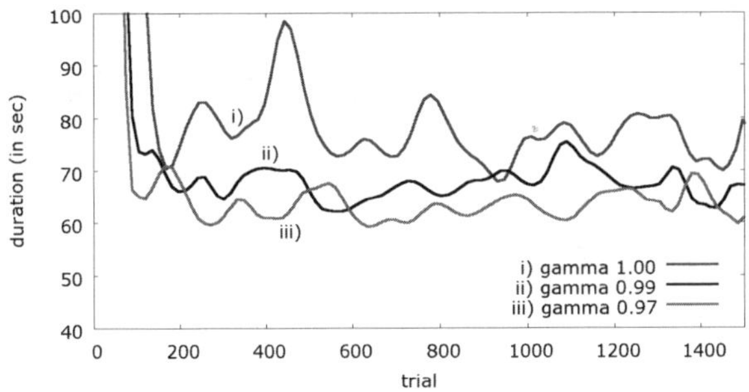

Fig. 3. Duration of successful episodes. A lower value of γ leads to a faster accomplishment of the task.

sequences of actions, as can be seen in Fig. 3. This figure shows the duration of successful episodes for different values of γ. The success rate (not shown in the plot) does not differ significantly for the different values. However, the goal is reached much faster with $\gamma = 0.97$ than without discounting the future ($\gamma = 1$). Thus, the discount factor has an important influence on the policy that is going to be learned. A lower value may be advantageous for time-sensitive tasks. On the other hand, quick solutions might involve a higher risk of failure, so that γ cannot be chosen arbitrarily small.

7.3 Results with the Real Robot

As a first approach to generate successful behavior for the real robot, we transfered the learned policy from the simulator onto the real robot. The result was that all ten out of ten episodes in the scoring test were completed successfully. The average duration was 153s. Thus, the policy learned in the simulator already outperforms the hard coded behavior by 20s [14].

In another experiment, we evaluated the learning process on the real robot. We used the proposed concept of imitation and memory as it was described by Setup iv) above. As can be seen in Fig. 4, the success rate increases quickly up to 70% after just 30 trials. The final performance is a success rate of 85%. Although the curve is not yet stable at this point, the results from the simulations suggest that this performance level can be maintained and maybe further improved.

8 Related Work

Machine learning has been widely investigated in recent works on robotics, intelligent agents or control systems. Classical approaches are increasingly combined with psychological mechanisms like imitation, curiosity, selective attention, and memory.

A well-known example for the successful application of RL techniques is the backgammon computer developed by Tesauro [2]. It consist of a feed-forward neural network, which is trained by playing against itself. TD-backgammon outperformed all commercial backgammon programs available at that time. The extensive training was

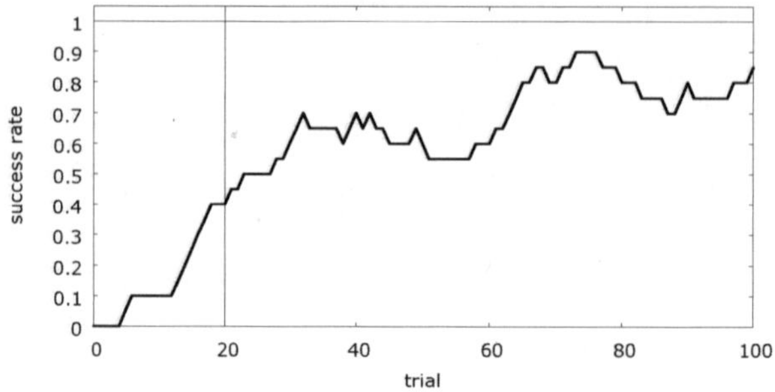

Fig. 4. Learning with the real robot. The proposed concepts of imitation and memory as well as function approximation lead to a quick acquisition of successful behavior.

essential part of this great success. The program was trained in up to 200,000 matches against itself.

Learning with memory can be used to overcome the hidden-state problem of non-Markovian environments. Algorithms like U-tree [17] or HQ-learning [18] integrate past information into the current state. As memory is only used within an episode, this can be seen as a form of short-term memory. Our approach does not aim at the solution of the hidden state problem within an episode but rather at preserving experience for later access. This corresponds to episodic memory. Our work is closely related to the fitted Q iteration algorithm [19], where RL is applied in batch mode to a large set of single observations. An observation is a four-tuple (s, a, s', r).

Different concepts of guided exploration have been proposed to accelerate RL methods. Reinforcement-driven information acquisition (RDIA) [20] combines knowledge from information theory with RL to model curiosity. Experiments with table-based Q-learning in a simulated environment show that transition probabilities can be learned much faster than with random exploration. Another form of guided exploration is coaching. The RATLE algorithm [21] uses Q-learning with a feed-forward neural network and allows to process external advice in form of rules. This is done by translating the rules into neural units and inserting these new units directly into the network. In [22] imitation is used to solve a maze problem.[2] The learning agent has an "innate" imitation behavior, which consists in following a teacher. Experiments in simulations show that increasingly complex mazes can be solved. The results are not compared to other approaches.

The successful application of RL methods to robotic soccer has been recently demonstrated by the team Brainstormers Osnabrück, the 2005 World Champion in the RoboCup 2D-Simulation-League. Their research focuses on the use of RL techniques for multi-agent systems [3]. Classical RL with feed-forward networks is used to learn basic skills that are combined to more complex behaviors [23]. The application of RL to real-world soccer robots is investigated in [24]. First, basic skills are learned via RL. Then, a policy is learned as well that chooses among the basic skills according to the

[2] The problem is called a maze problem in the original publication. However, it is rather a corridor. So the problem is not finding the exit, but to follow a given path without collisions.

current situation. Thus, a two-level hierarchy is used. The agent successfully learns good behavior in a simulated environment. Applied to the real robot, the behavior does not reach the performance of an explicitly programmed solution. The results are compared to a robot, whose top speed has been reduced to $\frac{1}{3}$ of the usual top speed.

9 Conclusions

In this paper, we presented several techniques to reduce the amount of training data for RL. The main idea was to build extensive knowledge from few experiences. This is crucial for the application of RL methods to real-world scenarios.

We use imitation to replace the random exploration of the huge state and action space with a guided exploration. In our approach, the agent has full access to experiences of a teacher, which has the same state and action space and gets identical rewards. Perceptions, actions, and rewards of the experienced agent are stored and can be accessed and reused later. Similarly, own experiences are stored and reevaluated later. This dramatically reduces the training expenses. Classical RL methods process the current observation and discard it immediately. This way, valuable information might be lost, since it cannot be correctly assessed at the moment of the experience. We let the agent repeatedly reprocess past experiences to avoid this problem.

In addition, the quick generalization of similar situations while preserving the possibility to distinguish between different situations, essentially contributes to the acceleration of the learning process. Coarse coding with binary features allows locally constant function approximation. In this case, inputs that activate identical features are treated identically. Our function approximation with continuous features is locally linear. Inputs that activate identical features remain discriminable by the different intensities of the activation. This way, generalization and discrimination can be better combined.

As the experimental results show, fundamental soccer skills can be learned using RL in simulation. The approach also works with a real humanoid robot on the soccer field. The given task is accomplished quickly and reliably. Although the training with the real robot requires more time than the training in simulation, it stays within a reasonable limit. We also showed that the learned behavior in the simulator can be directly used by the real robot and yields good results.

Acknowledgment

This project is supported by the German Research Foundation (DFG), grant BE2556/2-1.

References

1. Russell, S., Norvig, P.: Artificial Intelligence: A Modern Approach, 2nd edn. Prentice-Hall, Englewood Cliffs, NJ (2003)
2. Tesauro, G.: Practical issues in temporal difference learning. In: Proc. of Conference on Advances in Neural Information Processing Systems, vol. 4, pp. 259–266. Morgan Kaufmann Publishers, San Francisco (1992)

3. Riedmiller, M., Merke, A., Nowak, W., Nickschas, M., Withopf, D.: Brainstormers 2003 - team description. In: Proceedings of Robocup 2003, Padua, Italy (2003)
4. Asada, M., Ogino, M., Matsuyama, S., Ooga, J.: Imitation learning based on visuo-somatic mapping. In: Proc. of International Symposium on Experimental Robotics (ISER) (2004)
5. Bentivegna, D.C., Atkeson, C.G., Cheng, G.: Learning tasks from observation and practice. Journal of Robotics & Autonomous Systems 47(2-3), 163–169 (2004)
6. Dillmann, R.: Teaching and learning of robot tasks via observation of human performance. Journal of Robotics & Autonomous Systems 47(2-3), 109–116 (2004)
7. Ito, M., Tani, J.: Joint attention between a humanoid robot and users in imitation game. In: Proc. of the Int. Conf. on Development and Learning (ICDL) (2004)
8. Mataric, M.J.: Sensory-motor primitives as a basis for imitation: Linking perception to action and biology to robotics. In: Dautenhahn, K., Nehaniv, C. (eds.) Imitation in Animals and Artifacts, MIT Press, Cambridge (2002)
9. Schaal, S.: Learning from demonstration. In: Proc. of the Conf. on Neural Information Processing Systems (NIPS) (1997)
10. Sutton, R.S.: Learning to predict by the methods of temporal differences. Machine Learning 3, 9–44 (1988)
11. Hinton, G.E.: Distributed representations. Technical Report CMU-CS-84-157, Carnegie-Mellon University, Computer Science Department, Pittsburgh, PA (1984)
12. Peng, J., Williams, R.J.: Incremental multi-step Q-learning. In: Proceedings of the 11th International Conference on Machine Learning, pp. 226–232 (1994)
13. Jaakkola, T., Jordan, M.I., Singh, S.P.: Convergence of stochastic iterative dynamic programming algorithms. In: Cowan, J.D., Tesauro, G., Alspector, J. (eds.) Proc. of 7th Conference on Advances in Neural Information Processing Systems, pp. 703–710. Morgan Kaufmann, San Francisco (1994)
14. Behnke, S., Müller, J., Schreiber, M.: Playing soccer with RoboSapien. In: Proceedings of 9th RoboCup International Symposium (2005)
15. Aloul, F.A., Markov, I.L., Sakallah, K.A.: Efficient symmetry breaking for Boolean satisfiability. In: International Joint Conference on Artificial Intelligence, vol. 3, pp. 271–282. AAAI, Stanford (2003)
16. Withopf, D., Riedmiller, M.: Effective methods for reinforcement learning in large multi-agent domains. Information Technology Journal 47(5) (2005)
17. McCallum, A.: Learning to use selective attention and short-term memory in sequential tasks. In: Maes, P., Matari, M., Meyer, J.A., Pollack, J., Wilson, S. (eds.) From Animals to Animats 4: Proceedings of the Fourth International Conference on Simulation of Adaptive Behavior, Berlin, pp. 315–324. MIT Press, Cambridge (1996)
18. Wiering, M., Schmidhuber, J.: HQ-learning. Adaptive Behavior 6(2), 219–246 (1997)
19. Ernst, D., Geurts, P., Wehenkel, L.: Tree-based batch mode reinforcement learning. Journal of Machine Learning Research 6, 503–556 (2005)
20. Storck, J., Hochreiter, J., Schmidhuber, J.: Reinforcement driven information acquisition in non-deterministic environments. In: Proc. of ICANN'95. vol. 2., Paris, pp. 159–164 (1995)
21. Maclin, R., Shavlik, J.W.: Incorporating advice into agents that learn from reinforcements. In: Proc. of 12th National Conference on Artificial Intelligence, pp. 694–699 (1994)
22. Demiris, J., Hayes, G.: A robot controller using learning by imitation. In: Proceedings of the 2nd International Symposium on Intelligent Robotic Systems, Grenoble, France (1994)
23. Riedmiller, M., Merke, A., Meier, D., Hoffmann, A., Sinner, A., Thate, O., Ehrmann, R.: Karlsruhe Brainstormers — A reinforcement learning approach to robotic soccer. Lecture Notes in Computer Science (2001)
24. Dietl, M.: Reinforcement-Lernen im Roboterfußball. Diplomarbeit (in German), Albert-Ludwigs-Universität Freiburg (2002)

The Chin Pinch: A Case Study in Skill Learning on a Legged Robot

Peggy Fidelman and Peter Stone

Department of Computer Sciences, The University of Texas at Austin
1 University Station C0500, Austin, Texas 78712-0233
{peggy,pstone}@cs.utexas.edu

Abstract. When developing skills on a physical robot, it is appealing to turn to modern machine learning methods in order to automate the process. However, when no accurate simulator exists for the type of motion in question, all learning must occur on the physical robot itself. In such a case, there is a high premium on quick, efficient learning (specifically, learning with low sample complexity). Recent results in learning locomotion have demonstrated the feasibility of learning fast walks directly on quadrupedal robots. This paper demonstrates that it is also possible to learn a higher-level skill requiring more fine motor coordination, again with all learning occurring directly on the robot. In particular, the paper presents a learned ball-grasping skill on a commercially available Sony Aibo robot, with no human intervention other than battery changes. The learned skill significantly outperforms our best hand-tuned solution. As the learned grasping skill relies on a learned walk, we characterize our learning implementation within the layered learning formalism. To our knowledge, the two learned layers represent the first use of layered learning on a physical robot.

Keywords: learning and adaptive systems, sensor-motor control.

1 Introduction

In order for robots to be useful for many real-world applications, they must be able to adapt to novel and changing environments. Ideally, a robot should be able to respond to a change in its surroundings by adapting both its low-level skills, such as its walking style, and the higher-level skills which depend on them. Because hand-coding is time-consuming and often leads to brittle solutions, this adaptation should occur as autonomously as possible. Machine learning promises a way to generate solutions with little human interaction, so that when the environment changes the solution can be revised with limited human effort. Machine learning can also lead to better solutions than hand-tuning, because humans are often biased toward exploring a small part of the space of possible solutions, whereas machine learning explores the space in a systematic way.

Current learning methods typically need a large amount of training data to be effective. Thus, an appealing approach to creating learning *robots* is to train behaviors first in simulation before implementing them in the real world [5].

G. Lakemeyer et al. (Eds.): RoboCup 2006, LNAI 4434, pp. 59–71, 2007.

However, especially when concerned with complex perception or manipulation tasks, we cannot assume an adequate simulator will always exist for a given robot. With no simulator, each trial requires interaction with the physical world in real time. In such cases, it is not possible to offset the costs of an inefficient learning algorithm with a faster processor. The learning algorithm must make efficient use of the information gained from each trial (i.e., it must have low sample complexity).

For this reason, until recently, most of the locomotion approaches for quadrupedal robots have centered around hand-tuning a parameterized gait. However, in recent years, there has been a spate of research on efficient learning algorithms for quadrupedal locomotion [2,4,9,11,12,13,14]. A common feature of these approaches is that the robots time themselves walking across a known, fixed distance, thus eliminating the need for any human supervision.

This paper demonstrates that it is possible to similarly learn a higher-level more fine-motor skill, again with all learning occurring directly on the robot. In particular, the paper presents a learned ball-grasping skill on a commercially available Sony Aibo robot, with no human intervention other than battery changes. We show that a learning algorithm that has proven effective for learning walks applies directly to this new task. However, due to the different task characteristics, significant changes to the training scenario are required. This paper contributes a full specification of a training scenario that enables autonomous learning of a ball-grasping skill. The learned skill significantly outperforms an extensively hand-tuned solution.

As the learned grasping skill relies on a learned walk itself, we characterize our learning implementation within the layered learning formalism. *Layered learning* [17] is a hierarchical machine learning paradigm that leverages a given task decomposition to learn complex tasks efficiently. A key feature is that the learning of each subtask directly facilitates the learning of the next-higher subtask layer. Layered learning has been used previously to generate complex, multi-layer behaviors in *simulated* environments [6,7,17,18]. To our knowledge, our two learned layers represent the first use of layered learning on a physical robot.

The remainder of this paper is organized as follows. Section 2 describes the background and motivation for this work. Section 3 specifies the tasks to be learned and how the layered learning paradigm can be used to relate them, as well as how the training scenario is set up for each task. Section 4 describes the primary machine learning algorithm used in the work. Section 5 details the results of the training, and Section 6 discusses the contributions of this work, as well as possible directions for the future.

2 Background

This section describes the robot hardware used in all experiments and introduces the target task towards which it is trained (Section 2.1). It also summarizes the layered learning formalism (Section 2.2) within which we frame our approach.

2.1 Ball Acquisition by a Legged Robot

Acquiring an object is a prerequisite for many types of manipulations in the world [1,8]. For example, in the case of a Sony Aibo robot playing soccer, one of our motivating testbed domains, it is much easier to design effective ways for the robot to kick the ball if we may assume that the ball starts in a specific position relative to the robot. Furthermore, if the robot can *grasp* the ball securely enough, it can move the ball into a better position relative to the objects in the robot's environment before executing a kick. (For example, the robot can turn with the ball until it is pointed at the opponent's goal.) Thus, as a representative high-level task for learning, we consider the aim of having a robot walk up to a ball and gain control of it. For the purposes of this paper, we define *control* to mean that the robot holds the ball under its chin in a way that allows it to turn with the ball as shown in Figure 1.

Fig. 1. An Aibo with control of a ball. Achieving this position without knocking the ball away in the process is a challenge; our learning method allows the Aibo to do this more reliably without sacrificing walking speed.

As the robot platform for this research, we use the commercially available Sony Aibo ERS-7, a quadruped robot [15]. The ERS-7 has four legs with three degrees of freedom in each, a head with three degrees of freedom, and a CMOS camera in the head. It has several pressure sensors and two infrared range sensors, as well as position sensors in each of its joints. The robot is able to capture frames from the camera at a rate of 30 Hz. From these images, our software recognizes objects such as the orange ball based on color segmentation and aggregation. This variety of sensors allows us to rely on local sensing alone. In addition, the 576 MHz 64 bit RISC processor allows all necessary processing to be done onboard. In this work, we use a system for vision processing, walking, and kicking that was developed as part of our larger robot soccer project [16].

2.2 Layered Learning

Layered learning is a general hierarchical machine learning paradigm that leverages a given task decomposition to learn complex tasks efficiently. Though it has been validated previously in simulation, this paper presents the first application of layered learning on a physical robot. Specifically, the robot first learns a fast walk, then uses that walk to approach the ball while learning to grasp it.

The main principles of layered learning are summarized in Table 1. A detailed description of these principles is given by Stone and Veloso[17].

We cast our learned behaviors within the formal layered learning framework as defined in the remainder of this section [17]. Consider the learning task of identifying a hypothesis h from among a class of hypotheses H which map a set of

Table 1. The key principles of layered learning

1. Learning a mapping directly from inputs to outputs is not tractable.
2. A bottom-up, hierarchical task decomposition is given.
3. Machine learning exploits data to train and/or adapt. Learning occurs separately at each level.
4. The output of learning in one layer feeds into the next layer.

state feature variables S to a set of outputs O such that, based on a set of training examples, h is most likely (of the hypotheses in H) to represent unseen examples.

When using the layered learning paradigm, the complete learning task is decomposed into hierarchical subtask layers $\{L_1, L_2, \ldots, L_n\}$ with each layer defined as

$$L_i = (\boldsymbol{F_i}, O_i, T_i, M_i, h_i)$$

where:

$\boldsymbol{F_i}$ is the input vector of state features relevant for learning subtask L_i. $\boldsymbol{F_i} = \langle F_i^1, F_i^2, \ldots \rangle$. $\forall j,\ F_1^j \in S$.

$\boldsymbol{O_i}$ is the set of outputs from among which to choose for subtask L_i. $O_n = O$.

$\boldsymbol{T_i}$ is the set of training examples used for learning subtask L_i. Each element of T_i consists of a correspondence between an input feature vector $\boldsymbol{f} \in \boldsymbol{F_i}$ and $o \in O_i$.

$\boldsymbol{M_i}$ is the ML algorithm used at layer L_i to select a hypothesis mapping $\boldsymbol{F_i} \mapsto O_i$ based on T_i.

$\boldsymbol{h_i}$ is the result of running M_i on T_i. h_i is a function from $\boldsymbol{F_i}$ to O_i.

Note that a layer describes more than a subtask; it also describes an approach to solving that subtask and the resulting solution.

As stated in the Decomposition principle of layered learning, the definitions of the layers L_i are given *a priori*. The Interaction principle is addressed as follows. $\forall i < n$, h_i directly affects L_{i+1} in at least one of three ways:

- h_i is used to construct one or more features F_{i+1}^k.
- h_i is used to construct elements of T_{i+1}; and/or
- h_i is used to prune the output set O_{i+1}.

It is noted above in the definition of $\boldsymbol{F_i}$ that $\forall j,\ F_1^j \in S$. Since $\boldsymbol{F_{i+1}}$ can consist of new features constructed using h_i, the more general version of the above special case is that $\forall i, j,\ F_i^j \in S \cup_{k=1}^{i-1} O_k$.

When training a particular component, layered learning freezes the components trained in previous layers, thereby adding additional constraints to the learning process. It also adds guidance, by training each layer in a special environment intended to prepare it well for the target domain.

The original implementation of the layered learning paradigm was on the full robot soccer task in the RoboCup soccer simulator [17]. First, a neural network was used to learn an interception behavior. This behavior was used to train a decision tree for pass evaluation, which was in turn used to generate the input representation for a reinforcement learning approach to pass selection.

A subsequent application of layered learning uses two layers, each learned via genetic programming, for a soccer keepaway task in a simplified abstraction of the TeamBots environment [7]. In the full TeamBots environment, four learned layers were used, also on a keepaway task [18]. To our knowledge, there has been no previous implementation of layered learning on a physical robot.

3 Layered Learning on a Physical Robot

The process of approaching a ball and then gaining control of it relies on the gait that allows the robot to move toward the ball. Thus, when both the gait and the grasping are individually learned, we have a layered learning hierarchy consisting of two layers. This section casts the recent research on gait learning within the layered learning formalism (L_1), and then builds upon it to learn ball grasping, a second, higher-level skill (L_2).

3.1 Learning a Gait

In recent years, several approaches to learning a gait on an Aibo have been studied. Among these approaches, most of the differences between gaits stem from the shape of the loci through which the feet pass and the exact parameterizations of those loci. For example, Kohl and Stone used elliptical loci to learn high-speed walks using a policy gradient learning approach [11], while simultaneously but independently, Quinlan et al. were able to generate high-velocity gaits using a genetic algorithm and loci of arbitrary shape [12], and Roefer created a flexible gait implementation that allows use of a variety of different shapes of loci [14]. They then used an evolutionary learning algorithm to optimize a novel fitness function based on proprioception to learn a fast gait [13]. Chernova and Veloso similarly used an evolutionary approach with good success [2] and Lee et al. refined Kohl and Stone's approach to estimate gait speeds more effectively [4].

This paper builds upon the successful approach of Kohl and Stone [11], in which the gait is defined by a set of 12 continuous parameters specifying, among other things, the shape of the trajectory through which each leg moves as well as the target heights of the front and rear of the body. Thus, gait learning is framed as a parameter optimization problem, with forward speed as the objective function. The learning is accomplished via the policy gradient algorithm summarized in Section 4.

The fitness of a policy, or set of values for the 12 parameters, is obtained by having one or more Aibos time themselves as they walk a fixed, known distance indicated by a pair of landmarks. To reduce the effect of noise, this evaluation process is performed three times for each policy, and the resulting times are averaged to get the fitness of the policy.

In the notation of layered learning, the gait layer (L_1) is thus defined as:

$\boldsymbol{F}_1$: $\emptyset$;
O_1: values for the 12 parameters defining a gait, plus the speed of the resulting gait;

T_1: the set of training examples obtained by recording the time it takes to walk back and forth across a fixed distance;
M_1: the policy gradient algorithm described in Section 4;
h_1: the parameters of the fastest discovered gait, and its speed.

The walks learned using this technique perform similarly to those reported by Stone and Kohl [11], who report that with three robots continually walking across the field more than 1000 total times for approximately 3 hours, they achieved the fastest known walk on the Aibo at the time. Notably, the robots learned without any human intervention other than battery changes approximately once an hour, and the walk speed was nearly doubled during training.

Though our learned walk itself is a reproduction of previous results, the formulation of the walking task within the framework of layered learning is novel to this paper. Next, Section 3.2 introduces this novel learned skill in full detail and similarly frames it within layered learning. As prescribed by layered learning, the new skill uses the learned walk (h_1) as a part of its training scenario.

3.2 Learning to Acquire the Ball

The task of learning to capture a ball under the robot's chin is motivated by the ongoing development of our four-legged robot soccer team [16]. The robot is only able to kick in certain directions, so it is useful to be able to capture the ball and turn with it before kicking. Our team adopted the following strategy for getting the ball into this position: when the Aibo is walking to a ball with the intent of kicking it and gets close enough, it first slows down to allow for more precise positioning, and then it lowers its head to capture the ball under its chin (the *capturing motion*).

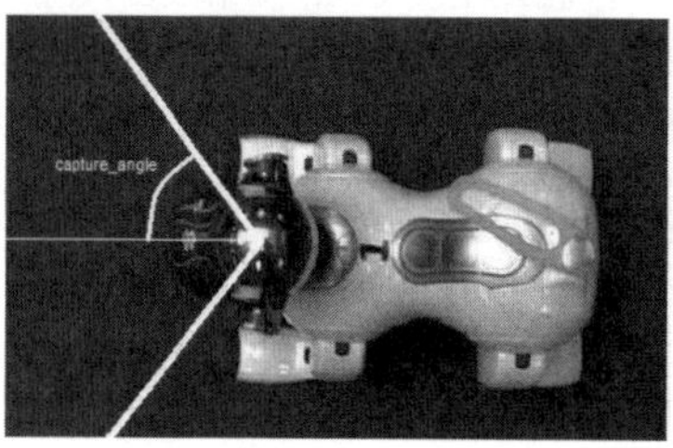

Fig. 2. Illustration of `capture_angle`. If the Aibo believes that the center of the ball is to the right of the thick white lines, then it will continue to turn toward the ball rather than beginning the capturing motion, even if the ball distance is believed to be less than `capture_dist`.

Executing the capturing motion without knocking the ball away is a challenge: if the head is lowered when the ball is too far away, the head may knock the ball away; but if it is not lowered in time, the body of the robot may bump the ball away. Furthermore, certain aspects of the acquisition motion interact, such as the perceived ball distance at which the head should be lowered and the amount that the robot slows down when close to the ball. Parameters like these must therefore be tuned simultaneously. This entire process is time-consuming to perform by hand.

The parameters that control the transition from walking to capturing the ball, as indicated in Figure 3, are as follows:

- `slowdown_dist`: the ball distance (in millimeters) at which slowing down begins;
- `slowdown_factor`: the (multiplicative) factor, in the range [0,1], by which the gait slows down at this point;
- `capture_angle`: the maximum ball angle (in degrees) at which the capturing motion may begin (see Figure 2);
- `capture_dist`: the ball distance (in millimeters) at which the capturing motion begins (if the ball is within the specified angle);
- `turn_cutoff`: the minimum ball angle (in degrees) at which the robot will not move directly toward the ball at all, but instead will turn in place to face the ball more directly. This parameter controls how straight the final part of the robot's approach will be.

```
totalscore ← 0
for j ∈ [1, n] do
  locate ball
  while ball farther than slowdown_dist do
    if ball angle more than turn_cutoff then
      turn toward ball
    else
      walk to ball at maxspeed
    end if
  end while
  while ball farther than capture_dist and outside of capture_angle do
    if ball angle more than turn_cutoff then
      turn toward ball
    else
      walk to ball at maxspeed*slowdown_factor
    end if
  end while
  lower head over ball
  if head tilt position sensor senses ball then
    totalscore ← totalscore + 1
    if center of field to robot's left then
      kick to left
    else
      kick to right
    end if
  end if
  turn 180°
end for
policy_score ← totalscore/n
```

Fig. 3. Method for evaluating policies while learning to approach the ball. n is the number of trials per policy; in our experiments, we used $n = 12$.

Given this parameterization, we are faced with a parameter optimization problem in five dimensions. Because our policies can be expressed in this way, and because our domain has the same efficiency constraints as that of learning fast locomotion for the Aibo, the policy gradient learning algorithm used to learn the gait (see Section 4) is again a natural choice.

However, there are new challenges in learning ball acquisition; specifically, i) defining an appropriate reward signal, and ii) defining an appropriate training scenario. The policy gradient algorithm relies on the magnitude of the fitness difference between policies. This magnitude is readily available in the learned walking scenario, because speed provides a natural and continuous measure of fitness. But in the case of ball acquisition, there is no straightforward way to rate a particular policy with regard to "how well" it captures the ball: it either does or it does not.

Therefore, we use a binary reinforcement signal: if the robot captures the ball, it receives a reward of 1; if not, it receives a reward of 0. The Aibo can determine autonomously whether it has captured the ball by trying to put its chin all the way down to its chest and then taking note of the value of the position sensor in its head tilt joint; if the ball is indeed under its chin, the head tilt motor will stop moving before getting to the requested position. During training, the score for a given policy is determined by running a fixed number of trials (12) with that policy and averaging the reinforcement signal over those trials (thus producing a discrete reinforcement signal). In other words, a policy's score is the number of times it successfully captures the ball over the course of 12 trials: an integer between 0 and 12 inclusive.

Each trial consists of the robot approaching the ball from a random location on the standard field used in the 2004 RoboCup competition, which is surrounded by a short wall designed to keep the ball from leaving the field. The training procedure is summarized in pseudocode in Figure 3.

One goal of the training procedure is to generate as many trials as possible in the open field, rather than with the ball starting against the wall. The latter trials are somewhat less informative because capturing the ball along the wall is considerably harder; even a good policy will fail much more frequently along the wall, which can lead to a smaller spread of scores among policies. In order to keep the ball in the open field, if the Aibo successfully captures it, it kicks it in whichever direction it estimates is away from the wall (lines 21–25 in Figure 3). Before starting the next trial, the Aibo turns around approximately 180° in place in order to knock the ball away from it if it is still close (line 27), so as to make the different trials as independent as possible. Once it has done this, it begins the next trial by searching for the ball and then approaching it with the parameters of the current policy (lines 3–17). Videos depicting the training process in action are available online[1].

In the notation of layered learning, we thus have the following definition of the acquisition layer (L_2):

$\boldsymbol{F_2}$: {BallAngle, BallDistance} $\in \{[-180, 180], [0, \infty)\}$. The five thresholds that comprise an acquisition policy (`slowdown_dist`, etc.) relate to these two sensor readings alone;

O_2: whether or not to lower the head at the current time;

T_2: evaluations of mappings from $\boldsymbol{F_i}$ to O_i, obtained by repeatedly trying to grasp the ball by the process described above and summarized in Figure 3. In particular, the learned walk (h_1) is used during training;

M_2: the policy gradient algorithm described in Section 4;

h_2: the final learned acquisition policy.

All learning is done on the Aibo itself, including all calculations necessary to execute the learning algorithm. Interruptions caused by dead batteries are of little consequence, since the learning algorithm we use has practically no state: if we resume from its last base policy, we will never lose as much as an entire iteration of the algorithm. With the algorithm parameters used in our experiments, a battery typically lasts for the amount of time necessary to complete two iterations, so on average a run requires about 4 battery changes.

---

[1] `http://www.cs.utexas.edu/~AustinVilla/legged/learned-acquisition/`

4 The Policy Gradient Algorithm

The learning algorithm common to both learned layers estimates the gradient of the policy's value function near the current policy via efficient experimentation. It then takes a step in the direction of the estimated gradient and repeats the process. We use the policy gradient algorithm presented and evaluated against alternatives for learned locomotion by Kohl and Stone [10]. This section summarizes the algorithm in task-independent terms and points out some of its advantages for the purpose of ball acquisition.

Table 2. Parameters for the policy gradient algorithm in the ball acquisition learning task

Parameter	Value
Policies per iteration (t)	8
Increment for `slowdown_dist` (ϵ_1)	10mm
Increment for `slowdown_factor` (ϵ_2)	0.1
Increment for `capture_angle` (ϵ_3)	5°
Increment for `capture_dist` (ϵ_4)	10mm
Increment for `turn_cutoff` (ϵ_5)	10°
Scalar step size (η)	2

Starting from a base policy $\{\theta_1, ..., \theta_N\}$, $t-1$ new policies are chosen by selecting one of $\{\theta_i - \epsilon_i, \theta_i, \theta_i + \epsilon_i\}$ randomly for each dimension i, where ϵ_i is a fixed increment particular to dimension i. These t policies (the base policy and the $t-1$ randomly selected policies) are then evaluated for their fitness. Their scores are used to estimate the partial derivative of fitness with respect to each of the N dimensions, which leads to a new base policy.

The estimation of partial derivatives works as follows. For each dimension i, the policies are divided into three sets according to the value of parameter i: if its value is $\theta_i - \epsilon_i$, the policy is in set $S_{-\epsilon,i}$; if it is θ_i, the policy is in set $S_{0,i}$; and if it is $\theta_i + \epsilon_i$, the policy is in set $S_{+\epsilon,i}$. Then the average score over all the policies in each set is computed and used to build an adjustment vector A of size N. For each i, if the average score over the set $S_{0,i}$ is greater than the average score over each of the other two sets, then $A_i = 0$; otherwise, A_i is the difference between the average scores over set $S_{+\epsilon,i}$ and set $S_{-\epsilon,i}$. A is then normalized and multiplied by a scalar step size η, so that the policy is adjusted by a fixed amount each time. The above process comprises one iteration of the algorithm. For the parameters used in learning ball acquisition, see Table 2.

5 Results

The success of layer L_1 at producing a significantly faster forward gait has been demonstrated previously [11]. In this paper, we demonstrate that, in the layered learning paradigm we present here, L_2 can build upon the gait improvement conferred by L_1. In particular, we hypothesize the ability to learn a significantly improved ball-acquisition skill to go with the significantly improved gait.

To test this hypothesis quantitatively, we learn ball acquisition using three gaits learned by separate runs of layer L_1. All three of these learned gaits represent significant improvements in speed over the initial hand-tuned gait. The initial (before learning) ball acquisition policy was hand-tuned for the initial hand-tuned gait from which gait learning began. The ball-acquisition learning paradigm described by layer L_2 was then applied to each of these gaits, and significantly improved acquisition policies were discovered for all three.

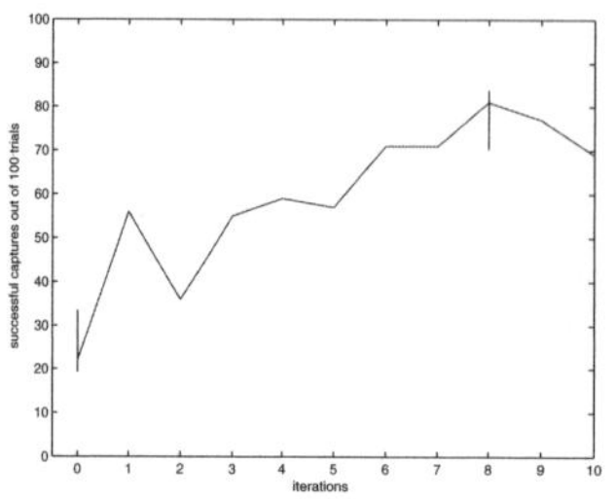

Fig. 4. Progress of acquisition learning on gait A. This learning curve was produced by running 100-trial evaluations on the base policy of each iteration. Error bars (showing the 95% confidence interval) are depicted for the initial and best learned policy; these were obtained by running five 100-trial evaluations on each policy.

Figure 4 shows the learning curve for one of these gaits, which we will refer to as gait A. For this gait, the initial ball acquisition policy acquires the ball roughly 26% of the time, whereas the best learned policy acquires the ball approximately 77% of the time. This improvement was reached in 8 iterations, which requires 768 attempted acquisitions (approximately 3 hours). The initial policy and the best learned policy were each subjected to five 100-trial evaluations, resulting in five approximations of the success rate of each. Statistical significance was then established by applying a t-test to these success rates.

Gait A has a speed of about 315mm/sec, whereas the initial hand-tuned gait from which it was learned has a speed of about 245mm/sec. The gait training process also requires roughly 3 hours. Therefore, with 6 hours of training, our robot's walking speed increased 29% and its reliability at acquiring the ball more than doubled[2] in comparison with the original hand-coded solution.

Table 3 summarizes the ball acquisition policies learned for all three gaits. It also shows the success rate of each when tested on the gait with which it was learned. These success rates were obtained by running 100-trial evaluations of the policy (except for gait A, where the data is the result of all 500 trials run to establish statistical significance on the data in Figure 4). The success rate of the initial hand-coded policy is 26% for gait A, 14% for gait B, and 14% for gait C.

Note that in all cases, the method learned not to slow down at all (`slowdown_factor` is 1). When `slowdown_factor` is 1, the parameter `slowdown_dist` has no effect on the robot's behavior, which is presumably why learning resulted in such a wide range of values for this parameter.

The fact that in all cases our method learns not to slow down demonstrates the advantage that machine learning can bestow because of its unbiased exploration of the space. In hand-tuning, we believed that slowing down would make the ball approach more reliable at the expense of speed, since the estimates of ball distance should change less rapidly if the robot is walking more slowly. Our system, however – which optimized only for reliability – found that slowing

[2] The initial ball-acquisition skill had a success rate of 36% with the initial gait, and was the result of extensive tuning involving the testing of dozens of parameter settings over the course of several days.

down is in fact a disadvantage: in all learning trials it actively increased the `slowdown_factor` parameter from its initial value of 0.8 to 1.0.

Table 3. Policy values learned for each gait, and the approximate success rate of each

Policy	slowdown_dist	slowdown_factor	capture_angle	capture_dist	turn_cutoff	Success rate
Initial	200	0.8	15	110	90	26%/14%/14%
Best: gait A	193	1	32	155	80	77%
Best: gait B	187	1	19	155	69	84%
Best: gait C	228	1	31	129	84	52%

We originally hypothesized that different gaits would require different acquisition policies. This hypothesis was supported by the fact that the initial ball acquisition policy dropped in effectiveness from approximately 36% on the gait for which it was hand-tuned to 14–26% on the learned gaits.

However, it turned out not to be the case with these learned gaits and their trained acquisition policies. Rather, upon testing the best acquisition policy learned with gait A on each of the other two learned gaits, there was no significant difference in performance — if anything, the acquisition policy learned on gait A performs better in each case, as shown in Table 4.

Table 4. Success rates of best natively learned acquisition policy and best acquisition policy learned on gait A

Gait	Natively learned	Best on gait A
B	84%	91%
C	52%	53%

Nonetheless, the layered learning paradigm enabled the separation of the learning for the walk and ball acquisition into two distinct phases. Given that they learn most efficiently in different training environments, such a hierarchical approach is an essential component of our successful skill learning.

6 Conclusion

This paper makes two main contributions: i) a significantly improved grasping skill achieved via fully autonomous machine learning with all training and computation executed on-board the robot, and ii) the first instantiation of layered learning on a physical robot.

The layered learning approach to locomotion and ball acquisition learning that we describe here is very useful in practice. Compared to manually tuning these skills, this method saves time and can generate better policies. Indeed, we used the described automated training paradigms for both the gait and the acquisition in our competitive team development for the RoboCup 2004 and 2005 robot soccer competitions, reaching the semifinals (out of 8) at the regional event and the quarterfinals (out of 24) at the international event both years [16].

In our ongoing research, we aim to identify additional skills and behaviors that can be learned in a similarly autonomous and efficient fashion. Several candidates for an L_3 that builds on the grasping skill learned in L_2 exist. Currently, for example, all design and tuning of kicks for our RoboCup team are done by hand. If this process could be automated, it would likely save time and might also lead to improved solutions. However, since most kicks begin by grasping the ball, autonomous learning of kicks would be intractable without a good grasping behavior. Another possible candidate for an L_3 that builds on the learned grasping skill is the tuning of walks that manipulate the ball, such as the one used in the turning-with-ball behavior which makes grasping so crucial in the first place (see Section 2.1). Eventually, these learned skills may feed into still higher-level learned decision-making behaviors (where to pass or when to shoot) based on the current learned skills. Indeed, an immediately realizable L_3 related to kicking is the *modeling* of hand-tuned kicks as accomplished via regression learning by Chernova and Veloso [3]. They use these models to demonstrably improve the robot's decision-making when choosing form among different kicks. Ultimately, we hope to characterize the full range of characteristics of tasks on a mobile robot that may be improved by these methods.

Acknowledgments

Thanks to Uli Grasemann and Bikram Banerjee for valuable comments, and also to the members of the UT Austin Villa team for their efforts in developing the software used as a basis for the work reported in this paper. Special thanks to Nate Kohl for sharing his machine learning infrastructure used for Aibo locomotion. This research was supported in part by NSF CAREER award IIS-0237699, ONR YIP award N00014-04-1-0545, and DARPA grant HR0011-04-1-0035.

References

1. Bicchi, A., Kumar, V.: Robotic grasping and contact: A review. In: Proceedings of the IEEE International Conference on Robotics and Automation (April 2000)
2. Chernova, S., Veloso, M.: An evolutionary approach to gait learning for four-legged robots. In: Proceedings of IROS'04 (September 2004)
3. Chernova, S., Veloso, M.: Learning and using models of kicking motions for legged robots. In: Proceedings of International Conference on Robotics and Automation (ICRA'04) (May 2004)
4. Cohen, D., Ooi, Y.H., Vernaza, P., Lee, D.D.: The University of Pennsylvania RoboCup 2004 legged soccer team. Available at URL `http://www.cis.upenn.edu/robocup/UPenn04.pdf`
5. Gat, E.: On the role of simulation in the study of autonomous mobile robots. In: AAAI-95 Spring Symposium on Lessons Learned from Implemented Software Architectures for Physical Agents. Stanford, CA (March 1995)
6. Gustafson, S.M.: Layered learning for a cooperative robot soccer problem. Master's thesis, Kansas State University (2000)

7. Hsu, W.H., Gustafson, S.M.: Genetic programming and multi-agent layered learning by reinforcements. In: Genetic and Evolutionary Computation Conference, New York,NY, pp. 764–771. Morgan Kaufmann, San Francisco (2002)
8. Kamon, I., Flash, T., Edelman, S.: Learning to grasp using visual information. Technical report, The Weizmann Institute of Science, Revhovot, Israel (March 1994)
9. Kim, M.S., Uther, W.: Automatic gait optimisation for quadruped robots. In: Australasian Conference on Robotics and Automation, Brisbane (December 2003)
10. Kohl, N., Stone, P.: Machine learning for fast quadrupedal locomotion. In: The Nineteenth National Conference on Artificial Intelligence, pp. 611–616 (July 2004)
11. Kohl, N., Stone, P.: Policy gradient reinforcement learning for fast quadrupedal locomotion. In: Proceedings of the IEEE International Conference on Robotics and Automation (May 2004)
12. Quinlan, M.J., Chalup, S.K., Middleton, R.H.: Techniques for improving vision and locomotion on the sony aibo robot. In: Proceedings of the 2003 Australasian Conference on Robotics and Automation (December 2003)
13. Röfer, T.: Evolutionary gait-optimization using a fitness function based on proprioception. In: Nardi, D., Riedmiller, M., Sammut, C., Santos-Victor, J. (eds.) RoboCup 2004. LNCS (LNAI), vol. 3276, Springer, Heidelberg (2005)
14. Rofer, T., Burkhard, H.-D., Duffert, U., Hoffman, J., Gohring, D., Jungel, M., Lotzach, M., Stryk, O.v., Brunn, R., Kallnik, M., Kunz, M., Petters, S., Risler, M., Stelzer, M., Dahm, I., Wachter, M., Engel, K., Osterhues, A., Schumann, C., Ziegler, J.: Germanteam robocup 2003. Technical report (2003)
15. Sony: Aibo robot (2004) `http://www.sony.net/Products/aibo`
16. Stone, P., Dresner, K., Fidelman, P., Jong, N.K., Kohl, N., Kuhlmann, G., Sridharan, M., Stronger, D.: The UT Austin Villa 2004 RoboCup four-legged team: Coming of age. Technical Report UT-AI-TR-04-313, The University of Texas at Austin, Department of Computer Sciences, AI Laboratory (October 2004)
17. Stone, P., Veloso, M.: Layered learning. In: López de Mántaras, R., Plaza, E. (eds.) ECML 2000. LNCS (LNAI), vol. 1810, pp. 369–381. Springer, Heidelberg (2000)
18. Whiteson, S., Kohl, N., Miikkulainen, R., Stone, P.: Evolving keepaway soccer players through task decomposition. Machine Learning 59(1), 5–30 (2005)

Half Field Offense in RoboCup Soccer: A Multiagent Reinforcement Learning Case Study

Shivaram Kalyanakrishnan, Yaxin Liu, and Peter Stone

Department of Computer Sciences
The University of Texas at Austin
Austin, TX 78712-0233
{shivaram,yxliu,pstone}@cs.utexas.edu

Abstract. We present half field offense, a novel subtask of RoboCup simulated soccer, and pose it as a problem for reinforcement learning. In this task, an offense team attempts to outplay a defense team in order to shoot goals. Half field offense extends keepaway [11], a simpler subtask of RoboCup soccer in which one team must try to keep possession of the ball within a small rectangular region, and away from the opposing team. Both keepaway and half field offense have to cope with the usual problems of RoboCup soccer, such as a continuous state space, noisy actions, and multiple agents, but the latter is a significantly harder multiagent reinforcement learning problem because of sparse rewards, a larger state space, a richer action set, and the sheer complexity of the policy to be learned. We demonstrate that the algorithm that has been successful for keepaway is inadequate to scale to the more complex half field offense task, and present a new algorithm to address the aforementioned problems in multiagent reinforcement learning. The main feature of our algorithm is the use of inter-agent communication, which allows for more frequent and reliable learning updates. We show empirical results verifying that our algorithm registers significantly higher performance and faster learning than the earlier approach. We also assess the contribution of inter-agent communication by considering several variations of the basic learning method. This work is a step further in the ongoing challenge to learn complete team behavior for the RoboCup simulated soccer task.

1 Introduction

RoboCup simulated soccer [2,4] has emerged as an excellent domain for researchers to test ideas in machine learning. Learning in the RoboCup soccer domain has to overcome several challenges, such as a continuous multi-dimensional state space, noisy sensing and actions, multiple agents (including adversaries), and the need to act in real-time. Machine learning techniques have been used in the past on a wide range of tasks in RoboCup soccer. For instance, the Brainstormers team [8,9] uses reinforcement learning to train both individual behaviors and team strategies. Several researchers have focused on specific subtasks like goal-shooting [3,6].

G. Lakemeyer et al. (Eds.): RoboCup 2006, LNAI 4434, pp. 72–85, 2007.

Keepaway. is a subtask of RoboCup soccer that has recently been proposed by Stone *et al.* [11] as a testbed for reinforcement learning methods. In keepaway, a team of *keepers* tries to keep possession of the ball away from the opposing team of *takers* within a small rectangular region. The task is episodic, and each episode ends when the takers gain possession, or when the ball goes outside the region of play. The keepers seek to maximize the duration of the episode, and are rewarded based on the time elapsed after every action. Stone *et al.* [11] provide a Sarsa-based reinforcement learning method to learn keeper behavior at a high level of abstraction.

In this paper, we extend keepaway to a more complex task of RoboCup soccer, **half field offense**. This task is played on one half of the soccer field, much bigger than the typical keepaway region. There are also typically more players on both teams. In each episode, the *offense* team needs to score, which involves keeping possession of the ball, moving up the field, and shooting goals. The *defense* team tries to stop it from doing so. Since the task realistically models the offense scenario in soccer, a policy learned for half field offense can be integrated quite naturally into full-fledged RoboCup simulated soccer games.

Both keepaway and half field offense have to cope with the usual difficulties associated with RoboCup soccer: continuous state space, noisy actions, and multiple agents. But several factors contribute to making half field offense a much harder multiagent reinforcement learning problem than keepaway. Maintaining possession of the ball is the main objective in keepaway, but it is only a subtask in half field offense. In order to succeed in half field offense, the offense players not only have to keep possession, but must also learn to pass or dribble to forge ahead towards the goal, and shoot whenever an angle opens up. With a larger state space and a richer action set than keepaway, a successful half field offense policy is therefore quite complex. A factor that makes learning in half field offense even more difficult is that the success of the task is evaluated simply based on whether a goal is scored or not at the end of an episode. Since goal scoring episodes are rare initially, it becomes necessary that the learning algorithm make the most efficient use of such information.

The learning method proposed for keepaway [11] only achieves limited success on the more difficult half field offense task. We analyze this method and propose a new method that overcomes many of its shortcomings. While reinforcement learning is indeed constructed to accommodate delayed updates and learning complex policies, we show that the learning process on a complex multiagent task like half field offense can be expedited by making better design choices. In particular, our algorithm uses inter-player communication to speed up learning and achieve better performance. We introduce the half field offense task in Section 2 and the learning method in Section 3. Section 4 presents emprical results of the performance of our method on the half field offense task. Section 5 discusses related work, and we conclude in Section 6.

2 Half Field Offense Task Description

Half field offense is an extension of the keepaway task [11] in RoboCup simulated soccer. In half field offense, an offense team of m players has to outsmart the defense team of n players, including a goalie, to score a goal. Typically $n \geq m$. The task is played over one half of the soccer field, and begins near the half field line with the ball close to one of the offense players. The offense team tries to maintain possession (keep the ball close enough for it to be kicked), move up the field, and score. The defense team tries to take the ball away from the offense team. The task is episodic, and an episode ends when one of three events occurs: 1. A goal is scored, 2. The ball is out of bounds, or 3. A defender gets possession of the ball (including the goalie catching the ball). Fig. 1 shows a screen-shot from a half field offense task, where four players are on the offense team and five players including a goalie are on the defense team. We denote this version of the task 4v5 half field offense, and discuss it in Section 2.2.

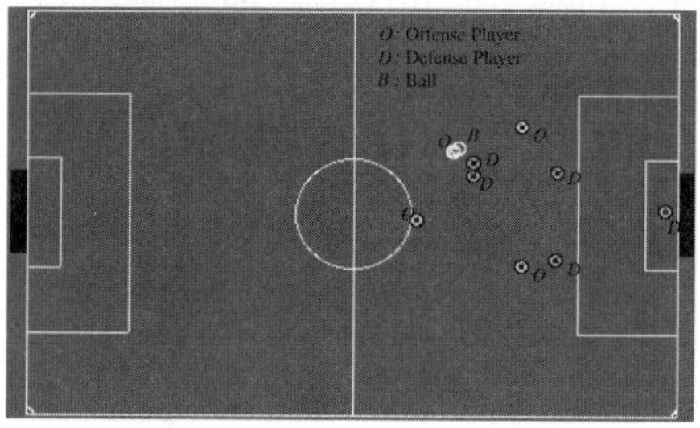

Fig. 1. Half field offense game in progress

In principle, it is possible to frame half field offense as a learning problem for either the offense team or the defense team (or both), but here we only focus on learning by the offense team. The objective is to increase the goal-scoring performance of the offense team, while the defense team follows a fixed strategy. A similar approach is also adopted for keepaway [11].

The offense team player who possesses the ball (and is hence closest to it) is required to take one of the following actions:[1]

- **Pass$k$.** This action involves a direct kick to the teammate that is the k-th closest to the ball, where $k = 2, 3, \ldots, m$. The representation used is indexical, since it is based on distances to the teammates, and not their actual jersey numbers.
- **Dribble.** In order to encourage the offense player with possession to dribble towards the goal, a cone is constructed with the player at its vertex and its axis passing through the center of the goal. The player takes a small kick within this cone in a direction that maximizes its distance to the closest defense player also inside the cone. The half angle of the cone is small (15°) when it is far away from the goal and opponents, but is progressively increased (up to 75°) as it gets closer to the goal or opponents. Thus it is

[1] They are in fact high-level skills, and are better described by the term "options," which are themselves composed of low-level actions over extended time periods. They nevertheless play the role of actions in the sense of reinforcement learning. We simply refer to them as actions for simplicity. For a more detailed discussion, see [11].

encouraged to forge ahead towards the goal whenever possible, but has room to move away from the defense players should they get too close.
- **Shoot.** By taking this action the player kicks the ball towards the goal in the direction bisecting the widest open angle available between the goalie, other defenders, and the goalposts.

When no offense player has possession of the ball, the one closest to the ball seeks to reach it by dashing directly towards it (**GetBall**). Offense players other than the one closest to the ball always try to to maintain a formation in order to take the attack forward (**GetOpen**). More precisely, any player from the offense team is constrained to behave as follows.

if I have possession of the ball **then**
  Execute some action from {**Pass**2, ..., **Pass**m, **Dribble**, **Shoot**}
else if I am the closest offense player to the ball **then**
  GetBall (Intercept the ball).
else
  GetOpen (Move to the position prescribed by the formation, see Section 2.2).

Therefore, the behavior of any offense player is fixed except when it has possession of the ball. Deciding which action to take when in possession of the ball precisely constitutes the learning problem, as is also the case in keepaway [11]. In principle, the player who has possession can benefit from a larger action set than the one we have described, but these actions are enough to achieve quite a high level of performance. More importantly, they allow us to focus on learning high-level strategies.

2.1 State Representation

In RoboCup simulated soccer [2], the server provides the players sensory information at regular intervals of time. Players typically process the low-level information thus obtained to maintain estimates of the positions and velocities of the players and the ball. For our task, we define the state using a set of variables involving distances and angles between players, which can be derived from information about their positions.These are listed below. The offense players are numbered according to their distances to the ball using an indexical representation. The offense player with the ball is always denoted O_1. Its teammates are $O_2, O_3, \ldots, O_m$. The defense players are also numbered according their distances to the ball; they are $D_1, D_2, \ldots, D_n$. The goalie, which may be any of the D_i, is additionally denoted D_g.

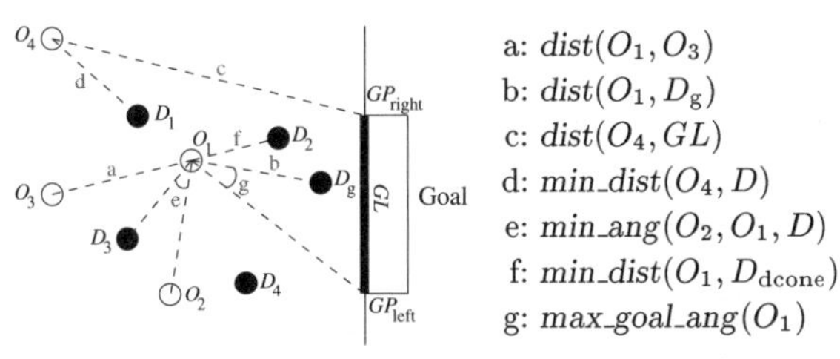

Fig. 2. Sample state variables for 4v5 half field offense

- $dist(O_1, O_i), i = 2, 3, \ldots, m$. The distances from O_1 to its teammates.
- $dist(O_1, D_g)$. The distance from O_1 to the goalie on the defense team.
- $dist(O_i, GL), i = 1, 2, \ldots, m$. For the offense player, the distance to the segment of the goal line GL between the goalposts.
- $min\_dist(O_i, D), i = 1, 2, \ldots, m$. For each offense player, the closest distance to any opponent, that is, $min\_dist(O_i, D) = \min_{j=1,\ldots,n} dist(O_i, D_j)$.
- $min\_ang(O_i, O_1, D), i = 2, 3, \ldots, m$. For offense players other than O_1, the smallest angle $\angle O_i O_1 D$ among all D, where D is a defense player, that is, $min\_ang(O_i, O_1, D) = \min_{j=1,\ldots,n} ang(O_i, O_1, D_j)$.
- $min\_dist(O_1, D_{\mathrm{dcone}})$. The distance from O_1 to the closest defender in the dribble cone or dcone. The dribble cone is a cone with half angle 60° with its vertex at O_1 and axis passing through the center of the goal. D_{dcone} is the set of defenders in the dribble cone.
- $max\_goal\_ang(O_1)$. The maximum angle with the vertex at O_1, formed by rays connecting O_1 and goalposts or defense players that are in the goal cone, which is the triangle formed by O_1 and the two goalposts GP_{left} and GP_{right}.

We adopt this set of state variables expecting them to be of direct relevance to the actions, although they are neither independent nor complete. We expect $dist(O_1, GL)$, $max\_goal\_ang(O_1)$, and $dist(O_1, D_g)$ to directly affect **Shoot**; $min\_dist(O_1, D)$ and $min\_dist(O_1, D_{\mathrm{dcone}})$ to affect **Dribble**; and the other variables to affect **Pass***k*. As in keepaway [11], the indexical representation based on distances is expected to help the players generalize better. We arrived at this set of state variables through experimentation, but did not expend much time optimizing the set. We note that the set of state variables is independent of the number of defense players, and has a linear relation with the number of offense players, therefore scaling to versions of the task with large numbers of players. The 4v5 version of the task uses 17 state variables.

2.2 4v5 Half Field Offense and Benchmark Policies

4v5 half field offense (see Fig. 1) is a version of the task involving four offense players and five defense players, including the goalie. We use this version of the task for all our experiments. In 4v5, the offense player wiith the ball must choose an action from the set {**Pass**2, **Pass**3, **Pass**4, **Dribble**, **Shoot**}, while the other offense players, following a fixed strategy, stay in an arc formation. The defense players also follow a static policy. Due to space limitations, the complete behaviors of the offense and defense players on the 4v5 task are specified on a supplementary web site[2]. The web site also lists examples of policies (including **Random**, in which actions are chosen randomly, and **Handcoded**, a policy we have manually engineered) for the 4v5 task and videos of their execution.

3 Reinforcement Learning Algorithm

Since half field offense is modeled as an extension of keepaway [11], they share the same basic learning framework. In fact, the learning method that has been

[2] `http://www.cs.utexas.edu/~AustinVilla/sim/halffieldoffense/index.html`

most successful on keepaway [11] can be directly used for learning half field offense. But while with keepaway the learning curve obtained using this method typically levels off after just 15 hours of simulated time, we find that with half field offense it continues to gradually rise even after 50 hours. Furthermore, when we visualize the execution of the policy thus learned,[2] it is quite apparent that it is sub-optimal. In order to ascertain whether indeed the learning can be improved, we analyze the task and the learning method in detail. We then proceed to introduce a new learning approach using inter-agent communication, which significantly improves the learning rate and the resulting performance. In this section, we explain how the reinforcement learning method is set up, the problems faced by multiagent reinforcement learning on this task, and how we handle them using explicit inter-agent communication.

3.1 Basic Setup

As in keepaway [11], the reinforcement learning problem is modeled as a semi-Markov decision process [1], where decisions are taken at unequal intervals of time, and only by the player with the ball. Each agent uses a function approximator to represent an action-value function or Q-function that maps state and action pairs (s, a) to real numbers, which are its current estimates of the expected long term reward of taking action a from state s. Each agent updates its Q-function using the Sarsa learning method. The main difference between the method from [11] and the one we propose in Section 3.3 is in how each agent obtains the experience, and how frequently updates are made.

Rewards for the reinforcement learning problem are defined in Table 1. A positive reward of 1.0 is given for an action that results in a goal, while small negative rewards are given when the episode ends unsuccessfully. It is conceivable to give small positive rewards (say 0.01) for successful passes, but we found that a zero reward was just as effective. Negative rewards are provided at the end of unsuccessful episodes to encourage keeping the ball in play. The ratios between different rewards can have a significant impact on the learning process [3]. We informally tested out different values and found this particular assignment effective. Since the task is episodic, we do not use discounting.

Table 1. Definition of rewards

Game Scenario	Notation	Reward
Goal	*goal_reward*	1.0
Ball with some offense player	*offense_reward*	0
Ball caught by goalie	*catch_reward*	−0.1
Ball out of bounds	*out_reward*	−0.1
Ball with some defense player	*defense_reward*	−0.2

3.2 Difficulty of Multiagent Learning

In the method used for keepaway [11], each agent learns independently. The only points in time when it receives rewards and makes updates to its Q-function are when it has possession of the ball, or when the episode ends. The reward itself is the length of the duration between when the action was taken and when possession was regained or the episode ended. Clearly, it is easy to apply this scheme to half field offense, using the reward structure specified in Table 1. We illustrate this method by tracing through a typical episode from half field offense (depicted in Fig. 3). The episode begins with O_A in possession of the ball.[3] O_A passes the ball to O_B, who takes three dribble actions before passing it to O_C. O_C then passes it back to O_B, who shoots the ball into the goal. O_D does not participate in this episode. Using the learning method from [11], O_A will receive *goal_reward* for its pass action (1); O_B will receive *offense_reward* for its dribble actions (2, 3, and 4) and pass action (5), and *goal_reward* for its shoot action (7); while O_C will get *goal_reward* for its pass action (6). O_A and and O_C will only make their learning updates at the end of the episode, while O_B will also make intermediate updates whenever it regains possession. O_D does not make any learning update in this episode.

We find shortcomings in the method from [11] that we have just described. First, consider another episode that differs from the above example only in that the final **Shoot** action results in the ball being caught by the goalie instead of finding the goal. In this case also, O_A, O_B, and O_C make corresponding updates to their Q-functions, but the reward used for the updates is *catch_reward*. Even though the reason for the failure to score on this episode is only perhaps a slightly flawed **Shoot** action, the reward assigned to O_A for its successful pass to O_B (and indeed actions 6 and 7) now becomes drastically different (negative instead of positive). This case illustrates that it is more desirable for O_A to update its Q-function for the pass to O_B right after O_B receives the ball, since the update then would be based on the Q-value of O_B in its current state, and the reward for a successful pass. While using the method from [11], the update is only based on how the episode ends. This can lead to a higher variance for updates to the Q-function, especially since the task is stochastic. The problem is that there is a long temporal delay between the execution of an action by a player and the corresponding learning update; because of this delay, the assigned reward and the next resulting state can change drastically. It is not too harmful in keepaway, since the rewards are themselves the time elapsed between the events, and do not change based on the next state. But in half field offense, even slight changes in the middle of an episode can lead to very different outcomes and very different rewards.

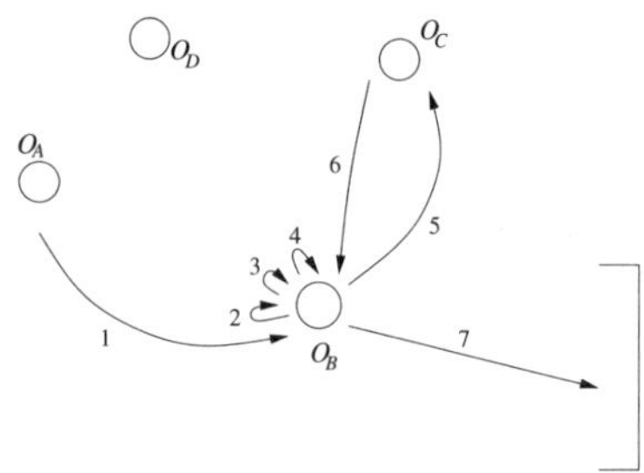

Fig. 3. Example episode: The numbers indicate the sequence of actions

Another evident shortcoming is the case of player O_D. Since each player learns solely based on its own experiences in the method from [11], O_D, which is not

[3] We use subscripts A, B, C, D to indicate fixed players since numerical subscripts indicate players according to how close they are to the ball.

involved in this episode, does not update its Q-function even once during this episode. Since the players are homogeneous, it seems conceivable that the players' experiences can be shared. For instance, O_D should be able to learn, based on O_B's experience, that **Shoot** action taken from close to the goal is likely to receive high reward. In fact, even among O_A, O_B, and O_C, only O_B records the information that the **Shoot** action resulted in a goal; O_A and O_C are only able to make updates to their respective **Pass** actions. Surely, they will also benefit by making O_B's update for the **Shoot** action. Sharing experiences can be particularly useful early in training, when successful episodes are rare. We next describe our learning method, which uses inter-agent communication to overcome the shortcomings described above.

3.3 Agent Communication

In the solution we propose, inter-agent communication is used to facilitate information sharing among the agents, and to enable frequent and more reliable updates. The protocol followed is similar to one used by Tan [12] for learning in an artificial predator-prey domain, where agents are able to communicate their experiences to their partners.

Table 2. Messages broadcast during the example episode

Number	Sender	State	Action	Reward
1	O_A	s_1	**Pass**3 (to O_B)	*offense_reward*
2	O_B	s_2	**Dribble**	*offense_reward*
3	O_B	s_3	**Dribble**	*offense_reward*
4	O_B	s_4	**Dribble**	*offense_reward*
5	O_B	s_5	**Pass**2 (to O_C)	*offense_reward*
6	O_C	s_6	**Pass**2 (to O_B)	*offense_reward*
7	O_B	s_7	**Shoot**	*goal_reward*

Since every action leads either to some offense player getting possession or the end of the episode, in our scheme, the appropriate reward for that action is provided as soon as one of these events occurs. Thus, in our example, O_A is given *offense_reward* as soon as its pass (action 1) to O_B succeeds, instead of having to wait until the end of the episode to receive *goal_reward*. As soon as it receives *offense_reward*, O_A broadcasts a message (see Table 2) to its teammates, describing the state in which it was when it took the pass action (s_1), the pass action itself (**Pass**3), and the reward received (*offense_reward*). In general, every time a player takes action a in state s and receives reward r, it broadcasts a message of the form (s, a, r) to the team. Since a Sarsa update is completely specified by a (s, a, r, s', a') tuple, each player records the messages received and makes an update as soon as enough information is available for it. Thus, when O_B broadcasts message 2 (see Table 2), *all* the players make a Sarsa update using the tuple (s_1, **Pass**3, *offense_reward*, s_2, **Dribble**.). It is quite clear that in this scheme, the SMDP step is designed to last only until some teammate gets possession (unless the episode ends before that), therefore keeping the updates more reliable. At the same time, communication permits each player to make an update

Algorithm 1. Reinforcement Learning with Communication

```
Initializations;
for all episode do
  s ← NULL;
  repeat
    // acting
    if I have possession of the ball then
      s ← getCurrentStateFromEnvironment();
      Choose action a using Q-function and ε-greedy selection;
      Execute action a;
      r ← waitForRewardFromEnvironment();
      broadcast(s, a, r);
    else
      if I am the closest offense player to the ball then
        GetBall;
      else
        GetOpen;
    // learning
    if I receive message (s_m, a_m, r_m) then
      if s is NULL then
        s, a, r ← s_m, a_m, r_m;
      else
        s', a', r' ← s_m, a_m, r_m;
        Perform Sarsa update based on (s, a, r, s', a'):
            Q(s, a) ← Q(s, a) + α (r + γQ(s', a') − Q(s, a));
        s, a, r ← s', a', r';
  until episode ends;
```

for every action that has been taken by any of the offense players during the episode. In fact, since all the players begin with the same initial Q-function and make the same updates, we can expect that their action-value functions will always be alike, thereby reducing an essentially distributed problem to one of centralized control (imagine a single "virtual" agent who resides at all times inside the player who currently has possession of the ball). However, in practice, the message passing is not completely reliable, so a small number of updates get missed.

Algorithm 1 provides the pseudocode of the algorithm the learning players implement. Each player stores its current action value function using a function approximator. When in possession of the ball, it decides which action to take based on an ϵ-greedy action selection scheme. After executing a and receiving a reward r, the player broadcasts the triple (s, a, r) to the team. Players not in possession of the ball simply follow the static policy. Each player uses the first message that is received during an episode to initialize values for the triple (s, a, r), and on every subsequent message (s', a', r') makes a learning update using the saved and received information.

We implement the inter-player communication using a "trainer," an independent agent that can communicate with all the players. The player broadcasting an (s, a, r) message actually sends it to the trainer, who then sends it to all the players. To be consistent, we assume that even to make an update corresponding to its own action, a player first sends a message to the trainer, and makes the update only on receiving it back from the trainer. The trainer sends a special (s, a, r) message to the players when the end of an episode is reached, so that they may make a final update for that episode and start afresh for the next.

We use Sarsa(0) as our reinforcement learning algorithm, along with CMAC tile coding for function approximation (as in [11]). The CMACs comprise 32 tilings for each feature. Distance features have tile widths of $3m$, while the tile width for angle features is fixed at $10°$. We use $\alpha = 0.125$, $\gamma = 1.0$, and $\epsilon = 0.01$.

4 Experimental Results and Discussion

In this section we present performance results of our learning algorithm. The graphs depict learning curves with the y-axis showing the fraction of successful episodes, and the x-axis the number of training episodes. The learning curves are smoothed using a sliding window of 1000 episodes. Each curve is an average of at least 30 independent runs. We have performed t-tests every $5,000$ episodes comparing the values of the curves, and we report the levels of significance for important comparisons.

To focus on learning while still perserving a high level of complexity in our experiments with the 4v5 half field offense task, we have modified a couple of RoboCup simulated soccer defaults. While the RoboCup default only allows players to "see" within a 90° cone, we allow for 360° vision, which removes hidden state, but still retains sensor noise. Also, we do not enforce the offsides rule in our task, even though our players get offsides only occasionally. These changes are enforced in all our experiments, in order to make meaningful comparisons between different offense team policies.

Fig. 4(a) plots the performance of our learning algorithm. Our learning algorithm using inter-agent communication achieves a long term success ratio of 32%, while one where the agents learn independently (as in keepaway [11]) only manages to register 23%. Beyond 5000 episodes, their order is significant ($p < 10^{-8}$). Clearly, the gain from using communication is substantial. This is particularly apparent when we compare it to the performance recorded by other static policies. Within 2000 episodes of training, our algorithm is able to learn a more successful policy than the **Handcoded** policy mentioned in Section 2.2. When we visualize the execution of the learned policy,[2] it is noticeably different from the **Handcoded**, suggesting that learning is able to capture behavior that is non-intuitive for humans to describe. Fig. 4(a) also plots the performance of the **Random** policy, which succeeds less than 1% of the time, thereby confirming that extended sequences of the right actions are required to score goals. The other curve in the graph shows the performance achieved by a set of four offense players (numbers 6, 7, 8, and 9) from the UvA 2003 RoboCup team [5], which won the RoboCup simulated league championship that year. The comparison between our players and the UvA offenders is not completely fair, because they have not been tuned specifically for the half field offense task. But the fact that our players are able to learn a policy that performs at least twice as well as the UvA players in this setting still gives some insight into the effectiveness of the policy they learn.

In order to get a clearer understanding of the impact of communication, we ran a set of experiments in which only subsets of players communicate among

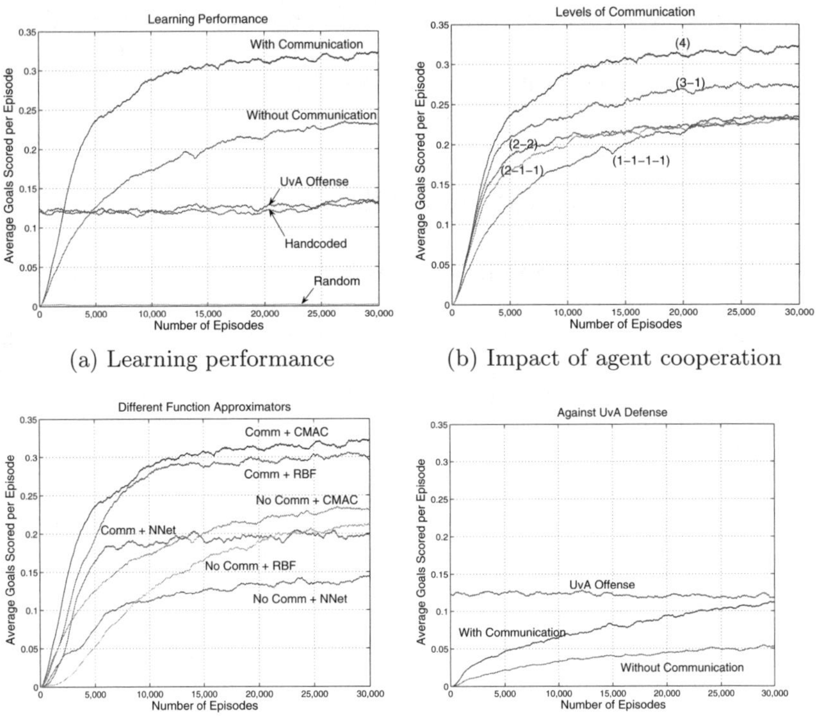

(a) Learning performance

(b) Impact of agent cooperation

(c) Different function approximators

(d) Against UvA defenders

Fig. 4. Experimental results

themselves, while some learn independently (as in keepaway [11]). Fig. 4(b) plots learning curves from experiments in which all four players communicate their updates (4), only three of them communicate while one learns independently (3-1), all players communicate but each only with a partner (2-2), two of them communicate while two learn independently (2-1-1), and they are all independent (1-1-1-1). At 5,000 and 10,000 episodes of training, the order specified above (except between 2-2 and 2-1-1) is significant with $p < 10^{-3}$, suggesting that increased communication results in a faster learning rate. After 30,000 episodes of training, the full-communication curve (4) remains ahead of all the others, with $p < 10^{-8}$. It is clear, therefore, that communication does make a significant difference to the performance. Through informal experimentation, we verified that communication results in a faster learning rate for the keepaway task too, though the final policy it learns does not perform significantly better than one learned with no communication [11].

While we use CMACs for function approximation in most of our experiments, we ran an additional set of experiments using different function approximators to see how this change affects the performance. The other function approximators we used are neural networks (NNets) and radial basis functions (RBFs). They have also been in the past used for learning keepaway [10]. Fig. 4(c) plots the performance obtained by using these function approximators both in the full-communication case and the no-communication case. In both cases, the order CMAC > RBF > NNet is preserved beyond 10,000 episodes of training (In keepaway the RBFs perform slightly better than the CMACs). But more importantly, we find that for each of the function approximators, significantly higher performance is achieved by the communication-based algorithm.

In order to verify to what extent our learning algorithm was robust to changes in the task, we ran a variation in which the offense team plays against a set of defense players from the UvA 2003 team [5] (players 5, 6, 2, 3, and 1). These players offer a far more defensive strategy than ours, and position themselves strategically to block passes between the offense players. After 30,000 episodes of learning, our players show an 11% success rate, which almost matches the performance achieved by the set of UvA offense players against this opposition. In fact, the learning curve still seems to be rising at this point. The UvA offense performs better than the learned policy, however with a low confidence ($p < 0.2356$). Therefore our learning method is able to achieve a high level of performance against a world-class opposition, despite having only been trained with a limited action set. Surely, the main reason for its success is inter-agent communication, as the no-communication algorithm only manages to achieve a success rate of 5% in the same number of episodes.

5 Related Work

Half field offense is a natural extension of the keepaway task introduced by Stone *et al.* [11]. It is a much harder problem to learn than keepaway, and we have shown that inter-agent communication can be effectively coupled with the algorithm that has so far been successful for keepway [11] to boost its performance significantly. Geipel [3] and Maclin *et al.* [6] have in the past applied reinforcement learning techniques to goal-shooting scenarios, but these have typically involved fewer players and a smaller field than 4v5 half field offense. The Brainstormers RoboCup team [8,9] has consistently applied reinforcement learning techniques to train different aspects of team behavior. They use reinforcement learning to learn high-level skills called “moves” in terms of low-level actions, and use these as primitives for learning high-level tactical behavior for attacking players [9]. For function approximation, they use neural networks, inputs to which are low-level state information. They also formulate simulated soccer as a Multi-agent Markov Decision Process (MMDP) [8] and discuss different models of agents based on their action sets and coordination. While their focus has been to develop a general architecture for learning team behavior, we, in this paper, address the specific problem of learning high-level behavior by the offense player with ball possession. For this reason, we use predefined high-level skills like **Pass***k* and **Shoot**. Since we mainly use CMACs, which cannot represent arbitrary non-linear functions, we design our state features to be at a high level of abstraction, in order to facilitate better generalization.

Multiagent reinforcement learning with inter-agent communication has been studied in the past. Whitehead [13] describes a *Learn-by-Watching* method similar to ours and obtains theoretical bounds for the speedup in learning by multiple Q-learning agents. Tan [12] empirically evaluates the effect of inter-agent communication on a much simpler problem than ours, a predator-prey scenario within a 10x10 discrete grid. The predators can cooperate by sharing their sensations, learning episodes, and whole policies. Of these, sharing episodes involves communicating extended series of state-action-reward messages between the predators, and has a direct correspondence with the method we have employed. Again, communication is is shown to be beneficial to learning. Mataric [7] uses reinforcement learning to train real robots on a box-pushing task. Communication is mainly used to share their sensations in order to form a complete state of the world, unlike in our algorithm, where communication directly impacts updates made to the agents' action-value functions.

6 Conclusions and Future Work

In this paper, we have introduced half field offense, a novel subtask of RoboCup simulated soccer. It extends an earlier benchmark problem for reinforcement learning, keepaway [11]. Half field offense presents significant challenges as a multiagent reinforcement learning problem. We have analyzed the learning algorithm that has been most successful for keepaway [11], and scaled it to meet the demands of our more complex task. The main feature of our new algorithm is the use of inter-agent communication, which allows for more frequent and reliable learning updates. We have presented empirical results suggesting that the use of inter-agent communication can increase the learning rate and the resulting performance significantly.

Learning half field offense is a step further in the ongoing challenge to learn complete team behavior for RoboCup simulated soccer. In this work we have only focused on learning the behavior of the offense player who has possession of the ball. It is in principle possible to pose as learning problems the behaviors of the other offense players, as well as the defense team. Also, high-level skills like **Pass***k* and **Shoot**, which we have directly used for learning here, may themselves be learned in terms of low-level actions like *turn* and *kick*. These are avenues for future research.

Acknowledgements

This research was supported in part by NSF CISE Research Infrastructure Grant EIA-0303609, NSF CAREER award IIS-0237699, and DARPA grant HR0011-04-1-0035.

References

1. Bradtke, S.J., Duff, M.O.: Reinforcement learning methods for continuous-time Markov decision problems. In: Advances in Neural Information Processing Systems 7 (NIPS-94) (1995)

2. Chen, M., Foroughi, E., Heintz, F., Huang, Z., Kapetanakis, S., Kostiadis, K., Kummeneje, J., Noda, I., Obst, O., Riley, P., Steffens, T., Wang, Y., Yin, X.: Users Manual: RoboCup Soccer Server — for Soccer Server Version 7.07 and Later. The RoboCup Federation (August 2002)
3. Geipel, M.M.: Informed and advice-taking reinforcement learning for simulated robot soccer. Master's thesis, Fakultät für Informatik, Forschungs- und Lehreinheit Informatik IX, Technische Universität München (2005)
4. Kitano, H., Asada, M., Kuniyoshi, Y., Noda, I., Osawa, E., Matsubara, H.: RoboCup: A challenge problem for AI. AI Magazine 18(1), 73–85 (1997)
5. Kok, J.R., Vlassis, N., Groen, F.C.A.: UvA Trilearn 2003 team description. In: Proceedings CD RoboCup 2003 (2003)
6. Maclin, R., Shavlik, J., Torrey, L., Walker, T., Wild, E.: Giving advice about preferred actions to reinforcement learners via knowledge-based kernel regression. In: Proceedings of the Twentieth National Conference on Artificial Intelligence (AAAI-05) (2005)
7. Matarić, M.J.: Using communication to reduce locality in distributed multi-agent learning. Journal of Experimental and Theoretical Artificial Intelligence 10(3), 357–369 (1998)
8. Merke, A., Riedmiller, M.: Karlsruhe Brainstormers — a reinforcement learning approach to robotic soccer II. In: Birk, A., Coradeschi, S., Tadokoro, S. (eds.) RoboCup 2001. LNCS (LNAI), vol. 2377, Springer, Heidelberg (2002)
9. Riedmiller, M., Merke, A., Meier, D., Hoffmann, A., Sinner, A., Thate, O., Ehrmann, R.: Karlsruhe Brainstormers — a reinforcement learning approach to robotic soccer. In: Stone, P., Balch, T., Kraetzschmar, G.K. (eds.) RoboCup 2000. LNCS (LNAI), vol. 2019, Springer, Heidelberg (2001)
10. Stone, P., Kuhlmann, G., Taylor, M.E., Liu, Y.: Keepaway soccer: From machine learning testbed to benchmark. In: Bredenfeld, A., Jacoff, A., Noda, I., Takahashi, Y. (eds.) RoboCup 2005. LNCS (LNAI), vol. 4020, Springer, Heidelberg (2006)
11. Stone, P., Sutton, R.S., Kuhlmann, G.: Reinforcement learning for RoboCup-soccer keepaway. Adaptive Behavior 13(3), 165–188 (2005)
12. Tan, M.: Multi-agent reinforcement learning: Independent vs. cooperative agents. In: Proceedings of the Tenth International Conference on Machine Learning (ICML-93), pp. 330–337 (1993)
13. Whitehead, S.D.: A complexity analysis of cooperative mechanisms in reinforcement learning. In: Proceedings of the Ninth National Conference on Artificial Intelligence (AAAI-91), pp. 607–613 (1991)

Autonomous Learning of Ball Trapping in the Four-Legged Robot League

Hayato Kobayashi[1], Tsugutoyo Osaki[2], Eric Williams[2], Akira Ishino[3], and Ayumi Shinohara[2]

[1] Department of Informatics, Kyushu University, Japan
[2] Graduate School of Information Science, Tohoku University, Japan
[3] Office for Information of University Evaluation, Kyushu University, Japan
{h-koba@i,ishino.uoc@mbox.nc}.kyushu-u.ac.jp
{osaki,ayumi}@shino.ecei.tohoku.ac.jp, eaw@ucla.edu

Abstract. This paper describes an autonomous learning method used with real robots in order to acquire ball trapping skills in the four-legged robot league. These skills involve stopping and controlling an oncoming ball and are essential to passing a ball to each other. We first prepare some training equipment and then experiment with only one robot. The robot can use our method to acquire these necessary skills on its own, much in the same way that a human practicing against a wall can learn the proper movements and actions of soccer on his/her own. We also experiment with two robots, and our findings suggest that robots communicating between each other can learn more rapidly than those without any communication.

1 Introduction

For robots to function in the real world, they need the ability to adapt to unknown environments. These are known as *learning* abilities, and they are essential in taking the next step in RoboCup. As it stands now, it is humans, not the robots themselves, that hectically attempt to adjust programs at the competition site, especially in the real robot leagues. But what if we look at RoboCup in a light similar to that of the World Cup? In the World Cup, soccer players can practice and confirm certain conditions on the field before each game. In making this comparison, should robots also be able to adjust to new competition and environments on their own? This ability for something to learn on its own is known as *autonomous learning* and is regarded as important.

In this paper, we force robots to autonomously learn the basic skills needed for passing to each other in the four-legged robot league. Passing (including receiving a passed ball) is one of the most important skills in soccer and is actively studied in the simulation league. For several years, many studies [1,2] have used the benchmark of good passing abilities, known as "keepaway soccer", in order to learn how a robot can best learn passing. However, it is difficult for robots to even control the ball in the real robot leagues. In addition, robots in the four-legged robot league have neither a wide view, high-performance camera, nor laser range finders. As is well known, they are not made for playing soccer. Quadrupedal locomotion alone can be a difficult enough challenge. Therefore, they must improve upon basic skills in order to solve these difficulties, all

G. Lakemeyer et al. (Eds.): RoboCup 2006, LNAI 4434, pp. 86–97, 2007.

before pass-work learning can begin. We believe that basic skills should be learned by a real robot, because of the necessity of interaction with a real environment. Also, basic skills should be autonomously learned because changes to an environment will always consume much of people's time and energy if the robot cannot adjust on its own.

There have been many studies conducted on the autonomous learning of quadrupedal locomotion, which is the most basic skill for every movement. These studies began as far back as the beginning of this research field and continue still today [3,4,5,6]. However, the skills used to control the ball are often coded by hand and have not been studied as much as gait learning. There also have been several similar works related to how robots can learn the skills needed to control the ball. Chernova and Veloso [7] studied the learning of ball kicking skills, which is an important skill directly related to scoring points. Zagal and Solar [8] studied the learning of kicking skills as well, but in a simulated environment. Although it was very interesting in the sense that robots could not have been damaged, the simulator probably could not produce complete, real environments. Fidelman and Stone [9] studied the learning of ball acquisition skills, which are unique to the four-legged robot league. They presented an elegant method for autonomously learning these unique, advanced skills. However, there has thus far been no study that has tried to autonomously learn the stopping and controlling of an oncoming ball, i.e. *trapping* the ball. In this paper, we present an autonomous learning method for ball trapping skills. Our method will enhance the game by way of learned pass-work in the four-legged robot league.

The remainder of this paper is organized as follows. In Section 2, we begin by specifying the actual, physical actions used in trapping the ball. Then we simplify the learning process for ball trapping down to a one-dimensional model, and finally, we illustrate and describe our training equipment used by the robots while training in solitude. In Section 3, we formalize a learning problem and show our autonomous learning algorithm for it. In Section 4, we experiment using one robot, two robots, and two robots with communication. Finally, Section 5 presents our conclusions.

2 Preliminary

2.1 Ball Trapping

Before any learning can begin, we first have to accurately create the appropriate physical motions to be used in trapping a ball accurately before the learning process. The picture in Fig. 1 (a) shows the robot's pose at the end of the motion. The robot begins by spreading out its front legs to form a wide area with which to receive the ball. Then, the robot moves its body back a bit in order to absorb the impact caused by the collision of the body with the ball and to reduce the rebound speed. Finally, the robot lowers its head and neck, assuming that the ball has passed below the chin, in order to keep the ball from bouncing off of its chest and away from its control. Since the camera of the robot is equipped on the tip of the nose, it actually cannot watch the ball below the chin. This series of motions is treated as single motion, so we can neither change the speed of the motion, nor interrupt it, once it starts. It takes 300 ms (= 60 steps $\times$ 5 ms) to perform. As opposed to grabbing or grasping the ball, this trapping motion is instead

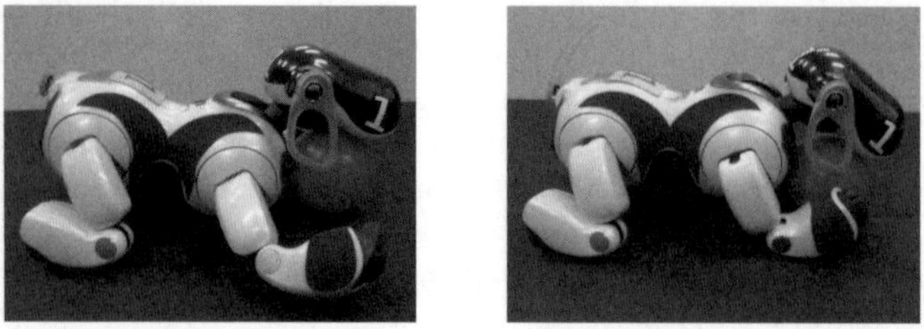

(a) trapping motion (b) pre-judgment motion

Fig. 1. The motion to actually trap the ball (a), and the motion to judge if it succeeded in trapping the ball (b)

thought of as keeping the ball, similar to how a human player would keep control of the ball under his/her foot.

The judgment of whether the trap succeeded or failed is critical for autonomous learning. Since the ball is invisible to the robot's camera when it's close to the robot's body, we utilized the chest PSD sensor. However, the robot cannot make an accurate judgment when the ball is not directly in front of their chest or after it takes a droopy posture. Therefore, we utilized a "pre-judgment motion", which takes 50 ms (= 10 steps $\times$ 5 ms), immediately after the trapping motion is completed, as shown in Fig. 1 (b). In this motion, the robot fixes the ball between its chin and chest and then lifts its body up slightly so that the ball will be located immediately in front of the chest PSD sensor, assuming the ball was correctly trapped to begin with.

2.2 One-Dimensional Model of Ball Trapping

Acquiring ball trapping skills in solitude is usually difficult, because robots must be able to search for a ball that has bounced off of them and away, then move the ball to an initial position, and finally kick the ball again. This requires sophisticated, low-level programs, such as an accurate, self-localization system; a strong shot that is as straight as possible; and a locomotion which utilizes the odometer correctly. In order to avoid additional complications, we simplify the learning process a bit more.

First, we assume that the passer and the receiver face each other when the passer passes the ball to the receiver, as shown Fig. 2. The receiver tries to face the passer while watching the ball that the passer is holding. At the same time, the passer tries to face the receiver while looking at the red or blue chest uniform of the receiver. This is not particularly hard to do, and any team should be able to accomplish it. As a result, the robots will face each other in a nearly straight line. The passer need only shoot the ball forward so that the ball can go to the receiver's chest. The receiver, in turn, has only to learn a technique for trapping the oncoming ball without it bouncing away from its body.

Ideally, we would like to treat our problem, which is to learn ball trapping skills, one-dimensionally. In actuality though, the problem cannot be fully viewed in one-dimension, because either the robots might not precisely face each other in a straight line, or because the ball might curve a little due to the grain of the grass. We will discuss this problem in Section 5.

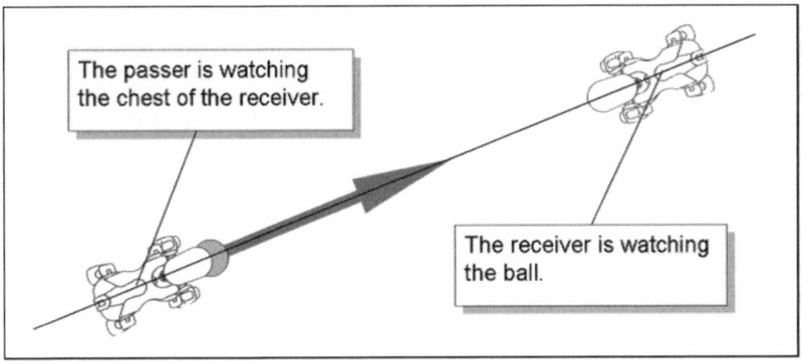

Fig. 2. One-dimensional model of ball trapping problem

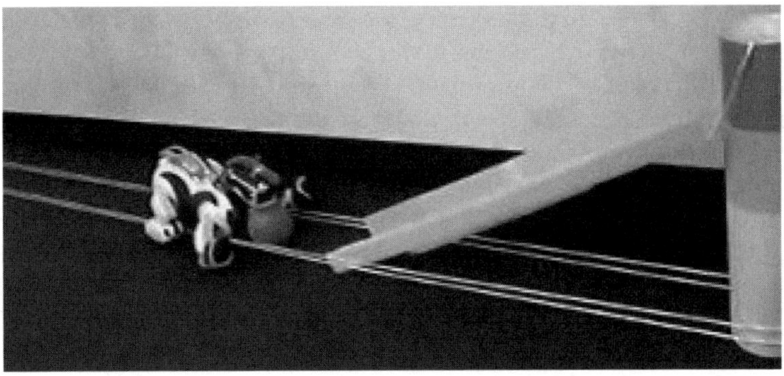

Fig. 3. Training equipment for learning ball trapping skills

2.3 Training Equipment

The equipment we prepared for learning ball trapping skills in one-dimensional is fairly simple. As shown in Fig. 3, the equipment has rails of width nearly equal to an AIBO's shoulder-width. These rails are made of thin rope or string, and their purpose is to restrict the movement of the ball, as well as the quadrupedal locomotion of the robot, to one-dimension. Aside from these rails, the robots use a slope placed at the edge of the rail when learning in solitude. They kick the ball toward the slope, and they can learn trapping skills by trying to trap the ball after it returns from having ascended the slope.

3 Learning Method

Fidelman and Stone [9] showed that the robot can learn to grasp a ball. They employed three algorithms: hill climbing, policy gradient, and amoeba. We cannot, however, directly apply these algorithms to our own problem because the ball is moving fast in our case. It may be necessary for us to set up an equation which incorporates the friction of the rolling ball and the time at which the trapping motion occurs if we want to view our problem in a manner similar to these parametric learning algorithms. In this paper, we apply reinforcement learning algorithms [10]. Since reinforcement learning requires no background knowledge, all we need to do is give the robots the appropriate reward for a successful trapping so that they can successfully learn these skills.

The reinforcement learning process is described as a sequence of states, actions, and rewards

$$s_0, a_0, r_1, s_1, a_1, r_2, \ldots, s_i, a_i, r_{i+1}, s_{i+1}, a_{i+1}, r_{i+2}, \cdots,$$

which is a reflection of the interaction between the learner and the environment. Here, $s_t \in S$ is a state given from the environment to the learner at time t ($t \geq 0$), and $a_t \in \mathcal{A}(s_t)$ is an action taken by the learner for the state s_t, where $\mathcal{A}(s_t)$ is the set of actions available in state s_t. One time step later, the learner receives a numerical reward $r_{t+1} \in \mathcal{R}$, in part as a consequence of its action, and finds itself in a new state s_{t+1}.

Our interval for decision making is 40 ms and is in synchronization with the frame rate of the CCD-camera. In the sequence, we treat each 40 ms as a single time step, i.e. $t = 0, 1, 2, \cdots$ means 0 ms, 40 ms, 80 ms, $\cdots$, respectively. In our experiments, the states essentially consist of the information on the moving ball: relative position to the robot, moving direction, and the speed, which are estimated by our vision system. Since we have restricted the problem to one-dimensional movement in Section 2.2, the state can be represented by a pair of scalar variables x and dx. The variable x refers to the distance from the robot to the ball estimated by our vision system, and dx simply refers to the difference between the current x and the previous x of one time step before. We limited the range of these state variables such that x is in [0 mm, 2000 mm], and dx in [-200 mm, 200 mm]. This is because if a value of x is greater than 2000, it will be unreliable, and if the absolute value of dx is greater than 200, it must be invalid in games (e.g. dx of 200 mm means 5000 mm/s).

Although the robots have to do a large variety of actions to perform fully-autonomous learning by nature, as far as our learning method is concerned, we can focus on the following two macro-actions. One is *trap*, which initiates the trapping motions described in Section 2.1. The robot's motion cannot be interrupted for 350 ms until the trapping motion finishes. The other is *ready*, which moves its head to watch the ball and preparing to *trap*. Each reward given to the robot is simply one of $\{+1, 0, -1\}$, depending on whether it successfully traps the ball or not. The robot can make a judgment of that success by itself using its chest PSD sensor. The reward is 1 if the *trap* action succeeded, meaning the ball was correctly captured between the chin and the chest after the *trap* action. A reward of -1 is given either if the *trap* action failed, or if the ball touches the PSD sensor before the *trap* action is performed. Otherwise, the reward is 0. We define the period from kicking the ball to receiving any reward other than 0 as one *episode*. For example, if the current episode ends and the robot moves to a random position with the ball, then the next episode begins when the robot kicks the ball forward.

In summary, the concrete objective for the learner is to acquire the correct timing for when to initiate the trapping motion depending on the speed of the ball by trial and error. Fig. 4 shows the autonomous learning algorithm used in our research. It is a combination of the episodic SMDP Sarsa(λ) with the linear tile-coding function approximation (also known as CMAC). This is one of the most popular reinforcement learning algorithms, as seen by its use in the keepaway learner [1].

```
while still not acquiring trapping skills do
    go get the ball and move to a random position with the ball;
    kick the ball toward the slope;
    s ← a state observed in the real environment;
    forall a ∈ A(s) do
        F_a ← set of tiles for a, s;
        Q_a ← Σ_{i∈F_a} θ(i);
    end
    lastAction ← an optimal action selected by ε-greedy;
    e⃗ ← 0;
    forall i ∈ F_lastAction do e(i) ← 1;
    reward ← 0;
    while reward = 0 do
        do lastAction;
        if lastAction = trap then
            if the ball is held then reward ← 1;
            else reward ← −1;
        else
            if collision occurs then reward ← −1;
            else reward ← 0;
        end
        δ ← reward − Q_lastAction;
        s ← a state observed in the real environment;
        forall a ∈ A(s) do
            F_a ← set of tiles for a, s;
            Q_a ← Σ_{i∈F_a} θ(i);
        end
        lastAction ← an optimal action selected by ε-greedy;
        δ ← δ + Q_lastAction;
        θ⃗ ← θ⃗ + αδe⃗;
        Q_lastAction ← Σ_{i∈F_lastAction} θ(i);
        e⃗ ← λe⃗;
        if player acting in state s then
            forall a ∈ A(s) s.t. a ≠ lastAction do
                forall i ∈ F_a do e(i) ← 0;
            end
            forall i ∈ F_lastAction do e(i) ← 1;
        end
    end
    δ ← reward − Q_lastAction;
    θ⃗ ← θ⃗ + αδe⃗;
end
```

Fig. 4. Algorithm of our autonomous learning (based on keepaway learner [1])

Here, F_a is a *feature set* specified by tile coding with each action a. In this paper, we use two-dimensional tiling and set the number of tilings to 32 and the number of tiles to about 5000. We also set the tile width of x to 20 and the tile width of dx to 50. The vector $\overrightarrow{\theta}$ is a *primary memory vector*, also known as a *learning weight vector*, and Q_a is a *Q-value*, which is represented by the sum of $\overrightarrow{\theta}$ for each value of F_a. The policy *ϵ-greedy* selects a random action with probability ϵ, and otherwise, it selects the action with the maximum Q-value. We set $\epsilon = 0.01$. Moreover, $\overrightarrow{e}$ is an *eligibility trace*, which stores the credit that past action choices should receive for current rewards. λ is a *trace-decay parameter* for the eligibility trace, and we simply set $\lambda = 0.0$. We set the *learning rate parameter* $\alpha = 0.5$ and the *discount rate parameter* $\gamma = 1.0$.

4 Experiments

4.1 Training Using One Robot

We first experimented by using one robot along with the training equipment that was illustrated in Section 2.3. The robot could train in solitude and learn ball trapping skills on its own.

Fig. 5(a) shows the trapping success rate, which is how many times the robot successfully trapped the ball in 10 episodes. It reached about 80% or more after 250 episodes, which took about 60 minutes using 2 batteries. Even if robots continue to learn, the success rate is unlikely to ever reach 100%. This is because the trapping motions, which force the robot to move slightly backwards in order to try and reduce the bounce effect, can hardly be expected to capture a slow, oncoming ball that stops just in front of it.

Fig. 6 shows the result of each episode by plotting a circle if it was successful, a cross if it failed in spite of trying to trap, and a triangle if it failed because of doing nothing. From the 1st episode to the 50th episode, the robots simply tried to trap the ball while it was moving with various velocities and at various distances. They made the mistake of trying to trap the ball even when it was moving away ($dx > 0$), because we did not give them any background knowledge, and we only gave them two variables: x and dx. From the 51st episode to the 100th episode, they learned that they could not trap the ball when it was far away ($x > 450$) or when it was moving away ($dx > 0$). From the 101st episode to 150th episode, they began to learn the correct timing for a successful trapping, and from the 151st episode to 200th episode, they almost completely learned the correct timing.

4.2 Training Using Two Robots

In the case of training using two robots, we simply replace the slope in the training equipment with another robot. We call the original robot the *Active Learner* (AL) and the one which replaced with slope the *Passive Learner* (PL). AL is the same as in case of training using one robot. On the other hand, PL differs from AL in that PL does not search out nor approach the ball if the trapping failed. Only AL does so. Other than this difference, PL and AL are basically the same.

We experimented for 60 minutes by using both AL and PL that had learned in solitude for 60 minutes using the training equipment. Theoretically, we would expect them

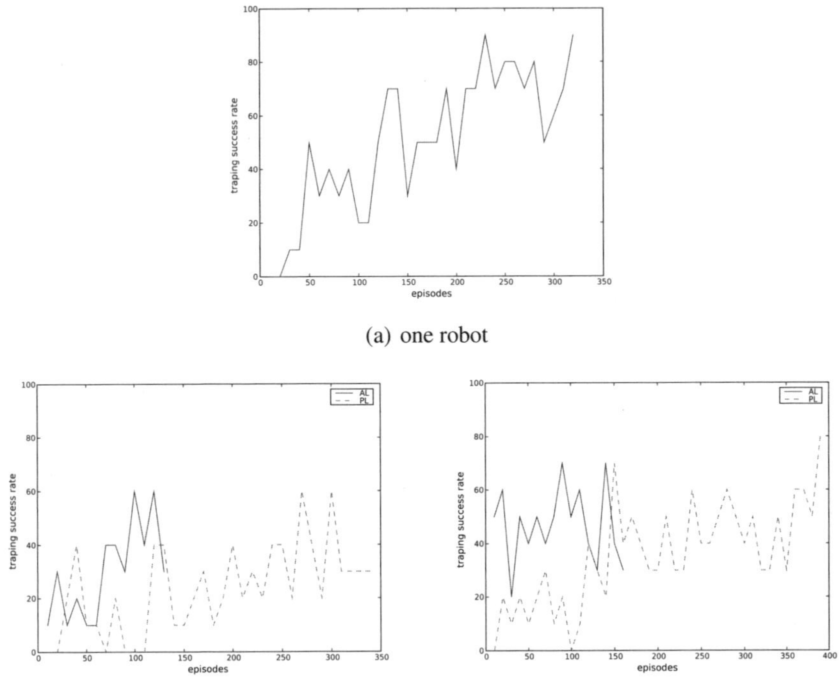

(a) one robot

(b) two robots

(c) two robots with communication

Fig. 5. Results of three experiments

to succeed in trapping the ball after only a short time. However, by trying to trap the ball while in obviously incorrect states, they actually failed repeatedly. The reason for this was because the estimation of the ball's distance to the robot-in-waiting became unreliable, as shown in Fig. 7. This, in turn, was due to the other robot holding the ball below its head before kicking it forward to its partner. Such problems can occur during the actual games, especially in poor lighting conditions, when teammates and adversaries are holding the ball.

Although we are of course eager to overcome this problem, we should not force a solution that discourages the robots from holding the ball first, because ball holding skills help them to properly judge whether or not they can successfully trap the ball. It also serves another purpose, which is to give the robots a nicer, straighter kick. Moreover, there is no way we can absolutely keep the adversary robots from holding the ball. Although there are several solutions (e.g. measuring the distance to the ball by using green pixels or sending the training partner to get the ball), we simply continued to make the robots learn without having made any changes. This was done in an attempt to allow the robots to gain experience related to irrelevant states. In fact, it turns out they should never try to trap the ball when $x \geq 1000$ and $dx \geq 200$. Moreover, they should probably not try to trap the ball when $x \geq 1000$ and $dx \leq -200$.

Fig. 5(b) shows the results of training using two robots. They began to learn that they should probably not try to trap the ball while in irrelevant states, as this was a

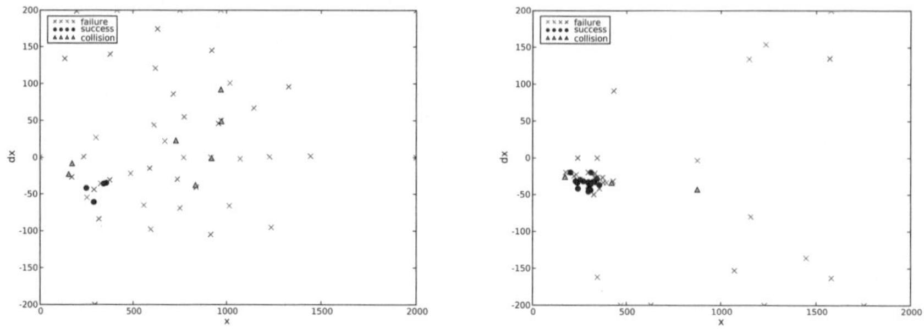

(a) Episodes 1–50 (b) Episodes 51–100

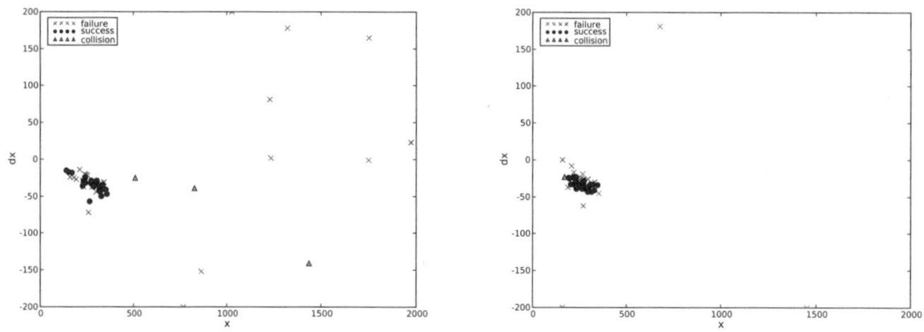

(c) Episodes 101–150 (d) Episodes 151–200

Fig. 6. Learning process from 1st episode to 200th episode. A circle indicates successful trapping, a cross indicates failed trapping, and a triangle indicates collision with the ball.

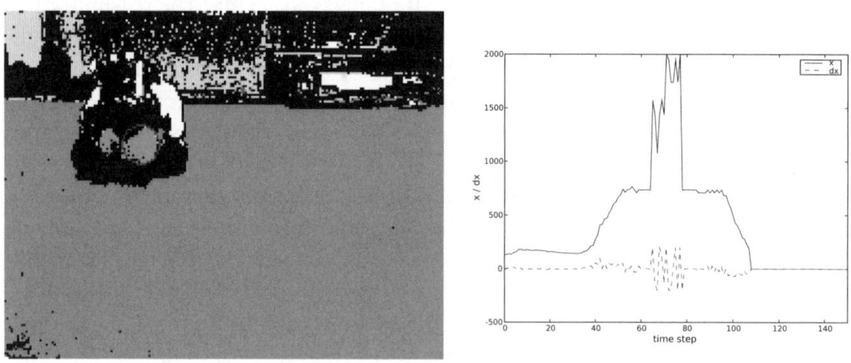

Fig. 7. The left figure shows how our vision system recognizes a ball when the other robot holds it. The ball looks to be smaller than it is, because a part of it is hidden by the partner and its shadow, resulting in an estimated distance to the ball that is further away than it really is. The right figure plots the estimated values of the both the distance x and the velocity dx, when the robot kicked the ball to its partner, the partner trapped it, and then the partner kicked it back. When the training partner was holding the ball under its head though (the center of the graph), we can see the robot obviously miscalculated ball's true distance.

likely indicator that the training partner was in possession of the ball. This was learned quite slowly though, because the AL can only learn successful trapping skills when PL itself succeeds. If PL fails, AL's episode is not incremented. Even if the player nearest the ball can go get it, the problem is not resolved because then they just learn slowly in the end, though simultaneously.

4.3 Training Using Two Robots with Communication

Training using two robots, like in the previous section, unfortunately takes a long time to complete. In this section, we will look at accelerating their learning by allowing them to communicate with each other.

First, we made the robots share their experiences with each other, as in [11]. However, if they continuously communicated with each other, they could not do anything else, because the excessive processing would interrupt the input of proper states from the real-time environment. Therefore, we made the robots exchange their experiences, which included what action a_t they performed, the values of the state variables x_t and dx_t, and the reward r_{t+1} at time t, but this was done only when they received a reward other than 0, i.e. the end of each episode. They then updated their $\overrightarrow{\theta}$ values using the experiences they received from their partner. As far as the learning achievements for our research is concerned, they can successfully learn enough using this method.

We also experimented in the same manner as Section 4.2 using two robots which can communicate with each other. Fig. 5(c) shows the results of this experiment. They could rapidly adapt to unforeseen problems and acquire practical trapping skills. Since PL learned its skills before AL learned, it could relay to AL the helpful experience, effectively giving AL about a 50% learned status from the beginning. These results indicate that the robots with communication learned more quickly than the robots without communication.

4.4 Discussion

The three experiments above showed that robots could efficiently learn ball trapping skills and that the goal of pass-work by robots can be achieved in one-dimension. In order to briefly compare those experiments, Fig. 8 presents a few graphs, where the x-axis is the elapsed time and the y-axis is the total number of successes so far. Fig. 8(a) and Fig. 8(b) shows the learning process with and without communication, respectively, for 60 minutes after pre-learning for 60 minutes by using two robots from the beginning. Fig. 8(c) and Fig. 8(d) shows the learning process with and without communication, respectively, after pre-learning for 60 minutes in solitude.

Comparing (a) and (c) with (b) and (d) has us conclude that allowing AL and PL to communicate with each other will lead to more rapid learning compared to no communication. Comparing (a) and (b) with (c) and (d), the result is different from our expectation. Actually, the untrained robots learned as much as or better than trained robots for 60 minutes. The trained robots seems to be over-fitted for slow-moving balls, because the ball was slower in the case of one robot learning than in the case of two due to friction on the slope. However, it is still good strategy to train robots in solitude at the beginning, because experiments that solely use two robots can make things more

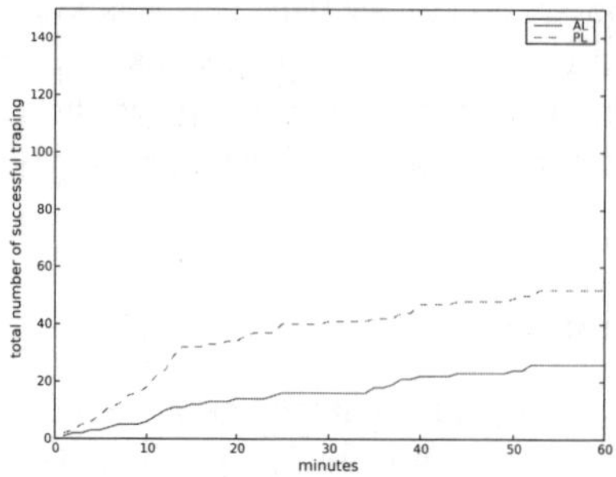

(a) without communication after pre-learning by using two robots

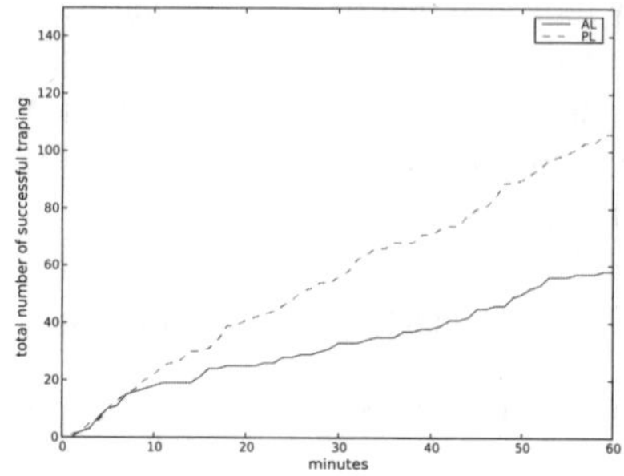

(b) with communication after pre-learning by using two robots

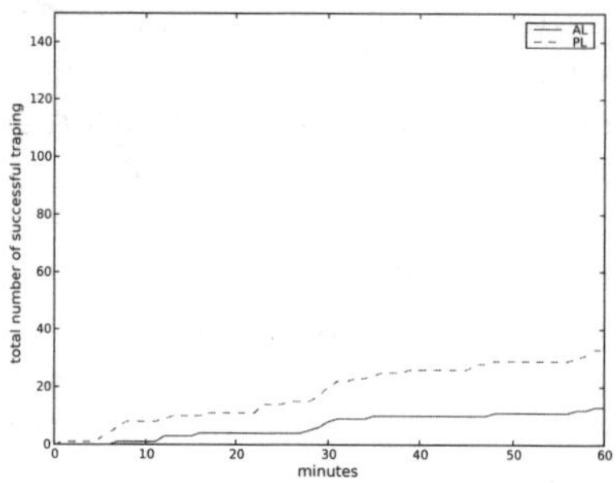

(c) without communication after pre-learning in solitude

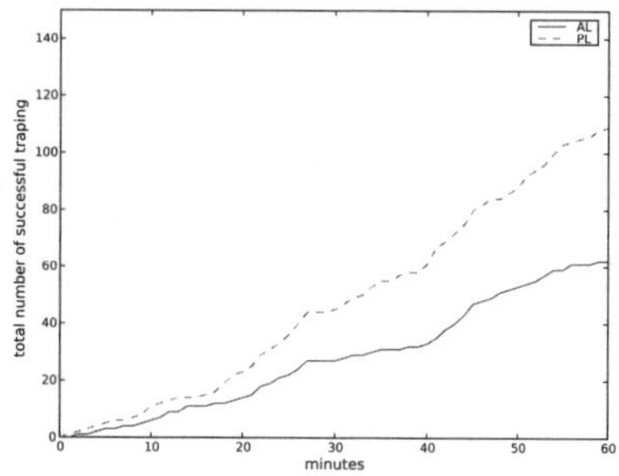

(d) with communication after pre-learning in solitude

Fig. 8. Total numbers of successful trappings with respect to the elapsed time

complicated. In addition robots should also learn the skills for a relatively slow-moving ball anyway.

5 Conclusions and Future Work

In this paper, we presented an autonomous learning method for use in acquiring ball trapping skills in the four-legged robot league. Robots could learn and acquire the skills without human intervention, except for replacing discharged batteries. They also successfully passed and trapped a ball with another robot and learn more quickly when exchanging experiences with each other. All movies of the earlier and later phases of our experiments are available on-line (http://www.jollypochie.org/papers/).

We also tried finding out whether or not robots can trap the ball without the use of the training equipment (rails for ball guidance). We rolled the ball to the robot by hand, and the robot could successfully trap it, even if the ball moved a few centimeters away from the center of its chest. At the same time though, the ball would often bounce off of it, or the robot did nothing if the ball happened to veer significantly away from the center point. In the future, we plan to extend trapping skills into two-dimensions using layered learning [12], e.g. we will try to introduce three actions of staying, moving to the left, and moving to the right into higher-level layers. Since two-dimensions are essentially the same as one-dimension in this case, it may be possible to simply use a wide slope. Good two-dimensional trapping skills can directly make keepers or goalies stronger. In order to overcome the new problems associated with a better goalie on the opposing team, robots may have to rely on learning better passing skills, as well as learning even better ball trapping skills. A quick ball is likely to move straightforward with stability, but robots as they are now can hardly trap a quick ball. Therefore, robots must learn skills in shooting as well as how to move the ball with proper velocity. It would be most effective if they learn these skills alongside trapping skills. This is a path that can lead to achieving successful keepaway soccer [1] techniques for use in the four-legged robot league.

References

1. Stone, P., Sutton, R.S., Kuhlmann, G.: Reinforcement learning for robocup soccer keepaway. Adaptive Behavior 13(3), 165–188 (2005)
2. Hsu, W.H., Harmon, S.J., Rodriguez, E., Zhong, C.: Empirical comparison of incremental reuse strtegies in genetic programming for keep-away soccer. In: Late Breaking Papers at the 2004 Genetic and Evolutionary Computation Conference (2004)
3. Hornby, G.S., Takamura, S., Yamamoto, T., Fujita, M.: Autonomous evolution of dynamic gaits with two quadruped robots. IEEE Transactions on Robotics 21(3), 402–410 (2005)
4. Kim, M.S., Uther, W.: Automatic gait optimisation for quadruped robots. In: Proceedings of 2003 Australasian Conference on Robotics and Automation, pp. 1–9 (2003)
5. Kohl, N., Stone, P.: Machine learning for fast quadrupedal locomotion. In: The Nineteenth National Conference on Artificial Intelligence, pp. 611–616 (2004)
6. Weingarten, J.D., Lopes, G.A.D., Buehler, M., Groff, R.E., Koditschek, D.E.: Automated gait adaptation for legged robots. In: IEEE International Conference on Robotics and Automation, IEEE Computer Society Press, Los Alamitos (2004)
7. Chernova, S., Veloso, M.: Learning and using models of kicking motions for legged robots. In: Proceedings of International Conference on Robotics and Automation (2004)
8. Zagal, J.C., Ruiz-del-Solar, J.: Learning to kick the ball using back to reality. In: Nardi, D., Riedmiller, M., Sammut, C., Santos-Victor, J. (eds.) RoboCup 2004. LNCS (LNAI), vol. 3276, pp. 335–347. Springer, Heidelberg (2005)
9. Fidelman, P., Stone, P.: Learning ball acquisition on a physical robot. In: 2004 International Symposium on Robotics and Automation (ISRA) (2004)
10. Sutton, R.S., Barto, A.G.: Reinforcement Learning: An Introduction. MIT Press, Cambridge (1998)
11. Kretchmar, R.M.: Parallel reinforcement learning. In: The 6th World Conference on Systemics, Cybernetics, and Informatics (2002)
12. Stone, P., Veloso, M.M.: Layered learning. In: de Mántaras, R.L., Plaza, E. (eds.) ECML 2000. LNCS (LNAI), vol. 1810, pp. 369–381. Springer, Heidelberg (2000)

Autonomous Learning of Stable Quadruped Locomotion

Manish Saggar[1], Thomas D'Silva[2], Nate Kohl[1], and Peter Stone[1]

[1] Department of Computer Sciences
[2] Department of Electrical and Computer Engineering
The University of Texas at Austin

Abstract. A fast gait is an essential component of any successful team in the RoboCup 4-legged league. However, quickly moving quadruped robots, including those with learned gaits, often move in such a way so as to cause unsteady camera motions which degrade the robot's visual capabilities. This paper presents an implementation of the policy gradient machine learning algorithm that searches for a parameterized walk while optimizing for both speed and stability. To the best of our knowledge, previous learned walks have all focused exclusively on speed. Our method is fully implemented and tested on the Sony Aibo ERS-7 robot platform. The resulting gait is reasonably fast and considerably more stable compared to our previous fast gaits. We demonstrate that this stability can significantly improve the robot's visual object recognition.

1 Introduction

In the robot soccer domain, a fast gait is an important component of a successful team. As a result a significant amount of recent research has been devoted to the problem of developing fast legged locomotion for Sony Aibo ERS-7 robots, leading to considerable improvement in gait speeds [1,2,3,4,5].

However, learned gaits optimized solely for speed tend to produce body motions that cause the camera to shake. Such unsteady gaits lead to camera images in which objects are rotated, translated, or blurred compared to camera images from a steady gait. These images make it difficult for the robot to identify objects. For example a pink over yellow beacon is usually identified as a pink blob over a yellow blob, however the pink does not appear above the yellow when the image is rotated. Thus, unstable gaits degrade a robot's object recognition and localization abilities which can cause problems during a game.

This paper proposes optimizing both gait speed *and* stability simultaneously, using a multi-criteria objective function. In addition, experiments are described that explore the idea of using active head movements to compensate for uneven body motion.

The remainder of this paper is organized as follows. Section 2 presents existing machine learning techniques that have been applied to optimize gait parameters for speed. Section 3 describes the parameterized Aibo gait, head motions, and the policy gradient algorithm used to train new gaits. Section 4 describes our

G. Lakemeyer et al. (Eds.): RoboCup 2006, LNAI 4434, pp. 98–109, 2007.

training experiments in detail and compares two different methods to offset unstable body movements. In Section 5, applications of stable gaits and future work are outlined, and Section 6 concludes.

2 Related Work

When generating quadrupedal robot gaits, the machine learning (ML) approach offers several advantages over hand-tuning of parameters. Using learning can reduce the amount of time required to find a fast gait and can be easily applied to different surfaces and different robots. ML techniques also do not suffer from the bias a human engineer might have when hand-tuning a gait. For example, there is evidence that when walking the actual joint angles of the Aibo differ considerably from requested joint angles, because of the force exerted by the ground [6]. ML techniques may be less susceptible to this problem than humans who often hand-tune gaits based on the locus of points the foot ideally moves through, as opposed to the actual locus the foot moves through.

Applying ML techniques to directly control an Aibo by manipulating joint angles is a difficult task. Evaluations on physical robots are noisy and take a long time compared to evaluation in simulation. Moreover, some of the intermediate exploratory gaits that ML algorithms generate may cause physical damage to the robot. The Aibo also does not have sensors that can be used during training that can provide closed loop feedback to the controller.

Nonetheless, reinforcement learning (RL) has been used to learn several similar control problems, not limited to Aibo locomotion. RL has been used to control a model helicopter than can hover while inverted in air [7]. Other ML techniques have been applied to directly control simulated bipedal robots: in [8] a central pattern generator was used for rhythm generation in the hips and knees of a simulated bipedal robot, and a dynamics controller was used to control the ankles of robot.

Similarly, previous work has shown that ML algorithms can excel at generating fast gaits for the Aibo by taking advantage of algorithms to optimize parameterized gaits for desirable characteristics. The earliest attempt to use ML algorithms to learn a gait used a genetic algorithm to optimize parameters describing joint velocities and body positions [9].

More recent approaches attempt to learn parameters for gaits that move the Aibo's four feet through a locus of points. In previous work, the policy gradient algorithm has been used, with a half-elliptical locus, to learn an Aibo gait that is optimized for speed [2,3]. Powell's method of multidimensional minimization has been used to optimize a parameterized gait with a rectangular locus [4]. A genetic algorithm that used interpolation and extrapolation for the crossover step was used to optimize a parameterized gait with a half-elliptical locus [1]. Odometry was used in order to evolve an omni-directional parameterized gait using a genetic algorithm by training the robot to move forward with its target orientation constantly changing [10]. In [5], a genetic algorithm and an acceleration model of the Aibo body was used to optimize a parameterized Aibo gait.

One of the fastest known forward Aibo gaits, which has a speed of 451 mm/s, was learned using a genetic algorithm and an overhead camera to quickly determine walk speeds [11].

To the best of our knowledge, all of these approaches have optimized exclusively for walk speed. This paper is based on the observation that the resulting gaits are often unstable, thus degrading the robot's visual capabilities. We demonstrate that this problem can be solved by optimizing the gait for both speed *and stability* by incorporating stability information into the objective function. This paper applies two different approaches to learning a stable walk. In the first approach, the objective function incorporates stability information. In the second approach, compensatory head movements are performed to counter the unstable body motions of a fast gait.

3 Background

The Sony Aibo ERS-7 robot is a quadruped with three degrees of freedom in each leg [12]. A controller must specify the set of twelve joint angles at each instant in order to specify a gait. Learning a controller for a fast gait by directly manipulating joint angles is a difficult non-linear control problem. One solution to this problem is parameterizing a gait by specifying the loci of points that the Aibo's feet moves through. Doing so can constrain the search space both to make it easier to search and to avoid gaits that can damage the robot. This paper uses a modified version of a half-elliptical parameterized gait modeled after that presented by Stone et al. [13]. Four additional parameters were added to this parameterization that govern compensatory head movements designed to improve head stability.

3.1 Parameterized Motion

The half-elliptical locus used by the fast gait is shown in Figure 1. Each foot moves through a half-elliptical locus with each pair of diagonally opposite legs in phase with each other and out of phase with the other two legs (a trot gait).

The four parameters that define the half ellipse are:

1. The length of the ellipse
2. The height of the ellipse
3. The position of the ellipse on the x axis
4. The position of the ellipse on the y axis

The symmetry of the Aibo is used to reduce the number of parameters that have to be optimized. The length of the ellipse is the same for all four legs to ensure a straight gait. The left and right sides of the body use the same parameters to describe the locus of the gait. The height, x position and y position of the elliptical loci of the front and back two legs use different parameters.

In addition to the leg movements, the head was allowed to make elliptical compensatory movements in order to cancel the effect of body motions that

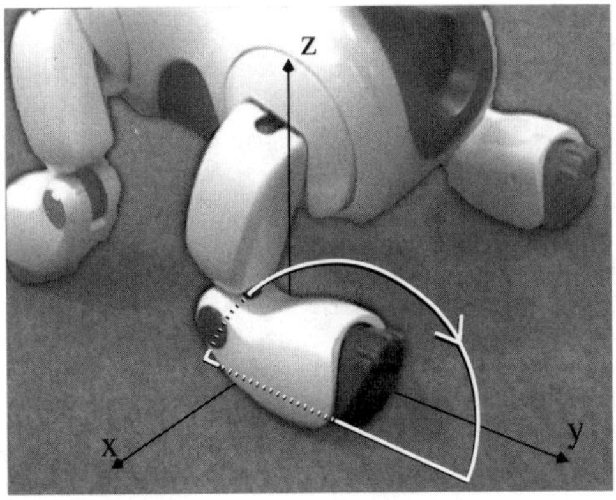

Fig. 1. The half-elliptical locus of each of the Aibo's feet is defined by length, height and position in the x-y plane

cause the camera to shake. Figure 2 depicts the two types of head movement that were used, which have the overall effect of moving the head in an ellipse. Two parameters were used to specify the head tilt angle limit and head tilt increment at each timestep. Similarly, two parameters describe the head pan motions. Initial values for these parameters were determined by testing just a few sets of values. We leave it to future work to determine how big of an effect these initial values have.

The 15 parameters that completely define the Aibo's movements are:

- The front locus: height, x position and y position (3 parameters)
- The rear locus: height, x position and y position (3 parameters)
- Locus length (same for all loci)
- Front body height
- Rear body height
- Time taken for each foot to move through locus
- The fraction of time each foot spends on the ground
- Head tilt limit and increment (2 parameters, with a limit from -10° to 10°)
- Head pan limit and increment (2 parameters, with a limit from -10° to 10°)

3.2 Policy Gradient Algorithm

This paper uses a policy gradient algorithm modeled after that presented by Kohl and Stone [2] to optimize the Aibo gait in the continuous 15-dimensional parameter space. The objective function F to be optimized is a function of the gait speed, acceleration and stability, and is described in detail in Section 4.

The policy gradient algorithm uses an initial parameter vector $\pi = \{\theta_1, ..., \theta_N\}$ and estimates the partial derivative of the objective function F with respect to each parameter. This is done by evaluating t randomly generated policies

Fig. 2. The Aibo can combine pan and tilt head movements (shown as solid lines) to move the head through an elliptical locus (shown as a dotted line). The center of ellipse is determine by the landmark the Aibo is looking at. The locus is defined by four variables: tilt limit, tilt increment, pan limit, and pan increment.

$\{R_1, ...R_t\}$ near π, such that each $R_i = \{\theta_1 + \delta_1, ..., \theta_N + \delta_N\}$ and δ_j is randomly chosen to be either $+\epsilon_j, 0, or - \epsilon_j$, where ϵ_j is a small fixed value relative to θ_j.

After evaluating each neighboring policy R_i on the objective function F, each dimension of every R_i is grouped into one of three categories to estimate an average gradient for each dimension:

- $Avg_{-\epsilon,n}$ if the nth parameter of R_i is $\theta_{n-\epsilon_n}$
- $Avg_{+0,n}$ if the nth parameter of R_i is θ_{n+0}
- $Avg_{+\epsilon,n}$ if the nth parameter of R_i is $\theta_{n+\epsilon_n}$

These three averages enable the estimation of the benefit of altering the nth parameter by $+\epsilon_n, 0$, and $-\epsilon_n$. An adjustment vector A of size n is calculated where $A_n \in$

- 0 if $Avg_{+0,n} > Avg_{+\epsilon,n}$and $Avg_{+0,n} > Avg_{-\epsilon,n}$
- $Avg_{+\epsilon,n} - Avg_{-\epsilon,n}$ otherwise

A is normalized and then multiplied by a scalar step size $\eta = 2$ to offset small ϵ_j. Finally A is added to π, and the process is repeated for the next iteration. Figure 3 describes the pseudocode for the policy gradient algorithm.

4 Empirical Results

The policy gradient algorithm described above was implemented and run on the Aibo as seen in Figure 4. In order to evaluate a particular gait parameterization,

```
π ← InitialPolicy
while !done do
  {R_1, R_2, ..., R_t} = t random perturbations of π
  evaluate( {R_1, R_2, ..., R_t} )
  for n = 1 to N do
    Avg_{+ε,n} ← average score for all R_i that have
                 a positive perturbation in dimension n
    Avg_{+0,n} ← average score for all R_i that have a zero
                 perturbation in dimension n
    Avg_{-ε,n} ← average score for all R_i that have a
                 negative perturbation in dimension n
    if Avg_{+0,n} > Avg_{+ε,n} and Avg_{+0,n} > Avg_{-ε,n} then
      A_n ← 0
    else
      A_n ← Avg_{+ε,n} − Avg_{-ε,n}
    end if
  end for
  A ← A/|A| * η
  π ← π + A
end while
```

Fig. 3. During each iteration t policies are sampled around π to estimate the gradient, then π is moved by η in the direction that increases the objective function the greatest

the Aibo was instructed to record various data while repeatedly walking back and forth between two landmarks.

In order to generate a gait that was both stable and fast, the learning algorithm had to be given an appropriate objective function. In previous work, the objective function was focused primarily on generating a fast gait. In this paper, since stability is desired, the objective function was modified. Figure 5 depicts the images a robot would see with a perfectly stable gait and with an unsteady gait. The image taken with the unstable gait is rotated and translated compared to the image taken with a stable gait.[1]

In order to find a stable gait, the original objective function (which was designed to optimize only for speed) was modified to include stability information. This modified objective function consists of four components:

1. M_t - The normalized time taken by the robot to walk between the two landmarks.
2. M_a - The normalized standard deviation (averaged over multiple trials) of the Aibo's three accelerometers
3. M_d - The normalized distance of the centroid of landmark from the center of an image.
4. M_θ - The normalized difference between the slope of landmark and the ideal slope (90°)

[1] Videos of a fast gait and a stable gait from the perspective of the robot can be found at `http://www.cs.utexas.edu/~AustinVilla/?p=research/learned_walk`

Fig. 4. The training environment during the gait parameter optimization experiment. The Aibo records how long it takes to move between two beacons. It also records the average accelerometer values, the average difference in the position of the centroid of the beacon and the center of the image, and the average slope of the beacon in the image.

These four components are combined to create a single objective function F:

$$F = 1 - (W_t M_t + W_a M_a + W_d M_d + W_\theta M_\theta) \quad (1)$$

The different components of the objective function are weighted by W_t, W_a, W_d, and W_θ, respectively, to optimize for desirable attributes. These weights are constrained such that their sum is equal to one. For example, if stability is more important than speed, the time taken to walk between landmarks W_t can be assigned a smaller value than the other three weights. The next section describes experiments that compared different weightings of this objective function.

4.1 Learning a Stable Gait

The first experiment we performed was designed to determine how best to train for stability while learning a gait. To do this, we used two different parameterizations for weighting the subcomponents of the objective function. The first parameterization used $W_t = 0.4$, $W_a = 0.1$, $W_d = 0.4$ and $W_\theta = 0.1$, which weighted speed slightly more than stability. The second parameterization used $W_t = 0.3$, $W_a = 0.3$, $W_d = 0.2$ and $W_\theta = 0.2$, which more evenly weighted all four components.

We used a relatively slow hand-tuned gait as a starting point for the policy gradient algorithm, since previous work suggested that starting from a faster gait could hinder learning [2]. This starting point was determined empirically after trying several different starting gaits. Learning performance was somewhat sensitive to the initial parameter settings, but we did not extensively optimize the initial values.

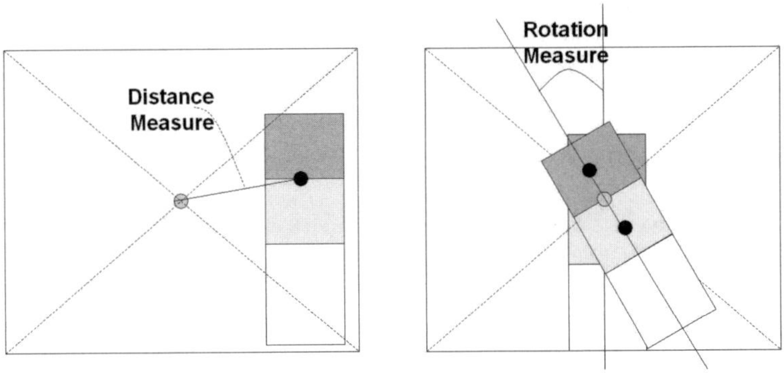

Fig. 5. Two visual clues that indicate an uneven gait. (a) shows the average displacement (M_d) of the centroid of the landmark with respect to the center of the image. (b) shows the average rotation (M_θ) of the landmark. If the camera is steady the average position difference should be zero and the average rotation should be 0°.

Table 1. Percentage reduction in the four objective function components for two different parameterizations without using compensatory head movements. In both cases the gait becomes more stable while only becoming slightly slower.

	Parameterization 1	Parameterization 2
M_t	-4.76	-4.5
M_a	34.7	32.6
M_d	60	57.14
M_θ	76.9	51.2

Figure 6 shows the progress of the policy gradient algorithm during training without head movements for the two different objective function parameterizations. The policy gradient algorithm generates 15 exploratory policies per iteration of the algorithm. In both parameterizations, the slope, distance and average acceleration measure of the objective function decrease considerably, while the time measure has a modest increase. This lead us to conclude that the weight parameters are not sensitive to smaller variations. Detailed results are shown in Table 1.

4.2 Adding Compensatory Head Movements

The previous results successfully demonstrate the ability of our robots to learn a stable gait while minimizing speed reduction. However, in that case, all of the learning was focused on the leg motion. Since the stability objective measures the robot's head motion, we hypothesized that allowing the robot to make compensatory head movements could effectively improve stability. To test this hypothesis, similar experiments to those described above were performed, but four

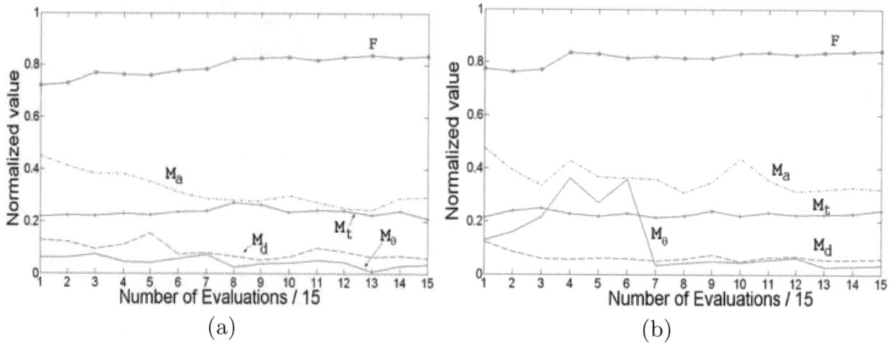

Fig. 6. (a) The overall fitness and fitness subcomponents (normalized to $[0, 1]$) for a single run with $W_t = 0.3$, $W_a = 0.3$, $W_d = 0.2$, and $W_\theta = 0.2$ without using head movements. The starting gait has an overall fitness of 0.72 and the final gait has an overall fitness of 0.83. (b) A similar plot, but with the parameters $W_t = 0.4$, $W_a = 0.1$, $W_d = 0.4$, and $W_\theta = 0.1$. The starting gait has an overall fitness of 0.78 and the final gait has an overall fitness of 0.83. Both speed and stability increase during learning.

additional parameters were added that governed compensatory head movements. For these experiments, the position of the landmark in the camera image was used to calculate the center of the ellipse that the Aibo's head moved through, and the tilt and pan angle limits and increments (set by the policy gradient algorithm) were used to calculate the length and height of the ellipse.

Figure 7 shows the progress of the policy gradient algorithm during training with head movements for two different objective function parameterizations. The policy gradient algorithm generated 19 exploratory policies per iteration of the algorithm. As in the experiment that learned a stable gait without head movements, the gait became more stable after learning. However, the results from this experiment were not as good as those from the previous experiment. This suggests that the addition of compensatory head movements does not significantly improve stability or speed.

Table 2 shows that gait parameters for the initial hand tuned gait, the final learned gait using head movements, the final learned gait without using head movements and the previously learned fast gait for comparison. The policy gradient algorithm was able to find a stable gait without much improvement in speed. These results demonstrate there is a tradeoff between gait speed and stability.

4.3 How Useful Is Stability?

The main premise of this paper is that walk stability is an important feature for robot gaits. In particular, we hypothesized that stable gaits would improve the robot's visual capabilities. The vision algorithm used for this work converts each image received from the camera into a pixel-by-pixel color-labeled image, then groups regions of similarly-colored pixels into bounding boxes. A variety of heuristics such as size, tilt, and pixel density are used to convert these bounding

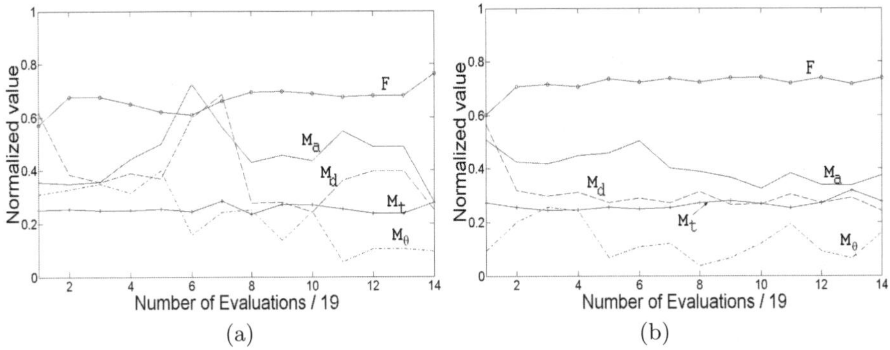

Fig. 7. (a) Scaled overall fitness and fitness subcomponents for a single run with $W_t = 0.3$, $W_a = 0.3$, $W_d = 0.2$, and $W_\theta = 0.2$ when compensatory head movements are enabled. The starting gait has an overall fitness of 0.57 and the final gait has an overall fitness of 0.76. (b) A similar graph, but with the parameters $W_t = 0.4$, $W_a = 0.1$, $W_d = 0.4$, and $W_\theta = 0.1$. The starting gait has an overall fitness of 0.60 and the final gait has an overall fitness of 0.74. In both cases, stability and speed increase, but the overall effect of compensatory head movements is negative.

boxes into high-level objects. If the robot is using an uneven gait, the camera will receive many images from an unexpected perspective, which can wreak havoc on the vision heuristics. The heuristics can always be improved, but this may take valuable processing time away from other components of the robot. Many vision algorithms employ such heuristics, making this a general problem for robotic vision [14,15].

In order to test whether the stable gaits learned above actually help vision, we conducted two experiments where the Aibo traversed the field while recording the objects that it saw. The number of objects that were correctly classified (averaged over four runs) is shown in Table 3. Using the learned stable walk, the Aibo displayed 39% more true positives and 54% fewer false positives. These results with statistically significant with $p < 0.05$.

5 Discussion and Future Work

The experiments detailed in this paper demonstrate that there is a tradeoff between gait speed and stability. Our version of the fast gait learned according to Stone et al. [13] achieves a speed of 340mm/s. When the objective function is changed to include stability information, the fastest walk that is learned has a speed of 259mm/s. Allowing the robot to make compensatory head motions to counterbalance for the body movements, reduced the speed marginally.

Even though the stable gait is not as fast as gaits optimized for speed, it could be used in situations where it is important not to lose sight of objects, for example if the robot has the ball and is near the opponent's goal, the stable gait can be used to ensure that the robot does not lose the ball from its vision and thus has a better chance at scoring. We leave deciding which gait to use when to future work.

Table 2. The parameterized starting gait, learned gaits with and without head movements and the learned fast gait. The policy gradient algorithm is able to find gaits that are considerably more stable than the learned fast gait while only sacrificing a small amount of speed. The final gait shows a small improvement in gait speed compared to the starting gait.

Parameter	Hand-tuned Gait	Stable gait	Stable gait with head movements	Fast gait
Front locus height	1.1	1.7	1.7	0.97
Front x position	-0.05	0.08	1.17	-0.04
Front y position	0.7	0.76	-0.08	0.3
Rear locus height	1.6	-0.45	1.54	1.61
Rear x position	0	1.54	1.7	-0.11
Rear y position	-0.4	0	0.66	-0.51
Locus length	0.4	0.5	0.68	0.57
Front body height	0.9	0.95	0.96	0.76
Rear body height	0.8	0.75	0.64	0.65
Time on ground	0.5	0.62	0.7	0.27
Time to move through locus	45	45.5	43.4	56
Tilt limit	n/a	n/a	4.93	n/a
Tilt increment	n/a	n/a	0.88	n/a
Pan limit	n/a	n/a	4.81	n/a
Pan increment	n/a	n/a	1.07	n/a
Gait speed	198 mm/s	259 mm/s	237 mm/s	340 mm/s

Table 3. The ratio of objects correctly and incorrectly classified by a vision algorithm using a learned fast gait and a learned stable gait. The stable gait leads to significantly ($p < 0.05$) better visual classification accuracy.

	True Positives	False Positives
Fast Gait	0.33	0.052
Stable Gait	0.46	0.028

Another interesting avenue for future work is to examine how different parameterizations for the gait and the head motion affect learning. Although the elliptical head motion described in this paper did not significantly increase head stability, other types of head motions might do better.

6 Conclusion

This paper presented results on using the policy gradient algorithm to learn a stable, fast gait. Experiments were performed using an objective function that optimizes for stability in addition to using head compensatory movements. In both cases, the policy gradient algorithm found a stable gait while sacrificing only a small amount of speed. Videos of a comparison between gaits optimized

for speed and gaits optimized for stability are available at:
`http://www.cs.utexas.edu/~AustinVilla/?p=research/learned_walk.`

Acknowledgments

The authors thank the Austin Villa team for their support and guidance. This research was supported in part by NSF CAREER award IIS-0237699, ONR YIP award N00014-04-1-0545, and DARPA grant HR0011-04-1-0035.

References

1. Rofer, T.: Evolutionary gait-optimization using a fitness function based on proprioception. In: Nardi, D., Riedmiller, M., Sammut, C., Santos-Victor, J. (eds.) RoboCup 2004. LNCS (LNAI), vol. 3276, pp. 310–322. Springer, Heidelberg (2005)
2. Kohl, N., Stone, P.: Machine learning for fast quadrupedal locomotion. In: The Nineteenth National Conference on Artificial Intelligence, pp. 611–616 (2004)
3. Kohl, N., Stone, P.: Policy gradient reinforcement learning for fast quadrupedal locomotion. In: Proceedings of the IEEE ICRA, IEEE Computer Society Press, Los Alamitos (2004)
4. Kim, M., Uther, W.: Automatic gait optimisation for quadruped robots. In: Australasian Conference on Robotics and Automation (2003)
5. Chernova, S., Veloso, M.: An evolutionary approach to gait learning for four-legged robots. In: Proceedings of IROS'04 (2004)
6. Stronger, D., Stone, P.: A model-based approach to robot joint control. In: Nardi, D., Riedmiller, M., Sammut, C., Santos-Victor, J. (eds.) RoboCup 2004. LNCS (LNAI), vol. 3276, pp. 297–309. Springer, Heidelberg (2005)
7. Ng, A.Y., Coates, A., Diel, M., Ganapathi, V., Schultel, J., Tse, B., Berger, E., Liang, E.: Autonomous helicopter flight via reinforcement learning. In: Advances in Neural Information Processing Systems 17, MIT Press, Advances in Neural Information Processing Systems (2004)
8. In, T.W., Vadakkepat, P.: Hybrid controller for biped gait generation. In: 2nd International Conference on Autonomous Robots and Agents (2004)
9. Hornby, G.S., Fujita, M., Takamura, S., Yamamoto, T., Hanagata, O.: Autonomous evolution of gaits with the sony quadruped robot. In: Genetic and Evolutionary Computation Conference, vol. 2, Morgan Kaufmann, San Francisco (1999)
10. Duffert, U., Hoffmann, J.: Reliable and precise gait modeling for a quadruped robot. In: Bredenfeld, A., Jacoff, A., Noda, I., Takahashi, Y. (eds.) RoboCup 2005. LNCS (LNAI), vol. 4020, Springer, Heidelberg (2006)
11. Rofer, T.: Germanteam robocup 2005. Technical report (2005)
12. Sony: Aibo robot (2005)
13. Stone, P., et al.: The UT Austin Villa 2004 RoboCup four-legged team: Coming of age. Technical Report UT-AI-TR-04-313, The University of Texas at Austin, Department of Computer Sciences, AI Laboratory (2004)
14. Sumengen, B., Manjunath, B.S., Kenney, C.: Image segmentation using multiregion stability and edge strength. In: The IEEE International Conference on Image Processing (ICIP) (2003)
15. Comaniciu, D., Meer, P.: Mean shift: A robust approach toward feature space analysis. IEEE Transactions on Pattern Analysis and Machine Intelligence 24(5), 603–619 (2002)

Using the Simulated Annealing Algorithm for Multiagent Decision Making

Jiang Dawei and Wang Shiyuan

Department of Computer Science and Technology,
Southeast University, P.R.China
jiang203@jlonline.com, desiree_wsy@yahoo.com.cn

Abstract. Coordination, as a key issue in fully cooperative multiagent systems, raises a number of challenges. A crucial one among them is to efficiently find the optimal joint action in an exponential joint action space. Variable elimination offers a viable solution to this problem. Using their algorithm, each agent can choose an optimal individual action resulting in the optimal behavior for the whole agents. However, the worst-case time complexity of this algorithm grows exponentially with the number of agents. Moreover, variable elimination can only report an answer when the whole algorithm terminates. Therefore, it is unsuitable in real-time systems. In this paper, we propose an anytime algorithm, called the simulated annealing algorithm, as an approximation alternative to variable elimination. We empirically show that our algorithm can compute nearly optimal results with a small fraction of the time that variable elimination takes to find the solution to the same coordination problem.

1 Introduction

A multiagent system (MAS) consists of a group of agents that interact with each other [1,2]. Research in MAS aims to provide theories and techniques for agents' behavior management. In this paper, we focus on the fully cooperative MASs in which the agents share a common goal. Examples are a team of robots who play football against another team or a group of agents who plan to build a house. A key aspect in such systems is *Coordination*: the procedure to ensure the individual actions of the agents generate optimal joint decisions for the whole group. RoboCup [3] provides a good platform for comparing and testing different coordination techniques.

To solve the above problem, previous research focuses on the use of game theoretic techniques [4], communication [5,6], social conventions or social lows [7], learning [8,9]. However, all these approaches need to exhaust the whole joint action space whose size grows exponentially with the number of agents. Thus, even in very small settings, they are infeasible.

A recent work to decrease the size of the joint action space uses a *coordination graph* (CG) [10,11,12]. The idea of CG is that in many situations, only a small number of agents need to coordinate their actions while the rest of others can

G. Lakemeyer et al. (Eds.): RoboCup 2006, LNAI 4434, pp. 110–121, 2007.

act individually. For example, in robotic soccer, only the ball owner and his surrounding players need to coordinate their actions to perform a pass while others can act individually. So the global joint payoff function, the representation of the global joint coordination dependencies between all agents, can be decomposed into a linear combination of local terms, each of which represents the local coordination dependencies between a small subgroup of the agents. Then each agent employs *variable elimination* (VE) algorithm to select an optimal individual action. The outcome results in optimal behavior for the whole group. However, the worst case time complexity of VE is the same with the aforementioned methods of exhausting all possibilities [13,14]. Moreover, although VE is an exact method which always reports the optimal joint action, it does not return any results until the entire algorithm terminates, which is not suitable for real-time systems. In [14], max-plus (MP) algorithm, which is analogous to the belief propagation algorithm [15] for Bayesian networks, was proposed as an approximate alternative to VE. MP can find optimal solutions for tree-structured coordination graphs and also the near optimal solutions in graphs with cycles, but it restricts each local payoff function involved at most two agents [14,15,16].

In this paper, we propose the *simulated annealing* (SA) algorithm as another approximation to VE. In our algorithm, agents repeatedly start independent tries. In an independent try, each agent tries to maximize the global payoff using his own action, while the actions of the other agents stay the same. If a better solution is found, accept it; otherwise, accept it with a certain probability.

We make the following contributions.

- The time complexity of our algorithm grows polynomially with the number of agents.
- Our algorithm is an anytime algorithm that reports result at any time.
- Our algorithm has no restrictions on the number of agents involved in local payoff functions.
- Experiments show that our algorithm can also find near optimal solution within only a small fraction of the time that VE takes to find the solution of the same coordination problem.

The paper is organized as follows. In section 2, we briefly describe the basic concepts of multiagent coordination problem and the process of finding the optimal joint action by VE and CG. Then we describe our proposed algorithm in section 3. Section 4 experimentally validate the correctness and efficiency of our algorithm, followed by conclusion and future work in section 5.

2 Variable Elimination and Coordination Graphs

In this section, we review the variable elimination (VE) algorithm. In a multiagent system, we have a collection of agents $\boldsymbol{G} = \{G_1, \ldots, G_n\}$[1]. Each agent G_i selects

[1] In this paper, we use upper case letters (*e.g.*, X) to denote random variables, and lower case (*e.g.*, x) to denote their values. We also use boldface to denote vectors of variables (*e.g.*, $\boldsymbol{X}$) or their values ($\boldsymbol{x}$).

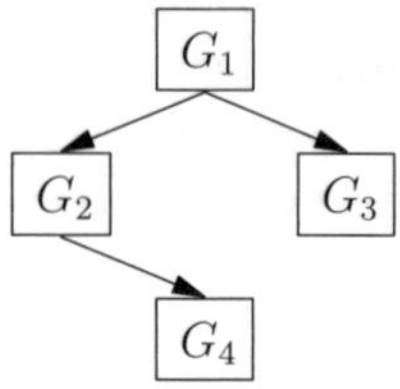

Fig. 1. Initial coordination graph

an individual action a_i from his own action set A_i. Their joint action space thus can be represented as $\boldsymbol{A} = \times_i A_i$. The global payoff function of the agents $\upsilon(\boldsymbol{a})$ maps each joint action $\boldsymbol{a}$ to a real value: $\upsilon(\boldsymbol{a}) \to \mathbb{R}$. The coordination problem is to find the optimal joint action $\boldsymbol{a}^*$ that maximizes $\upsilon(\boldsymbol{a})$, i.e., $\boldsymbol{a}^* = \text{argmax}_{\boldsymbol{a}} \upsilon(\boldsymbol{a})$. In a naive way, we may consider all possible joint actions and select the one that maximizes $\upsilon(\boldsymbol{a})$. Unfortunately, this approach is infeasible in even the simplest settings, for the number of joint actions grows exponentially with the number of agents (It is called "*curse of dimensionality*" [13]).

This "curse of dimensionality" may be solved by exploiting the structure of the problem to define a compact representation for the global joint payoff function [11,12]. In this way, the global joint payoff function is decomposed into a linear combination of a set of local payoff functions, each of which is only related to a part of system controlled by a small number of agents. For example, in RoboCup, only players that are close to each other have to coordinate their actions to perform a pass or a defend, thus we can use the sum of local payoff functions of subgroup agents to approximate the whole team's payoff. In some situations, this approach can get a very compact representation for coordination dependencies among agents. Furthermore, such representation can be mapped onto a coordination graph $G = (V, E)$ according to the following rules [11]: each agent is mapped to a node in V, and each coordination dependency is mapped to an edge in E. Then the agents can use VE which is identical to variable elimination (or bucket elimination) [17] in a Bayesian network on such CG to determine the optimal joint action.

We show how VE works as follows. Suppose we have 4 agents with each one having 4 different actions, then the number of joint actions is $4^4 = 256$, and global joint payoff function can be decomposed as:

$$\upsilon(\boldsymbol{a}) = \upsilon_1(a_1, a_2) + \upsilon_2(a_2, a_4) + \upsilon_3(a_1, a_3) \tag{1}$$

Fig. 1 shows the initial corresponding coordination graph. The key idea in VE is that, rather than enumerating all possible joint actions and summing up all functions to do maximization, we maximize over variables once at a time. Let us begin with optimization for agent 1. Agent 1 collects all local payoff functions including himself, i.e., υ_1 and υ_3 then does maximization. Hence, we obtain:

$$\text{max}_{\boldsymbol{a}} \upsilon(\boldsymbol{a}) = \text{max}_{a_2, a_3, a_4} \{\upsilon_2(a_2, a_4) + \text{max}_{a_1} [\upsilon_1(a_1, a_2) + \upsilon_3(a_1, a_3)]\} \tag{2}$$

After enumeration of possible action combinations of his neighbors, i.e., agent 2 and agent 3, agent 1 conditionally returns his best response and yield a new function $e_1(a_2, a_3) = \max_{a_1}[v_1(a_1, a_2) + v_3(a_1, a_3)]$ whose value at the point a_2, a_3 is the value of the internal max expression in equation (2). At this time, agent 1 is eliminated from CG. The global joint payoff function is rewritten as:

$$\max_{\boldsymbol{a}} v(\boldsymbol{a}) = \max_{a_2,a_3,a_4}\{v_2(a_2, a_4) + e_1(a_2, a_3)\} \tag{3}$$

Now fewer agents remain. Next, agent 2 does the same procedure. After collecting $v_2(a_2, a_4)$ and $e_1(a_2, a_3)$, agent 2 produces a conditional strategy based on the possible actions of agent 3 and agent 4, and returns his choice, i.e., $e_2(a_3, a_4) = \max_{a_2}[v_2(a_2, a_4)+e_1(a_2, a_3)]$ to the system, then is eliminated. The global payoff function only contains 2 agents now:

$$\max_{\boldsymbol{a}} v(\boldsymbol{a}) = \max_{a_3,a_4}\{e_2(a_3, a_4)\} \tag{4}$$

Agent 3 begins to do optimization. Enumerating actions of agent 4, he reports his own choice and gives a conditional payoff $e_3(a_4) = \max_{a_3} e_2(a_3, a_4)$. Finally, the only remaining agent 4 can simply choose his optimal action: $a_4^* = \text{argmax}_{a_4} e_3(a_4)$.

In the second pass, all agents do the entire process in reverse elimination order. To fulfill agent 4's optimal action a_4^*, agent 3 must select $a_3^* = \text{argmax}_{a_3} e_3(a_4^*)$. Then agent 2 can make a decision $a_2^* = \text{argmax}_{a_2} e_2(a_3^*, a_4^*)$. Finally, agent 1 does $a_1^* = \text{argmax}_{a_1} e_1(a_2^*, a_3^*)$ to choose his optimal action appropriately. The whole procedure needs only $4 \times 4 + 4 \times 4 + 4 = 36$ iterations which is much smaller than 256 iterations of the whole joint action space.

The outcome of VE is independent of the elimination order and always gives the optimal joint action [13]. However, the running speed of VE is depended on the elimination order and exponential in the *induced width* of the coordination graph [11,17]. Finding the optimal elimination order for VE is a well known NP-complete problem [18,19]. Thus, in some cases and especially in the worse case, the time consumed by VE grows exponentially with the number of agents. Furthermore, VE can not give any useful results until the termination of the complete algorithm, therefore it is not suitable for RoboCup 2D simulation league for the robot player has to send actions to server every 100ms. We aim to find an alternative approache that can circumvent such limitations.

3 The Simulated Annealing Algorithm

In many real-world applications, especially in limited computing time cases such as RoboCup, we should make tradeoff between the optimality of the actions and running time. Thus a sub-optimal or nearly optimal solution would be sufficient. In [14], max-plus (MP) algorithm was proposed as an approximation to VE. MP is essentially an instance of Perl's *belief propagation* (BP) algorithm [15] in Bayesian network. It can converge to optimal joint action in tree-structured CGs and find nearly optimal result in graphs with cycles [14,15]. However, MP limits the number of agents in local coordination dependencies not exceeding two.

In this section, we describe the *simulated annealing* (SA) algorithm proposed as an approximate alternative to VE without MP's limitation. The simulated annealing algorithm [2], inspired by statistical mechanics, is very popular for combinatorial optimization [20,21,22]. In this area, efficient techniques are developed to find minimum or maximum values for a function of a number of independent variables [22]. The simulated annealing process executes by "melting" the system being optimized at a high effective temperature at first, and then lowering the temperature by slow stages until the system "freezes" and no further change occurs.

We decide to apply SA to our multiagent decision making problem, since our problem also needs to optimize the global joint payoff function via a number of independent action variables of the agents. The key idea in our approach is rather similar to CG. We decompose the global joint payoff function into a sum of local terms, and then do optimization. Given n agents (defined in section 2), the global joint payoff function can be decomposed as follows:

$$v(\boldsymbol{a}) = \sum_{i \in \boldsymbol{G}} v_i(a_i) + \sum_{i,j \in \boldsymbol{G}} v_{ij}(a_i, a_j) + \sum_{i,j,k \in \boldsymbol{G}} v_{ijk}(a_i, a_j, a_k) + \cdots \quad (5)$$

Here, $v_i(a_i)$ represents the payoff that an agent contributes to the system when acting individually, e.g., dribbling with the ball. $v_{ij}(a_i, a_j)$ denotes the payoff of a coordination action, e.g., a coordination pass between agent i and agent j, and $v_{ijk}(a_i, a_j, a_k)$, depicts another coordination action involving three agents, e.g., pass from i to j, then j to k. Coordination dependencies with more players can be added if needed. Our decomposition is different from MP in that there is no limitation on the number of robot players involved in local terms. In MP algorithm, the global joint payoff function can only be decomposed into $\sum_{i \in \boldsymbol{G}} v_i(a_i) + \sum_{i,j \in \boldsymbol{G}} v_{ij}(a_i, a_j)$.

Now the goal is to find the optimal joint action, i.e., $\boldsymbol{a}^* = \text{argmax}_{\boldsymbol{a}} v(\boldsymbol{a})$. The pseudo-code of SA algorithm is presented in Alg.1.. The SA algorithm is implemented in a centralized version and performed by the agents in parallel, without assuming the availability of communication. The idea behind it is very straightforward. In each iteration (called an independent try), the algorithm starts with a random choice of joint action for the agents, then loop over all agents. Each agent optimizes the global payoff function with his own action while the actions of all the others stay the same. If the agent's local optimization can yield a better joint action than the initial one, we accept it, otherwise accept the solution with a probability of $\epsilon = \frac{1}{1+e^{-(\Delta/T)}}$. The looping continues until the temperature T decayed from T_{max} to a predefined threshold T_{min}. Then we select a new random starting position and repeat the whole process. When an agent should send action to the server, he returns his own action from the optimal joint action found so far.

Basically, what the SA algorithm does is to seek the global maximum of the global joint payoff function. The SA algorithm has some important differences

[2] The simulated annealing algorithm is also called monte carlo annealing or probabilistic hill-climber.

from VE. Firstly, SA is an anytime algorithm that can report an answer at any time, while VE reports until the whole algorithm terminates. Secondly, in each independent try, agent i only has to iterate his own actions instead of all combinatorial actions of his neighbors, thus makes the algorithm tractable. Finally, the SA is essentially a stochastic algorithm that can not guarantee to find the optimal joint action, [3] while VE is an exact and deterministic algorithm that always report the optimal result. As an approximation algorithm, SA is also different from MP in that SA has no limitation on the form of decomposed local functions while the latter has.

SA has a feature of stochastic movement from one solution to another, which helps it jump away from local maxima and improve the answer's quality [21,22,23]. Although SA can not guarantee the convergence to optimal joint action, we shall see that it can find an approximately optimal solution in a rather short time.

4 Experiments

In this section, we evaluate the simulated annealing algorithm by comparing it with other algorithms, especially with variable elimination. The experiments run in two stages. In the first stage, we fix the number of agents and the number of different actions per agent to test the scalability of the two algorithms when the number of neighbors per agent grows. In the second stage, we compare the relative payoff SA returned with the optimal payoff produced by VE.

Since multiagent system is such a large field that there is no standard problem one can test against, it is important to generate the proper test sets. In this paper, we use a random generator (RG) to produce all test sets. The inputs of the random generator are values of the number of agents $|\boldsymbol{G}|$, the number of different actions per agent $|A|$, maximum number of neighbors per agent Nr_{ne}, and the number of value rules each agent has Nr_{ρ}. We believe that these aspects are sufficient to show the difficulty of the coordination problem. The output of the random generator is a set of value rules, each of which is in the form $\langle \rho : v \rangle$. The value rule is introduced in [11] and proved suitable for plenty of real-world applications such as RoboCup. The global joint payoff function is thus represented by the sum of value rules of all agents. Table 1 depicts a sample output of the random generator (RG) with $|G| = 4$, $|A| = 4$, $Nr_{ne} = 3$, $Nr_{\rho} = 1$.

Here, the integer value of a_i is an action index. In a real RoboCup 2D simulation program, such an action index is finally mapped to a real predefined action (or skill, i.e., dribbling, pass, etc.) and sent to the server. We ignore the details of the specific applications and only focus on the performance of the decision algorithm.

In the first experiment, we generate 120 coordination problems and assign them to 4 test sets based on different actions of each agent. For the problem of each test set, the settings are as follows. The number of the agents is fixed

[3] Technically the SA can also find the optimal solution if the annealing process is very very slow [22]. However this will cause the algorithm to run too long time so that it has no practical use.

Algorithm 1. Pseudo-code of the simulated annealing algorithm

```
Define: G = {G_1, ..., G_n} the agents who want to coordinate their actions
Define: v(a) the global joint payoff function defined by section 3
Define: a* the optimal joint action found so far
Define: a_i the action of agent i
Define: a_i* the optimal action of agent i found so far
Define: a_{-i} the actions of all agents but agent i
  g ← 0
  t ← 0
  while t < MaxTries do
    a = random joint action
    T ← T_max
    repeat
      for each agent i in G do
        a' = argmax_{a_i} v(a_{-i} ∪ a_i)
        Δ ← v(a') − v(a)
        if Δ > 0 then
          a ← a'
        else
          a ← a' with probability 1/(1+e^{-(Δ/T)})
        end if
        if v(a) > v(a*) then
          a* ← a
          g ← v(a*)
          choose a_i* from a*
        end if
        if should send action to server then
          send a_i* to server
        end if
      end for
      T ← T · decay
    until T < T_min
    t ← t + 1
  end while
```

to $|\boldsymbol{G}| = 15$, while each agent has $Nr_\rho = 8$ value rules with different number of neighbors. The payoff of each value rule is generated from a uniform random variable $v \sim U[1, 10]$. The number of neighbors k in each value rule is in the range $k \in [1,\ Nr_{ne}]$. Each value has a chance of $\binom{Nr_{ne}}{k}/2^{Nr_{ne}}$. All the programs are implemented in C++, and the results are generated on a 2.2GHz/512MB IBM notebook computer.

When applying variable elimination algorithm, we accelerate the running time by eliminating the agent with the minimum number of neighbors. When running simulated annealing algorithm, we set $MaxTries$ to 10, the highest temperature $T_{max} = 0.3$, and lowest temperature $T_{min} = 0.05$. The temperature $decay$ of the algorithm is in proportion to Nr_{ne}, i.e., $decay \propto Nr_{ne}$. So if the coordination problem contains value rules involving large amounts of agents, we will do a

Table 1. Sample output of RG

$\langle \rho : v \rangle$
$\langle\ a_1 = 3 \wedge a_3 = 3 \wedge a_4 = 4 : 7.19085\rangle$
$\langle\ a_2 = 4 \wedge a_3 = 4 : 4.67774\rangle$
$\langle\ a_1 = 1 \wedge a_2 = 1 \wedge a_3 = 2 \wedge a_4 = 2 : 4.67774\rangle$
$\langle\ a_1 = 4 \wedge a_3 = 2 \wedge a_4 = 1 : 4.67774\rangle$

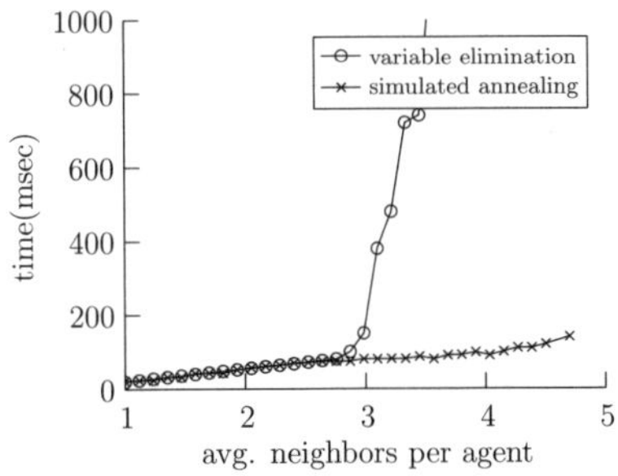

(a) Timing comparisons for VE and SA (4 actions per agent).

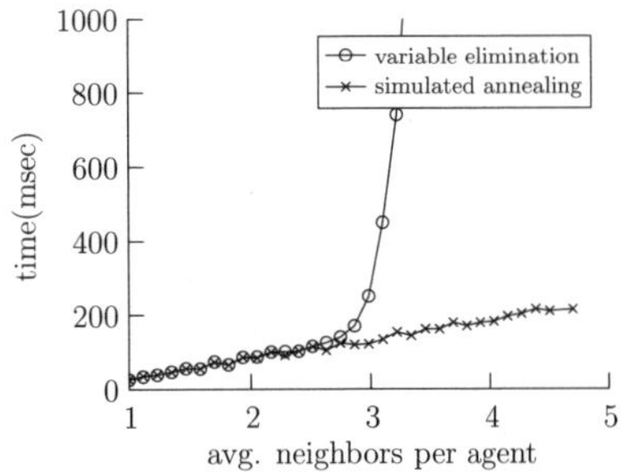

(b) Timing comparisons for VE and SA (6 actions per agent).

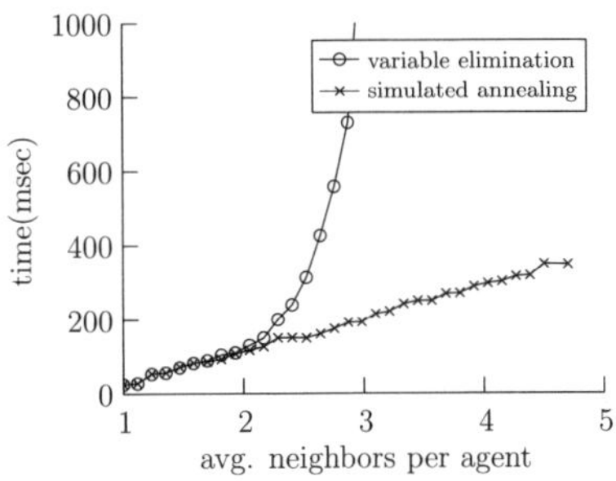

(c) Timing comparisons for VE and SA (8 actions per agent).

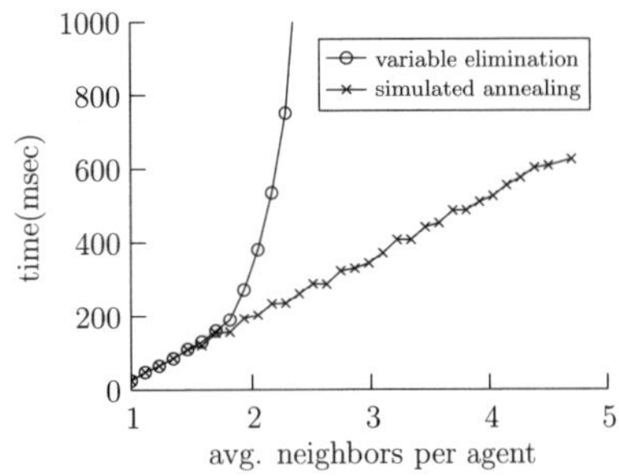

(d) Timing comparisons for VE and SA (10 actions per agent).

Fig. 2. Average timing comparisons for both VE and SA for testing the scalability when the number of neighbors per agent grows with 15 agents

deep search in an independent try, vice versa. The experiment repeats 10 times to weaken the effect of hardware and operation system.

In the second experiment, we produce 6 coordination problems, each of which has its own settings such as number of agents, different actions per agent, etc. VE and SA are both evaluated. When applying SA, instead of starting from a random choice for all agents, in ith independent try, we let the agent select action according to the ith highest value rule if he is involved, otherwise select action randomly. We also set $MaxTries = 200$ to ensure sufficient time to run. Other settings are the same as in the first experiment.

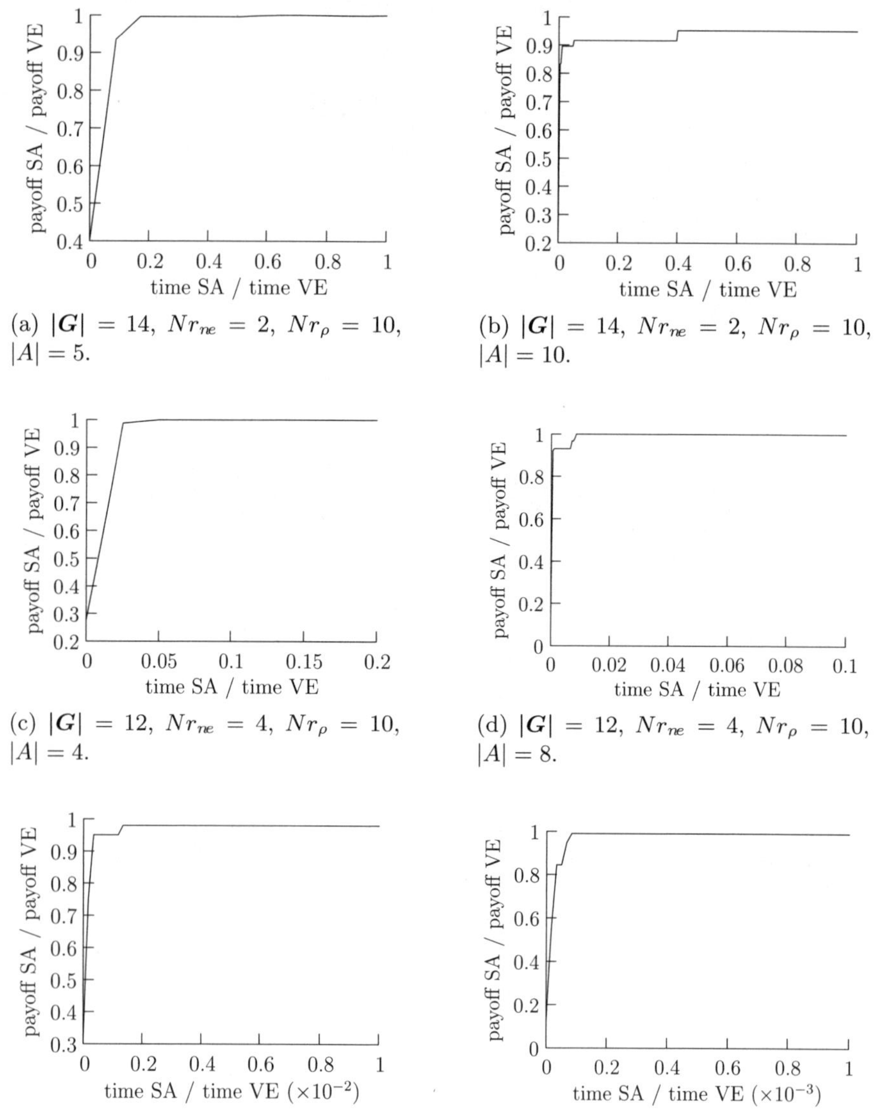

(a) $|\boldsymbol{G}| = 14$, $Nr_{ne} = 2$, $Nr_\rho = 10$, $|A| = 5$.

(b) $|\boldsymbol{G}| = 14$, $Nr_{ne} = 2$, $Nr_\rho = 10$, $|A| = 10$.

(c) $|\boldsymbol{G}| = 12$, $Nr_{ne} = 4$, $Nr_\rho = 10$, $|A| = 4$.

(d) $|\boldsymbol{G}| = 12$, $Nr_{ne} = 4$, $Nr_\rho = 10$, $|A| = 8$.

(e) $|\boldsymbol{G}| = 10$, $Nr_{ne} = 8$, $Nr_\rho = 10$, $|A| = 4$.

(f) $|\boldsymbol{G}| = 10$, $Nr_{ne} = 8$, $Nr_\rho = 10$, $|A| = 8$.

Fig. 3. Relative payoff found by SA with respect to VE

In order to give a clear image of VE and SA, we scale the payoff axis so that the global maximum payoff is 1. The time axis is also scaled so that the time it takes the whole VE to terminate is 1. Thus the points in the figure can be seen as the fraction of the payoff and the running time of VE. The results of SA will be scaled to its VE companion. Again, the experiment is also repeated 10 times to reduce hardware and software's side effects.

Fig. 2(a)–2(d) give the timing results for the four test sets in the first experiment. It can be seen that the running time of the SA algorithm grows linearly as the number of the neighbors per agent increases. The running time of VE grows exponentially, since it must enumerate all neighbor's possible combination actions in each iteration. Furthermore, when the average number of neighbors per agent was more than 3.5, VE can not always compute the optimal joint action, so these tests were removed from the test sets.

The relative payoff found by the SA with respect to VE are plotted in Fig. 3(a)–3(f). In all the plots, we see that the SA algorithm performed very well. It is obvious that we found approximately optimal results in all problems. In loosely connected coordination problem with few actions, i.e., Fig. 3(a), SA algorithm can converge to the maximum payoff while only using the 60% time of variable elimination [4]. However, if the number of actions is large (Fig. 3(b)), SA can not reach the optimal result, although it can find approximately optimal solution (96% payoff) quickly. Further experiments show that if the joint action space is huge (more than 15 agents, and each agent has more than 10 actions), we should increase the acceptable probability ϵ accordingly to speed up the convergence to optimal result. This is because in such situations, a little higher acceptable probability can increase the chance of stochastic solution movement for simulated annealing algorithm. This technique helps SA jump away from local optimizations and cover the joint action space as possible as it can. But the exact relationship between acceptable probability and the convergency speed are still not very clear. For the medium connected problems (Fig. 3(c)–3(d)), SA can compute the optimal policy with a little fraction of time (2%–6%) that variable elimination needs to solve the same problem. Fig. 3(e) and Fig. 3(f) give us a strong impression that SA can compute above 98% payoff within the time ranges between 0.015% to 0.2% of the time variable elimination takes in the densely connected problems. We also show that in these 2 experiments, although the SA can find near optimal solution very quickly, it still needs to take plenty of time to approximate the optimal result.

In our internal unpublished tests, we also compare SA with max-plus algorithm informally. The experiment shows that when reaching the same relative payoff, the time difference between the two algorithms is at most 5%. Although our algorithm is not faster than max-plus, we believe that our approach is more appropriate for complex coordination problems in which the coordination dependencies in the value rule is often more than two. Thus, max-plus can not be applied directly.

5 Conclusion

In this paper, we have described and investigated the use of the simulated annealing algorithm for cooperative action selection as an approximate alternative to variable elimination algorithm. As above-mentioned, Variable elimination is an exact approach that always reports the optimal joint action. It is also an

[4] Note that SA does not know even the maximum payoff has been found due to its stochastic property.

efficient algorithm in loosely connected coordination graphs. However, it is very slow in densely connected coordination graphs and unable to produce results at anytime. The simulated annealing algorithm repeats independent tries. In each try, each agent tries to maximize the global payoff using his own choice without influencing the actions of all other agents. Based on the result quality in each maximization, the algorithm accepts a solution with a certain probability. We have provided empirical evidences to show: 1) this method is almost optimal with a small fraction of the time that VE takes to compute the policy of the same coordination problem; 2) the running time of SA grows linearly with the increasing of the number of neighbors per agent; 3) it is an anytime algorithm to return result at any time. For above reasons, we believe that simulated annealing is an feasible approach for action selection in large complex cooperative autonomous systems such as RoboCup.

As for future research, we plan to implement the simulated annealing algorithm in our SEU_T 2D simulation team. Last year, we tried to use VE for our player's cooperative action selection framework, but the computational constraints made us only use a small set of value rules with each rule involving at most 3 agents [24]. Applying simulated annealing algorithm, we should produce more advanced coordination actions to involve much more agents.

Finally, we will figure out a appropriate setting of the acceptable probability, especially the decay rate in simulated annealing algorithm. Some recent work shows that neural network algorithm can produce a good decay rate for larger problems [23]. We would like to try to employ such techniques in our multi-agent decision making problem. Furthermore, we want to investigate whether reinforcement learning algorithms can be applied to automatic learning of the payoff instead of hand tuning.

References

1. Weiss, G. (ed.): Multiagent Systems: a Modern Approach to Distributed Artificial Intelligence. MIT Press, Cambridge, MA, USA (1999)
2. Woolridge, M., Wooldridge, M.J.: Introduction to Multiagent Systems. John Wiley & Sons, Inc., New York, NY, USA (2001)
3. Kitano, H., Asada, M., Kuniyoshi, Y., Noda, I., Osawa, E.: Robocup: The robot world cup initiative. In: AGENTS '97. Proceedings of the first international conference on Autonomous agents, Marina del Rey, California, United States, pp. 340–347. ACM Press, New York, NY, USA (1997)
4. Osborne, M.J., Rubinstein, A.: A Course in Game Theory. MIT Press, Cambridge (1999)
5. Carriero, N., Gelernter, D.: Linda in context. Communications of the ACM 32(4), 444–458 (1989)
6. Gelernter, D.: Generative communication in Linda. ACM Transactions on Programming Languages and Systems 7(1), 80–112 (1985)
7. Boutilier, C.: Planning, learning and coordination in multiagent decision processes. In: TARK '96. Proceedings of the 6th conference on Theoretical aspects of rationality and knowledge, The Netherlands, pp. 195–210. Morgan Kaufmann Publishers Inc., San Francisco, CA, USA (1996)

8. Tan, M.: Multi-agent reinforcement learning: Independent vs. cooperative learning. In: Huhns, M.N., Singh, M.P. (eds.) Readings in Agents, pp. 487–494. Morgan Kaufmann, San Francisco, CA, USA (1997)
9. Claus, C., Boutilier, C.: The dynamics of reinforcement learning in cooperative multiagent systems. In: AAAI/IAAI, pp. 746–752 (1998)
10. Guestrin, C., Koller, D., Parr, R.: Multiagent planning with factored MDPs. In: 14th Neural Information Processing Systems (NIPS-14) (2001)
11. Guestrin, C., Venkataraman, S., Koller, D.: Context specific multiagent coordination and planning with factored MDPs. In: The Eighteenth National Conference on Artificial Intelligence (AAAI-2002), Edmonton, Canada, July 2002, pp. 253–259 (2002)
12. Guestrin, C., Koller, D., Parr, R., Venkataraman, S.: Efficient solution algorithms for factored MDPs. Accepted in Journal of Artificial Intelligence Research (JAIR) (2002)
13. Guestrin, C.: Planning Under Uncertainty in Complex Structured Environments. PhD thesis, Stanford University (2003)
14. Kok, J.R., Vlassis, N.: Using the max-plus algorithm for multiagent decision making in coordination graphs. In: Bredenfeld, A., Jacoff, A., Noda, I., Takahashi, Y. (eds.) RoboCup 2005. LNCS (LNAI), vol. 4020, Springer, Heidelberg (2006)
15. Pearl, J.: Probabilistic Reasoning in Intelligent Systems: Networks of Plausible Inference. Morgan Kaufmann Publishers Inc., San Francisco, CA, USA (1988)
16. Wainwright, M., Jaakkola, T., Willsky, A.: Tree consistency and bounds on the performance of the max-product algorithm and its generalizations. Statistics and Computing 14, 143–166 (2004)
17. Dechter, R.: Bucket elimination: a unifying framework for reasoning. Artificial Intelligence 113(1-2), 41–85 (1999)
18. Arnborg, S., Corneil, D.G., Proskurowski, A.: Complexity of finding embeddings in a K-tree. SIAM J. Algebraic Discrete Methods 8(2), 277–284 (1987)
19. Bertelé, U., Brioschir, F.: Nonserial dynamic programming. Academic Press, London (1972)
20. Michalewicz, Z., Fogel, D.B.: How to solve it: modern heuristics. Springer, New York, NY, USA (2000)
21. Johnson, D.S., McGeoch, L.A.: The Traveling Salesman Problem: A Case Study in Local Optimization (Draft of November 20, 1995) In: Aarts, E.H.L., Lenstra, J.K. (eds.) To appear as a chapter in The book Local Search in Combinatorial Optimization, John Wiley & Sons, Inc., New York (1995)
22. Kirkpatrick, S., Gelatt, C.D., Vecchi, M.P.: Optimization by simulated annealing. Science 220(4598), 671–680 (1983)
23. Spears, W.M.: Simulated annealing for hard satisfiability problems. DIMACS Series in Discrete Mathematics and Theoretical Science 26, 533–558 (1996)
24. Dawei, J.: SEU_T 2005 team description (2D). In: Proceedings CD RoboCup 2005, Osaka, Japan, July 2005, Springer, Heidelberg (2005)

From RoboLab to Aibo: A Behavior-Based Interface for Educational Robotics

Rachel Goldman[1], M.Q. Azhar[2], and Elizabeth Sklar[3]

[1] Google, Inc.
1440 Broadway, New York, NY 10018 USA
rjg@google.com
[2] Dept of Computer Science
Graduate Center, City University of New York
365 5th Avenue, New York, NY 10016 USA
mazhar@gc.cuny.edu
[3] Dept of Computer and Information Science
Brooklyn College, City University of New York
2900 Bedford Avenue, Brooklyn, NY 11210 USA
sklar@sci.brooklyn.cuny.edu

Abstract. This paper describes a framework designed to broaden the entry-level for the use of sophisticated robots as educational platforms. The goal is to create a low-entry, high-ceiling programming environment that, through a graphical behavior-based interface, allows inexperienced users to author control programs for the Sony Aibo four-legged robot. To accomplish this end, we have extended the popular *RoboLab* application, which is a simple, icon-based programming environment originally designed to interface with the LEGO Mindstorms robot. Our extension is in the form of a set of "behavior icons" that users select within RoboLab, which are then converted to low-level commands that can be executed directly on the Aibo. Here, we present the underlying technical aspects of our system and demonstrate its feasibility for use in a classroom.

1 Introduction

Many teachers have an interest in introducing robots into their classrooms for teaching a variety of subjects other than specifically robotics, from traditional technical topics such as programming and mechanical engineering to other areas such as mathematics and physical science, where robots are used to demonstrate the concepts being taught. It has long been recognized that hands-on methods have a powerful impact on student learning and retention [1,2,3]. The growing field of *educational robotics*—the use of robots as a vehicle for teaching subjects other than specifically robotics [4]—has employed that approach, using the *constructionist* [5] process of designing, building, programming and debugging robots, as well as collaboration and teamwork, as powerful means of enlivening education [2,6,7]. Even very young children have been successfully engaged in hands-on learning experiences that expose them to some of the basic principles

G. Lakemeyer et al. (Eds.): RoboCup 2006, LNAI 4434, pp. 122–133, 2007.

of science and engineering, widening their horizons and preparing them for life in a highly automated, technically challenging world.

For the past several years, we have been designing and helping to implement educational robotics curriculum in inner-city primary and middle school classrooms, after-school programs and summer schools, undergraduate introductory programming courses and international robotic competitions [8,9,7,10]. To support these curricula, we have employed RoboLab [11], a widely used graphical programming environment developed at Tufts University, for operation on the LEGO Mindstorms Invention System robot [12]. RoboLab runs on a Mac, Windows or Unix computer. The environment is highly visual and provides a good first experience with procedural programming concepts. Entities such as motors and sensors are represented as rectangular icons on the screen, and users drag and drop them with the mouse onto a canvas to create "code". The icons are strung together using "wires", and all programs are downloaded from RoboLab onto the LEGO robot via a "communication tower", connected to the computer's USB or serial port, that transmits the program to the robot using an infra-red signal. A simple RoboLab program is illustrated in Figure 1.

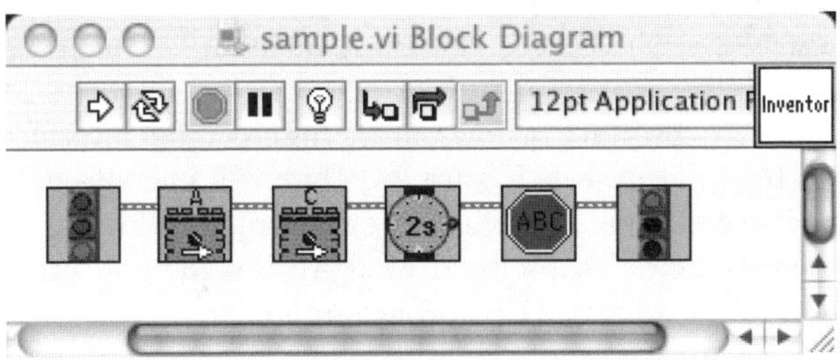

Fig. 1. A basic RoboLab program

If the robot has wheels and motors attached to two of its ports (labeled A and C), this program will make the robot go forward for 2 seconds and then stop.

RoboLab's graphical programming environment tiers the levels of programming, which allow a novice to produce results quickly and acquire skills without having to read a sophisticated text or complicated application manual. Ranging from "Pilot" to "Inventor" levels, RoboLab is broad enough to ensure the success of both beginners and advanced users. The initial Pilot levels are completely graphical so even users who cannot read can be successful. The more advanced "Inventor" levels incorporate advanced programming concepts and features to add power, flexibility and complexity to programs.

Our experiences working with classroom teachers and young students have raised several issues that have motivated us to pursue the development of a behavior-based interface, which abstracts away the low-level motor and sensor commands that often confuse inexperienced programmers or deter techno-phobic students [13]. Our longterm goal is to create a standard middle ground that can act as a sort of "magic black box", for current and future robotic platforms, following several design criteria:

- *ease of use:* programmers only have to deal with high-level icons—novices will not get discouraged with low-level text-based syntax;
- *disappearing boundaries:* programmers are able to test and run the same behaviors on multiple agent platforms—running the same RoboLab program on a LEGO robot and on an Aibo will help students understand about abstraction and behavior-based control;
- *interoperability:* a standard behavior language is used for multiple platforms—students do not need to learn different languages in order to use a variety of robot platforms, or our simulator (currently under development [14,15]); and
- *flexibility:* students from a wide range of backgrounds and teachers with a broad range of goals can use the system effectively, accommodating different levels, curricular needs, academic subjects and physical environments for instruction.

Because of RoboLab's popularity in the classroom and its icon-based programming style, it is well suited as the front-end for our interface. We have three underlying educational goals, each supporting different pedagogical needs:

- First, we want to produce a low-entry, high-ceiling interface to the Sony Aibo robot [16] (pictured in Figure 2a) that will encourage non-traditional computer science students to learn programming, allowing us to capitalize on the "cuteness factor" associated with Aibo while still providing students with a serious, first adventure in programming.
- Second, we want advanced students to gain an appreciation of the modularity of both a robot's controlling interface and its underlying hardware. Just as Java and JavaScript are platform independent, providing a robot programming environment that is platform independent will give students hands-on experience with *code abstraction*, witnessing the same code executing on both the LEGO robot and the Aibo (and other platforms).
- Third, we want to create a motivating, hands-on environment with which to introduce more advanced students to behavior-based concepts.

This paper is organized as follows. We begin by reviewing several different programming interfaces designed for the Sony Aibo. Then we describe some technical details about RoboLab and explain how our behavior-based programming interface operates. Finally, we end with a brief summary and mention directions for future work.

2 Background

OPEN-R is the programming interface designed for the Aibo, provided for free via download from Sony's web site [17]. OPEN-R gives programmers the ability

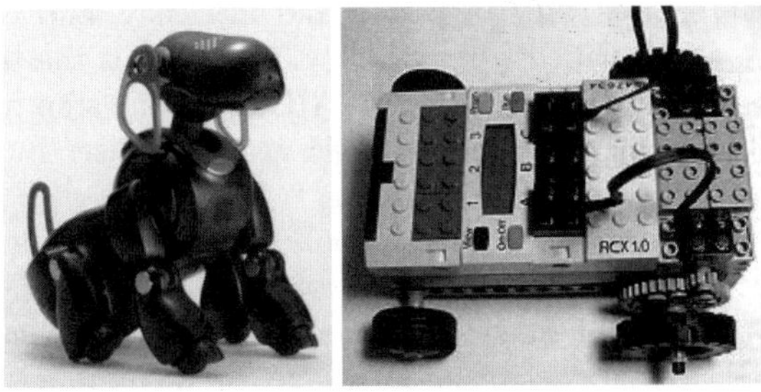

(a) Aibo ERS-7M2/B (b) LEGO Mindstorms

Fig. 2. Robot platforms

to develop software to control the low-level hardware of the robot. Some of the features of OPEN-R include: modularized hardware, modularized software and networking support. Because the hardware is modularized, each module is connected by a high-speed serial bus and can be exchanged for a new module. Each software module in OPEN-R is either a "data object" or a "distributed object" and is implemented in C++[1]. OPEN-R programs are built as a collection of concurrently running OPEN-R objects, which have the ability to communicate with each other via message passing. Due to the modularity of the software, individual objects can be easily replaced and each object is separately loaded from a Sony Memory Stick. Furthermore, OPEN-R supports Wireless LAN and TCP/IP network protocol.

Sony's formalism for describing the working cycle of OPEN-R objects resembles layered *finite state automata* [18]. Each OPEN-R object can have numerous states and must include an IDLE state. At any given point, an object can be in only one state, and objects move from state to state using *transitions*. In order to switch states, an *event* must activate a transition to a new state and the pre-condition of the new state must be satisfied.

The low-level functionality of the robot can be controlled using the OPEN-R API [17]. With the API, programmers can experiment with image processing, sensory feedback information and robot motion in order to develop (original) sets of behaviors for the Aibo. However, the OPEN-R API can be quite hard to use for novice and even intermediate programmers. As a result, several interfaces and abstraction languages have been developed to sit on top of OPEN-R in an attempt to hide the low-level complexity from the inexperienced end-user. These include:

- *R-CODE* [19],
- *YART* (Yet Another R-CODE Tool) [20],
- *Tekkotsu* [21], and
- *URBI* (Universal Robotic Body Interface) [22,23].

[1] C++ objects and OPEN-R objects are not the same and should not be confused with each other [18].

These languages/interfaces vary in power and intricacy, and each has its own goals and features, as described briefly below. OPEN-R is the basis upon which R-CODE, Tekkotsu, and URBI are built. YART goes one step further, abstracting R-CODE one level by providing a basic graphical user interface (GUI) to create and manipulate R-CODE programs.

The R-CODE SDK provides a set of tools that allow users to program the Aibo using the *R-CODE scripting language*, offering higher-level commands than traditional programming languages such as C, C++ and even OPEN-R. The benefits of R-CODE being a scripting language are its simplicity (to both learn and use) and its lack of compilation; however, programmers have less control than with other lower-level languages. With only a few lines of R-CODE, users can program the Aibo to perform complex behaviors such as dancing, kicking or walking. Because R-CODE does not require compilation, it can be written in a plain text file on any operating system and saved directly on a memory stick that has the R-CODE virtual machine pre-loaded on it. R-CODE commands can be viewed as OPEN-R macros, where the degree of precision and control depends solely on the underlying OPEN-R code. Although R-CODE is powered by OPEN-R functions, the developer does not have access to the underlying OPEN-R subroutines. R-CODE is best suited for performing actions and various behavior sequences.

YART [20] is an R-CODE front-end developed by a hobbyist. It provides a text-based GUI with simple drag-and-drop functionality and produces files of R-CODE commands. This tool can also be used to generate customized behaviors via pre-existing YART-compatible Aibo personalities. YART is an easy place to start programming simple behavior patterns. However, unlike RoboLab, YART is a text-based interface, so the objects that the user drags with the mouse are bits of text—whereas in RoboLab, the user drags graphical icons.

Tekkotsu [21], developed at Carnegie Mellon University by Touretzky et al., is an application development framework for intelligent robots that is compatible with the OPEN-R framework [17]. It is based on an object-oriented and event-passing architecture that utilizes some of the key features of C++, like templates and inheritance. Although Tekkotsu was originally created for the Sony Aibo, the current version can be compiled for multiple operating systems. Tekkotsu simplifies robotic application development by supplying basic visual processing, forward and inverse kinematics solvers, remote monitoring, teleoperation tools, and wireless networking support. Tekkotsu handles low-level tasks so the developer can focus on higher-level programming. It provides primitives for sensory processing, smooth control effectors, and event-based communication. Some of the higher-level features include a hierarchical state machine formalism used for control flow management and an automatically maintained world map. Additionally, Tekkotsu includes various housekeeping and utility functions and tools to monitor various aspects of the robot's state.

The Universal Robotics Body Interface (URBI) [22,23], developed at École Nationale Supérieure de Techniques Avancées (ENSTA), is an attempt to provide a standard way to control the low-level aspects of robots while providing the high-level capabilities of traditional programming languages. URBI is based on a client-server architecture where the server is running on the robot and is typically accessed by the client via TCP/IP. The client can be virtually any system or any other kind of computer, thereby adding flexibility to URBI. The URBI language is a high-level scripting language capable of controlling the joints and accessing the sensors, camera, speakers or other hardware on the robot. URBI has been primarily applied to entertainment robots because they tend to provide the most interesting interfaces and capabilities.

Each of these interfaces have informed our approach. We selected RoboLab as the front-end for several reasons. RoboLab is already quite widely used in classrooms worldwide. Teachers and students are comfortable with the interface and will not feel like they need to learn yet another programming environment in order to expand their use of robot platforms in the classroom. In addition, schools do not need to purchase another software package; our methodology is a free extension to RoboLab. Finally, as detailed below, RoboLab is constructed on top of a framework designed to support a wide range of hardware devices, thus the concept of expanding its use to interface with a range of robot devices is a natural fit.

3 Our Approach

This section describes the approach to our implementation, outlining the steps required for progressing from writing programs in RoboLab to generating code that is executed on the Aibo. Given that one of our objectives with this work is to develop a behavior-based programming interface for controlling Aibo, we have designed a set of generic low-level behaviors that can be graphically represented in RoboLab. As part of this process, we analyzed the main differences between the Aibo platform and the LEGO Mindstorms platform to ensure that our behaviors are suitable for both. The two platforms are physically quite different, not only in terms of processor configuration but also in regard to the types of sensors and effectors provided. The LEGO Mindstorms is typically constructed as a wheeled robot, as depicted in Figure 2b (though legged structures can be built). The kit comes with two touch sensors and a light sensor and has the ability to support numerous other LEGO and commercial sensors. The Aibo, a four-legged robot, comes with a variety of built-in sensors including: multiple touch sensors, distance sensors and a camera; and it cannot support any other sensors (without being dismantled).

With the capabilities of both platforms in mind, we defined a set of prototype behaviors and control structures suitable for our behavior-based "palette" in RoboLab (illustrated in Figure 3). RoboLab is implemented on top of National Instruments' LabVIEW [24]. Individual icons and full programs are saved as "VIs" (virtual instruments). Each icon or program can be seen as imitating an actual instrument [25]. We used RoboLab's built-in feature to create "subVIs" (VI modules or subroutines) in order to construct our customized behaviors.

These behaviors act as macros for sets of lower-level RoboLab commands. Although the ability to expand RoboLab's current set of icons is a remarkably powerful feature, we are constrained by the necessity to use a set of pre-defined icons as the underlying basis for each new icon. For example, as illustrated in Figure 4, our forward behavior icon is in reality a macro for the set of icons: motor A forward, motor C forward, and wait for (some amount of time).

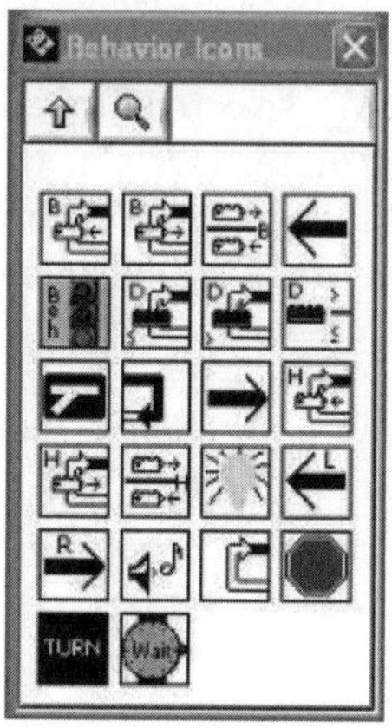

The behaviors icons, shown in the palette to the left, can be used just like any other RoboLab icon. As with basic RoboLab function icons, wiring together a sequence of behavior icons creates a well-formed RoboLab program. The meanings of the first eight icons are (from top left to right): loop while back sensors are pressed, loop while back sensors are not pressed, branch according to state of back sensor, move backwards, begin behavior, loop while distance sensor is less than or equal to parameter, loop while distance sensor is greater than parameter, branch according to state of distance sensor. See [13] for a complete and detailed description of the behavior icons.

Fig. 3. Behavior Icon Palette

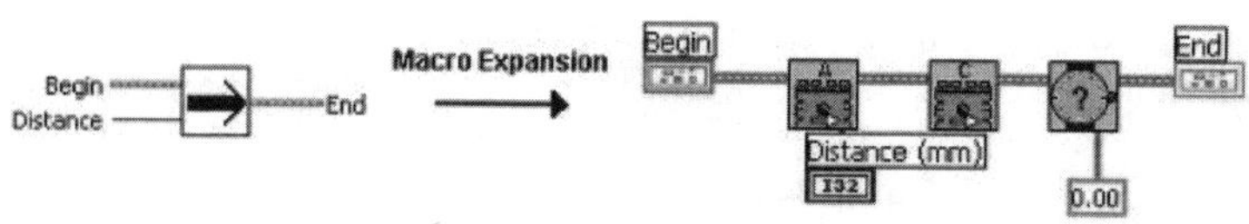

Fig. 4. Forward behavior icon and its underlying set of basic RoboLab icons

The first step in our approach is for a user to write a program in RoboLab and save the contents of the program to a file, in a manner that will preserve its functionality while allowing translation to Aibo commands to occur outside of RoboLab. The default output from RoboLab is *LASM*, or LEGO Assembly Language. We use this default, saving the LASM commands in an output file — whereas normally, users send the LASM commands directly to the LEGO robot via the communication tower. Once the LASM file is saved, we invoke our translation program, called `lasm2aibo`, that converts the LASM commands into Aibo commands. After examining the Aibo languages and interfaces discussed in section 2, we decided to use R-CODE to implement Aibo commands in our system because R-CODE is easy to use and does not require compilation, and because R-CODE offers a more natural mapping to RoboLab. With R-CODE,numerous behaviors can be prototyped quickly and tested efficiently.

Both Tekkotsu and URBI are better suited for applications that require complex solutions and greater computational power.

A key challenge in designing `lasm2aibo` was to determine which LASM command(s) and what parameters are generated for each RoboLab icon. Our behavior icons are macros comprised of multiple low-level built-in RoboLab icons, which complicates the translation process, as detailed below. Taking a file of LASM commands as input, our translater recognizes tokens that match relevant LASM commands, numbers, white space, and delineators (commas and new line characters), and "compiles" (or translates) these into R-CODE sequences. We designed and implemented our translator using the UNIX tools Lex [26] and Yacc [27].

Both Lex and Yacc greatly simplify compiler writing, or translating between two programming language representations. Lex generates the C code for building a lexical analyzer or lexer. Any given lexer takes an arbitrary input stream and divides it into a sequence of tokens based on a set of regular expression patterns. The Lex specification refers to the set of regular expressions that Lex matches against the input. A deterministic finite state automaton generated by Lex performs the recognition of the expressions. Lex allows for ambiguous specifications and will always choose the longest match at each input point. Each time one of the patterns is matched, the lexer invokes user-specified C code to perform some action with the matched token. Yacc is responsible for generating C code for a syntax analyzer or a parser. Yacc uses the grammar rules to recognize syntactically valid inputs and to create a syntax tree from the corresponding lexer tokens. A syntax tree imposes a hierarchical structure on tokens by taking into account elements such as operator precedence and associativity. When one of the rules has been recognized, then the user-provided code for this rule (an "action") is invoked. Yacc generates a bottom-up parser based on shift-reduce parsing. When there is a conflict, Yacc has a set of default actions. For a shift-reduce conflict, Yacc will shift. For reduce-reduce conflicts, Yacc will use the earlier rule in the specification.

In our case, Yacc reads the grammar description and the token declarations from `lasm2aibo.y` and generates a parser function `yyparse()` in the file `y.tab.c`. Running Yacc with the `-d` option causes Yacc to generate definitions for the tokens in the file `y.tab.h`. Lex generates a lexical analyzer function `yylex()` in the file `lex.yy.c` by reading the pattern descriptions from `lasm2aibo.l` and including the header file `y.tab.h`. The lexer and parser are compiled and linked together to form the executable `lasm2aibo`. From the `main()` function in `lasm2aibo`, the `yyparse()` function is called, which in turn calls the `yylex()` function to obtain each token.

Our Yacc parser is generated from the `lasm2aibo` grammar specification. Through a series of grammar productions, the parser attempts to group sequences of tokens into syntactically correct statements. Associated with each production is a set of semantic actions. Given that many of the LASM command sequences for varying groups of behaviors are the same, often behavior identification becomes part of the semantic analysis. For instance, all the motion behaviors including forward, backward, right and left produce the same LASM command output. Furthermore, there is no direct way to distinguish between activating an LED or a motor because they are both viewed as output devices. Only by examining the values of some of the parameters and by making certain assumptions can they be differentiated. Having to perform specific behavior recognition during the semantic analysis is one of the limitations involved in using LASM as our source program. Ideally, each behavior should be represented by a unique syntax, allowing for quicker and cleaner parsing.

When a behavior is recognized, the semantic actions involve calling the corresponding behavior function. These pre-defined behavior functions are used to generate the equivalent behavior in R-CODE and help flag the R-CODE subroutines that need to be included in the R-CODE source file. The final output of the translator includes the file `R-CODE.R`, the R-CODE program that matches the original RoboLab program, and a `behaviors.txt` file used for debugging purposes to ensure that behaviors, sensors, and control structures are appropriately classified. Running the `lasm2aibo` executable file with a LASM text file as input generates the file `R-CODE.R`. This file should be placed on a R-CODE-ready memory stick in the `OPEN-R/APP/PC/AMS` directory [28]. To execute the program, insert the memory into Aibo, turn on the power and watch Aibo come to life!

Throughout the process, there are multiple places where errors could occur; this includes lexical, syntactic and semantic errors. The `lasm2aibo` translator attempts to handle errors gracefully. If the error is fatal, a blank `R-CODE.R` file will be generated; however, if the error is non-fatal then the translator will continue to process the source file. In both cases, a descriptive error message and the line number where the error occurred is displayed on the console via standard output.

3.1 Example

A detailed example follows. First, the RoboLab behavior-based program (shown in Figure 5a) is constructed. The resulting LASM is contained in Figure 5b. Second, our `lasm2aibo` module is executed, taking the LASM file as input and creating an R-CODE equivalent, illustrated in Figure 5c. This gets written to a Sony memory stick, which also contains the R-CODE virtual environment, and is ready to be executed on the Aibo.

4 Summary

We have presented the design and implementation of our prototype framework for providing a simple, graphical, behavior-based interface to the Sony Aibo robot. Built into the popular RoboLab graphical programming environment and connected to the Aibo via a translater, `lasm2aibo`, this framework represents the proof-of-concept for a longterm project aimed at bringing educational robotics into a broad range of classrooms. The successful implementation of the direct

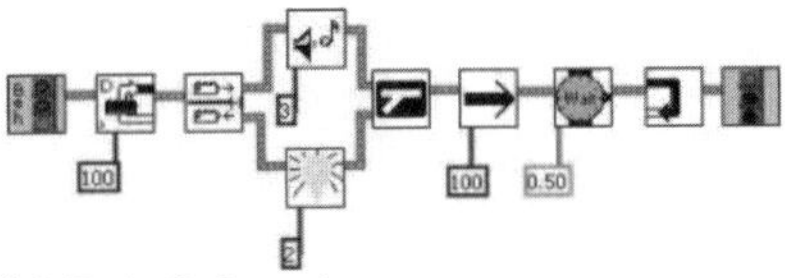

(a) RoboLab code

```
delt 0

task 0
sent 0,3
senm 0,4,0
sent 1,1
senm 1,1,0
Label1002:
chkl 2,100,1,9,0,Label1003
chkl 2,1,0,9,1,Label1004
plays 2
wait 2,100
wait 2,0
jmpl Label1005
Label1004:
pwr 2,2,2
dir 0,2
out 2,2
wait 2,0
Label1005:
pwr 1,2,100
dir 2,1
out 2,1
pwr 4,2,7
dir 2,4
out 2,4
wait 2,0
wait 2,50
ping
jmpl Label1002
Label1003:
endt

plays 5
```

(b) LASM code

```
:Start
PLAY:ACTION:STAND
WAIT
WHILE:100:<:Distance
IF:1:>:Head_ON:THEN
PUSH:2471
CALL:Play_Sound:1
PUSH:1000
CALL:Wait_For_Time:1
WAIT
ELSE
PUSH:2260
CALL:Display_LED:1
WAIT
ENDIF
CLR:SENSORS
PUSH:100
CALL:Forward:1
WAIT
PUSH:500
CALL:Wait_For_Time:1
WEND
CLR:SENSORS
EXIT

****************************************************
*                  BEHAVIOR FUNCTIONS              *
****************************************************

/**********************************************
 * walk forward for a specified distance in mm
 **********************************************/
 :Forward //pass a distance
  ARG:distance
  PLAY:ACTION:WALK:0:distance
  RETURN

/**********************************************
 * wait for a specified amount of time
 **********************************************/
 :Wait_For_Time
  ARG:time
  WAIT:time
  RETURN

/**********************************************
 * plays a specified sound
 **********************************************/
 :Play_Sound
  ARG:sound
  PLAY:MWCID:sound
  RETURN

/**********************************************
 * plays a specified led pattern
 **********************************************/
 :Display_LED
  ARG:pattern
  PLAY:MWCID:pattern
  RETURN
```

(c) R-CODE

Fig. 5. Example

This program will test whether the head sensor is pressed or not. If the head sensor is released, it will play sound number 3, otherwise if the head sensor is pressed it will display LED pattern number 2. The Aibo will then move forward 100mm and wait half asecond. This process will repeat while the distance sensor reads a value greater than100mm.

translation from RoboLab to Aibo demonstrates the feasibility and viability of the process. A user study is planned for Summer 2006.

Although our direct translation is appropriate for small-scale solutions, current work on this project is exploring a more abstract, generalized framework for linking RoboLab (or other graphical programming interfaces) to a variety of robot platforms [15]. Recent press releases have revealed that the LEGO Mindstorms will be succeeded in August 2006 by a more sophisticated platform called NXT [29], and Sony has announced that production of Aibo halted in March 2006 [30]. Given the changing face of consumer robotics, a flexible framework such as ours will help classrooms ease transitions from one robot platform to another by providing teachers and students with a familiar interface and an intuitive behavior-based methodology for programming, no matter what hardware lies beneath.

References

1. Piaget, J.: To Understand Is To Invent. The Viking Press, Inc., New York (1972)
2. Papert, S.: Mindstorms: Children, Computers, and Powerful Ideas. BasicBooks (1980)
3. Resnick, M.: Technologies for lifelong kindergarten. Educational Technology Research and Development 46(4) (1998)
4. Sklar, E., Parsons, S.: RoboCupJunior: a vehicle for enhancing technical literacy. In: Proceedings of the AAAI-02 Mobile Robot Workshop (2002)
5. Papert, S.: Situating constructionism. Constructionism (1991)
6. Slavin, R.: When and why does cooperative learning increase achievement? theoretical and empirical perspectives. In: Hertz-Lazarowitz, R., Miller, N. (eds.) Interaction in cooperative groups: The theoretical anatomy of group learning, pp. 145–173. Cambridge University Press, Cambridge (1992)
7. Sklar, E., Eguchi, A., Johnson, J.: RoboCupJunior: Learning with Educational Robotics. In: Kaminka, G.A., Lima, P.U., Rojas, R. (eds.) RoboCup 2002. LNCS (LNAI), vol. 2752, pp. 238–253. Springer, Heidelberg (2002)
8. Goldman, R., Eguchi, A., Sklar, E.: Using educational robotics to engage inner-city students with technology. In: Kafai, Y., Sandoval, W., Enyedy, N., Nixon, A.S., Herrera, F. (eds.) Proceedings of the Sixth International Conference of the Learning Sciences (ICLS), pp. 214–221 (2004)
9. Sklar, E., Eguchi, A.: RoboCupJunior – Four Year Later. In: Nardi, D., Riedmiller, M., Sammut, C., Santos-Victor, J. (eds.) RoboCup 2004. LNCS (LNAI), vol. 3276, Springer, Heidelberg (2004)
10. Sklar, E., Parsons, S., Stone, P.: RoboCup in Higher Education: A preliminary report. In: Polani, D., Browning, B., Bonarini, A., Yoshida, K. (eds.) RoboCup 2003. LNCS (LNAI), vol. 3020, Springer, Heidelberg (2004)
11. Tufts University: RoboLab (accessed January 16, 2006), `http://www.ceeo.tufts.edu/robolabatceeo/`
12. LEGO: Mindstorms robotics invention kit (accessed February 1, 2006), `http://www.legomindstorms.com/`
13. Goldman, R.: From RoboLab to Aibo: Capturing Agent Behavior. Master's thesis, Department of Computer Science, Columbia University (2005)

14. Chu, K.H., Goldman, R., Sklar, E.: Roboxap: an agent-based educational robotics simulator. In: Agent-based Systems for Human Learning Workshop at AAMAS-2005 (2005)
15. Azhar, M.Q., Goldman, R., Sklar, E.: An agent-oriented behavior-based interface framework for educationa robotics. In: Agent-Based Systems for Human Learning (ABSHL) Workshop at Autonomous Agents and MultiAgent Systems (AAMAS-2006) (2006)
16. Sony: AIBO (accessed January 16, 2006), `http://www.us.aibo.com/`
17. OPEN-R: SDE (accessed January 16, 2006), `http://openr.aibo.com/`
18. Serra, F., Baillie, J.C.: Aibo Programming Using OPEN-R SDK Tutorial (2003), `http://www.cs.lth.se/DAT125/docs/tutorial_OPENR_ENSTA-1.0.pdf`
19. R-CODE: SDK (accessed January 16, 2006), `http://openr.aibo.com/openr/eng/no_perm/faq_rcode.php4`
20. YART: Yet Another R-CODE Tool (accessed January 16, 2006), `http://www.aibohack.com/rcode/yart.htm`
21. Touretzky, D.S., Tira-Thompson, E.J.: Tekkotsu: A framework for AIBO cognitive robotics. In: Proceedings of the Twentieth National Conference on Artificial Intelligence (AAAI-05), Menlo Park, CA, AAAI Press, Stanford (2005)
22. Baille, J.C.: URBI: Towards A Universal Robotic Body Interface. In: Proceedings of the IEEE/RSJ International Conference on Humanoid Robots, Santa Monica, CA USA (2004)
23. Baille, J.C.: URBI: Towards a Universal Robotic Low-Level Programming Language. In: Proceedings of the IEEE/RSJ International Conference on Intelligent Robots and Systems, Edmonton, Canada (2005)
24. National Instruments: LabVIEW (accessed January 16, 2006), `http://www.ni.com/labview/`
25. LabVIEW: User Manual (accessed January 16, 2006), `http://www.ni.com/pdf/manuals/320999e.pdf`
26. Lesk, M.E., Schmidt, E.: Lex – A Lexical Analysis Generator. Bell Laboratories, Murray Hill, NJ (1975)
27. Johnson, S.C.: Yacc – Yet Another Compiler-Compiler. Bell Laboratories, Murray Hill, NJ (1975)
28. Téllez, R.: R-CODE SDK Tutorial (v1.2) (2004)
29. LEGO: What's NXT? LEGO Group Unveils LEGO MINDSTORMS NXT Robotics Toolset at Consumer Electronics Show (January 4, 2006), `http://www.lego.com/eng/info/default.asp?page=pressdetail&contentid=17278&countrycode=2057&yearcode=&archive=false&bhcp=1`
30. Duffy, J.: What happened to the Robot Age? BBC News Magazine (January 27, 2006)

The Robotics and Mechatronics Kit "qfix"

Stefan Enderle

Neural Information Processing Department
University of Ulm, Germany
enderle@neuro.informatik.uni-ulm.de

Abstract. Robot building projects are increasingly used in schools and universities to raise the interest of students in technical subjects. They can especially be used to teach the three mechatronics areas at the same time: mechanics, electronics, and software. However, it is hard to find reusable, robust, modular and cost-effective robot development kits in the market. Here, we present *qfix*, a modular construction kit for edutainment robotics and mechatronics experiments which fulfills all of the above requirements and receives strong interest from schools and universities. The outstanding advantages of this kit family are the solid aluminium elements, the modular controller boards, and the programming tools which reach from an easy-to-use graphical programming environment to a powerful C++ library for the GNU compiler collection.

1 Introduction

Robot building projects are a good means to bring the interesting field of robotics to schools, high-schools, and universities. Studying robotics the students learn a lot about mechanics, electronics, and software engineering. Additionally, they can be highly motivated and learn to work in a team.

Performing a lot of robot building labs with pupils and students, we found that there is a gap between the relatively cheap toy-like kits, like LEGO Mindstorms or Fischertechnik Robotics and the quite expensive off-the-shelf robots. The toy kits offer a good opportunity to start building robots, but they mostly support the control of only 2 or 3 motors and the same number of sensors. Off-the-shelf robots (see e.g. [1,4,7]) are completely built up, so typically only the programming of the robot can be studied.

Alternatively, there exist a number of controllers, like the 6.270 board or the HandyBoard [3] which come without mechanical parts and so must be used in combination with other toy kits, like RC-controlled cars, or custom-built robots. However, these boards, can control only small motors and are not very expandable.

After building RoboCup robots from scratch [6,9,2,11] and supporting schools developing their own RoboCupJunior robot [10], the authors gained a lot of experience about reasonable mechanical concepts and controller architectures for a usable robot development kit. Thus, we decided to develop the robot kit familiy *qfix* and to provide it to schools and universities.

G. Lakemeyer et al. (Eds.): RoboCup 2006, LNAI 4434, pp. 134–145, 2007.

2 The *qfix* Approach: Modularity

The main concept behind *qfix* ist modularity in the following dimensions:

- **Mechanics:** The mechanical parts are aluminium parts including rods, plates and holders for different sensors and actuators. These parts are the building blocks for constructing mechanical and mechatronic systems, like cars, walking robots, etc. Most parts contain threads and can easily be screwed together, so very robust models can be build.
- **Electromechanics and Electronics:** There already exist many compatible electromechanical and electronical parts including a variety of sensors, actuators, and controller boards. With these components it is possible to make the mechanical models *move* (by DC motors, servo-motors, stepper motors), *sense* (by tactile, infrared and ultrasonic sensors), and *think* (by powerful controller boards which can be programmed on the PC).
- **Software:** In the software area, modularity is no big deal. The *qfix* software comes with the powerful free GNU C++ toolchain (WinAVR for windows, respective libraries or RPMs for linux). Additionally, it contains an easy-to-use C++ class library for accessing all *qfix* electronics components.

  Since beginners, say, of an age from 12, have problems going directly into C or C++ programming, we developed a graphical programming environment called GRAPE in order to simplify the programming of self-built robots. This software directly produces C++ code from the graphical description and thus supports the beginner in learning object-oriented programming.

2.1 Mechanics

The basic building blocks of the *qfix* system are anodized aluminium rods with $\phi 6$ holes along all four sides and two M6 threads on the front and back side (see Fig. 1). Currently, there are rods from 20mm to 100mm including 45° rods.

Other basic elements are a variety of plates with holes and threads. These plates can be bolted to the rods using a screw and a nut or only a screw exploiting the rods' frontal threads. Like the rods, the plates are given in different lengths and widths, currently up to 200mm x 200mm (see Fig. 2 for an exemplary plate).

All mechanical parts use holes and threads according to DIN/ISO standards and have a grid of 10mm.

Fig. 1. Basic elements: rods

Fig. 2. Basic elements: plates (Here: 200x200mm with 400 threads)

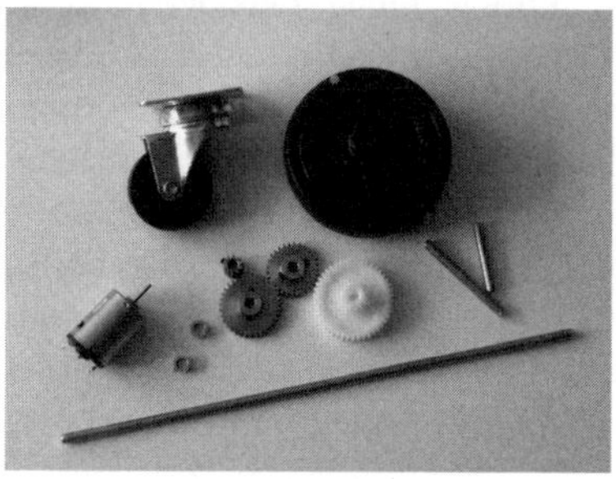

Fig. 3. Wheels, axes, and gears

In Figure 3, some additional mechanical elements can be seen: wheels, casterwheels, gears, and axes. They are usually used to implement dynamic models which then can be driven by different motors as shown in the next section.

2.2 Electromechanics/Electronics

Motors: In order to make a model move, motors are needed. Typical robotics applications often use different kinds of motors for different tasks: DC motors, servo motors, and stepper motors. *qfix* supports these different categories by providing the respective motor bearings (see Fig. 4) and electronics components for driving motor and wheel encoders.

Sensors: When building robots, it is also necessary to make them able to gather information about their environment. This can be done by mounting simple switches signalling bumps into obstacles, or by adding distance measuring devices like infrared or sonar sensors. As with motors, *qfix* supports numerous sensors by providing the respective bearing (see Fig. 5) for mounting the sensor to the model.

Controllers: Obviously, the motors and sensors must be driven by an electronics component. For *qfix*, we developed a new, modular controller board arhitecture which is both powerful and easy-to-handle. The board with the smallest controller is the "BobbyBoard" (see Figure 6) which uses the Atmel ATmega32 controller and supports the following I/Os:

Fig. 4. Exemplary motor bearing for a DC motor

Fig. 5. Exemplary sensor bearing for an IR distance sensor

Fig. 6. "BobbyBoard": controller board with ATmega32

- 2 DC motor controllers (battery voltage, 1A)
- 4 digital inputs (0/5V)
- 4 analog inputs (0-5V)
- 8 digital outputs (battery voltage, 100mA)
- 4 LEDs
- 4 buttons
- I^2C-bus for extensions

Further existing main boards are the "CAN128Board" which shows the same I/O capabilities but uses an Atmel AT90CAN128 controller with CAN interface, more memory and more speed. And, the "SoccerBoard" with 8 analog and 8 digital inputs, 8 digital outputs, 6 motor drivers, and optional CAN and USB interface (see Figure 7). This board is specifically designed for the requirements

Fig. 7. "SoccerBoard": controller board with ATmega128

Fig. 8. "LC-display board": expansion board with LC-display

of RoboCupJunior, where often omnidrive platforms with three driven wheels plus a kicker and a "dribbler" are used combined with multiple sensor systems.

All controller boards are programmed ("flashed") from the PC via a serial, parallel or USB link and then run autonomously without the host computer.

Extension Boards: The main idea behind the *qfix* boards is their flexible modular architecture: The main controller board runs the main program and communicates with expansion boards for setting actuator values and getting sensor data. The expansion boards themselves are responsible for controlling the attached devices, so the main processor does not have to perform expensive tasks, like feedback motor control, etc.

The controller boards contain an I^2C-bus and optionally a CAN bus which both allow to chain dozens of boards of the same or different kinds to a large controller network. So, it is possible to either control more I/Os or even to implement distributed applications with decentralized control (see e.g. [5]).

The following extension boards based on I^2C-bus are currently available:

- **Servo board 1:** The servo-board uses a Atmel mega8 for controling 4 servo motors independently.
- **Servo board 2:** This servo-board is designed for humanoid robots and can control 24 servo motors independently. It contains a mega128 controller and a Xilinx FPGA for fast I/O control.
- **Stepper-board:** The stepper-board can control 4 stepper motors independently. Both, full and half step mode are supported.

- **DC-power board:** The DC-power-board is capable of controlling two DC motors with 4A each. It also contains two encoder input lines for each motor.
- **LC-display board:** An LC-display with 4 lines of 20 characters each (see Figure 8).
- **Relais boards:** There are two relais boards: one to be connected to the digital output of the controller board and one to be connected via the I^2C-bus.

Further expansion boards, e.g. for Polaroid sonar sensors [8] and a camera board are currently under development.

2.3 Software

With *qfix* we provide the free GNU C++ toolchain including generic tools for downloading programs to the controller boards. Additionally, we provide a C++ class library supporting all *qfix* boards. On Windows, the generic tools mainly consist of the WinAVR GCC environment for Atmel controllers which includes the extensible editor *programmers notepad* and powerful download tools, like *avrdude*. All tools also run on Linux/Unix and Mac, so cross-platform development is fully supported.

The easy-to-use *qfix* C++ class library hides the low-level hardware interface from the programmer and supports the complete *qfix* extension board family. The main idea is to provide a specific C++ class for each *qfix* module. Therefore, the library provides the classes `BobbyBoard`, `SoccerBoard`, `LCD`, `SlaveBoard`, `StepperBoard`, `ServoBoard`, `RelaisBoard`, etc. For example, when building an application with the BobbyBoard and the LCD you use the respective classes, like the following:

```
#include "qfixBobbyBoard.h"        // include BobbyBoard library
#include "qfixLCD.h"               // include LCD library

int main()
{
  BobbyBoard board;                // construct object "board"
  LCD        lcd;                  // construct object "lcd"

  board.ledOn(0);                  // turn on LED 0
  board.waitForButton(0);          // wait until button 0 is pressed
  board.motor(0,255);              // turn on motor 0 to full speed
  lcd.print("Engines running");    // print a text on the LCD
}
```

As can be seen from the comments of the code, two instances of two classes are constructed: `board` and `lcd`. Their methods are called in order to let the main board turn on a LED and a motor, wait for a button press, and output text on the LCD.

A lot of the functionality is hidden in the constructors of both classes. When constructing the object `board` for instance, the constructor initializes all I/O pins and starts an interrupt routine for motor PWM control. When constructing `lcd`, the constructor opens an I^{2}C-bus channel and starts communicating with the physically connected LC-display. This mechanism works perfectly as long as expansion boards of different types are used only.

When using multiple expansion boards of the same type, the extended construction syntax can be used in order to connect the objects to the correct physical boards. Imagine you have a controller board and three identical LC-display boards:

```
#include "qfixBobbyBoard.h"      // include BobbyBoard library
#include "qfixLCD.h"             // include LCD library

int main()
{
  BobbyBoard board;              // construct object "board"
  LCD        lcd0(0);            // construct object "lcd0"
  LCD        lcd1(1);            // construct object "lcd1"
  LCD        lcd2(2);            // construct object "lcd2"

  board.waitForButton(0);        // wait until button 0 is pressed
  lcd0.print("Hallo");           // print a text on LCD 0
  lcd1.print("World!");          // print a text on LCD 1
  lcd2.print("Engines running"); // print a text on LCD 2
}
```

In this example, each of the three `lcd`X objects is connected to the physical LCD board with the respective ID. This ID can be hardcoded to the LCD by flashing the LCD board, or it can be dynamically changed by calling the method `lcd.changeID(newID)`.

For those who want to connect multiple controller boards but do not want to go into detail with programming the I^{2}C-bus, we provide a class `SlaveBoard` which can be used as a "remote control" for connected BobbyBoard main boards:

```
#include "qfixBobbyBoard.h"      // include BobbyBoard library
#include "qfixSlaveBoard.h"      // include SlaveBoard library

int main()
{
  BobbyBoard master;             // construct a master board object
  SlaveBoard slave0(0);          // construct a slave board object
  SlaveBoard slave1(1);          // construct a slave board object

  master.motor(0,100);           // turn on motor on master board
  slave0.motor(0,100);           // turn on motor on slave board 0
  slave0.waitForButton(0);       // wait for button on slave board 0
  slave1.ledOn(0);               // turn on LED on slave board 1
}
```

3 Graphical Programming Environment GRAPE

In addition to the C++ environment, we developed a new software system called *GRAPE* (which stands for GRAphical Programming Environment). With GRAPE it is possible to program the *qfix* controller boards in an object oriented way without having experience in C++.

The GRAPE application consists of three tabbed windows which are used sequentially: In the first tab, the desired classes (e.g. BobbyBoard and LCD) are loaded. Each class can then be instantiated by one or more objects. The object names can be freely chosen. The second tab holds the main window for graphical programming. Here, symbolic blocks are arranged intuitively in order to get a flow chart with the desired program flow (see Figure 9).

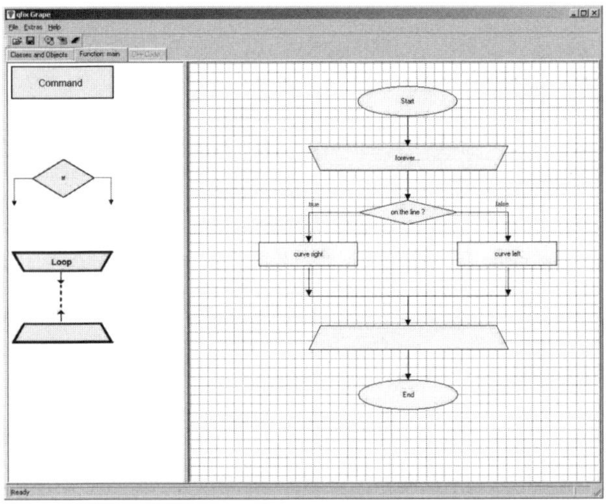

Fig. 9. Graphical program in GRAPE

For each symbolic icon, a properties dialog can be opened to define the semantics of the icon in a semi-graphical way: For *commands*, the user can select an object from the list of instantiated objects, then chose a method from the object's possible methods, and then select the desired parameters from the list of possible parameters for the chosen method. This selection defines all parts of a typical object-oriented method call: <object>.<method>(<parameters>).

After filling all graphical blocks with their respective meaning, the flow chart can be saved as a XML description file. This makes it possible to perform, e.g. in an individual tool, the translation to any object-oriented (or even classically procedural) programming language. In GRAPE, this translation is already integrated and the flow chart (or internally, the XML representation) is automatically translated to C++ code (see Figure 10).

With this approach, the basic concepts of a procedural programming language can easily be learned: commands, sequences of commands, if-clauses, while loops.

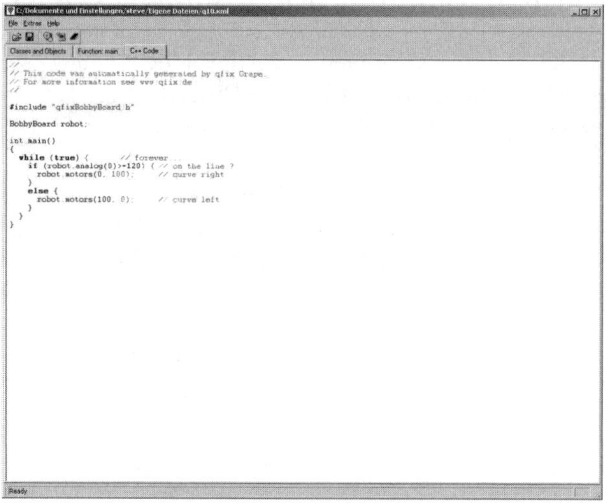

Fig. 10. Respective code in GRAPE

And, it can be studied how these concepts are translated to C++ or another programming language. In addition to that, the users learn to use given class libraries.

4 Experiments

In order to demonstrate the feasibility of the *qfix* parts and controller boards, we developed some typical robot and mechatronic applications.

4.1 Differential Drive Robot

The first mobile robot is a car with two independently driven wheels and a caster wheel, all mounted on a 10cm x 10cm base plate (see Fig. 11). The BobbyBoard drives the two motors as well as three infrared distance sensors (Sharp GP2D120) which are used for a simple collision avoidance behaviour. An improved version also uses bumpers, a line sensor for moving along a line and an LCD for displaying messages like "front blocked" or general status information.

4.2 Offroad Robot

Figure 12 shows an "offroad" robot which was built in order to test the power of the motor controllers (L293D).

For this robot the same mainboard as in the differential drive robot above is used and drives four stronger motors, where the left and the right ones are connected in parallel. The complete platform is much bigger (main plate of 20x20cm) than the above one and includes a boxed version of the LC-display.

Fig. 11. Differential drive robot with two driven wheels and three IR distance sensors

Fig. 12. Robot arm with three DOFs

Fig. 13. Omnidrive platform with three omnidirectional wheels

4.3 Soccer Robot

As a third application, a specialized soccer robot was built in order to demonstrate the flexibility of both mechanics and electronics components. As main platform we used a round plate of about 21cm diameter with three omnidirectional wheels (see Figure 13).

In order to control the three motors, two controller boards were connected via the I^2C-bus and communicate with each other to establish a reliable movement coordination. Additionally, the resulting soccer robot uses a kicker device and a so called "dribbler" to hold the ball near the robot. As sensors, infrared light sensors are used for detecting a RoboCupJunior ball. For obstacle avoidance,

Fig. 14. Soccer robots

infrared or ultrasonic distance sensors can be attached. The complete soccer robot including a trendy skin or "tricot" is shown in Figure 14.

5 Conclusion

We presented *qfix*, a construction kit for developing autonomous mobile robots and other mechatronics applications. *qfix* was mainly developed for educational and edutainment purposes. The kit consists of solid mechanical and electromechanical parts, powerful modular controller boards with several extension boards, and a complete C++ class library for easy support of all functionality.

Since the kits are often used in the RoboCupJunior area, where the users are only 12 or even less years old and have no programming experience, we developed the graphical programming environment GRAPE. This tool supports object oriented programming on a graphical level but directly generates C++ code which can be studied and edited.

The complete *qfix* robot kit family proves to be an appropriate tool for robot development. It is already used in educational classes and labs in schools and at universities. Additionally, the open architecture encourages the robotics community to help improving the kits.

Ackowledgement

This work is sponsored by KTB mechatronics GmbH, Germany (www.ktb-mechatronics.de).

We thank Bostjan Bedenik who mainly implemented the qfix Grape software.

References

1. ActivMedia: Pioneer 1 Operation Manual. RWI, Jaffrey, NH (1996)
2. Enderle, S.: The Sparrow-99 robot. Technical report, University of Ulm, Internal report (1999)

3. HandyBoard: (1998)
`http://lcs.www.media.mit.edu/groups/el/Projects/handy-board/index.html`
4. K-Team.: Khepera – user manual (1999)
5. Kaiser, J.: Real-time communication on the CAN-bus for distributed applications with decentralized control. In: 4th IFAC International Symposium on Intelligent Components and Instruments for Control Applications, Buenos Aires, Argentina (September 2000)
6. Kitano, H., Asada, M., Kuniyoshi, Y., Noda, I., Osawa, E., Matsubara, H.: RoboCup — a challenge problem for AI. AI magazine, pp. 73–85 (Spring 1997)
7. Nomadic: `http://www.robots.com`
8. POLAROID: Ultrasonic Ranging System. Polaroid Corporation, 784 Memorial Drive, Cambridge, MA 02139 (1991)
9. RoboCup: `http://www.robocup.org`
10. RoboCupJunior: `http://www.robocupjunior.org/de`
11. Utz, H., Sablatnög, S., Enderle, S., Kraetzschmar, G.K., Palm, G.: Miro – Middleware for mobile robot applications. IEEE Transactions on Robotics and Automation, Special issue of on Object Oriented Distributed Control Architectures, 2002 (submitted)

Cooperative Visual Tracking in a Team of Autonomous Mobile Robots

Walter Nisticò, Matthias Hebbel, Thorsten Kerkhof, and Christine Zarges

Institute for Robot Research
Universität Dortmund
Otto-Hahn-Str.8, 44221 Dortmund
forename.surname@uni-dortmund.de

Abstract. Robot soccer is a challenging domain for sensor fusion and object tracking techniques, due to its team oriented, fast-paced, dynamic and competitive nature. Since each robot has a limited view about the world surrounding it, the sharing of information with its teammates is often crucial in order to be ready to react to situations which might involve it in the near future. In this paper we propose a Particle Filter based approach that addresses the problem of cooperative global sensor fusion by explicitly modeling the uncertainty concerning the robots' positions, the data association about the tracked object, and the loss of information over the network.

1 Introduction

The tracking of fast moving objects has received constant attention in the context of autonomous robots interacting with highly dynamic and potentially hostile environments, and the robot soccer domain is particularly suitable to this purpose, as a robot has to interact cooperatively and competitively with a set of moving objects such as teammates, opponents and the ball.

1.1 The Platform

This work has been developed on the robot Sony Aibo ERS-7 [1], which is the only allowed hardware platform on the 4-Legged RoboCup League [2]. This limitation poses several interesting challenges for the task of object tracking, due to the limited computational resources available (576MHz MIPS CPU, 64MB of RAM) and the presence of a single exteroceptive sensor, a low-power CMOS camera with a maximum resolution of 208×160 pixel. Even the localization and tracking algorithms have to consider the running time as a serious issue, and it is highly desirable to be able to process the sensory information at the maximum rate provided by the camera, 30Hz, since this device has a very limited field of view (57° horizontal, 45° vertical) and consequently it has to be moved at high angular velocities to scan the environment. This year, the problem has been further complicated by the new rules of the league, which nearly doubled the

G. Lakemeyer et al. (Eds.): RoboCup 2006, LNAI 4434, pp. 146–157, 2007.

field size (now is 6m×4m) and removed any border or fence which would prevent the robots from observing unknown objects outside of the field. At last, legged locomotion makes it impossible to accurately know the height of the camera relative to the field, as this changes continuously as the robot walks, and the camera is mounted on the snout of the robot, which can rotate with 3 degrees of freedom; the uncertainty in the camera pose relative to the ground adds severe noise levels to the measurements.

1.2 Related Work

The most significant achievement in the field of individual tracking has been presented in [3], where the authors have used a sophisticated Rao-Blackwellised Particle Filter to efficiently model strong non linearities in the ball motion due to the interactions with the environment, such as bouncing on borders or being kicked by a robot. Cooperative object tracking is instead still in its infancy in this context, as the additional problems of the uncertainty on the robots' own positions, and the sharing of information over an unreliable network further complicate the problem. In [4] the authors have compared several sensor fusion techniques applied in the context of the RoboCup Middle-Size League, including Bayesian Filtering techniques [5], simple techniques such as arithmetic or weighted mean of percepts, and an anchoring approach. Not surprisingly, the Kalman Filter and the Particle Filter resulted to be the top performers, with the former as the solution of election due to its limited computational requirements, but it has to be noted that on this platform the uncertainty over the robot location is very small, due to the availability of omni-directional cameras and range sensors such as laser scanners. At last, in [6] has been proposed an Extended Kalman Filter based approach for global sensor fusion in the Four-Legged League, which takes into account the localization problem. However, this paper does not consider the data association problem, assuming the ball to be unique on the field, an assumption which is not true anymore since the rule changes in the league which have removed the protective fence from around the field.

2 Tracking the Ball with a Particle Filter

Even if the RoboCup rules allow only a single ball to be present on the field, there are still several situations where ambiguities may arise. The camera is the only exteroceptive sensor of this robot, and the ball is mainly recognized for its color (orange), its spherical shape (although it is frequently incomplete due to occlusions), and the fact that it lays on a green surface (the soccer field), however there are situations where objects around the field, such as clothing or shoes in the audience, can appear as valid ball candidates. Furthermore, the official red jerseys used to distinguish one team of robots have also a rounded shape, and due to limitations in the camera hardware, lighting, and blur, they can appear as potential balls. For all these reasons, we feel that a Kalman Filter based approach as described in [6] is not robust enough for our needs, as it does not deal very well with sensor ambiguity since it cannot deal with multi-modal probability distributions; consequently, we have decided to use a Particle Filter approach similar to what is described in [3], which can track multiple ball hypotheses. To avoid adding the robot localization uncertainty to the ball tracking own uncertainty, the ball position and velocity will be represented in a robot-centric reference system.

2.1 Particle Filters

The *Particle Filter* [5] is a non-parametric implementation of the general *Bayes Filter*, which is a recursive algorithm that calculates the belief or *posterior* $bel(x_t)$ at time t of the state x_t of a certain process, by integrating measurement observations z_t and control actions u_t over the belief of the state at time $t-1$. The Bayes Filter is based on the Markov assumption or *complete state* assumption which postulates the conditional independence of past and future data given the current state x_t. A Particle Filter represents an approximation of the posterior $bel(x_t)$ in the form of a set of samples randomly drawn from the posterior itself; such a representation has the advantage, compared to closed form solutions of the Bayes Filter such as the Kalman Filter [7], of being able to represent a broad range of distributions and model non-linear processes, whereas parametric representations are usually constrained to simple functions such as Gaussians. Given a set of N samples or *particles* $\Pi_t := x_t^1, x_t^2, \ldots, x_t^N$, at time t each particle represents an hypothesis of the state of the observed system; obviously, the higher the number of samples N, the better the approximation, however [8] has shown how to dynamically adjust N. An estimate of $p(x_t|u_t, x_{t-1}^i)$ is called

---

Algorithm 1. Particle Filter

Require: particle distribution Π_{t-1}, control action u_t, measurement observation z_t
  for $i = 1$ to N **do**
    Process update: update the particle state as the result of control action u_t: $x_t^i \propto p(x_t|u_t, x_{t-1}^i)$
    Measurement update: calculate the particle importance factors $w_t^i = p(z_t|x_t^i)$ from the latest observation
    Add $\langle x_t^i, w_t^i \rangle$ to the temporary set $\overline{\Pi}_t$
  end for
  Resampling: create Π_t from $\overline{\Pi}_t$ by drawing the particles x_t^i in a number proportional to their importance w_t^i.

Process Model, while $p(z_t|x_t^i)$ is known as *Sensor Model*. In the context of robot localization and object tracking, particle filters are often referred to as *Monte Carlo Localization* [9].

2.2 Sensor Model

Unlike other robot platforms, an important source of noise in the camera measurements is the uncertainty about the camera pose relative to the robot-centered reference system: the camera can rotate with 3 degrees of freedom, and its height relative to the ground changes dynamically as the robot walks. Furthermore, the on-board camera has only 3 shutter speed settings, with a minimum exposure time of $\frac{1}{200}s$; as a result, images are affected by blur, which gets more noticeable as the robot and camera speed increase. Lastly, the camera captures an image sequentially from the top scanline to the bottom, with a frequency of 30fps, so that a time delay exists between the top of the image and the bottom of $\approx \frac{1}{30}s$, which distorts the image and the percepts especially in case of a fast camera panning motion (for a description of the problem and a possible solution see [10]). The uncertainty about the measurement is modeled as a 2-dimensional gaussian, with one axis oriented as the distance between the robot and the ball (σ_ρ) and the other perpendicular to it ($\sigma_\perp$); the variances of such gaussian are dependent on the following factors:

1. Percept confidence $p_c(z) \in [\frac{1}{5}, 1]$ calculated from the image processor based on several criteria used to identify the ball, such as color, shape, sharpness of the contour; its reciprocal is multiplied by both axes
2. Distance to the percept f_ρ; is proportional to σ_ρ
3. Camera panning velocity $f_{\dot{\alpha}}$; is proportional to $\sigma_\perp$
4. Robot speed f_{v_R}, which affects the amount of "head bobbing", causing motion blur and inaccuracy in the camera pose; is multiplied by both axes

In order to estimate f_ρ, $f_{\dot{\alpha}}$ and f_{v_R}, we used an external camera[1] mounted on the ceiling above the soccer field to compare the robot's own measurements with the true ball position. Based on the observed data, we have modeled f_ρ and $f_{\dot{\alpha}}$ as second order polynomials, while for f_{v_R} we have chosen a piecewise linear approximation; the results have been represented in Figure 1. In case of several objects in the image which might look like a ball, the vision system provides a list of candidate ball hypotheses, each one with a certain percept confidence $p_c(z)$, and they are all used to perform the measurement update of the particle filter.

2.3 Process Model

Since the ball is tracked in a robot-centric reference system, the robot's own motion is an apparent speed relative to the ball. At each time instant, an estimate of the robot motion is represented by the odometry vector: $o_t = \left[o_{vx}^t \; o_{vy}^t \; \omega_t \right]$ where (o_{vx}^t, o_{vy}^t) is the robot translation speed and ω_t is its angular velocity at time t. In addition, we use a *constant speed model*, which propagates the position of the ball $\vec{s}(t)$ at time t based on the speed $\vec{v}(t)$ at time $t-1$; this because

[1] The average measurement error of such a vision system is below 0.5cm.

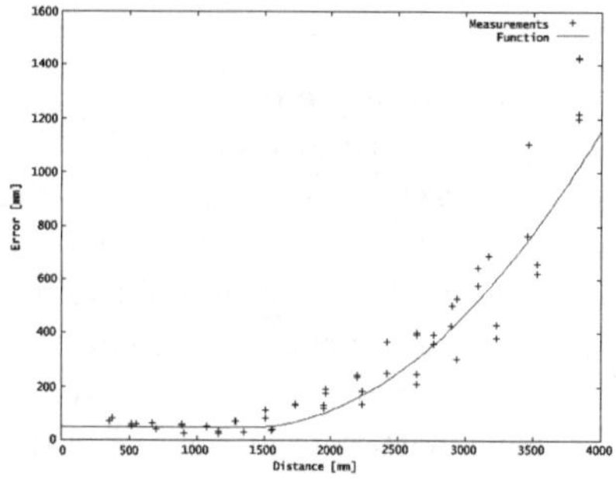

(a) Distance Error

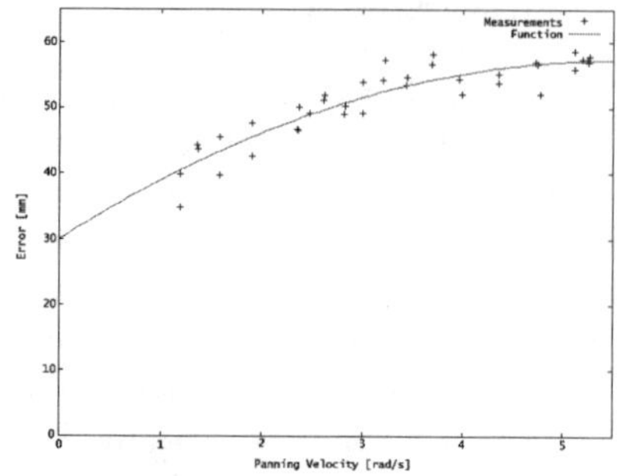

(b) Panning Velocity Error

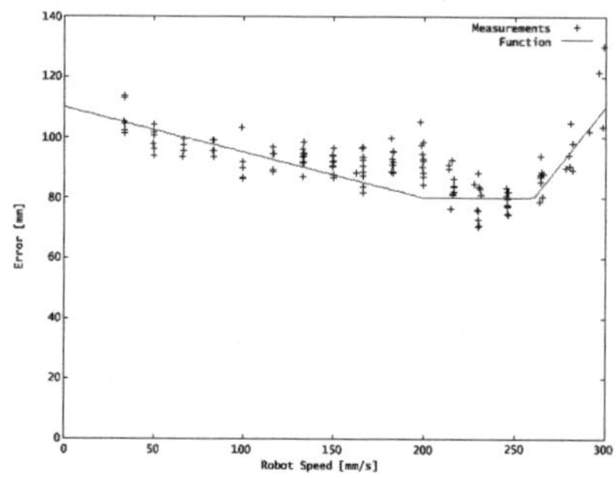

(c) Robot Speed Error

Fig. 1. Ball measurement error functions learned from experimental data

accelerations are very difficult to measure with highly noisy sensors like this camera. Consequently, the time-update is performed as follows[2]:

$$\begin{bmatrix} s_x^t \\ s_y^t \\ v_x^t \\ v_y^t \end{bmatrix} = \begin{bmatrix} \cos(\omega_t \Delta t) & -\sin(\omega_t \Delta t) & \Delta t & 0 \\ \sin(\omega_t \Delta t) & \cos(\omega_t \Delta t) & 0 & \Delta t \\ 0 & 0 & \cos(\omega_t \Delta t) & -\sin(\omega_t \Delta t) \\ 0 & 0 & \sin(\omega_t \Delta t) & \cos(\omega_t \Delta t) \end{bmatrix} \cdot \begin{bmatrix} s_x^{t-1} \\ s_y^{t-1} \\ v_x^{t-1} \\ v_y^{t-1} \end{bmatrix} + \begin{bmatrix} 0 \\ 0 \\ o_{vx}^t \\ o_{vy}^t \end{bmatrix} \tag{1}$$

Since the constant speed model is only an approximation and the odometry vector is noisy itself, in the time update we add to the probability distribution a constant amount of gaussian noise ($\sigma = 0.25$ empirically derived).

3 Multi-robot Tracking

As can be seen in Figure 1(a), the measurement error grows very quickly with the distance to the ball (error $\approx$ 1m at a distance of 3.5m), making it very difficult to track a ball which rolls in the opposite side of the field. Further, it

[2] In this equation the sign of the odometry vector has to be reversed, here the negative sign is omitted to keep compact the graphical representation.

happens very frequently that one or more robots in a team have the line of sight to the ball completely occluded by teammates or opponents, but still they need to know where the ball is in order to position themselves strategically on the field. At last, it is desirable that the robots share a common estimate about the ball position, so that they can coherently take strategical decisions such as who should go to the ball and try to get in control of it. Thereto, merging the sensor information of all the robots in a team can often result in a better estimate than what would be possible for each robot alone. However, the cooperative tracking of an object must be performed in a global coordinate system, and this severely complicates the problem compared to local sensor fusion techniques, since the uncertainty on the robots' positions adds to the ball measurement errors. For example, it happens that a robot following the ball focuses its attention for too long on the ground, without looking at global landmarks: since field features such as the field lines, the kick-off circle and the line crossings are symmetrical on the field, the robot location probability distribution tends to concentrate around symmetric modes, and consequently the localization state can jump frequently from a peak to another. When the ball probability distribution of that robot is transformed in global coordinates using the current robot pose, it will jump along with it providing contradictory information to the whole team estimate of the ball position and this is not desirable, especially if such robot is very close to the ball and has a good view of it. For the robot localization, we have implemented a *Monte Carlo* algorithm similar to what is described in [11], with a sample set of 100 particles.

3.1 Multi-robot Belief Merging

Since each particle in the self locator represents a candidate pose, ideally each robot should calculate the global position of its ball probability distribution **for each particle of the self locator**: with 40 particles to represent the (robot centric) ball probability $\pi^l_B(i), i \in [0, 40)$, and 100 particles for the localization probability $\pi_{SL}(i), i \in [0, 100)$, the final set of ball particles in the global reference system would be 4000 particles per robot $\pi^g_B(i), i \in [0, 4000)$. The probability of each ball particle should be also multiplied by the probability of the corresponding localization particle:

$$p(\pi^g_B(i \cdot j)) = p(\pi^l_B(i)) \cdot p(\pi_{SL}(j)) \tag{2}$$

Then such particles have to be sent to the teammates, merged with their particles, and clustered to find the expected "team ball" position. Algorithm 2 can deal with situations where the localization distribution of some robot presents strong ambiguities, because such ambiguities can be resolved by the information provided by the teammates. The main problem with this approach is the huge amount of particles that have to be computed and sent over the network: if each particle is represented by the values $\left[s_x \; s_y \; p(\pi^g_B(i)) \right]$ and each value is stored in 32 bit precision, even when using broadcasts we would still use 1.5Mbit of traffic per iteration of the algorithm just for particle data. An alternative to

Algorithm 2. "Naive" Merging

for all i, j such that $i \in [0, 40), j \in [0, 100)$ **do**
  calculate the position of $\pi_B^g(i)$
  calculate $p(\pi_B^g(i))$
end for
send $\pi_B^g(i)$ to teammates
receive $\pi_B^g(i)$ from teammates
cluster the joint particle set and compute the expected ball position

save on network traffic would be to send $\pi_B^l(i)$ and $\pi_{SL}(j)$ and multiply the two sets at the destination; however this would result in even greater computational costs, and our platform is already not suitable to process particle sets of such dimensions.

3.2 Reducing the Joint Particle Set Size

In most game situations, the particle distribution is not spread uniformly accross the field (this normally happens only when the robot is placed on the field for the first time) but it is concentrated in a very limited number of clusters. This is because low probability particles are replaced with new samples in the positions calculated from the latest observation, following the sensor-resetting [12] / Mixture Monte Carlo idea [13], so even when the robot is teleported or "kidnapped" by the referee, a new cluster forms very quickly at the new position of the robot. In our experiments, in a typical match over 90% of the total probability is concentrated in at most 3 clusters, so we calculate the 3 robot pose hypotheses with the highest probabilities to generate the ball distribution in global coordinates. Further, to keep the running time and network traffic low, we subsample the ball global probability distribution to obtain up to 12 *"representative particles"* out of the set of 120. Since each robot provides at most 12 particles to his teammates, the global ball estimate is calculated out of 48 particles as the cluster with the highest probability. To efficiently calculate the representative particles, the soccer field is recursively split into cells to form a quad-tree, with a maximum depth $\delta = 7$:

- *Basic Cell*: a cell which contains one particle or none
- *Composite Cell*: a cell which contains 4 Basic Cells

While it might appear that 12 particles are too few to represent the belief of a single robot about the ball position, it has to be noted that Algorithm 3 is applied to the particle set *before* its normalization / resampling. As such, a small number of particles can carry the same amount of information of a much larger set after the normalization, because such a process replaces high probability particles with several copies having all the same importance factor.

Algorithm 3. Representative Particles Computation

1. The whole field is initialized as a Composite Cell
2. **if** a Basic Cell contains more than one particle
   - transform it in a Composite Cell by subdividing it into 4 Basic Cells
   - particles are inserted into each new Basic Cell depending on their position on the field
3. apply recursively step 2 until a maximum depth $\delta = 7$ is reached
4. representative particles are generated out of the 12 cells which contain the highest probability; if a chosen cell contains more than one particle, the representative particle position is calculated as a weighted average of the particles there contained

3.3 Loss of Information over the Network

To keep network utilization and latency to low levels, in our system a robot exchanges data with his teammates through UDP broadcasts [14]. However, UDP does not guarantee that the packets will reach their destination, and it is quite common to have network performance problems in crowded places or at the competition sites, since 802.11 networks are now so widely popular. Since our ball tracking runs at a fairly high rate (the same as the vision system, 30Hz), it is not so unlikely that for a frame or two no particle is received from a certain teammate. In such unfortunate case, it is wiser to use older information from the corresponding robot instead of immediately discarding all the particles of the previous iteration, since in such a short interval of time (up to 100ms) the ball state cannot change too much. Therefore the current implementation of our ball tracker stores the particles received from all teammates. If in the new frame, at least one particle is received from a certain robot, all its old particles are discarded and substituted by the new ones. Otherwise, the old particles can be used, but random noise has to be added to reflect the increased uncertainty due to the unmodeled ball motion, and their reliability has to be lowered. That is achieved by replacing all old particles with two new ones, each new particle carrying half of the original probability. If this results in more than 12 particles for that particular robot, only the 12 particles with the highest validities are retained. Afterwards all particles are spread by the addition to their position of gaussian noise, to represent a probabilistic search around their original position, since the direction of movement of the ball is unknown. The standard deviation of such noise is a function of the number of frames where no particles were received from the teammate. Finally, the validity of these particles is decreased by a factor which is also a function of their age. In our tests, this approach has worked better than propagating the old state by using the speed, because the speed estimation is very noisy in itself, the ball motion is often non-linear, and the constant-speed model is a valid approximation only for very short periods of time.

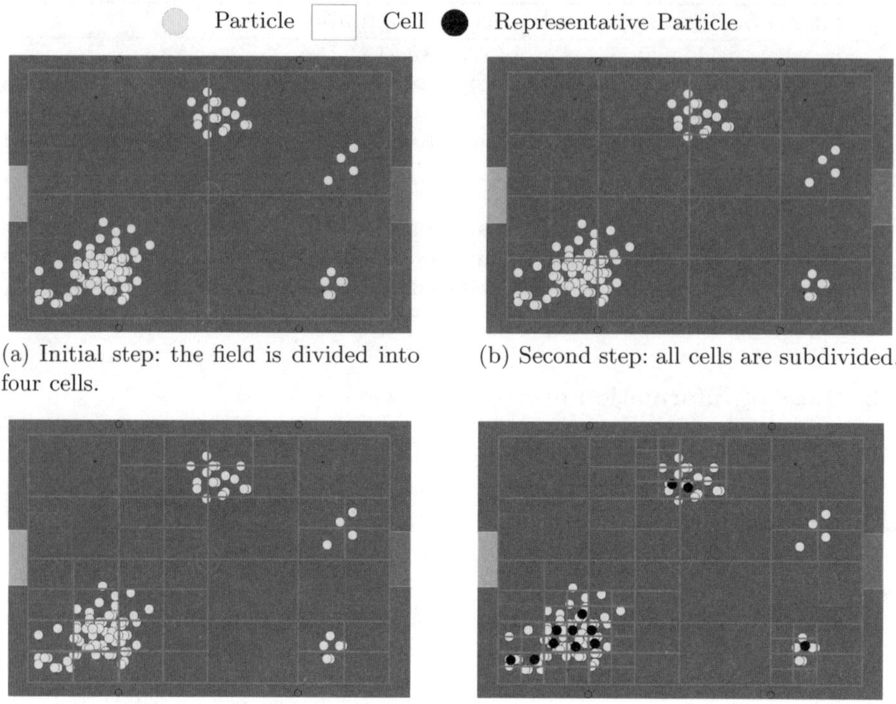

(a) Initial step: the field is divided into four cells.

(b) Second step: all cells are subdivided.

(c) Third step: cells which are empty or contain only a single particle are not further subdivided. The other are again subdivided.

(d) Last step: in the 12 cells containing the highest probability, a representative particle is calculated.

Fig. 2. Visualization of Algorithm 3 with an example

4 Experiments and Results

Finding a suitable reference system to compare against our proposed approach has not been an easy task, as global sensor fusion is still in its infancy on our development platform, and results from other leagues such as [4] cannot be directly compared, due to the vast difference of computational resources and sensor capabilities. A good candidate for our comparison has been found in [15], since this approach is adopted by 5 different teams on this hardware platform and its source code is publicly available.

4.1 Reference System

The approach described makes use of a Kalman Filter [7] to track the ball position and velocity in a robot-centric reference system. The robots in the team exchange their localization and local ball estimates, and the global team ball is calculated from the information provided by the robot with the highest confidence in its own localization.

4.2 Experimental Results

Our experiments have been performed by running in parallel on the same robots the reference system and our new solution, processing exactly the same data. The results of both systems are compared with the ground truth obtained from a ceiling camera global vision system as described in Section 2.2. In all of the following scenarios, the environment surrounding the soccer field is unstructured and unknown, and the robots might incorrectly identify false landmarks and false balls in it.

Scenario 1. In our first test scenario, 4 robots are placed on the field, without opponents or obstacles which might occlude their sight. Their vision systems are perfectly calibrated for the lighting conditions, and all the robots move freely on the field. The "observing robot" can never see the ball; the other 3 can, but they also have to periodically distract their attention from it in order to localize themselves. This test represents a "best case" scenario to evaluate the performance in a condition where the sensor information is relatively accurate and reliable. The results are shown in Figure 3(a).

Scenario 2. In this scenario, the conditions are similar to the first test, but one of the robots which can see the ball has a problem in its vision system, so that it consistently detects "ghost balls" inside the yellow goal. Such a problem is not infrequent during the competitions, and can severely penalize the performance of the whole team. The results are shown in Figure 3(b).

Scenario 3. This scenario is based on a real game situation. All robots are free to move and look at the ball, but the presence of opponents can occlude their sight to the ball and to the landmarks. Even worse, the opponents struggle against the observing team for getting control over the ball, compromising the localization state of the robots, as such collisions cannot be detected since the robots do not have any range or contact sensors. The results are shown in Figure 3(c).

Scenario 4. This test is similar to the previous scenario, but the frequency and entity of the collisions is greater. The results are shown in Figure 3(d).

4.3 Performance

Our goal was to be able to process all the sensory information at the native camera frame rate of 30Hz, this because of the limited field of view of the camera, which forces the robot to look around continuously at high angular velocities. The tracking system composed of the individual robot-centric tracker and the global tracker requires about $\approx$ 1ms to execute, being on par in terms of run-time with Kalman Filter based approaches, and many times faster than other Particle Filter based implementations. On average, each robot broadcasts 5 particles per frame, 12 bytes per particle, 30 times per second, for a total network traffic (team of 4 robots) of $\approx$ 56Kbit/s, while in the worst case, this value reaches

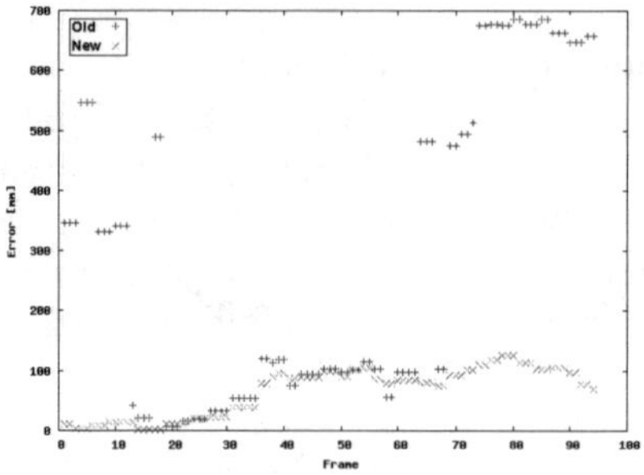

(a) Scenario 1: the x axis of the graph represents the temporal axis, where 30 frames = 1s. The old system performs particularly poorly after frame 70

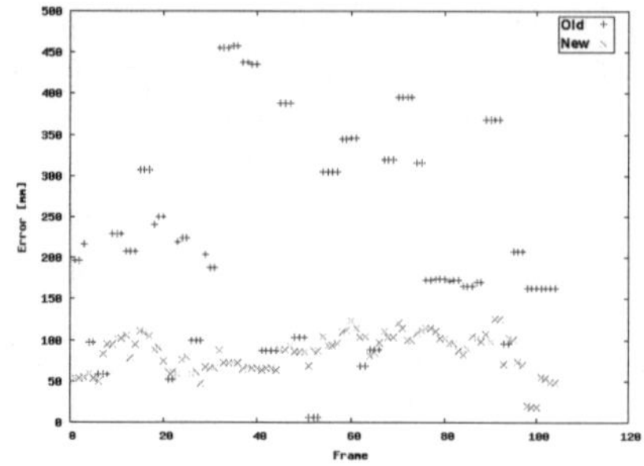

(b) Scenario 2: the old system cannot cope with the ghost balls, and is consistently outperformed

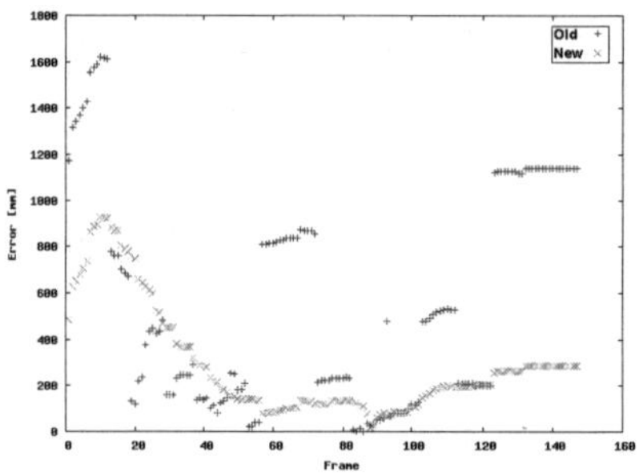

(c) Scenario 3: in the beginning of the test there is a strong collision. The new system performs constantly better.

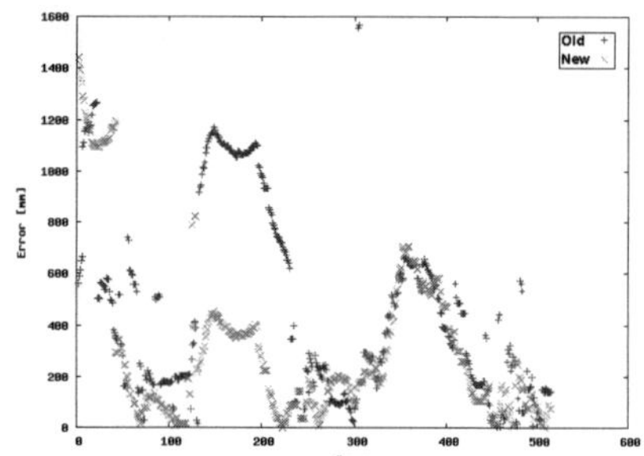

(d) Scenario 4: several collisions, around frame 150 the new system is 3 times more accurate than the reference

Fig. 3. Test scenarios

135Kbit/s; this is compatible with the competition constraints, which limit to 512Kbit/s the total bandwidth available to a team that has to be used also for other communication tasks such as role assignments and strategical data exchange.

5 Conclusion

It has been presented a Particle Filter based approach that tackles the problem of global sensor fusion in presence of high uncertainty concerning the robot positions, the data association about the tracked object, and the loss of information over the network. The system meets all the performance constraints set by the platform, and is competitive in terms of running time with simpler approaches which do not deal with all the aforementioned problems. In the future, we plan to investigate the possibility to make use of the speed information in the global tracker, by building a better sensor model with the help of our ceiling camera application and machine-learning techniques.

Acknowledgment

We would like to thank Microsoft MSDNAA for their support which made it possible for us to take part in the fascinating world of RoboCup.

References

1. Sony Aibo ERS-7: (2005), http://www.aibo.com
2. The RoboCup Federation: In: Bredenfeld, A., Jacoff, A., Noda, I., Takahashi, Y. (eds.) RoboCup 2005. LNCS (LNAI), vol. 4020, Springer, Heidelberg (2006), http://www.robocup.org
3. Kwok, C., Fox, D.: Map-based Multiple Model Tracking of a Moving Object. In: Nardi, D., Riedmiller, M., Sammut, C., Santos-Victor, J. (eds.) RoboCup 2004. LNCS (LNAI), vol. 3276, Springer, Heidelberg (2005)
4. Ferrein, A., Hermanns, L., Lakemeyer, G.: Comparing Sensor Fusion Techniques for Ball Position Estimation. In: Bredenfeld, A., Jacoff, A., Noda, I., Takahashi, Y. (eds.) RoboCup 2005. LNCS (LNAI), vol. 4020, Springer, Heidelberg (2006)
5. Fox, D., Hightower, J., Liao, L., Schulz, D., Borriello, G.: Bayesian Filtering for Location Estimation. PERVASIVE computing, 10–19 (2003)
6. Karol, A., Williams, M.A.: Distributed Sensor Fusion for Object Tracking. In: Bredenfeld, A., Jacoff, A., Noda, I., Takahashi, Y. (eds.) RoboCup 2005. LNCS (LNAI), vol. 4020, Springer, Heidelberg (2006)
7. Kalman, R.E.: A new approach to linear filtering and prediction problems. Transactions of the ASME - Journal of Basic Engineering 82, 35–45 (1960)
8. Fox, D.: Adapting the sample size in particle filters through kld-sampling. I. J. Robotic Res. 22(12), 985–1004 (2003)
9. Fox, D., Burgard, W., Dellaert, F., Thrun, S.: Monte carlo localization: Efficient position estimation for mobile robots. In: Proceedings of the Sixteenth National Conference on Artificial Intelligence (AAAI'99) (1999)
10. Nistico, W., Röfer, T.: Improving percept reliability in the Sony Four-Legged League. In: Bredenfeld, A., Jacoff, A., Noda, I., Takahashi, Y. (eds.) RoboCup 2005. LNCS (LNAI), vol. 4020, Springer, Heidelberg (2006)
11. Röfer, T., Jüngel, M.: Fast and robust edge-based localization in the sony four-legged robot league. In: Polani, D., Browning, B., Bonarini, A., Yoshida, K. (eds.) RoboCup 2003. LNCS (LNAI), vol. 3020, Springer, Heidelberg (2004)
12. Lenser, S., Veloso, M.: Sensor resetting localization for poorly modeled mobile robots. In: Proc. of the IEEE International Conference on Robotics and Automation (ICRA), IEEE Computer Society Press, Los Alamitos (2002)
13. Thrun, S., Fox, D., Burgard, W.: Monte carlo localization with mixture proposal distribution. In: Proc. of the National Conference on Artificial Intelligence, pp. 859–865 (2000)
14. Postel, J.: RFC 768 - User Datagram Protocol (1980), http://www.ietf.org/rfc/rfc768.txt
15. Röfer, T., et al.: German Team RoboCup 2004. Technical report (2004) Available online: http://www.germanteam.org/GT2004.pdf

Selective Visual Attention for Object Detection on a Legged Robot

Daniel Stronger and Peter Stone

Department of Computer Sciences, The University of Texas at Austin
{stronger,pstone}@cs.utexas.edu
http://www.cs.utexas.edu/~{stronger,pstone}

Abstract. Autonomous robots can use a variety of sensors, such as sonar, laser range finders, and bump sensors, to sense their environments. Visual information from an onboard camera can provide particularly rich sensor data. However, processing all the pixels in every image, even with simple operations, can be computationally taxing for robots equipped with cameras of reasonable resolution and frame rate. This paper presents a novel method for a legged robot equipped with a camera to use selective visual attention to efficiently recognize objects in its environment. The resulting attention-based approach is fully implemented and validated on an Aibo ERS-7. It effectively processes incoming images 50 times faster than a baseline approach, with no significant difference in the efficacy of its object detection.

1 Introduction

Processing a stream of visual images is an important but time-consuming task. One technique that has been used to speed up vision processing is that of selective visual attention [1]. This technique is based on the idea that not all areas in a given visual scene are relevant to the task at hand. Therefore by restricting one's attention to the relevant parts of the scene, the agent can greatly increase its visual processing speed. This intuition is corroborated by work in cognitive science confirming that human vision processing takes advantage of selective attention. For example, Sprague et al. [2] present a model of visual attention and compare it to human eye-tracking data.

In robotic vision, selective attention can take two main forms. One is gaze direction, in which a robot moves its camera so that its field of view is faced towards the important information [3,4,5]. That approach is analogous to human eye saccading, but does not address the question of how to process each image, an often time-consuming process.

The other main approach to selective attention, which is taken in this paper, involves only processing the areas of the image that are likely to have relevant features. Because of the large amount of data in every image, processing each image in its entirety is difficult to do at frame rate, and where it is possible, it severely limits the amount of time the robot can spend processing the interesting areas of the image. By restricting its attention to the parts of the image that

G. Lakemeyer et al. (Eds.): RoboCup 2006, LNAI 4434, pp. 158–170, 2007.

are most likely to contain the important information, the robot can dramatically speed up its image processing. This approach raises the challenge of identifying the useful areas of the image.

One common way to find the useful areas of the image is to first compute a saliency map [6], which represents the conspicuity at each point in the image. The most salient regions of the image can then be processed in detail. This approach has been applied to tasks such as face and handwritten digit recognition [7] and recognizing an object in a cluttered visual field [8]. Unfortunately, constructing the saliency map still requires processing the entire image, a time-consuming task in the context of trying to process a video stream at frame rate.

Another method for focusing attention on the important parts of the image is feature tracking [9,10]. For example, Shi and Tomasi [9] present a method for identifying the optimal features in the image for tracking, i.e. the ones that are most likely to correspond to points in the real world. In feature tracking, a specific area of the image is analyzed between two consecutive images to characterize the movement of the feature. This method requires that corresponding points in consecutive images be close to each other, so that the tracking mechanism can properly identify the differences between consecutive frames as *movement*. However, there are sometimes cases in which objects in the field of view can move across a large portion of the image between consecutive frames. For example, in the case of a legged robot, the jerky motion caused by walking can lead to very sharp motion in the image. In these cases, feature tracking is not applicable.

This paper considers the situation of an autonomous legged robot equipped with a camera that moves within an environment with fixed landmarks. The robot's goal is to visually detect the landmarks as efficiently as possible. In this context, we present a novel technique for applying selective visual attention to the task of object detection. Like feature tracking, the technique is based on the idea that the robot can use prior information to predict the *expected location* in the image of each object. However, unlike previous work, it does not assume that an object's expected location in one frame is necessarily close to its location in the previous frame. The robot uses the objects' expected locations to direct its visual search towards the most likely areas of the image, enabling it to perform object detection very efficiently, despite the sharp motion caused by walking.

This approach presents three main challenges. First, predicting the location of an object in the image requires having an accurate estimate of the camera's *pose*, its position and orientation in space. On a legged robot, meeting this requirement is particularly challenging because as the robot walks, its body, and thus also its camera, rock quickly from side to side. Second, when preliminary analysis suggests that the object is not present at the expected location, the robot must have a strategy for continuing its search of the image for the target object. Third, the fact that most of the image is not processed at all presents a new challenge for object recognition: When processing a few pixels suggests the presence of an object, the robot must decide which pixels to process next to complete the object detection.

Although each image on our test platform contains over 33000 pixels, our technique allows the robot to examine fewer than 1200 of them on average as it identifies landmarks. This reduced processing enables the robot to process images 50 times faster than a baseline approach. Nonetheless, there is no significant drop in the success rate of object detection. The technique is implemented and validated on a popular mobile robot platform, the Sony Aibo ERS-7.

The remainder of this paper is organized as follows. The following section presents an overview of our technique. Sections 3-5 present solutions to the three challenges raised by this approach. Section 6 presents experimental results and Section 7 concludes and discusses future work.

2 Overview

In this paper we consider the task of a vision-based autonomous robot operating in a known environment. The robot has access to a series of images that are generated by a camera located on board the moving robot, which arrive at video frame rate. We assume that there is a fixed set of *objects* relevant to the robot's decision making (e.g., landmarks). The goal of the robot's vision module is to identify these objects when they are present in the image. The robot also maintains an estimate of its own pose in the environment over time, based on its visual observations and its odometry estimate. One popular approach to this self-localization problem, which we use in the experiments reported in this paper, is Monte-Carlo localization, or particle filtering [11,12].

At each time step, the robot estimates its camera's pose. The details of how this is accomplished while the robot is walking are presented in Section 3. Then, given the camera's pose, the robot can loop through the fixed objects in the environment and, for each one, predict whether or not and where it is expected to appear in the robot's field of view. This prediction is achieved by projecting the object location onto the image plane (correcting for distortion if necessary).

For each object, if it is expected to be behind the image plane and therefore invisible to the robot, it is discarded. Similarly, if the object's projection onto the image plane is outside the field of view of the camera, it is discarded. Otherwise, the resulting image location is considered the object's expected location in the captured visual frame. If the expected location is examined and the object is not found, then the robot continues to search for it throughout the remainder of the image. This process is described in Section 4. Notably, a key challenge compared to previous approaches is that an object's location in the image plane may not be close to its location in the previous frame.

When preliminary analysis of an image location suggests that the object is present there, the robot must analyze that region of the image in detail. The goal of this processing is to accurately and efficiently determine whether or not the object is present in that location, and if so, how large it is in the image plane. A solution to this problem in our test-bed domain is presented in Section 5.

Finally, once the objects in the image have been identified, they can be used as landmarks for the purposes of localization. This process consists of converting the size and location of objects in the image into the corresponding distances and angles from the robot.

The entire method is summarized in Algorithm 1. The variable *Obj* loops through all of the environmental landmarks. Each one is projected onto the image plane if possible, initializing *TestLocation* to be the expected location of the object. This location is advanced by the routine *SeededSearch*, specified in Section 4. If the object is found, it is then used to inform localization. The underlined procedures in the algorithm will be described in the following sections.

Algorithm 1. Algorithm Summary

```
ComputeCameraPose
for all (objects Obj in the environment) do
  transform location of Obj into camera reference frame
  if (Obj is in front of the image plane) then
    project Obj onto ExpectedLocation in image plane
    if (Obj is in the camera's field of view) then
      TestLocation ← ExpectedLocation
      repeat
        Examine TestLocation for match with Obj
        advance TestLocation according to SeededSearch
      until (Obj is found) OR (entire image searched)
      if (Obj is found) then
        determine extent of Obj in the image
        project Obj back into global reference frame
        compute distance and angle from Obj
        incorporate information in localization
      end if
    end if
  end if
end for
```

3 Computing the Camera Pose

In order to accurately predict the location of the objects in the image, the robot needs to have an accurate estimate of its camera's pose. On a legged robot, this is particularly difficult because of the jagged motion caused by walking. The details of the method for finding the camera pose are necessarily dependent on the configuration and sensors of the specific robot being used. Nevertheless, the principles used here can be extended to apply to any robotic platform.

The experiments reported in this paper were performed on a Sony Aibo ERS-7.[1] The robot is roughly $280mm$ tall and $320mm$ long. It has 20 degrees of freedom: three in each of four legs, three in the neck, and five more in its ears, mouth, and tail. At the tip of its nose there is a CMOS color camera that captures images at 30 frames per second in YCbCr format. The images are 208×160 pixels giving the robot a field of view of 56.9° horizontally and 45.2° vertically. The robot's processing is performed entirely on-board on a 576 MHz processor.

Preliminary experiments have shown that an object's location in the robot's field of view can change by as much as 80 pixels between two consecutive video frames. This distance is a large fraction of the 208-pixel width of the image. The

[1] http://www.aibo.com

fact that a fixed point in the robot's view can move so far in one thirtieth of a second demonstrates the instability of the camera's pose. A video depicting the world from the robot's point of view is available online.[2]

The camera's pose can be estimated from the robot's joint angles and accelerometer values. This computation occurs in three phases. First, the height of the robot's body is estimated based on the back legs' joint angles. Second, the body's tilt and roll are estimated based on the accelerometer values. Finally, the head and neck angles are used to complete the computation.

Ideally, the pose of the camera with respect to the ground plane could be computed by multiplying a series of homogeneous transformation matrices based only on the robot's joint configuration and angles [13]:

$$T_{foot}^{cam} = T_{foot}^{hip} \cdot T_{hip}^{body} \cdot T_{body}^{neck} \cdot T_{neck}^{cam} \tag{1}$$

where T_B^A represents the transformation from coordinate system A to coordinate system B. However, as the robot walks, the coordinate system of any given foot is not constrained to be either parallel to the ground or in contact with the ground. To compensate for this problem, we separately estimate the height of one of the rear hips and use the robot's internal accelerometers to estimate the body's tilt and roll. The resulting equation is

$$T_{ground}^{cam} = T_{ground}^{hip} \cdot T_{hip}^{body} \cdot T_{body}^{neck} \cdot T_{neck}^{cam} \tag{2}$$

In this equation, the *hip* coordinate frame is parallel to the ground. Then T_{ground}^{hip} is only a vertical translation, whose magnitude is determined by the rear legs' joint angles (based on the assumption that the more outstretched leg is the one touching the ground). The *body* coordinate frame is attached to the robot's body, so that the transformation T_{hip}^{body} must account for the body's tilt and roll. These quantities are estimated by using the robot's accelerometers. The three accelerometers report the component of gravity (combined with the body's acceleration) in the direction of each of the three cardinal axes in the body's reference frame. By taking the average of a rolling window of accelerometer values for each direction, the robot is able to filter out the effects of noise and acceleration and isolate the body's tilt and roll. The final transforms from the body to the neck and camera are done via standard DH-transforms, as described by Schilling [13]. The full details of our implementation are presented in our technical report [14].

Once the camera pose has been determined, it can be combined with the robot's prior body pose estimate in its environment to compute expected locations for objects in the robot's field of view. These locations can then be used to seed the visual search for those objects.

4 Seeded Visual Search

After the camera pose has been used to compute an object's expected location in the image, the robot must search for the object, as per Algorithm 1. This section

[2] http://www.cs.utexas.edu/~AustinVilla/?p=research/selective-vision/

describes our solution to the seeded visual search problem. That is, given that the robot has an expected location for an object in the image, how can the robot best take advantage of this knowledge to find the object as quickly as possible? We assume that, for each object, the robot has an *object decision mechanism* that can start at any pixel and eventually determine whether or not that pixel is part of the object in question. Ideally, the mechanism should usually reject pixels that are not part of the object after only some quick preliminary processing. For example, in a color-coded domain, the object decision mechanism can *segment* the pixel in question into a color category, and only for pixels that are the correct color, examine the surrounding region in more detail. In other settings the preliminary processing may require examining multiple pixels. The object decision mechanisms we use in our test-bed domain are described in Section 5.

Once the robot has identified an expected location for a given object, it first applies the object decision mechanism starting at the expected location. If the mechanism fails to find the object in its first application, the robot continues by applying the decision mechanism repeatedly, starting at a different pixel each time. If the mechanism starts at a pixel and successfully identifies the target object, the search can terminate. Otherwise, the searching process should continue starting the decision mechanism at different points, gradually moving outwards from the original expected location. In the worst case, the search could expand to cover the entire image. But in practice this rarely happens as will be demonstrated in Section 6.

To accomplish this goal, we use a series of starting points that follows a square spiraling pattern that expands outwards. The pattern used is depicted in Figure 1, and the points examined are all on a square lattice. The distance between lattice points depends on the minimum possible size of the target object. This property allows the robot to process as little of the image as possible while still ensuring that if the object is in the image it will be found. The spiral expands until either the target object is found or the entire image is filled. Because the image is filled in the worst case, the method is able to recover from a completely incorrect prior localization estimate.

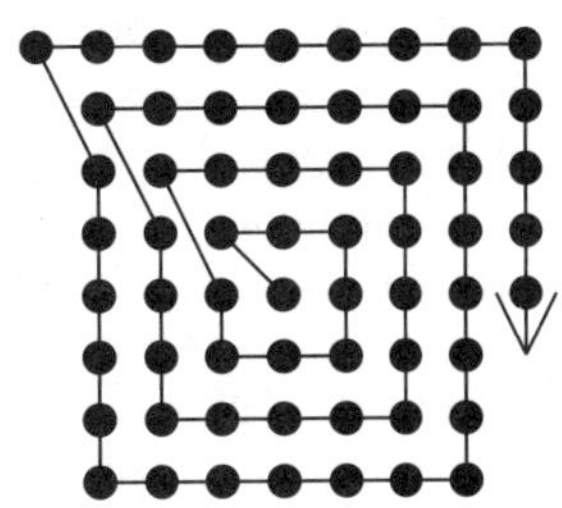

Fig. 1. The circles represent the locations in the image that are analyzed, starting from the expected object center. The distance between adjacent circles depends on the size of the target object.

5 Object Detection

The previous section describes how to process the image in a manner that enables the target object to be found quickly, given an object decision mechanism. This section presents our object decision mechanism, completing the description of the attention-based approach to object detection presented in this paper. First, however, it is necessary to specify the types of objects that are detected in the experiments reported in this paper.

The Aibo's environment used in this paper is a legged-league field measuring 4.4 meters by 2.9 meters. It has one blue goal, one yellow goal, and four visually distinct cylindrical beacons each with two colors: one is pink, and the other is blue or yellow. The robot is faced with the task of accurately identifying the goal and beacons that appear in the image.

Since the RoboCup field is color-coded, the object decision mechanism discussed in Section 4 analyzes pixels in the image by segmenting them and observing whether or not they are the color corresponding to the desired object: blue or yellow for a goal, or pink for a beacon. Color segmentation is carried out via a *color table*, a three-dimensional array with a color label for each possible combination of Y, Cb, and Cr values. The color table is created off-line, by manually labeling a large suite of training data and using a Nearest Neighbor learning algorithm to learn the best label for each YCbCr combination. The full details of our color segmenting algorithm are presented in our technical report [14]. After segmentation, a goal appears to the robot as a blue or yellow rectangle and a beacon appears as a pink square above or below a blue or yellow square of the same size. Whether the pink is above or below, and whether the other color is blue or yellow, uniquely determine which of the four beacons it is.

For each object, a single point is projected onto the image plane, as specified in Section 3, to yield an expected location. For goals, the center of the rectangle is used, while for beacons the center of the pink square is used. After the pink square is identified, the rest of the the beacon can be found easily.

As discussed in Section 4, an object decision mechanism is needed that examines the image starting at a given pixel and decides whether or not that pixel is part of the target object. The mechanism begins by segmenting the given pixel. If it is not the color corresponding to the target object, the mechanism rejects the pixel. Otherwise, the point is *expanded* into a maximal approximate rectangle of pixels all the same color. Since it starts with only one pixel segmented, it segments further pixels as they are needed.

To expand a point into a rectangle of a given color, we first expand the point into a line, by expanding to the right and then to the left. Expanding to the right consists of segmenting consecutive pixels, each one to the right of the previous one. This process continues until a sufficiently long string of consecutive pixels are all not the given color, marking the right end of the line. The threshold used for consecutive wrong-colored pixels, denoted as $ConsecThresh$, is three.

The pixel is expanded to the right and left, and then the process is repeated on the pixel directly above. This process continues upward until a pixel is found that cannot expand in either direction, which we consider to signify the top of the rectangle. Similarly the robot proceeds downward from the root pixel until the bottom of the rectangle is found. The rightmost right edge of an expansion line, and the leftmost left edge, are taken to be the left and right edges of the rectangle. Pseudocode for this point expansion routine is given in Algorithm 2. $BaseX$ and $BaseY$ are the image coordinates of the starting point for the expansion routine. The returned values represent the boundary coordinates of the rectangle.

Once a rectangle of the appropriate color has been found, the object decision mechanism decides whether or not the target object has been found, as opposed to some spurious pixels of the target color. For goals, the blue or yellow rectangle found is assumed to be the goal, as long as it is sufficiently large. For beacons, depending on the beacon's identity, the robot expects a blue or yellow square, either directly above or below the pink square that has been found. The location of the expected square center is based on the size and location of the pink square that has already been found. This new center is expanded into a blue or yellow rectangle in accordance with Algorithm 2. If this operation is successful and the resulting combined beacon rectangle is sufficiently large, the object decision mechanism registers a success. The resulting goal and beacon rectangles are the output of the vision processing.

Algorithm 2. Color Expanding Routine

```
Given: ConsecThresh, ColorTable, DesiredColor
Given: BaseX, BaseY
Given: ColorTable[Pixel[BaseX, BaseY]] = DesiredColor
x ← BaseX, y ← BaseY
repeat
  y goes up, then down, from BaseY, one pixel at a time
  repeat
    x goes right, then left, from BaseX, one pixel at a time
    Compare ColorTable[Pixel[x,y]] to DesiredColor
    if equal then
      ConsecutiveMisses ← 0
    else
      ConsecutiveMisses ← ConsecutiveMisses +1
    end if
  until on each side, ConsecutiveMisses reaches ConsecThresh OR x reaches image edge
until on top and bottom, y reaches a row where no DesiredColor is found or the image edge
MinX, MaxX, MinY, MaxY ← lowest and highest values of x and y, respectively to segment
to DesiredColor
Return: MinX, MaxX, MinY, MaxY
```

Although the description in this section of the object detection mechanism is specific to our test-bed environment, the overall algorithm (presented in Section 2) does not depend on the details of the object detection mechanism. As long as the robot is equipped with a local method for determining whether a pixel or set of pixels depict an object of interest, that object decision mechanism can be plugged into Algorithm 1.

6 Experimental Validation

The goal of the method presented in this paper is to obtain a high object-detection accuracy without much computational complexity. As such, the mark of success is to perform roughly as well at object detection as a state-of-the-art approach that processes the entire image. In this section, we evaluate the technique presented in this paper by comparing it to a common baseline approach [14,15,16]. Section 6.1 describes our implementation of this baseline approach, and Section 6.2 presents experiments comparing the two methods.

6.1 Baseline Method

Our group has previously solved the problem of object detection on the Aibo ERS-7 on the RoboCup field using the baseline method mentioned above. However, this solution involves processing all of the pixels in every image, and despite the fact that we have optimized it aggressively, it consumes almost all of the robot's available processing time. The methods used are summarized in this section, while the full details of this baseline approach can be found in our technical report [14]. The robot executes the following three steps.

1. Color Segmentation: Classify every image pixel as one of a small set of distinct colors.
2. Region Merging: Collect adjacent pixels of the same color into monochromatic regions.
3. Object Detection: Identify the monochromatic regions that correspond to specific objects in the environment.

During the robot's image processing, it loops through every pixel in each image, classifying it according to the color table. Note that each image measures 208×160 pixels, for over 33000 pixels total.

As each pixel is segmented, it is also incorporated into a run-length encoding of the image. That is, each maximal horizontal string of consecutive pixels that are the same color is stored as a run-length. These run-lengths comprise a highly compressed version of the segmented image. Next, vertically adjacent run-lengths of the same color are combined into a *bounding box*, a rectangular structure consisting of the rectangle's coordinates and the color inside. The robot continues to merge bounding boxes that are adjacent and of the same color. Heuristics are used to determine if some adjacent boxes should not be merged, and also if some boxes should be deleted because they contain a density of the desired color that is too low. These bounding boxes are the input to object detection.

The robot first attempts to detect goals in the image, because they are generally the largest objects found. Thus the blue and yellow bounding boxes are sorted from largest to smallest, and tested for being the goal in that order. The goal tests consist of heuristics to ensure that the bounding box in question is close enough to the right height to width ratio, and that there is a high enough density of the desired color inside the bounding box for a goal, as well as other heuristics. To find the beacons, the robot finds bounding boxes of pink and blue or yellow that satisfy appropriate beacon heuristics, and then combines them into beacon bounding boxes, labeled by which beacon is inside. The resulting goal and beacon bounding boxes are the output of the vision process.

Once goals and beacons have been identified in the image, the robot uses them to update its localization estimate. This update takes place in two stages: translating the bounding box information into object distances and angles and incorporating these distances and angles into the robot's pose estimate.

The robot first computes its distance and horizontal angle to any objects it has identified. We have found it to be more effective to use the left and right edges of each goal instead of goal centers as landmarks for localization.

To compute its distances and angles to the landmarks in the image, the robot takes into account the camera pose and the location and size of the object in the image. The distances and angles yielded by this process are often quite noisy, and it is rare to see enough objects in one frame to triangulate one's pose uniquely. To alleviate these factors, the robot maintains an estimate of its pose through particle filtering [11,12], which gathers the robot's visual information and odometry into a coherent pose estimate over time.

The details in this section summarize our own baseline implementation for object detection that has been used successfully for a couple of years. Though there are many other object detection implementations within this domain that differ in their details [15,16,17], none of the approaches that we know of on the legged robot process the image based on the projected locations of objects.

6.2 Results

The baseline and attention-based methods were evaluated and compared based on their rate of success identifying landmarks in the image. In order to measure these success rates, we had the robot save a series of representative images to its memory stick with the identified landmarks marked. Then the images were viewed on a monitor and the objects were labeled manually.

In order to generate a series of representative images, the robot walked to a sequence of fixed poses on the field while continually moving its head from side to side, stopping at each pose for 15 seconds. The points and their order are shown in Figure 2. They were specifically chosen to represent a wide range of difficulty for object detection. Every ten seconds, the robot saved an image for evaluation. This ensured that a representative and varied sample of images was used for the evaluation.

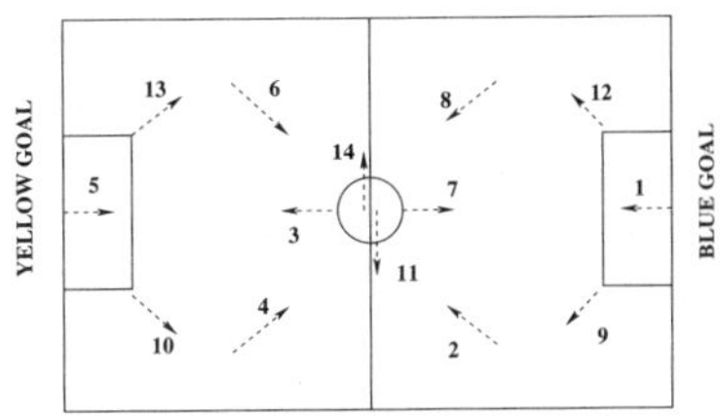

Fig. 2. The points on the field visited by the Aibo

Both methods were evaluated for ten trials. In each trial, the robot walked between the poses until it had taken 50 images total. The robot's rate of success within each trial was recorded. This rate is defined as the number of correctly identified landmarks (goal edges and beacons) divided by the number of landmarks identified by the manual labeling. This quantity represents the fraction of the objects that were actually in the image that the object detection successfully identified. We also counted the number of false positives: instances in which the robot reported a landmark that was not actually there. The false positive rate was extremely low for both methods. There were none at all recorded with the baseline method and two with the attention-based method over all ten trials. Compared to the total number of objects in the images captured while testing the attention-based method, 597, these errors represent a false positive rate of only 0.34 percent.

Furthermore, we recorded the amount of time taken to process each image. That is, for each image frame the robot receives, the amount of time taken to

Table 1. Comparison of the baseline and new, prediction-based methods

Method	Baseline	New Method
Avg Detection Rate	**77.54 ± 7.32**	**74.33 ± 10.63**
Avg Pixel Count	**33280**	**1138**
Avg Time/Frame	**35.049** ms	**0.695** ms

identify the vision objects in the image is recorded. These times do not include the time taken for particle filtering, nor for behavior and motion processing, which also occur in real time on-board the robot. The vision processing times are averaged over the full extent of all ten trials. We also measured the average number of pixels examined by the attention-based method on a typical run. This is compared to the number of pixels examined by the baseline method, which is the total number of pixels in the image. The average processing times, accuracies, and numbers of pixels processed attained by the baseline method and the new method are shown in Table 1.

The prediction-based method took an average of 0.695 ms to process each image, compared to the baseline method taking 35.049 ms on average. This represents a speed up of a factor of 50.4. To process at frame rate, there is 33 ms of computation time available per frame. Using the baseline method, vision processing takes 35 ms of computation, so that there is no time for additional processing while continuing to operate at roughly 30 Hz. In fact, even with minimal computation performed for localization and decision-making, the baseline process leads to an ability to process at only 24 Hz. Thus the robot completely ignores information from 20% of the available frames. In contrast, when using our method, the vision processing of stationary objects takes only 0.7 ms, leaving more than enough time to perform localization and decision-making while operating at 30Hz, and in fact freeing up much additional processing time. This time could potentially be used for precise tracking of mobile objects, as well as enabling the exploration of completely new approaches that would not have been tractable previously, perhaps such as detailed behavior planning or teammate/opponent modeling.

At the same time, the new method achieved a very similar quality of vision accuracy to the baseline method. Over the ten trials, the attention-based method attained an average detection rate of 77.54 ± 7.32 percent, while the baseline method attained an average rate of 74.33 ± 10.63. Based on the variances in these measurements, the difference in their means is statistically insignificant ($p > 0.4$ in a two-tailed t-test). These tests demonstrate that the attention-based method sped up the robot's vision processing by a factor of 50, without any significant effect on the success rate of object detection.

7 Conclusion and Future Work

This paper presents a technique for a legged robot equipped with a camera to use visual selective attention to efficiently recognize objects in its environment.

Given the inapplicability to this domain of previous techniques, a novel approach is presented based on taking into account the robot's prior knowledge about its camera's pose. This approach presents three general challenges, which are particularly difficult in the context of a legged robot. The paper presents solutions to each of those challenges in that context.

When these solutions are combined, the resulting technique enables a legged robot to process its visual data stream very efficiently. We have fully implemented and validated this technique on an Aibo ERS-7. This attention-based approach effectively processes incoming images 50 times faster than a baseline approach, with no significant difference in the efficacy of its object detection.

This work has a number of interesting possibilities for future extensions. One enhancement would be to extend the method so that it could also find mobile objects. Achieving this goal would require having the ability to predict the objects' locations in advance. Such an ability could derive from a predictive model of the objects' locations based on their velocities, communication with other robots, or a combination thereof. Another possibility for future work would be for the robot to draw cues from objects that have already been processed in a given image to decide which regions of the same image will be most likely to contain other objects. Furthermore, these methods could be extended so that they influence gaze direction as well as the focus of attention within each image. Finally, this technique is designed to be generally applicable and can therefore be applied in different domains to ascertain the range of possible environments and types of objects with which it can be effective.

Using the camera's pose to speed up vision is an example of enhancing vision processing by incorporating high-level information. This general idea is promising, based on the notion that in order to effectively perceive its environment, an agent should take advantage of as much prior knowledge as possible about that environment. This work represents progress towards the long-term goal of enabling autonomous agents to make effective use of their knowledge of the world in their perceptual processing.

Acknowledgements

This research was supported in part by NSF CAREER award IIS-0237699, ONR YIP award N00014-04-1-0545, and DARPA grant HR0011-04-1-0035. The authors thank Mohan Sridharan and Aniket Murarka for helpful suggestions.

References

1. Yarbus, A.L.: Eye movements during perception of complex objects. In: Riggs, L.A. (ed.) Eye movements and vision, ch. VII, pp. 171–196. Plenum Press, New York (1967)
2. Sprague, N., Ballard, D., Robinson, A.: Modeling attention with embodied visual behaviors (2005), `http://www.cs.rochester.edu/`*dana/WalterTheory25.pdf*

3. Mitsunaga, N., Asada, M.: Sensor space segmentation for visual attention control of a mobile robot based on information criterion. In: Proceedings of the IEEE International Conference on Intellegent Robots and Systems, IEEE Computer Society Press, Los Alamitos (2001)
4. Kwok, C., Fox, D.: Reinforcement learning for sensing strategies. In: Proceedings of the IEEE International Conference on Intelligent Robots and Systems, IEEE Computer Society Press, Los Alamitos (2004)
5. Najemnik, J., Geisler, W.: Optimal eye movement strategies in visual search. Nature 434, 387–391 (2005)
6. Itti, L., Koch, C., Niebur, E.: A model of saliency-based visual attention for rapid scene analysis. IEEE Transactions on Pattern Analysis and Machine Intelligence 20(11), 1254–1259 (1998)
7. Salah, A.A., Alpaydin, E., Akarun, L.: A selective attention-based method for visual pat- tern recognition with application to handwritten digit recognition and face recognition. IEEE Transactions on Pattern Analysis and Machine Intelligence 24(3) (March 2002)
8. Walther, D., Rutishauser, U., Koch, C., Perona, P.: On the usefulness of attention for object recognition. In: The 2nd Workshop on Attention and Performance in Computer Vision (2004)
9. Shi, J., Tomasi, C.: Good features to track. In: IEEE Conference on Computer Vision and Pattern Recognition, IEEE Computer Society Press, Los Alamitos (1994)
10. Baluja, S., Pomerleau, D.: Expectation-based selective attention for visual monitoring and control of a robot vehicle. Robotics and Autonomous Systems 22(3–4) (1997)
11. Dellaert, F., Fox, D., Burgard, W., Thrun, S.: Monte carlo localization for mobile robots. In: Proceedings of the IEEE International Conference on Robotics and Automation (ICRA), IEEE Computer Society Press, Los Alamitos (1999)
12. Sridharan, M., Kuhlmann, G., Stone, P.: Practical vision-based monte carlo localization on a legged robot. In: IEEE International Conference on Robotics and Automation, IEEE Computer Society Press, Los Alamitos (April 2005)
13. Schilling, R.: Fundamentals of Robotics: Analysis and Control. Prentice-Hall, Englewood Cliffs (2000)
14. Stone, P., Dresner, K., Fidelman, P., Jong, N.K., Kohl, N., Kuhlmann, G., Sridharan, M., Stronger, D.: The UT Austin Villa 2004 RoboCup four-legged team: Coming of age. The University of Texas at Austin, Department of Computer Sciences, AI Laboratory, Tech. Rep. UT-AI-TR-04-313 (October 2004)
15. Bunting, J., Chalup, S., Freeston, M., McMahan, W., Middleton, R., Murch, C., Quinlan, M., Seysener, C., Shanks, G.: Return of the NUbots! the 2003 NUbots team report (2003), `http://robots.newcastle.edu.au/publications/NUbotFinalReport2003.pdf`
16. Mitsunaga, N., Toichi, H., Izumi, T., Asada, M.: Babytigers 2003: Osaka legged robot team (2003), `http://www.er.ams.eng.osaka-u.ac.jp/robocup/BabyTigers/BabyTigers-TechR eport-2003.pdf`
17. Roefer, T., et al.: German team: Robocup 2004 (2004), `http://www.germanteam.org/GT2004.pdf`

Towards Probabilistic Shape Vision in RoboCup: A Practical Approach

Sven Olufs, Florian Adolf, Ronny Hartanto, and Paul Plöger

Department of Computer Science, Bonn-Rhein-Sieg University of Applied Sciences, D-53757 St. Augustin, Germany
sven@olufs.com, paul.ploeger@fh-brs.de

Abstract. This paper presents a robust object tracking method using a sparse shape-based object model. Our approach consists of three ingredients namely shapes, a motion model and a sparse (non-binary) subsampling of colours in background and foreground parts based on the shape assumption. The tracking itself is inspired by the idea of having a short-term and a long-term memory. A lost object is "missed" by the long-term memory when it is no longer recognized by the short-term memory. Moreover, the long-term memory allows to re-detect vanished objects and using their new position as a new initial position for object tracking. The short-term memory is implemented with a new Monte Carlo variant which provides a heuristic to cope with the loss-of-diversity problem. It enables simultaneous tracking of multiple (visually) identical objects. The long-term memory is implemented with a Bayesian Multiple Hypothesis filter. We demonstrate the robustness of our approach with respect to object occlusions and non-Gaussian/non-linear movements of the tracked object. We also show that tracking can be significantly improved by using compensating ego-motion. Our approach is very scalable since one can tune the parameters for a trade-off between precision and computational time.

1 Introduction

The ability of knowing "where is what?" seems to be easy for humans while it is a cognitive challenge for a machine. The knowledge about *object tracks* (e.g., position, movement) is one of the key capabilities for a high level behaviour. In a typical RoboCup MSL scenario this means that we want to detect and track static and dynamic objects, e.g. a ball, goals, corner-posts, other robots or humans. Additionally we must cope with the object dynamics and a rapidly changing environment.

We can distinguish two major components in a typical visual tracker: (1) *Target Representation and Localisation* and (2) *Filtering and Data Association.* (1) is mostly a bottom-up process which has to cope with changes in the appearance of the target, like for example Mean-Shift [5,3], Particle filtering [10,11] or iterative error-minimising [8] approaches. Those kinds of approaches typically keep the position of the object of interest in an image sequence. (2) is mostly

G. Lakemeyer et al. (Eds.): RoboCup 2006, LNAI 4434, pp. 171–182, 2007.

a top-down process dealing with the *dynamics* of the tracked object, e.g. like Probabilistic Data Association filters (PDAF) [1], Sample-Based Joint Probabilistic Data Association Filters (SJPDAF) [15] or Multiple Hypothesis Tracking (MHT) [14,6] approaches. Those kind of filters are typically dealing with an abstract (anonymous) observation and solve the *tracking and data association problem* [1].

Solely *bottom-up* approaches [5,3,10,11,8] provide a limited robustness with respect to high dynamics (e.g. a bouncing ball), occlusions, visually hidden or disappearing gone objects. Usually, it assumes that the initial position of the object of interest is known a-priori and no visually identical objects appear in the scene. In fact, this assumption does not hold true for a typical RoboCup scenario where we have two goals and multiple players on the field. Recovering lost objects and tracking of multiple objects are typically the domains of *top-down* approaches using (primitive) classifiers. The idea of using a binary homogeneity criterion (e.g. binary colour segmentation) as the primitive classifier is still popular within the RoboCup community [9], though it is bound to the RoboCup soccer domain and not directly applicable to real world applications.

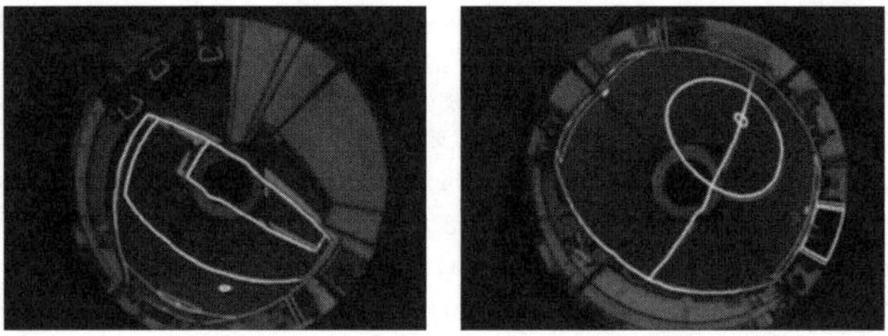

(a) (b)

Fig. 1. Contours of the detected and tracked objects and the projected believe pose of the robot. The tracker has no knowledge about the initial object poses.

Our new probabilistic approach is a hybrid solution of both concepts using a Bayesian *dynamic state space* formulation, see [7] for theoretical issues. Note that we are not interested in solving the *data association* problem. Basically our approach is inspired by the idea of a *short-term* and a *long-term memory*. The objective of the short term-memory is to track objects through image sequences. An observation is immediately lost if the object is no longer detected in the image. The objective of the long-term memory is to maintain tracks of the recognised object coming from the short-term memory. Its role is to re-detect an object if it is no longer detected by the short-term memory. The component gives a feedback to the short term memory in case of a "missing" object. Technically the long-term memory is implemented using a Bayesian *Multiple Hypothesis Tracker*. The short-term memory is implemented by using a new extended method of Bayesian Monte Carlo filters, namely the *extended Particle Filter* method. This extension allows obtaining multiple independent states estimation. We show that our tracking approach is more robust for fast moving objects than exclusively used (standard) approaches. We also show that ego motion compensation can improve the performance and robustness of the tracker considerably. In order to be applicable to the RoboCup domain we assume all motions of robots and objects are non-Gaussian/non-linear. Furthermore we assume that any initial position of the object of interest is a-priori unknown and that visibility is not known. Finally we assume that multiple visually identical objects can appear in the image.

The current implementation of our approach was tested extensively for coloured objects in various RoboCup soccer competitions as well as on real-world soccer playgrounds. Furthermore our method was successfully applied to other tasks like mobile vehicle surveillance and multi-target tracking of real-world objects in cluttered scenes.

The paper is organised as follows. Section 2 introduces the probabilistic *object model* used by the short-term memory. In section 3 a practical approach is presented. Experimental results are shown in Section 4. Finally we discuss the results and the approach in Section 5.

2 Object Model

The **pose** of a tracked object is defined as $\Phi(x\ y\ \theta\ \lambda\ \dot{x}\ \dot{y})$ where x, y is the position, θ is the relative orientation, λ is a scale factor of the observed object in the image and $\dot{x}, \dot{y}$ denotes the relative motion of the object of interest in one frame to the subsequent frame. Let q be an a-priori known target model of the object of interest. Let $\text{CONTOUR}(\Phi)$ be a function that is generated by a 3D projection of a known 3D shape of the object a contour model for all Φ. Our implementation

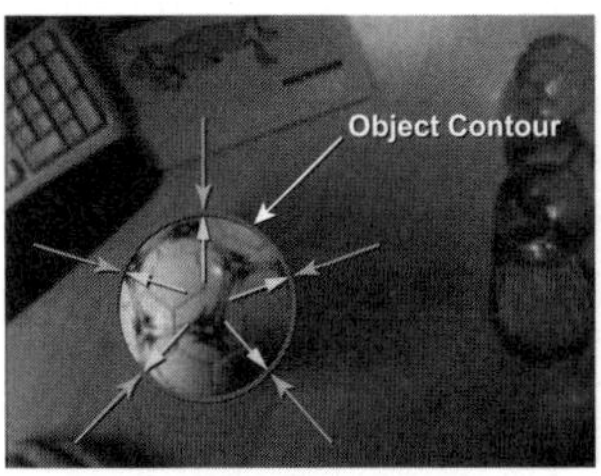

(a) inner and outer force

(b) areas

Fig. 2. Principle of our tracking approach

defines this function for ball, goals, corner-posts and obstacles. The main idea of our approach is to keep the believed contour of an object of interest as close as possible to its background, instead of keeping it only close to the object's contour alone. This is illustrated in fig. 2(a), the arrows head for the two virtual forces that "push" the contour of the object to its correct position. The outer arrows indicate forces that maximise the likelihood of the statistics of the beliefs of the objects. The inner arrows indicate forces, which result from the function that maximises the ratio of the ability to distinguish between object and background. This method prevents contours from collapsing because of low-contrast objects, like e.g. a white ball.

In practice, tracking of the exact contour of the object is difficult due to motion blur. Thus we track the region that is close to the object and its background region as shown in fig. 2(b). The real contour lies in the middle of both regions. We choose the $m - bin$ histogram for the feature space representation of the object region and the background region. Using histograms is not an optimal non-parametric density estimate method [5] but it suffices for our purposes. Both histograms are generated by function $h(x_i)$, which assigns the colour at location x_i to the corresponding histogram bin. In our experiments we use an illumination invariant colour space $r'g'I$ that was introduced in [12]. The conversion of RGB values is given by $(r', g', I) = (\frac{r}{g}, \frac{g}{b}, \frac{r+g+b}{3})$. Our experimental setup uses a $12 \times 12 \times 4$ bin histogram. The usage of a 4 bin illumination per 12×12 bin colour information provides a smaller sensitivity to illumination changes. Our experiments have shown that this also works by using only 16×16 colour bins without illumination information. However, in this case it becomes impossible to distinguish black from white objects. The colour distribution $p_\Phi = {y_u^{(u)}}_{u=1\ldots m}$ with the pose Φ is calculated as

$$p_\Phi^{(u)} = f \sum_{i=1}^{l} \delta[h(x_i) - u] \tag{1}$$

where l is the number of pixels in the area, δ is the Kronecker delta function and f is a normalisation constant ensuring $\sum_{u=1}^{m} p_\Phi^{(u)} = 1$. The probability distribution functions of both histograms are calculated separately. For the sake of simplicity we assume that all corresponding objects and background areas are known a-priori for all Φ. Note that both histogram bins are not binary e.g. we do not apply traditional segmentation. Next we need a similarity measure of $p_{object}(u)$ and $q_{object}(u)$ to enable tracking and target localisation. The bin-likelihood is calculated using the Bhattacharyya metric [5,13]:

$$\rho[p, q] = \sum_{u=1}^{m} \sqrt{p_{obj}(u) \cdot q_{obj}(u)} \tag{2}$$

In theory a perfect match will result in $\rho[p, q] = 1$. A non perfect match will result in a value between zero and one. To avoid certain observations, like for example a "white ball" on a "white wall", we can assume that the object is always distinguishable from the background. We extend (2) to

$$\rho[p, q] = \left\lfloor \left\lfloor \sum_{u=1}^{m} \sqrt{p_{obj}(u) \cdot q_{obj}(u)} - \sqrt{p_{bck}(u) \cdot q_{obj}(u)} \right\rfloor - \vartheta \right\rfloor \tag{3}$$

The background is extracted bin-wise instead of subtracting the summed likelihood. This leads to a higher stability in case of the background being alike to the object scheme. In general, probabilistic frameworks output the "most believed" state, for example they will yield the "best fit". In practice this becomes often a problem when the object of interest is not visible in the image, e.g. due to partial or full occlusion, or when it is no longer in the image. In this case the next most believed state output will be considered as a false positive. In order to overcome that problem we apply thresholding using a lower limit $\vartheta \in \mathbb{R}$. Also, in the probabilistic framework of [10] a lower limit was used to detect the loss of an object.

Tracking whole regions yields relatively expensive computations, e.g. when we apply sequential Monte Carlo filters. Hence, we apply an approximation that uses structured *sample points* [8] near the contour, where the sample points can be considered as both, object and background. Using this technique allows us to have an adjustable trade-off between runtime and precision.

3 Probabilistic Object Tracking

In this section we consider the two co-operating memories of our approach. First we give a brief introduction to Particle Filters and the related *loss of diversity* problem in particle filtering. Next we explain the technical details of both memory models.

3.1 Particle Filters

Particle Filtering [10] or Monte Carlo [7] (MC) was developed to track objects in clutter. It is a Bayesian probabilistic method. The position of an object is represented by using a set of n weighted particles. Each particle contains a "believed" position with an assigned probability π. The pose of the object is estimated by using the observations as a function of the likelihood of believed poses/states while the particle filter attempts to maximise the likelihood of the beliefs. The usual MC algorithm works recursively in four different stages: (1) first the *prediction* stage of the motion model that is used to integrate the *actions* u to all particles e.g. the particles are simply moved. In the following stage (2) the observations are used to *update* the weight π of the particles. As next (3) the weight of all particles are normalised to one. Finally (4) the particles are *re-sampled* to get the posterior distribution. Technically, re-sampling discards particles with low weights and moves their weights to specific (random) particles with higher weights. In our implementation we move to the position of a new "offspring" particle with respect to the weight of the parent particle. For example a low weight of the parent particle will result in a high transformation.

3.2 Extended Particle Filtering (ePF)

Particle filters approximate probability density functions using a discrete set of n particles. Filters like the Bootstrap algorithm typically approximate the density by maximising the probability of particles by re-sampling particles with

procedure EXTENDPARTICLEFILTER
    // ...
    NORMALISEIMPORTANCEWEIGHTS($\tilde{\pi}_t^{(0)}, ..., \tilde{\pi}_t^{(n)}$)
    // Selection & Resampling step (Bayes Prediction Step)
    // **generate m clusters from all particles**
    $c^{(j)}{}_{j=1...m} \leftarrow$ OBTAINCLUSTERS($\tilde{x}_{0...t}^{(i)}$)
    // **calculate the weight of the clusters**
    $\varpi^{(j)}{}_{j=1...m} \leftarrow$ NORMALISECLUSTERWEIGHTS($\varpi'^{(j)}_{j=1...m}$)
    for $l \leftarrow 1$ **to** m **do**
        $N'^{(l)}_t \leftarrow$ SELECTCLUSTER($c^{(l)}$)
        $d \leftarrow$ NUM($c^{(l)}$) // **Apply traditional resampling for each cluster..**
    end for
    // ...
end procedure

Fig. 3. Extended Particle Filter, see [7] for the original algorithm

a "high" probability. The terms "high" and "low" are relative in the theory of probability density functions: The weights of all particles are usually normalised such that the overall sum is one. This means that particles with a "high" probability are considered more important than those with a relatively low probability. For example "high" ranked particles can generate more offspring in the re-sampling step.

The discrete approximation can lead to unwanted side effects caused by the discrete set of particles. For example, it can be the case when we apply particle filters to semi-bimodal distributions (one short dominant peak and a wide little peak): This can result in relatively many particles with a low probability and relatively few particles with a high probability. According to the probability function it can happen that low weighted particles gain a lot of attention of the filter compared to the few particles with a high probability, like e.g. the 'particle clinging ' effect when the system needs several time-steps to converge. With visual object tracking we face the problem that we have several areas with very low probabilities, e.g. if the background has a minimal similarity of the object. This is exactly the case with our object model (section 2). It is a relatively liberal model ensuring that lost objects are re-detected in short time. In the case of a lost object the particles are distributed randomly. Indeed the usage of a strict probability can reduce the required time to converge, however it has a high impact on the robustness of the system. In eq. (3) we introduced a lower limit to reduce the chance of this effect to occur. In literature this problem is sometimes referred as the "loss of diversity" (in particle filters) problem. In practise the probability density is unknown; hence we have to expect "particle clinging ". The "particle clinging " effect is also noticeable in the case of Monte Carlo localisation. With bi-modal distributions we face the problem that all particles will always converge to the highest probability [15] and may lose the object.

The main reason for the particle clinging effect is that we consider the particles piece-wise during the re-sampling step. We introduce a new *extended Particle filtering* (ePF) method. It is based on the idea that we consider groups of particles (clusters) during re-sampling. This means that we treat clusters like ordinary particles. For example we assign each cluster a count of offspring particles. Each cluster generates its own offspring while we use the same method as in traditional methods, e.g. the traditional re-sampling step is applied to each cluster. We use an implementation based on the mean shift metric [2, pp 790]. This method builds clusters of the state space using a weighted kernel function and a kernel size. In contrast to the popular k-means clustering method the number of clusters is determined by the algorithm itself. Using the clusters we are able to obtain multiple states from one particle set by applying the MC estimation methods separately on each cluster. In practice 80% of the entire time only one cluster appears. In the worst case this extended method performs equally to the non-extended method.

3.3 Short-Term Memory

The short term memory deals with frame-to-frame changes of Φ_t in image sequences using the extended particle filters. We assume that the tracked object undergo translation or rotation in the image of the camera. Additionally we assume that only the object size changes while the shape remains unchanged. A straightforward model of such motion applied to the state is Φ_{t-1} to obtain Φ_t using:

$$\Phi_t = \Phi_{t-1} + \Delta_{mov} \tag{4}$$

To simplify matters we assume that the process noise is contained within Δ_{mov}. Based on the idea of the *Mass Inertia Model* we use the particle filtering method to estimate the translative motion. The relative motion is expressed by $\dot{x}$ and $\dot{y}$ through the translation:

$$\Phi_{xt} = \Phi_{xt-1} + \Phi_{\dot{x}t-1} \quad \Phi_{y_t} = \Phi_{y_{t-1}} + \Phi_{\dot{y}_{t-1}} \tag{5}$$

The theory of particle filters states that particles with a "good" proposal will survive in the sample set and populate in the resample set. The parameters $\dot{x}, \dot{y}$ are propagated in the resample step of the filtering process according to their prior weights. A "good" proposal will generate more but concentrated particles, while a "bad" proposal will generate fewer but highly varied particles. In theory the set will convert to "good" values after a few (> 3) iterations. So we obtain

$$\Phi_t = \Phi_{t-1} + \begin{pmatrix} \Phi_{\dot{x}t-1} \\ \Phi_{\dot{y}_{t-1}} \\ 1.1 - {\pi'}_{t-1} \\ 0 \\ 1.1 - {\pi'}_{t-1} \\ 1.1 - {\pi'}_{t-1} \end{pmatrix}^T \cdot \begin{pmatrix} 1 \\ 1 \\ \rho_\omega \\ 0 \\ \rho_{move} \\ \rho_{move} \end{pmatrix}^T \cdot \begin{pmatrix} 1 \\ 1 \\ \text{RND}() \\ 0 \\ \text{RND}() \\ \text{RND}() \end{pmatrix}^T + \Delta_{Noise} \tag{6}$$

where ρ_{move}, ρ_ω is the expected (maximum) relative movement and orientation change made in one time step, respectively. $\text{RND}() : [\text{-}1{:}1] \in \mathbb{R}$ is a non Gaussian and non linear random function. π' represents the normalised weight ranging from $[0:1] \in \mathbb{R}$. Note that the (usual) π is normalised such that the sum of all π is one. To prevent a local minima we use 1.1 instead of 1.0 as normalisation value and we add uniform distributed noise Δ_{Noise} to the particles. Note that initially all parameters in Φ_t are set to zero.

3.4 Long-Term Memory

Although our particle filter is an efficient and robust visual object tracker, it only provides limited robustness to occlusions. In such a case the particle filter will converge to a uniform distribution of its particles. In order to keep objects for a "longer time" we use a well-tuned "Multiple Hypothesis Tracking"(MHT) filter. We apply a MHT algorithm that assigns an ID to the observed tracks of the particle filter. The original implementation given in [14], later extended in [6], provides already a full framework to suffice our needs. It is able to cope with missing measurements and predicts the motion of a-priori known unobserved objects. In our experimental setup, an unobserved object is "forgotten" after 2 seconds. Note that the original implementation is bound to linear/Gaussian systems.

We use a straightforward technique to give a feedback to the short-term memory: We add b additional particles to the particle filter using the positions of the predicted hypothesis of missing objects. The position of additional particles (per hypothesis) depends on its "age", e.g. the count of iterations until the last update step to a found observation. An older "age" will result in a larger spread of the additional particles. Note that the MHT filter is used only as a "long term memory" e.g. its only feeding "hints" to the short-term memory. The estimation of the believed state is done in the short-term memory, e.g. the most dominant observation is output. In the case of visually gone objects these "hints" are ignored by the particle filtering.

3.5 Ego-Motion Compensation

The main idea of ego-motion compensation is to apply the inverse motion of the robot to the tracker. In our experimental setup we use the self-localisation of the robot system. We use a simple heuristic to apply the *Ego-Motion Compensation* to the two tracker sensors: First we calculate the relative movement of the robot by using the difference of the current and previous poses, like e.g. the pose obtained from the current and previous iteration. We use the robot's pose probability as weight factor for the observed motion, which prevents "artificial motion" caused by non explicit poses, like for example during the global localisation. The weighted deltas are applied to the particles and the MHT hypothesis by shifting them. Note that the particles are only rotated. We do not shift the particles translatively because it can lead to unwanted behaviours. The MHT hypotheses are shifted translatively and rotatory according to their deltas. The motion models of both trackers remain untouched since we assume that the object itself moves (approaching the robot).

4 Experimental Results

In this section we will evaluate our approach and compare it to standard tracking methods like the *Mean Shift* tracker proposed by [5] and the *Colour-Based probabilistic tracking* proposed by [7,13]. The *Colour-Based probabilistic*approach is closely related to ours while *Mean Shift* differs from it. The *Mean Shift* is usually used in combination with a Kalman filter.

4.1 Test Environment

We use test sequences from the RoboCup Middle Size context to measure the performance. Our robot is equipped with an omni-directional camera system (omni-vision) which is the only sensor of the system. These kind of camera systems lead to the effect that *ego-motion* is more intense in the image as in ordinary pan and tilt setups.

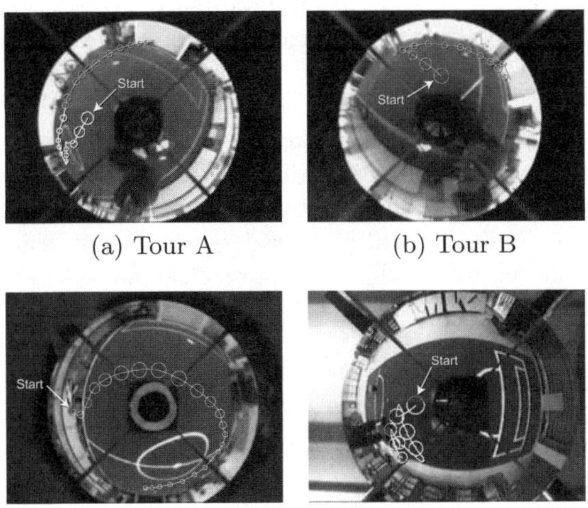

(a) Tour A (b) Tour B

(c) Tour C (d) Tour D

Fig. 4. Sample Tours used for Evaluation, the unit of the values are frames

First, tour "A" tests the object tracking performance over a large range which is a typical situation in a RoboCup tournament: The ball can be far away from its observer while it is possible that the object moves rapidly towards the observer within short time. In tour "B" we assess the 'trackers' performance in case of highly dynamic object movements. The ball is pushed to the centre of the field and shot (very fast) towards the yellow goal. The shots are repeated three times and the ball bounces back. The objective of tour "C" is to test the stability w.r.t. ego motion which represents a typical behaviour of a RoboCup robot in the role of a striker. Tour "D" tests the stability of the trackers w.r.t. a background highly

Table 1. Accuracy of the trackers, the unit of all values is pixel

	Mean Shift		Mean Shift + KL		MC Tracker		our approach	
Tour	Avr.	Max	Avr.	Max	Avr.	Max	Avr.	Max
Tour A	21.125	168.848	5.510	48.664	40.253	348.319	4.651	62.047
Tour B	109.807	291.239	74.895	223.243	42.254	290.805	4.883	115.242
Tour C	153.664	350.957	168.434	513.469	26.951	238.484	9.273	221.215
Tour D	87.963	310.499	7.950	64.315	21.350	215.758	5.438	63.523

similar to the object of interest. Note that we use an ordinary white soccer ball which contrasts the RoboCup rule-set (2006).

4.2 Results

We use the following settings to parameterise our tracker: 200 particles for the *extended Particle Filter* module, 50 "hint" particles for the MHT module and 50 random particles. During the test runs the *ego-motion compensation* is not used. For the other contestant we use values that were proposed by [5,11]: an ellipsoid area is used for the *Mean Shift* and *MC*. We measure the Euclidean error $e1$ of the believed pose of the tracker with respect to the ground truth position. In tours with high dynamics the *MC* tracker performance is better than the *Mean Shift* tracker. Results are vice versa in tours with low dynamics. One reason for this is the incapability of a Kalman filter to adapt to non-linear/non-Gaussian dynamics. In tour C the Kalman filter degrades even the performance of the *Mean Shift* tracker. Altogether we see that our approach shows the lowest average error of all contestants.

4.3 Benchmarking

First we analyse the influences of the number of particles on the performance of different configurations of the tracker, see fig. 5(a). "MC" denotes a tracker using traditional Monte Carlo filtering. All configurations converge while the number

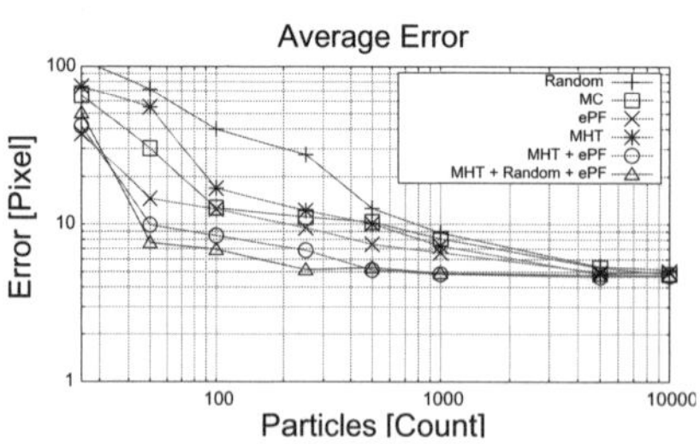

(a) different number of particles

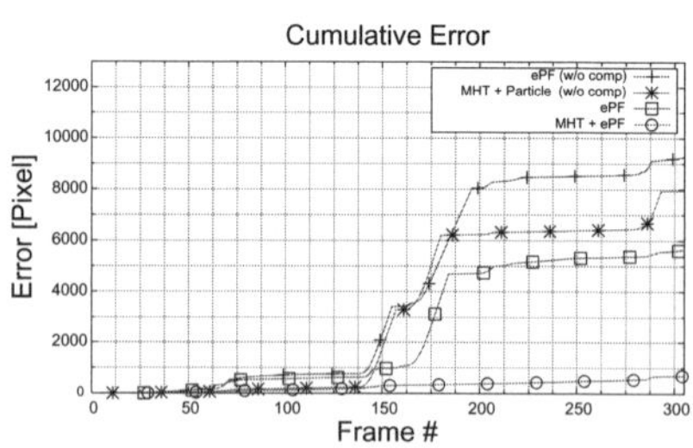

(b) ego motion compensation

Fig. 5. Using different configurations of our approach

of particles increases. The *MC* tracker performs better than the MHT tracker. In cased where MHT loses the object, all particles are set randomly to recover the position of the object. That's why MHT converges with an increasing number of particles. The extended Particle Filter method performs better than the traditional Monte Carlo. In fact the combination of MHT, random particles and extended Particle Filter shows the best results. The main reason for the improvement is that the MHT filters adapt better to "long term" motion (of the robot and objects) such as the extended Particle Filter, while the MHT filter fails to adapt to "short term" motion (e.g. if the ball is shot or is bounces back). It is also shown in [16] that random particles can improve the performance of particle filters. Next we consider the influence of the *ego motion compensation* on the accuracy of the *hybrid tracker.* Figure 5(b) shows the cumulative error of tour C using different configurations. Similar to fig. 5(a) an extended Particle Filter can be improved using additional particles resulting from the MHT "long term" memory. We see that *Ego motion compensation* significantly improves the performance. Turnings of the robot have only a slight influence on the performance.

5 Discussion

We showed that our new tracking technique can robustly track objects with high and low dynamics. Moreover we showed that the usage of our extended Particle Filter yields better performance compared to traditional Monte Carlo Filters. It allows us to obtain multiple (disjoint) independent states. The major drawback of our implementation is that we have to assume a *kernel size* for the object in the state space. The usage of too high or too low values can lead to multiple observations of the same object which degrades the performance to the level of traditional Monte Carlo filters. [4] proposed a parameter-free version of the *Mean shift* algorithm by analysing the density of the state space. We cannot apply this method because we use too few particles (≈ 1000 particles are needed). One advantage of our approach is that it only requires a minimum of initial knowledge about an object's motion dynamics since we assume that an object can move in any direction.

Furthermore we faced the problem of the dynamic scale adoption. In our approach we propagate these parameters using particles, where [5] applies a heuristic by using the best match probing the scales $\lambda, \lambda 0.9, \lambda 1.1$ in every iteration. Both solutions are sub-optimal and can lead to inaccuracy in case of poor illuminated environments. In fact the inaccuracy is caused by the false size adaptation which leads to a "shift" of the position of the believed track. See [5,13] for more details. In our approach we observed two bottlenecks: (1) the generation and evaluation of the colour scheme candidates and (2) the generation of the sample points of the approximated shape model. The computational time on the evaluation of the colour scheme highly depends on the number of used bins of the sensor model.

The *Mean Shift* is the fastest method of all tested trackers. This explains why it became popular in the visual tracking community. Theoretically the runtime of our approach and the *MC* tracker is identical. Practically m of our approach

Table 2. Computational complexity

	Mean Shift	MC Tracker	our Approach
Complexity	$O(m \log m)$	$O(nm \log m)$	$O(nm \log m)$
Average Runtime	$< 1ms$	$\approx 20ms$	Ball:$\approx 6ms$, Goal $\approx 11ms$
typical n	-	150	300
typical m	30x30	30x30	16x8

is much smaller than MC due to the fact that we use sample pixels instead of the pixel areas. The runtime was measured on a 1.1GHz PentiumM notebook.

References

1. Bar-Shalom, Y., Fortmann, T.: Tracking and data association. Mathematics in science and engineering, 1st edn., vol. 179, Academic Press Inc., London (1988)
2. Cheng, Y.: Mean shift, mode seeking, and clustering. IEEE Transactions on Pattern Analysis and Machine Intelligence 17, 790–799 (1995)
3. Collins, R., Liu, Y., Leordeanu, M.: On-line selection of discriminative tracking features. IEEE Transaction on PAMI 27(10), 1631–1643 (2005)
4. Comaniciu, D., Meer, P.: Mean shift: A robust approach toward feature space analysis. IEEE Transaction on PAMI 24, 603–619 (2002)
5. Comaniciu, D., Ramesh, V., Meer, P.: Kernel-based object tracking. IEEE Transaction on PAMI 25, 564–575 (2003)
6. Cox, I., Hingorani, S.: An efficient implementation and evaluation of reid's mht algorithm for visual tracking. In: ICPR94, pp. 437–442 (1994)
7. Doucet, A., Freitag, N., Gordon, N.: Sequential Monte Carlo Methods in Parctise, ch. 2, pp. 4–16. Springer, New York (2001)
8. Hanek, R., Schmitt, T., Buck, S., Beetz, M.: Towards robocup without color labeling. In: RoboCup International Symposium 2002 (2002)
9. Hundelshausen, F., Rojas, R.: Tracking regions. In: Polani, D., Browning, B., Bonarini, A., Yoshida, K. (eds.) RoboCup 2003. LNCS (LNAI), vol. 3020, Springer, Heidelberg (2004)
10. Isard, M., Blake, A.: Condensation – conditional density propagation for visual tracking. International Journal of Computer Vision 29(1), 5–28 (1998)
11. Nummiaro, K., Koller-Meier, E., Gool, L.: An adaptive color-based particle filter. Image and Vision Computing 21(1), 99–110 (2003)
12. Olufs, S.: Realtime color-segmentaion of fast moving objects (in german). Master's thesis, University of Applied Sciences Bonn-Rhein-Sieg (2002)
13. Prez, P., Hue, C., Vermaak, J., Gangnet, M.: Color-based probabilistic tracking. In: Heyden, A., Sparr, G., Nielsen, M., Johansen, P. (eds.) ECCV 2002. LNCS, vol. 2350, pp. 661–675. Springer, Heidelberg (2002)
14. Reid, D.: An algorithm for tracking multiple targets. IEEE Transaction on Automatic Control 24(6), 843–854 (1979)
15. Schulz, D., Burgard, W., Fox, D., Cremers, A.: People tracking with a mobile robot using sample-based joint probabilistic data association filters. Journal of Robotics Research (IJRR) (2003)
16. Thrun, S., Fox, D., Burgard, W., Dellaert, F.: Robust monte carlo localization for mobile robots. Artificial Intelligence 128(1-2), 99–141 (2000)

Parabolic Flight Reconstruction from Multiple Images from a Single Camera in General Position

Raúl Rojas, Mark Simon, and Oliver Tenchio

Institut für Informatik
Freie Universität Berlin
Takustr. 9, 14195 Berlin, Germany

Abstract. This paper shows that it is possible to retrieve all parameters of the parabolic flight trajectory of an object from a time stamped sequence of images captured by a single camera looking at the scene. Surprisingly, it is not necessary to use two cameras (stereo vision) in order to determine the coordinates of the moving object with respect to the floor. The technique described in this paper can thus be used to determine the three-dimensional trajectory of a ball kicked by a robot. The whole calculation can be done, at the limit, with just three measurements of the ball position captured in three consecutive frames. Therefore, this technique can be used to forecast the future motion of the ball a few milliseconds after the kick has taken place. The computation is fast and allows a robot goalie to move to the correct blocking position. Interestingly, this technique can also be used to self-calibrate stereo cameras.

1 Introduction

The years 2004/2005 constitute a turning point in the small-size RoboCup league. Several of the competing teams used chip kickers to lift the ball above 15 cm high robots placed at a distance of 30 to 50 cm from the kicking robot. Unfortunately for the defending team, it is very difficult to position the goalie when such a kick occurs. The projection of the parabolic flight of the ball on the video images looks like a curve. It is not immediately clear, whether the ball is flying high or is just rolling along a curved trajectory. Fig. 1 shows the apparent trajectory of a ball kicked high from a corner to the opposite corner of the field. Fig.1 is the combined result of observing the field with two cameras. In this paper, we discuss only the mathematics for a single camera overlooking the field. Generalizing to several cameras whose fields of view do not overlap is straightforward.

It would be very useful for a defending team to be able to determine the parameters of the parabolic ball trajectory (that is, the direction of the kick and its height) immediately after the kick occurs. As we show, any three points

G. Lakemeyer et al. (Eds.): RoboCup 2006, LNAI 4434, pp. 183–193, 2007.

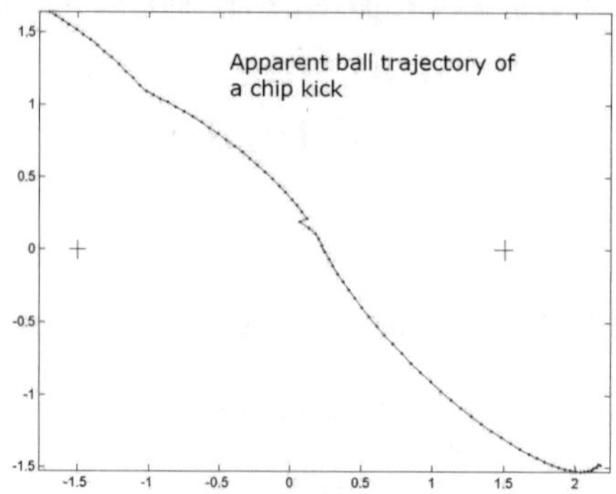

Fig. 1. Chip kick as seen with two global cameras in a small-size field (the kick starts in the lower right). The axis dimensions are given in meters. The cameras are located at the coordinates $(1.5, 0)$ and $(-1.5, 0)$. Their fields of view overlap only slightly in the middle of the field.

measured at any time in the apparent trajectory of the ball can be used to reconstruct the three-dimensional parabolic path. We do not need to apply methods from stereoscopic vision [1], [2].

Kim et al. [3] have studied a related problem: the reconstruction of the parabolic flight of a ball from a video of a soccer game. However, their method is based on using the two extremes of the parabola (when the ball touches the ground, at the start and at the end of the ballistic motion), and is not suitable for predicting future parabolic motion using a minimal number of points just after a kick. A variation of their method, in which they adjust a quadratic function to many alternative planes of motion, searching among them for an optimal fit, is too cumbersome and inefficient. The method described here, by contrast, is direct, does not require any search driven computation, and can be used for forecasting future motion using only three video frames just after the kick.

2 Projective Transformation and Reconstructed Path

We adopt the following conventions. The origin of world coordinates is on the floor. The camera is situated above the field. Figure 2 shows how the two coordinate systems (field and camera) are related. Without any corrective computation, the flying ball is interpreted as a "virtual ball" rolling on the ground along a curved path. In the general case, the three axis of the camera's system of coordinates are rotated relative to the field's coordinate axis. Let us denote by R the rotation matrix needed for transforming from world to camera coordinates, and by ℓ the translation vector from the origin of the world system to the camera coordinate system. Therefore, a point with world coordinates $p = (x, y, z)^{\mathrm{T}}$ has coordinates $q = R(p - \ell)$ in the camera's coordinate system. The inverse transformation, from camera to world coordinates, is therefore $p = R^{-1}q + \ell$.

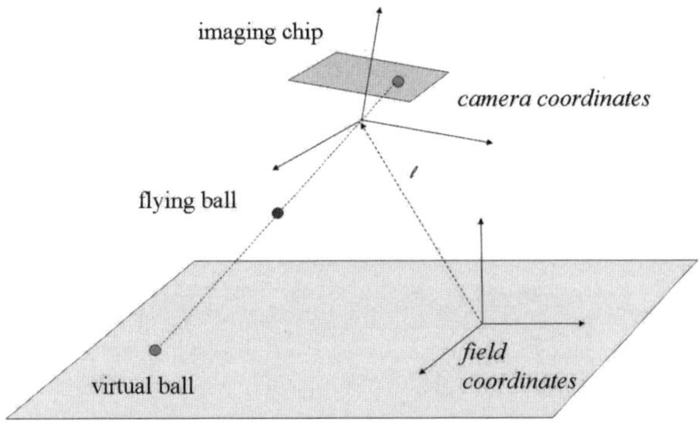

Fig. 2. Coordinate systems of the field and camera

Our computer vision system is self-calibrating. Just by selecting four points on the field, whose world coordinates are known, such as the corners of the field itself, the software computes automatically the rotation matrix R and the translation vector ℓ mentioned above (by solving a system of linear equations, see [2], [5]). Given the position of an object in a pixel of the imaging chip, the computer vision can then determine its position in world coordinates on the field.

In what follows, we assume that the successive projections of the ball on the field, with world coordinates $p_i = (x_i, y_i, 0)$, for $i = 1 \ldots$, have been transformed to the camera coordinate system using the transformation $R(p_i - \ell)$. This is a straightforward step since the matrix R and the vector ℓ are known from the initial calibration step. This also means that the camera can have any angle and displacement in relation to the world coordinates, and that the method described next is completely general.

We assume also, for the sake of the computation, and without losing generality, that the camera imaging chip is at a unit distance from the pinhole. The point where parabolic flight starts to be measured has camera coordinates (x_0, y_0, z_0). The velocity of the ball, after the kick, is given by the vector $v = (v_x, v_y, v_z)$. The parabolic flight of the ball is then described by the following parameterized path, in the camera's coordinate system:

$$(x, y, z) = \left(x_0 + v_x t + \frac{1}{2} g_x t^2,\ y_0 + v_y t + \frac{1}{2} g_y t^2,\ z_0 + v_z t + \frac{1}{2} g_z t^2 \right)$$

where t is the time elapsed, starting with the first data point ($t = 0$ for that point), and (g_x, g_y, g_z) are the components of the earth's acceleration in the camera's coordinate system (which can be tilted with respect to the vertical).[1]

[1] Gravitational acceleration varies with the latitude, because the earth is not perfectly spherical. The Geodetic Reference formula of 1967, used by geographers, is given by $g = -9.7803185(1 + 0.005278895 \sin^2\phi - 0.000023462 \sin^4\phi)$ m/s. Surface features of the earth are not considered in the formula.

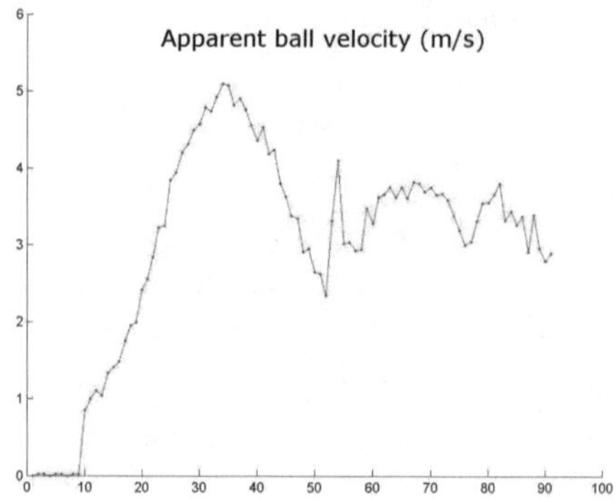

Fig. 3. Apparent velocity of the "virtual" ball from frame to frame in m/s

The components of the earth's acceleration with respect to the camera system can be computed using the rotation matrix R:

$$(g_x, g_y, g_z)^{\mathrm{T}} = R(0, 0, -9.8)^{\mathrm{T}}$$

Let us assume that the kick is never so high as to reach the camera plane (that is, $z \neq 0$). The projection of the position of the ball in the image plane of the camera (at a unit distance from the pinhole) is then

$$\left(\frac{x_0 + v_x t + \frac{1}{2} g_x t^2}{z_0 + v_z t + \frac{1}{2} g_z t^2}, \frac{y_0 + v_y t + \frac{1}{2} g_y t^2}{z_0 + v_z t + \frac{1}{2} g_z t^2}\right)$$

Assume now that m points in m images are given, where the "virtual" ball has been detected at times $t_1, t_2, \ldots, t_m$ (setting $t_1 = 0$). Let us denote the coordinates of the m points on the ground with respect to the camera system by $(x'_1, y'_1, z'_1), \ldots, (x'_m, y'_m, z'_m)$. Then, since "virtual ball" and real ball have the same projection on the camera chip, we have in general:

$$(\frac{x'_i}{z'_i}, \frac{y'_i}{z'_i}) = \left(\frac{x_0 + v_x\, t_i + \frac{1}{2} g_x t_i^2}{z_0 + v_z t_i + \frac{1}{2} g_z t_i^2}, \frac{y_0 + v_y\, t_i + \frac{1}{2} g_y t_i^2}{z_0 + v_z t_i + \frac{1}{2} g_z t_i^2}\right) \qquad \text{Eq.1}$$

We start processing the video sequence only when the ball is moving (which is easy to check). We denote by α_i the ratio x'_i/z'_i and by β_i the ratio y'_i/z'_i. From the expression above (and for the i-th point) we can derive two linear equations:

$$z_0\alpha_i + v_z\alpha_i t_i - x_0 - v_x t_i + 0 \cdot y_0 + 0 \cdot v_y t_i = -\frac{1}{2} g_z \alpha_i t_i^2 + \frac{1}{2} g_x t_i^2$$

and

$$z_0\beta_i + v_z\beta_i t_i + 0 \cdot x_0 + 0 \cdot v_x t_i - y_0 - v_y t_i = -\frac{1}{2} g_z \beta_i t_i^2 + \frac{1}{2} g_y t_i^2$$

We have two linear equations for six variables. Therefore, three points on the parabolic flight path provide enough equations (six) which can be used to solve

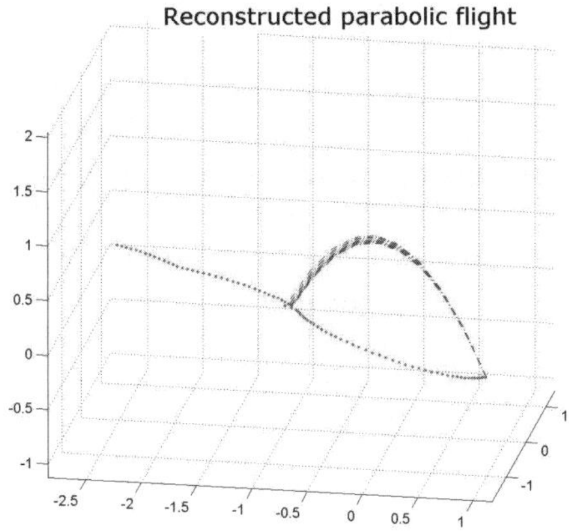

Fig. 4. Reconstruction of the parabolic flight (20 to 30 points on the trajectory were used)

the system. If we use more than 3 points, let us say m, then the general form of the system of equations we obtain is

$$D(z_0, v_z, x_0, v_x, y_0, v_y)^{\mathrm{T}} = d$$

where D (for data) is a $2m \times 6$ matrix and d is a $2m$-dimensional vector. Using the pseudoinverse D^+ of D we find the solution

$$(z_0, v_z, x_0, v_x, y_0, v_y)^{\mathrm{T}} = D^+ d$$

where $D^+ = (D^T D)^{-1} D^T$. The pseudoinverse allows us to use as many points for the calculation as we have already measured, because we are interested in producing an estimate of the flight trajectory of the ball as soon as possible, but also as precise as possible.

The computed initial position of the ball (x_0, y_0, z_0) and the velocity vector (v_x, v_y, v_z) must be transformed now from the camera's to the world's frame of reference. That is, we compute

$$(x_0^w, y_0^w, z_0^w)^{\mathrm{T}} = R^{\mathrm{T}}(x_0, y_0, z_0)^{\mathrm{T}} + \ell$$

and

$$(v_x^w, v_y^w, v_z^w)^{\mathrm{T}} = R^{\mathrm{T}}(v_x, v_y, v_z)^{\mathrm{T}}$$

where the superindex w means "world coordinates". Since R is a rotation matrix, its inverse is the transpose of R.

A word of caution: since we start the clock t_i when we get hold of the first data frame, it is necessary to start the calculation only when the ball is really moving fast on the field (a ball being pushed by a robot is usually not so fast). Fig. 3 shows the velocity of the ball on the ground, as computed from a series of frames of a real game. The moment after the ball has been kicked (frame 10) is readily identifiable. Therefore, it is easy to avoid performing erroneous calculations for a ball which has not been kicked (more about this below).

3 Resetting the Clock

We assumed in the previous sections that the clock starts running with the first available data point (after we detect that the ball was kicked). If we miss the exact frame in which the ball was kicked, we can start the clock later with any frame in which we see the "virtual" ball still on the air. When we compute z_0^w, we can compare it to 0 (points on the field have z coordinate equal to zero). If both are different, we can compute the time T it took the ball to reach the height z_0^w(relative to the field plane). Having this elapsed time, we can go "back in time" and find the exact point where the ball was kicked.

Let v_{z0}^w be the initial vertical velocity of the ball, at the moment of the kick. Then, if v_z^w is the vertical velocity in the frame we started recording data, we know that

$$v_z^w = v_{z0}^w - gT$$

where g is the downward earth acceleration, and

$$z_0^w = v_{z0}^w T - \frac{1}{2} g T^2 = v_z^w T + \frac{1}{2} g T^2.$$

Solving this quadratic equation, we find T. With T, we compute v_{z0}^w and the following corrections for all other parameters (the superindex f means *final*):

$$\begin{aligned} x_0^f &= x_0^w - v_x^w T & \qquad v_x^f &= v_x^w \\ y_0^f &= y_0^w - v_y^w T & v_y^f &= v_y^w \\ z_0^f &= z_0^w - (v_z^w T - \tfrac{1}{2} g T^2) & v_z^f &= v_z^w + gT \end{aligned}$$

4 Experimental Results

No robotic vision system is perfect. There is an intrinsic error in all measurements of the ball position. This noise, which is even larger when the ball is flying high, makes necessary to use more than three data points in order to make the best future projection of the ball's path. We have experimented with several different numbers of points. The best results were obtained (for our computer vision software) when more than 20 points were used for the ball trajectory. The camera was running at 52.7 frames/second. The ball was in flight for around 60 frames, so that we had to use one third of the data points to get a good reconstruction of the parabolic path, with subcentimeter accuracy.

Fig. 4 shows the result of the inverse computation described in this paper. Given the path of the ball projected on the ground, we computed parabolas based on 20 to 30 data points from as many video frames. The ball is in flight for around 50 frames. All parabolas are very near to each other.

Fig.5 shows a reconstruction computed with only ten frames, starting the computation at different moments (between frame 10 and frame 20). The figure is a screenshot of our vision software for the small-size league [4],[5], enriched with this new feature.

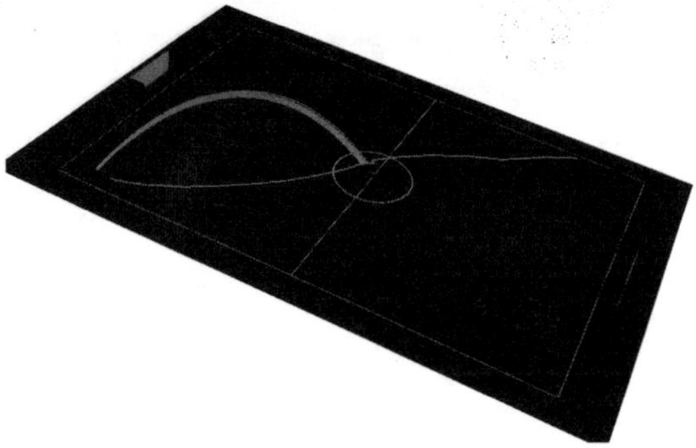

Fig. 5. Reconstruction of the parabolic flight (10 points on the trajectory were used, several different start frames were selected). The color screenshot is from our our small-size computer vision program.)

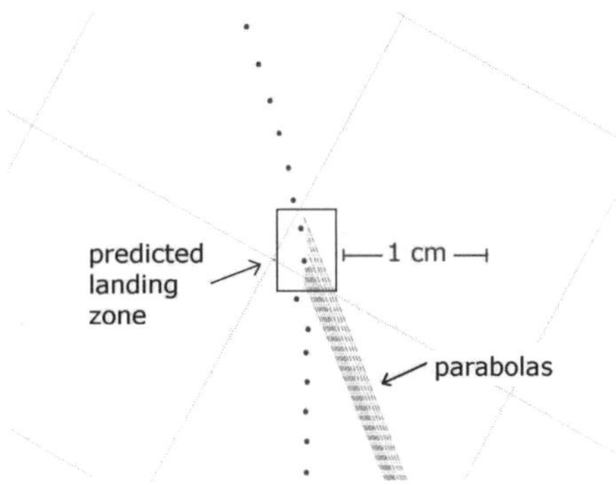

Fig. 6. Predicted impact position computed with 5 to 14 points from the trajectory. The predicted impact position is at the end of each parabolic path line. The isolated points are "virtual" images of the ball, before and after impact.

Fig.6 shows the predicted impact position of the ball after a chip kick, using different numbers of data points (from 5 to 14 frames). Each prediction is the tip of the predicted flight line. As can be seen, the spread of the prediction is smaller than 1 cm. Such an error, after a 2m chip kick, is almost negligible.

Fig.7 shows how the three components of the 3D-velocity of the ball converge to very precise values, when the number of points along the parabola is increased from 5 to 25. Already ten points produce a very good estimate of the velocity vector. The time needed, when using ten frames for the computation, is about one fifth of a second.

5 Reaction Time

The parabolic flight reconstruction would be only of theoretical interest if the goalie had not time to react. We show now that in many cases the goalie can

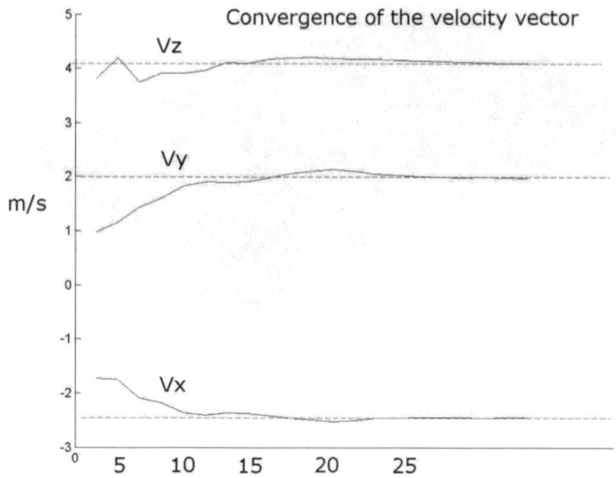

Fig. 7. Convergence of the three components of the velocity vector for different number of data points on the parabola

indeed intercept the ball. Whatever the distance from the goal line is, the ball has to be lifted above the goalie, that is, above 0.15 m. The time the ball is in the air in parabolic flight is given by $t = 2\sqrt{2h/g}$. Given a minimum height of 0.15 m, and the acceleration constant g, we obtain a minimum time of flight of $t = 0.35$ seconds. When the ball falls (for half of the flight-time) starting from rest at the highest point of a parabola just high enough to fly above a robot, the final velocity is $v_z = -0.175\,s \times g$, which is around -1.7 m/s. This is also the initial vertical velocity of the ball, when it is chip-kicked (with positive sign). Therefore, any slower ball will not fly high enough to go above the goalie or any robot on the field. This can be used as a fast check for a "dangerous" flying ball. Fig. 8 shows how fast the "virtual" ball is accelerated by our chip kick. The maximum apparent acceleration value is around 40 g's. The apparent acceleration remains positive as long as the ball is approaching the camera objective and its apparent speed continues to increase. When the ball starts falling, it seems as if the ball falls being accelerated by more than one g.

Now we can consider if a chip kick is stoppable. The delay from our vision system is about 100 ms (that is the time it takes an image to be accepted and processed)[4]. Assume that we use 5 frames for predicting the parabolic flight (first provisional estimation). At 52,7 frames per second, those 5 frames correspond to 95 ms. The robot has just $350 - 195 = 155$ ms for moving to the approximate blocking position. If the distance is not very large, the robot could reach the correct position. Our robots travel at around 2 m/s. They can move 31 cm in 155 ms, but only after having been accelerated. The question becomes therefore how fast can the robots start moving from scratch.

First we would like to have a table of the flight time of the ball for different distances on the field. Let us assume that the kick is optimal: the ball goes only as high as it needs to go, and no higher. The ball can go forward as fast as desired, but the real limit for the ball is to go above the goalie and below the

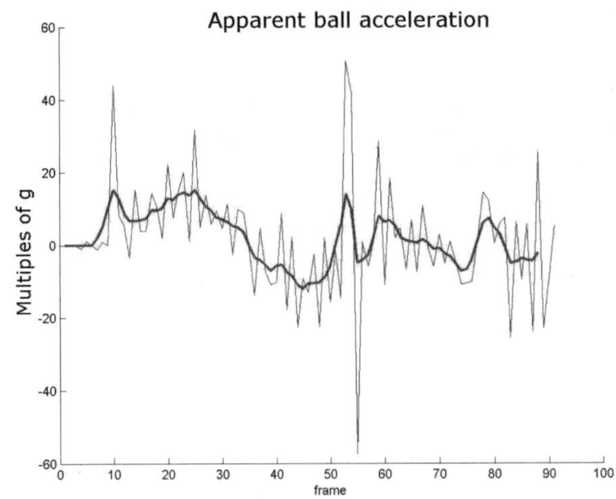

Fig. 8. Apparent acceleration of a ball kicked high (virtual ball acceleration)

goal bar (positioned at 15 cm). We do the following: we position the goalie at a distance r from the goal line. The goalie has a width of 18 cm and a height of 15 cm. The ball has to go under the bar. The ball has a diameter of 4 cm (in reality 4.3 cm, but let us simplify). We then compute the best parabolic shot that clears the robot and enters the goal box going below the bar. The time of flight is then plotted according to the coordinates on the field of the point where the ball is kicked.

The results obtained with this method can be seen in Fig.9. As stated before, there is a minimum time of flight of around 0.35 seconds. When the goalie is 10 cm away from the goal line, the minimum time for a chip kick over the goalie from far away (5 meters) rises to about 0.7 seconds, double the minimal time. However, teams cannot shoot so impressively from far away. They cannot adjust the angle of the shoot, which is fixed, and this produces suboptimal trajectories. Therefore, in practice, the real time of flight will be always much larger than the lower and optimal bound computed here.

We measured the velocity of some of our robots. Due to initial inertia, our goalie needs around 0.2 seconds to move 10 cm. An optimal shot from midfield requires around 0.5 seconds flight time. If we collect 10 data points before moving the robot (in 0.2 seconds), and if the vision delay is 0.1 seconds, we have 0.2 seconds left to position the goalie. If we need to move the robot less than 10 cm, stopping a chip kick is possible. Parabolic flight prediction has been incorporated now in our vision software. It is possible to stop all chip kicks coming from far away. Stopping nearer chip kicks depends on the position of the robot. Chip-kicks from anywhere near the goal area are unstoppable, if the goalie is not touching the line.

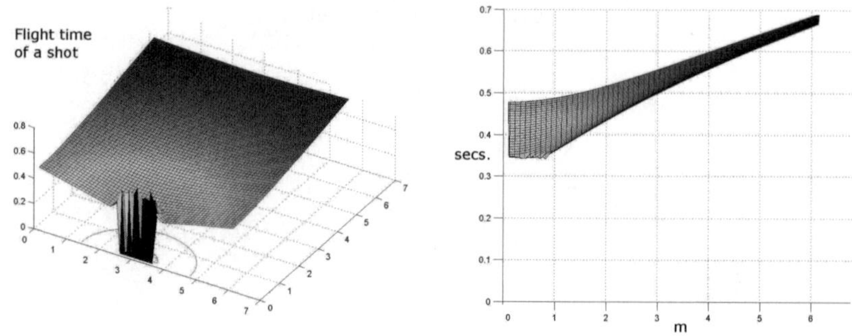

Fig. 9. Time of flight in seconds for optimally lifting the ball above the goalie and scoring (left image). The goalie has been positioned 10 cm away from the goal line. Lateral view of the same surface (right figure). The maximum distance is 6 meters (x-coordinate 6). The time of flight was set to zero 50 cm from the center of the goal box (a graphic artifact arises). Chip kicks from far away require 0.7 seconds. The minimum flight time for lifting the ball above the goalie is 0.35s.

6 Conclusions

This paper has shown that after a chip kick it is possible to reconstruct the three-dimensional flight path of the ball using a single camera overlooking a field. We have shown that the minimal velocity required for clearing a small-size robot is 1.7 m/s. This criterion can be used as a trigger for starting the code which computes three-dimensional paths. Hard flat shots can be discriminated by just looking at the "straightness" of the ball's path. A false alarm triggered by a robot pushing the ball in a curved path can be handled easily since no robot can be moving right behind a ball in parabolic flight. It can also be checked if the path corresponds to an acceleration compatible with the gravity constant. If not, the parabolic path can be discarded.

The method described in this paper can be used also in the mid-size league with omnidirectional or frontal cameras. Since the computer vision software for omnicameras produces an estimation of the position of the ball on the floor, as if we had a camera overlooking the field, the method described here can be used without any modification (provided the ball is visible under the horizontal plane of the omnicam).

There are other applications for our technique, for example, for obtaining the orientation of a camera overlooking a featureless area. If the camera is not pointing straight down, its orientation can be computed by tracking a ball thrown in front of the camera. The method described in this paper must be just extended to deal with a gravitational acceleration with unknown components (g_x, g_y, g_z) in the tilted frame of reference of the camera. In that case, Eq. 1 transforms into

$$(\alpha_i, \beta_i) = \left(\frac{x_0 + t_i\, v_x + \frac{1}{2} g_x t_i^2}{z_0 + t_i v_z + \frac{1}{2} g_z t_i^2}, \frac{y_0 + t_i\, v_y + \frac{1}{2} g_y t_i^2}{z_0 + t_i v_z + \frac{1}{2} g_z t_i^2} \right)$$

where (α_i, β_i) are the ball coordinates in the camera's chip. This expression provides two linear equations for each detected ball in each frame. For the nine unknowns we need at least five points to compute the value of (g_x, g_y, g_z), and from this, the angle of the camera axis with the vertical.

There is also an application for stereo-vision. If two stereoscopic cameras overlook a common field of view, and if their pose is unknown, an object thrown in parabolic flight can be used to determine the direction of gravity relative to each camera. The cameras do not need to be triggered simultaneously. Bringing the reconstructed parabolic flight curves for each camera in correspondence can then be used to calibrate the stereoscopic vision system. The numeric for such stereoscopic calibration without camera parameters is much simpler than that used by other groups [6], [7], and does not require matching points, nor the world coordinates of the corresponding points in space. This result is more general than Luong and Faugeras technique [8], and will be published elsewhere.

References

1. Howard, I.P., Howard, A.P., Rogers, B.: Binocular Vision and Stereopsis. Oxford University Press, Oxford (1995)
2. Forsythe, D., Ponce, J.: Computer Vision: A Modern Approach. Prentice-Hall, Englewood Cliffs (2003)
3. Kim, T., Seo, Y., Hong, K.-S.: Physics-based 3D position analysis of a soccer ball from monocular image sequence. In: ICCV 1998, 1998th edn., Bombay, India, pp. 721–726 (January 1998)
4. Simon, M., Behnke, S., Rojas, R.: Robust Real Time Color Tracking. In: Stone, P., Balch, T., Kraetzschmar, G.K. (eds.) RoboCup 2000. LNCS (LNAI), vol. 2019, pp. 62–71. Springer, Heidelberg (2001)
5. Simon, M., Wiesel, F., Egorova, A., Gloye, A., Rojas, R.: Plug & Play: Fast Automatic Geometry and Color Calibration for Tracking Mobile Robots. In: Nardi, D., Riedmiller, M., Sammut, C., Santos-Victor, J. (eds.) RoboCup 2004. LNCS (LNAI), vol. 3276, Springer, Heidelberg (2005)
6. Hartley, R., Gupta, R., Chang, T.: Stereo from Uncalibrated Cameras. In: IEEE Conference on Computer Vision and Pattern Recognition, pp. 761–764. IEEE Computer Society Press, Los Alamitos (1992)
7. Hartley, R.I., Zisserman, A.: Multiple View Geometry in Computer Vision. Cambridge University Press, Cambridge (2000)
8. Luong, Q-T., Faugeras, O.: Self-calibration of a stereo rig from unkown camera motions and point correspondences. In: Bensoussan, A., Verjus, J.-P. (eds.) Future Tendencies in Computer Science, Control and Applied Mathematics. LNCS, vol. 653, Springer, Heidelberg (1992)

On the Calibration of Non Single Viewpoint Catadioptric Sensors

Alberto Colombo[1], Matteo Matteucci[2], and Domenico G. Sorrenti[1]

[1] Università degli Studi di Milano Bicocca, Dipartimento di Informatica, Sistemistica e Comunicazione
[2] Politecnico di Milano, Dipartimento di Elettronica e Informazione

Abstract. A method is proposed for calibrating Catadioptric Omnidirectional Vision Systems. This method, similarly to classic camera calibration, makes use of a set of fiducial points to find the parameters of the geometric image formation model. The method makes no particular assumption regarding the shape of the mirror or its position with respect to the camera. Given the camera intrinsic parameters and the mirror profile, the mirror pose is computed using the projection of the mirror border and, eventually, the extrinsic parameters are computed by minimizing distance between fiducial points and back-projected images.

1 Introduction

A catadioptric imaging system is built combining reflective (catoptric) and refractive (dioptric) elements. They allowed building omni-directional sensors [1] at an affordable price, which made catadioptric systems become of widespread use also in fields like video-surveillance and teleconferencing; in contrast omnidirectional dioptric-based systems usually cost much more.

The choice of a specific mirror shape is based mainly on required resolution, manufacturing cost, assembly cost and geometric properties. With respect to these properties, mirror shapes can be classified in two categories: *Single Viewpoint* (SVP, also called *central*) imaging systems, and *Non-SVP* (*non-central*) imaging systems [2]. In SVP mirror systems, the lines of those incoming rays, which are reflected to the image plane, intersect at a single point, called *Effective Viewpoint* (see Fig. 1). It has been proved that the only useful mirror shapes with this property are hyperboloid mirrors (coupled with a perspective camera) and paraboloid mirrors (coupled with an orthographic camera). The uniqueness of the viewpoint allows a mapping of every scene point on the image plane, similarly to what happens with perspective cameras. The result of the catadioptric projection is equivalent to a panoramic image taken with a camera rotating around its viewpoint, and the resulting catadioptric image can be rectified to obtain an omni-directional undistorted perspective image called *panorama* [3,4]. On the other hand, SVP imaging systems suffer from a number of problems. Firstly, the SVP constraint leaves little room for changing other optical parameters that could be useful to tune because the relative pose of the mirror and

G. Lakemeyer et al. (Eds.): RoboCup 2006, LNAI 4434, pp. 194–205, 2007.

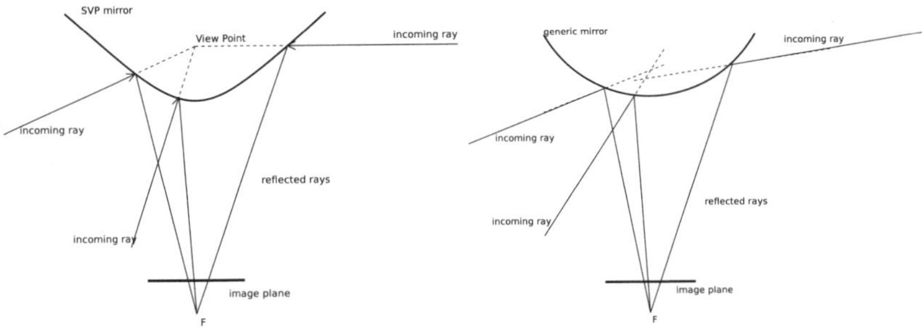

(a) central sensor (b) non-central sensor

Fig. 1. Comparison between a central catadioptric system (featuring an hyperbolic mirror) and a non-central one. Note that in *a* all rays intersect at the hyperboloid focus, while in *b* there is no unique intersection for the rays.

the camera are strictly bound. Due to the curved vertical section of the mirrors, spatial resolution is badly distributed: typically the ground and the camera fill the central part of the omni-directional image, leaving little room, on the peripheral part of the image, for the (interesting) surrounding environment. Regarding assembly costs, cameras based on hyperboloid and paraboloid mirrors are more expensive needing very accurate assembly, for the viewpoint to be unique.

The work on non-SVP sensors aimed at overcoming these limitations. Early non-SVP systems were based on a spheric mirror [5] or on a conical mirror [1] due to the low cost of such mirrors and the mathematical simplicity of the projection. In general, however, non-SVP systems pose no constraints on mirror shape and position. Catadioptric images taken with such devices cannot be rectified though, i.e., there is no function that takes a non-SVP omni-directional image and unwraps it into a perspective image. The freedom in the mirror shape, however, can be exploited for designing optimal mirrors for each specific application. Examples include single-camera stereo-vision [6], multi-part mirrors [7] and many kind of non-distorting mirrors [8].

This paper contribution is an algorithm to calibrate non-SVP systems using the projection of the mirror border and a set of points whose scene and image coordinates are known. By *calibration* we mean to estimate the mirror pose (with respect to the camera), as well as the camera pose (with respect to a *world reference frame*). We will be assuming that the camera intrinsics are known. We will therefore call *intrinsic parameters* of the catadioptric system the mirror pose; *extrinsic parameters* of the catadioptric system the camera pose in the world reference frame. By knowing the calibration parameters it will be possible to map every pixel onto its interpretation line. These lines can then be exploited for metrology. The assumption we make is that the mirror shape is known.

Most of the scientific literature regarding catadioptric system calibration deals with SVP systems, probably because of their availability off-the-shelf. Since SVP systems assume that the mirror pose is known, their calibration consists of finding 1) the intrinsic parameters of the mirror-and-camera pair (seen as a single indivisible object) and 2) the extrinsic parameters, similarly to what happens with traditional cameras. There are even some ready-to-use calibration tools available on the Internet, e.g. [9], [10], [3].

The calibration of non-SVP systems, on the other hand, is a surprisingly unstudied problem. The only exception is the particular case of non-SVP systems featuring a conic mirror, like the COPIS [1]. In this case, if the cone axis matches the camera optical axis, it can be proved that there is one viewpoint for each direction the optical rays come from. The set of these viewpoints, sometimes referred to as *view-circle*, allows the conic projection to be described through relatively simple equations; it also allows to calibrate the camera intrinsic parameters together with the rest of the catadioptric system, like one would do with a SVP system. In our experiments, for the sake of ease, we used a conic mirror, but this will not necessarily be coaxial to the camera. In fact, the algorithm we develope (see Sect. 2.2) is equally fit to any mirror shape.

Strelow's method [11] for calibrating generic catadioptric systems separates the perspective projection from the catadioptric part and deals only with the latter. The mirror can be of any shape as long as the surface equation is known. The method consists of calculating the mirror position w.r.t. the camera (i.e., the catadioptric system intrinsic parameters) using the image of one or more targets whose three-dimensional coordinates (w.r.t. the camera) are known, and then minimizing the difference between the observed position on the image and their respective predicted back-projections, where these depend on the roto-translation describing the mirror pose. This requires solving a non-linear system of equations. This method obtains a precision of about 1%, but it suffers from two problems. Firstly, the calibration points coordinates must be known w.r.t. the camera reference frame, i.e., the camera extrinsic parameters must be known *before* the calibration. The second problem is tied to the intrinsic complexity of a non-SVP projection model. In order to back-project a scene point onto the image one has to find the line, exiting from the scene point, which, once reflected by the mirror, passes through the camera pin-hole. To minimize the distance between image points and back-projected scene points this method requires for each minimization iteration, and for each point, to numerically resolve a non linear system.

2 Our Method

In the first place, we shall start by looking at the geometric model adopted to describe catadioptric imaging systems; we shall then examine the two phases of our method: mirror localization and extrinsic parameters computation. In Fig. 2 we see a schematic drawing of a typical catadioptric system. The cube in the middle, indicated with C, represents the camera (*camera reference frame* $x_C y_C z_C$). Note that this reference frame has a precise physical meaning, as its origin is the center of projection. It is not defined arbitrarily, but depends on the camera intrinsic parameters. In front of the camera there is a solid of revolution (SOR) mirror. Solids of revolution feature an *axis of symmetry* and a *profile*. We

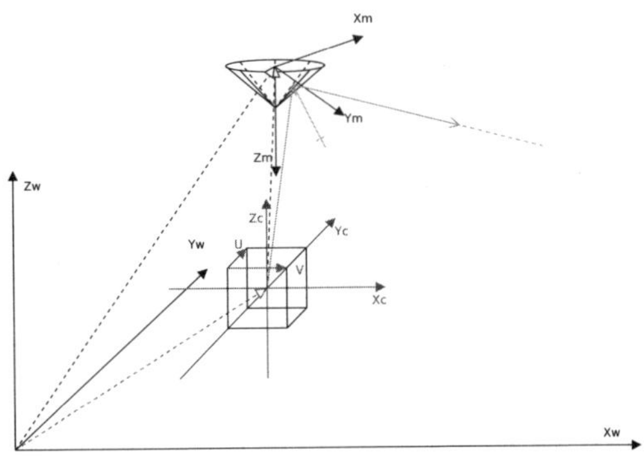

Fig. 2. Geometric model of a catadioptric imaging system

can define a reference frame local to the mirror whose z-axis corresponds to the SOR axis and whose origin lies in the center of the mirror base (*mirror reference frame* $x_M y_M z_M$). The mirror is in a generic position with respect to the camera: its axis is not parallel to the optical axis and it does not pass through the pinhole. The unknown roto-translation from the camera reference frame to the mirror is $\mathcal{T}_{C \to M}$: we will see how to compute it in Sect. 2.1. In Fig. 2 we see a third reference frame, $x_W y_W z_W$, called *world reference frame*; all distances measured by the user are defined in this frame. Its z-axis is vertical upward and the origin is usually located in a easily recognizable point. In Sect. 2.2 we will see how to compute the roto-translation $\mathcal{T}_{W \to C}$ and, as a consequence, the roto-translation $\mathcal{T}_{W \to M}$. In analogy with the terminology of traditional camera calibration, we will call the coefficients of $\mathcal{T}_{W \to C}$ *extrinsic parameters* of the catadioptric system. Because all rigid transformations are invertible, we will refer to the inverse of the above mentioned roto-translations, i.e., $\mathcal{T}_{M \to C}$, $\mathcal{T}_{M \to W}$ e $\mathcal{T}_{W \to C}$, whenever this would help readability.

2.1 Mirror Localization

The method exposed here is based on the analysis of the conic on the image plane corresponding to the projection of a circle. *The perspective projection of a circle is an ellipse.* The mirror pose can be estimated by this circle-to-ellipse correspondence. The idea of using the projected ellipse to estimate the circle pose is not new and it has found various applications (e.g. [12]); an introduction to this problem can be found in [13]. From the observed perspective projection of a circle with known radius, the circle support plane and the position of its center can be inferred analytically. Let the real, symmetric, non-zero matrix Q be the ellipse corresponding to the projection of a circle of radius r in the normalized camera reference frame (i.e., where the camera pinhole is in the origin and the image plane has equation $z = 1$). Because the matrix representation of quadrics is unique, up to an arbitrary scale factor, we choose a scale κ such that $\det \kappa Q = 1$. From matrix theory, we know $\kappa = \sqrt[3]{-\frac{1}{\det Q}}$. In the rest of this section, Q will refer to the normalized matrix. Let λ_1, λ_2 e λ_3 be the eigenvalues of Q, where $\lambda_3 < 0 < \lambda_1 \leq \lambda_2$, and let $\hat{u}_1$, $\hat{u}_2$ and $\hat{u}_3$ be the (orthonormal) eigenvectors corresponding to λ_1, λ_2 and λ_3 respectively. The normal $\hat{n}$ to the circle support plane is given by:

$$\hat{n} = \hat{u}_2 \sin\theta + \hat{u}_3 \cos\theta \tag{1}$$

where $\sin\theta = \sqrt{\frac{\lambda_2-\lambda_1}{\lambda_2-\lambda_3}}$, $\cos\theta = \sqrt{\frac{\lambda_1-\lambda_3}{\lambda_2-\lambda_3}}$. The distance of the support plane from the pinhole is

$$d = \lambda_1^{\frac{3}{2}} r\,. \tag{2}$$

To compute the position of the circle centre one only needs to multiply the inverse of the ellipse with the normal to the support plane:

$$\boldsymbol{c} = Q^{-1} \cdot \hat{n}\,. \tag{3}$$

Note that the sign of eigenvectors $\hat{u}_{1,2,3}$ are arbitrary. Since both $\hat{n}$ and $-\hat{n}$ represent the same plane orientation, from (1) we deduce that

1. If $\lambda_1 = \lambda_2$, then $\hat{n} = \sqrt{\frac{\lambda_1-\lambda_3}{\lambda_2-\lambda_3}}\hat{u}_3$, therefore there exists only one solution regardless of the sign of $\hat{u}_3$.
2. If $\lambda_1 \neq \lambda_2$, then there exist two interpretations.

In the second case, both interpretations are physically plausible, but of course only one is real. Unfortunately, a single image is not enough to solve this ambiguity, therefore an implementation of this algorithm will need more information in order to choose one of the two solutions. We will bring both hypothesis foreward up to the solution of the extrinsic parameters step. Only at that point we will have the ambiguity solved, by choosing the solution that gives the best estimate (with the smallest residual). In other applications, where two concentric circles are used, the solution can be directly disambiguated [14]; this could be useful for multi-part mirrors, like the one in [7].

2.2 Calibration of the Extrinsic Parameters

The extrinsic parameters are the coefficients of the transformation $\mathcal{T}_{C\rightarrow W}$ that converts a point coordinates from the camera reference frame $x_C y_C z_C$ to the world reference frame $x_W y_W z_W$. To compute these parameters we take a set of N *calibration points* or *fiducial points* whose scene coordinates (w.r.t. the world reference frame) and image coordinates are known. If we know the camera intrinsic parameters we can use the fiducial points image coordinates to compute their interpretation lines, with respect to the camera system. If we know the pose of the mirror, with respect to the camera system, we can compute the reflected interpretation lines. We also know that these lines come from their corresponding scene points. We know the reflected interpretation lines with respect to the camera system, while the fiducial points scene coordinates are known with respect to the world system, but we do not know the transformation between the two. We first see how to compute the reflection onto a surface of revolution, then we use the reflected interpretation lines for the estimation of the extrinsic parameters.

Reflection of the Interpretation Lines. If the mirror profile can be represented by a second degree equation, then the mirror shape is a quadric (a cone, a paraboloid or a hyperboloid) and the reflected line can be determined analytically. The same holds if the profile is a spline[1] of second degree or less, as long as the reflection is computed for each segment of the spline. On the other hand, when the profile is described by a third degree equation, the mirror shape is represented by a sixth degree equation that cannot be solved analytically [15]. In this work, without loss of generality, we implemented the reflection for quadric-shaped mirrors and we only applied it to conic mirrors. As explained in the following section, the same theory can be applied to mirrors of any shape using a local spline approximation of the profile.

Let $Q \in \Re^{4\times4}$ be a symmetric, non-zero matrix representing the quadric. The locus of points belonging to the quadric surface is given by $F = \boldsymbol{x}^{\mathrm{T}} Q \boldsymbol{x} = 0$ where $\boldsymbol{x} = [x,\ y,\ z,\ 1]^{\mathrm{T}}$. Given a point $\boldsymbol{p} = [p_x, p_y, p_z]^{\mathrm{T}}$ and a direction $\hat{d} = [d_z, d_y, d_z]^{\mathrm{T}}$, $(||\hat{d}|| = 1)$, let $\boldsymbol{x}(t) = \boldsymbol{p} + \hat{d} \cdot t$ be the parametric equation of the line passing through $\boldsymbol{p}$ with direction $\hat{d}$, whose intersection with the quadric we want to compute. By substituting the line into the quadric, grouping and rewriting as a function of t we obtain $A \cdot t^2 + B \cdot t + C = 0$ which is a second degree polynomial in t. We solve it to find t_1, t_2 and substitute them into the line to find the intersection points $\boldsymbol{P}_1$, $\boldsymbol{P}_2$ of the line with the quadric. Only one of the two intersections has a physical meaning, while the other is a point either behind the camera or outside the mirror or also occluded by the mirror itself (Fig. 4a). Once the intersection point $\boldsymbol{P}_i$ is known, we compute the normal to the quadric in $\boldsymbol{P}_i$ as

$$\hat{n}_i = \left[\left.\frac{\partial F}{\partial x_i}\right|_{\boldsymbol{P}_i} \left.\frac{\partial F}{\partial y_i}\right|_{\boldsymbol{P}_i} \left.\frac{\partial F}{\partial z_i}\right|_{\boldsymbol{P}_i} \right]^{\mathrm{T}}, \tag{4}$$

and then we apply the law of reflection to the line $\hat{d}$: $\hat{r} = 2\left(\hat{n}^{\mathrm{T}}\hat{d}\right)^{\mathrm{T}} \hat{n} - \hat{d}$ (Fig. 3).

Reflection on Mirrors of any Shape. In genral, mirrors are defined by the points of their profile, which can therefore approximated by a linear spline. This means that the mirror surface is a stack of conic trunks. To intersect a line with a conic trunk, one must first intersect the line with the full cone, and then check whether this intersection falls within the trunk. As we have a stack of trunks, the intersection must be computed separately for each trunk. Typically, only one intersection will survive the interval test; if more than one does, the one closest to the origin of the ray is chosen. In Fig. 4 ray r intersects all cones $(m_{1..3})$, but only the one with m_2 falls within the validity interval of the trunk.

[1] A *spline* is a curved line formed by two or more *vertices*, or *control points*, and a mathematical formula describing the curve(s) between them; if this formula is an n-th degree polynomial, then we get an n-th degree spline.

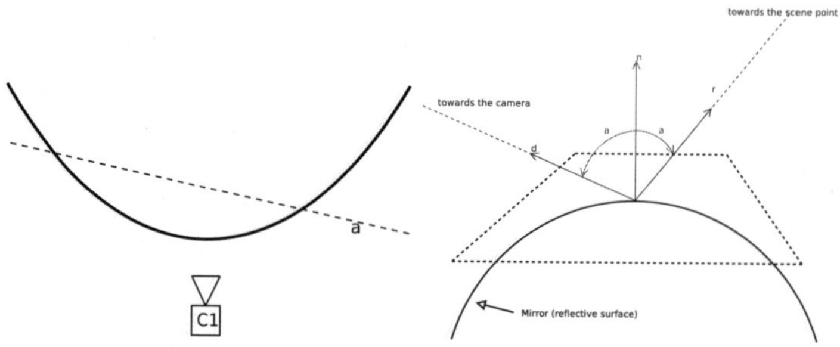

(a) Intersection of incoming ray with the mirror surface

(b) Reflection of a ray on a curved surface

Fig. 3. The reflection model for the interpretation lines

Computing the Roto-translation. The reflected interpretation lines pass through their corresponding scene points, i.e., the distance between a line and its point is zero. We look for the roto-translation $\mathcal{T}_{C\to W}$ that zeroes the distances between the fiducials and their reflected interpretation lines. Actually, because of noise, the lines will only pass *close* to the fiducials, i.e., we can only find a roto-translation M that minimizes the sum of the squares of the fiducial distances to their interpretation lines. If we call $\boldsymbol{p}_i$ and $\boldsymbol{q}_i(t)$, $(i = 1...N)$, respectively the fiducials and the parametric equations of the reflected interpretation lines. We look for a M that minimizes

$$\mathcal{J} = \sum_{i=1}^{N} \|\boldsymbol{p}_i - M\boldsymbol{q}_i(t_i)\| \tag{5}$$

where t_i is the value of the parameter t that minimize the distances between the lines $\boldsymbol{q}_i(t)$ and the fiducials $\boldsymbol{p}_i$, for a given M. M is computed by solving a nonlinear system of equations. The non-linearity comes from the orthonormality constraint on the rotational part of M.

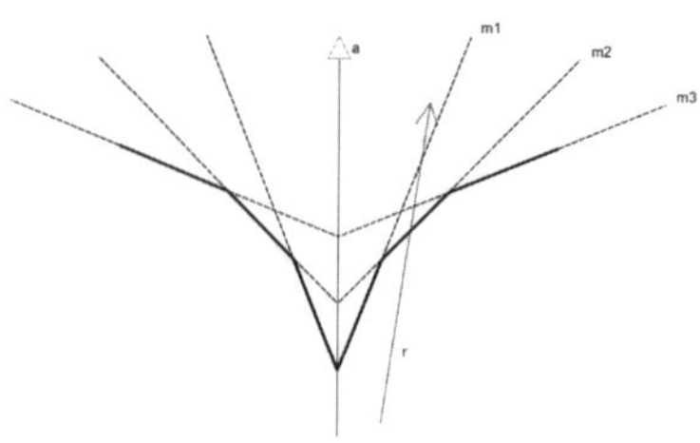

Fig. 4. Intersection of a line with a mirror profile approximated by a linear spline

3 Experimental Results

In order to test the correctness of the algorithm we tried it on synthetic images, so as to compare the results computed by our method against a ground truth. We performed the intrinsic and extrinsic calibrations separately, to avoid errors in the first phase to corrupt the results in the second phase. At the end of this section, we provide a complete calibration for a COPIS system.

3.1 Intrinsic Parameters

For space limits we present only two tests, using the same conic mirror, tilted by $\theta = 0°, 10°$ respectively. For each test we present the expected value (ground truth, GT) and the values obtained for the mirror position ($\boldsymbol{c}$, $\boldsymbol{c}_1$, $\boldsymbol{c}_2$ respectively), for the projection of the mirror centre ($\boldsymbol{p}$, $\boldsymbol{p}_1$, $\boldsymbol{p}_2$, in pixels) and for the normal to the support plane ($\hat{n}$, $\hat{n}_1$ and $\hat{n}_2$, where $\hat{n} = [\sin\theta, 0, \cos\theta]^{\mathrm{T}}$). We also computed the angles ϕ, ϕ_1, ϕ_2 respectively formed by $\hat{n}$, $\hat{n}_1$ and $\hat{n}_2$ with the z-axis of the camera reference frame. We present the differences $\Delta_1 = \phi - \phi_1$ and $\Delta_2 = \phi - \phi_2$ (expressed in degrees) between the expected angles and the computed ones. This difference is a good estimate for the mirror orientation error.

1. $\theta = 0°$

$$\hat{c} = \begin{bmatrix}0\ 0\ 100\end{bmatrix}^{\mathrm{T}}, \ \boldsymbol{c}_1 = \begin{bmatrix}0.066\ {-0.146}\ 100.003\end{bmatrix}^{\mathrm{T}} \ \boldsymbol{c}_2 = \begin{bmatrix}0.01\ {-0.312}\ 100.003\end{bmatrix}^{\mathrm{T}}$$

$$\boldsymbol{p} = \begin{bmatrix}319.5\ 239.5\end{bmatrix}^{\mathrm{T}}, \ \boldsymbol{p}_1 = \begin{bmatrix}319.681\ 240.703\end{bmatrix}^{\mathrm{T}}, \ \boldsymbol{p}_2 = \begin{bmatrix}319.521\ 238.499\end{bmatrix}^{\mathrm{T}}$$

$$\hat{n} = \begin{bmatrix}0\ 0\ 1\end{bmatrix}^{\mathrm{T}}, \ \hat{n}_1 = \begin{bmatrix}0.001\ 0.010\ 0.999\end{bmatrix}^{\mathrm{T}}, \ \hat{n}_2 = \begin{bmatrix}0.000\ {-0.008}\ 0.999\end{bmatrix}^{\mathrm{T}}$$

$$\Delta_1 = 0.632, \ \Delta_2 = 0.547$$

1. $\theta = 5°$

$$\hat{c} = \begin{bmatrix}0\ 0\ 100\end{bmatrix}^{\mathrm{T}}, \ \boldsymbol{c}_1 = \begin{bmatrix}3.492\ 0.097\ 100.22\end{bmatrix}^{\mathrm{T}} \ \boldsymbol{c}_2 = \begin{bmatrix}0.174\ 0.069\ 100.281\end{bmatrix}^{\mathrm{T}}$$

$$\boldsymbol{p} = \begin{bmatrix}319.5\ 239.5\end{bmatrix}^{\mathrm{T}}, \ \boldsymbol{p}_1 = \begin{bmatrix}303.092\ 239.389\end{bmatrix}^{\mathrm{T}}, \ \boldsymbol{p}_2 = \begin{bmatrix}319.169\ 239.66\end{bmatrix}^{\mathrm{T}}$$

$$\hat{n} = \begin{bmatrix}0.087\ 0\ 0.996\end{bmatrix}^{\mathrm{T}}, \ \hat{n}_1 = \begin{bmatrix}0.048\ 0.000\ 0.998\end{bmatrix}^{\mathrm{T}}, \ \hat{n}_2 = \begin{bmatrix}0.085\ 0.001\ 0.996\end{bmatrix}^{\mathrm{T}}$$

$$\Delta_1 = 2.222, \ \Delta_2 = 0.127$$

2. $\theta = 10°$

$$\hat{c} = \begin{bmatrix}0\ 0\ 100\end{bmatrix}^{\mathrm{T}}, \ \boldsymbol{c}_1 = \begin{bmatrix}7.003\ 0.092\ 100.029\end{bmatrix}^{\mathrm{T}} \ \boldsymbol{c}_2 = \begin{bmatrix}0.102\ 0.074\ 100.274\end{bmatrix}^{\mathrm{T}}$$

$$\boldsymbol{p} = \begin{bmatrix}319.5\ 239.5\end{bmatrix}^{\mathrm{T}}, \ \boldsymbol{p}_1 = \begin{bmatrix}285.694\ 238.972\end{bmatrix}^{\mathrm{T}}, \ \boldsymbol{p}_2 = \begin{bmatrix}319.51\ 239.645\end{bmatrix}^{\mathrm{T}}$$

$$\hat{n} = \begin{bmatrix}0.173\ 0\ 0.984\end{bmatrix}^{\mathrm{T}}, \ \hat{n}_1 = \begin{bmatrix}0.105\ 0.000\ 0.994\end{bmatrix}^{\mathrm{T}}, \ \hat{n}_2 = \begin{bmatrix}0.175\ 0.001\ 0.984\end{bmatrix}^{\mathrm{T}}$$

$$\Delta_1 = 3.997, \ \Delta_2 = 0.067$$

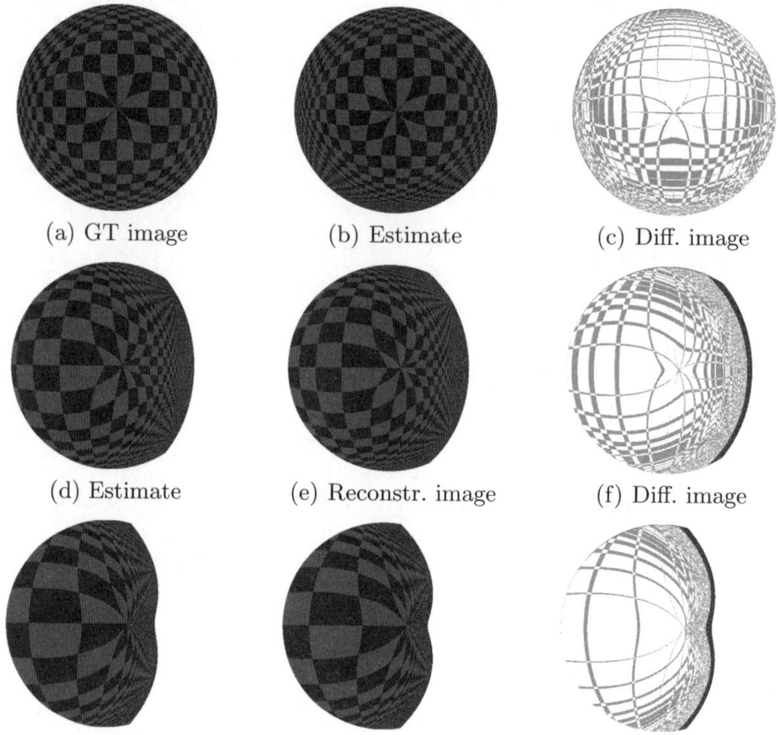

(a) GT image (b) Estimate (c) Diff. image

(d) Estimate (e) Reconstr. image (f) Diff. image

(g) GT image (h) Estimate (i) Diff. image

Fig. 5. Mirror not tilted (a,b,c); tilted 5° (d,e,f); tilted 10° (g,h,i)

Note how an increase in the mirror tilt improves the precision in the (correct) pose estimate and increases also the difference between the two solutions.

3.2 Extrinsic Parameters

We generated a set of images where the world reference frame matched the camera reference frame, in order to have a trivial ground truth. We then applied the calibration algorithm keeping the fiducials set fixed but varying the initial estimate. This test allowed us to verify when, and how quickly, the algorithm converged. As a measure of the quality of the result we used (5), normalized with respect to the number N of fiducials (ξ_N). For each test, we present the initial estimate for rotation angle $\tilde{\theta}$ (in degrees, around the world z-axis), the initial estimate for translation $\tilde{T}$, the goodness of the initial estimate $\tilde{\xi}_N$ (before a solution is computed) and the goodness of the calibration ξ_N (after a solution is computed), and the number n of iterations performed by the algorithm. The results are reported in Tab. 1.

3.3 Calibration of a Real COPIS System

Before the catadioptric calibration can begin, the *intrinsic* parameters of the camera must be known. We used *camcal* [16] to calibrate our 8mm Sony XC711, which yielded the following intrinsic parameters matrix:

$$K = \begin{bmatrix} -626.438 & 0 & 317.218 \\ 0 & -604.007 & 231.424 \\ 0 & 0 & 1 \end{bmatrix} .$$

The mirror base is well visible and, once the ellipse equation is extracted, the following values are found for the normal $\hat{n}_{1,2}$, the centre position $\boldsymbol{c}_{1,2}$, and its projection $\boldsymbol{p}_{1,2}$:

$$\hat{n}_1 = \begin{bmatrix} 0.133 \; 0.0059 \; 0.991 \end{bmatrix}^{\mathrm{T}}, \quad \hat{n}_2 = \begin{bmatrix} -0.0072 \; -0.0265 \; 0.999 \end{bmatrix}^{\mathrm{T}}$$

$$\boldsymbol{c}_1 = \begin{bmatrix} 6.388 \; -1.603 \; 123.501 \end{bmatrix}^{\mathrm{T}}, \quad \boldsymbol{c}_2 = \begin{bmatrix} 9.216 \; -0.951 \; 123.329 \end{bmatrix}^{\mathrm{T}}$$

$$\boldsymbol{p}_1 = \begin{bmatrix} 284.815 \; 239.268 \end{bmatrix}^{\mathrm{T}}, \quad \boldsymbol{p}_2 = \begin{bmatrix} 270.028 \; 235.754 \end{bmatrix}^{\mathrm{T}}$$

Because the two solutions are so similar, we cannot solve the ambiguity yet and so we estimate the extrinsic parameters for both. We will eventually choose the solution with the smallest residual. The next step is the collection of fiducial points. We are currently developing a new version of the tool, whose calibration pattern is a checkerboard to ease fiducials collection. That not available, we choose 40 among the most visible dots of the grid pattern (the ones in red in Fig. 6*a*). Having collected the fiducials, an initial estimate is needed to compute the extrinsic parameters. This initial estimate consists of a rotation matrix $\tilde{R}$ and a translation vector $\tilde{T}$. The latter can be measured by hand, while the rotation may be input as the angle $\tilde{\theta}$ around the world z-axis, assuming that the camera is approximately pointed vertically upward. In this case, the initial estimate was

$$\tilde{R} = I, \quad \tilde{T} = \begin{bmatrix} -1105 \; -780 \; -20 \end{bmatrix}^{\mathrm{T}}$$

which yielded the following residuals (for the two possible mirror poses): $\tilde{\xi}_{N_1} = 80924$, $\tilde{\xi}_{N_2} = 2306.78$. After the extrinsic parameters are computed, we get the following results (the subscripts indicate either the first or the second pose):

$$R_1 = I, \quad T_1 = \begin{bmatrix} -1105 \; -780 \; -20 \end{bmatrix}^{\mathrm{T}}, \quad \xi_{N_1} = 80924$$

Table 1. Convergence speed and quality of the results w.r.t. the initial estimate

$\tilde{\theta}$	0	0	0	0	0	0	1	2	10	1	1
$\tilde{T}$	$\begin{bmatrix}0\\0\\0\end{bmatrix}$	$\begin{bmatrix}0\\0\\1\end{bmatrix}$	$\begin{bmatrix}0\\0\\2\end{bmatrix}$	$\begin{bmatrix}0\\0\\10\end{bmatrix}$	$\begin{bmatrix}10\\0\\0\end{bmatrix}$	$\begin{bmatrix}10\\10\\10\end{bmatrix}$	$\begin{bmatrix}0\\0\\0\end{bmatrix}$	$\begin{bmatrix}0\\0\\0\end{bmatrix}$	$\begin{bmatrix}0\\0\\0\end{bmatrix}$	$\begin{bmatrix}0\\0\\10\end{bmatrix}$	$\begin{bmatrix}10\\10\\10\end{bmatrix}$
$\tilde{\xi}_N$	0	1.232	4.926	123.143	33.222	470.88	48.321	1753.97	5109.54	205.589	808.635
ξ_N	0	0	0	0	0	0	0	0	112.241	0	0
n	1	7	14	68	35	116	82	161	490	107	225

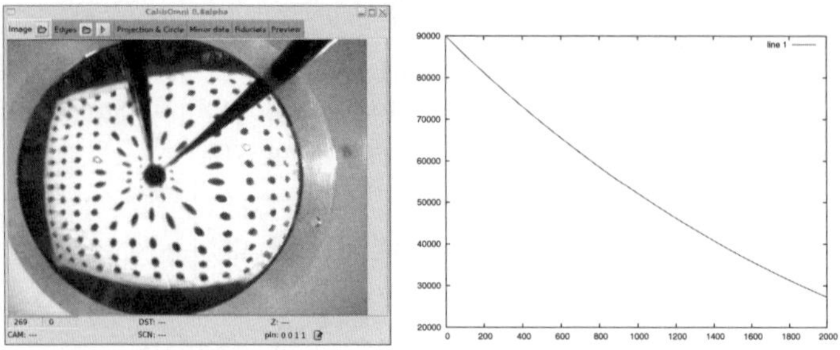

(a) Calibration pattern (b) Convergence speed

Fig. 6. Calibration of a COPIS system

$$R_2 = \begin{bmatrix} 0.999 & 0.019 & -0.02 \\ -0.019 & 0.999 & -0.021 \\ 0.019 & 0.022 & 0.999 \end{bmatrix}, \quad T_2 = \begin{bmatrix} -1106.29 \\ -760.09 \\ -87.88 \end{bmatrix}, \quad \xi_{N_2} = 699.438\,.$$

Note how the residual of option 1 is much greater than the residual of option 2. Also note that a solution could not be computed for option 1. This means that the correct mirror position is option 2. The convergence graph can be seen in Fig. 6*b*. Using these 40 fiducials, an AMD 2400+ can compute approximately 100 iterations per second.

A comparison with our method is unfortunately unfeasable. The only other tool that matches ours in flexibility is Strelow's [11], but it is based on a different setup and no software is available for download. All other tools are not generic w.r.t. mirror shape or pose.

4 Conclusions

We described a mathematical model for catadioptric image formation which is generic with respect to mirror shape and pose and to the camera optics. Then, we introduced a calibration method that allows to estimate the model parameters. The experimental session proves the method to be valid and gives a quantitative measure of the precision The main flaw of decomposing the calibration process in mirror localization and extrinsic parameters estimate is that an error in the first phase propagates to the second phase. A way to overcome this problem is to add a third phase where a global post-optimization is performed. This phase consists of solving a non-linear system where all parameters (both intrinsic and extrinsic) are unknown, and where the results of the previous two phases are used as an initial estimate; this is what we are currently developing.

References

1. Yagi, Y., Kawato, S.: A panorama scene analysis with conic projection. In: Proc. IROS90, vol. 1, pp. 181–187 (1990)
2. Baker, S., Nayar, S.: A theory of catadioptric image formation. In: Proc. ICCV98
3. Geyer, C., Daniilidis, K.: Paracatadioptric Camera Calibration. IEEE T-PAMI 25(5), 687–695 (2002)
4. Kaidan Incorporated: `http://www.kaidan.com/`
5. Charles, J.: How to build and use an all-sky camera. Astronomy Journal (1987)
6. Conroy, T.L., Moore, J.B.: Resolution invariant surfaces for panoramic vision systems. In: Proc. ICCV99
7. Blind for the review: 36(2-3), 87–102
8. Gaspar, J., Decco, C., Okamoto, J.J., Santos-Victor, J.: Constant resolution omnidirectional cameras. In: Proc. OmniVis02, pp. 27–34 (2002)
9. Mei, C.: Omnidir. Calibr. Toolbox Extension (2005), `http://www-sop.inria.fr/icare/personnel/Christopher.Mei/ChristopherMeiPhDStudentToolbox.html`
10. Scaramuzza, D., Martinelli, A., Siegwart, R.: A Flexible Technique for Accurate Omnidirectional Camera Calibration and Structure from Motion. In: Proc. of ICVS06 (2006)
11. Strelow, D., Mishler, J., Koes, D., Singh, S.: Precise Omnidirectional Camera Calibration. IEEE Computer Society Conference on Computer Vision and Pattern Recognition (CVPR'01) 1, 689 (2001)
12. Wang, J.G., Sung, E.: Eye gaze estimation from a single image of one eye. Ninth IEEE International Conference on Computer Vision 1, 136–143 (2003)
13. Kanatani, K.: Geometric Computer Vision. Oxford Science Publications, Oxford (1993)
14. Fremont, V., Ryad, C.: Direct camera calibration using two concentric circles from a single view. In: Proc. ICAT02 (2002)
15. Van Wijk, J.: Ray Tracing Objects Defined by Sweeping Planar Cubic Splines. In: ACM Transaction on Graphics, ACM Press, New York (1984)
16. Tarel, J.P., Vezien, J.M.: CamCal v.10 Manual - A Complete Software Solution for Camera Calibration. Technical report, INRIA (1996)

An Automated Refereeing and Analysis Tool for the Four-Legged League*

Javier Ruiz-del-Solar, Patricio Loncomilla, and Paul Vallejos

Department of Electrical Engineering, Universidad de Chile

Abstract. The aim of this paper is to propose an automated refereeing and analysis tool for robot soccer. This computer vision based tool can be applied for diverse tasks such as: (i) automated game refereeing, (ii) computer-based analysis of the game, and derivation of game statistics, (iii) automated annotations and semantic descriptions of the game, which could be used for the automatic generation of training data for learning complex high-level behaviors, and (iv) automatic acquisition of real game data to be used in robot soccer simulators. The most attractive application of the tool is automated refereeing. In this case, the refereeing system is built using a processing unit (standard PC) and some static and/or moving video cameras. The system can interact with the robot players and with the game controller using wireless data communication, and with the human spectators and human second referees by speech synthesis mechanisms or using visual displays. We present a refereeing and analysis system for the RoboCup Four-Legged League. This system is composed by three modules: object perception, tracking, and action analysis. The camera placement issue is solved by human controlled camera's placement and movement. Some preliminary experimental results of this system are presented.

1 Introduction

One of the RoboCup main goals is allowing robots play soccer as humans do. One interesting extension of this idea is having automated refereeing in the soccer games. The long-term goal could be developing robot referees, but an interesting intermediate step, to be developed in the next few years, is building automated refereeing using static and/or moving video cameras. Such a refereeing system can interact with the robot players and with the game controller using wireless data communication, and with the human spectators and human second referees by speech synthesis mechanisms or using visual displays.

Video-based refereeing employs computer vision methods that allow automating tasks like tracking, refereeing (taking refereeing decisions), annotation, indexing and generation of semantic descriptions. Therefore, an automated soccer referee is an analysis tool that can also be employed for performing a computer-based analysis of the game, an automated summary of the game, or for obtaining game statistics (% of the time that the ball is in each half of the field, % of ball possession of each team, number of goals of each team player, number of direct kicks to the goal by each team,

---

* This research was partially supported by FONDECYT (Chile) under Project Number 1061158.

G. Lakemeyer et al. (Eds.): RoboCup 2006, LNAI 4434, pp. 206–218, 2007.

number of faults of each team and each team player as ball holding, illegal defender, etc.). This statistical analysis can be also carried out offline, that means that the tool can be employed for analyzing game videos obtained using different recording systems.

Moreover, the same analysis tool can be used for obtaining annotations and semantic descriptions of the game, which could be used for the automatic generation of training data for learning complex high-level behaviors, as for example robot passing or team strategy planning. Furthermore, taking into account the availability of realistic simulators in some of the RoboCup soccer leagues (e.g., SimRobot [1] and UCHILSIM [2] in the Four-Legged league), the obtained annotations and semantic descriptions of the game can be used for the acquisition of real game data to be used in simulations. In this way, a robot controller can be adapted in a simulation of a real game situation, previously acquired by the automated analysis tool.

In this context, the aim of this paper is to propose such an automated refereeing and analysis system for the RoboCup Four-Legged League. To the best of our knowledge a similar system has not been proposed in the literature.

This paper is organized as follows. In section 2 some related work is presented. The here-proposed refereeing and analysis system is presented in section 3. In section 4 some experimental results of the application of this system are presented. Finally, in section 5 some conclusions of this work are given.

2 Related Work

Computer vision based analysis of sport videos has been addressed by many authors (e.g. [3]-[23]), and nowadays is a hot topic within the multimedia video analysis community. There are also successful video analysis programs that are being used by TV sport channels (e.g. Hawkeye [24] and Questec [25]). Applications have been developed in almost all massive sports such as tennis ([13][14][16][20]), soccer ([4][6][7][17][18][22][23]), baseball ([10][15]), basketball ([12]), and American football ([11]). However, to the best of our knowledge the automatic analysis of robot sports, as for example robot soccer, has not been addressed in the literature.

The major research issues for the automatic analysis of human sports include (see a survey in [20]): ball and players tracking, landmarks detection (goal mouth, oval, side lines, corners, etc.), tactic analysis for providing training assistance, highlight extraction (ace events in tennis, score events in soccer and basketball, etc.), video summarization (automatic generation of game summaries), content insertion (e.g. commercial banner replacement in the field), and computer-assisted refereeing (e.g. offside detection). Some of these research issues are still open as for example fully autonomous tactic analysis, soccer offside detection considering player's intention, or video summarization; current solutions are semi-automatic and provide assistance to human referees or video operators.

We believe that many of the accumulated knowledge in the automatic analysis of human sports can be employed for the automatic analysis of robot sports. Probably the main obvious difference being that in the robot sport case, robot players need to be detected and identified. The main problem for identifying robots is the fact that all players look the same. However, the robots can be individualized using the team uniform color [26], the player's number [28], or both.

The here-proposed automated refereeing and analysis system for the RoboCup Four-Legged league makes use of the accumulated knowledge in automatic analysis of human sports, and the image analysis tools developed in the Four-Legged league in the last years.

3 Proposed Automated Refereeing and Analysis System

3.1 General Description

The block diagram of the proposed refereeing and analysis system is shown in figure 1. The system is composed by three main subsystems *Object Perception*, *Tracking*, and *Motion & Action Analysis*, and makes use of two databases *Game Rules* (input) and *Game Statistics* (output).

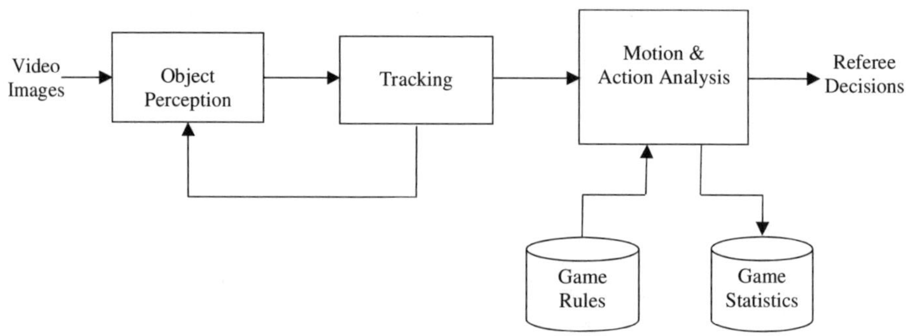

Fig. 1. Block diagram of the refereeing and analysis tool

The *Object Perception* module has three main functions: object detection, object identification, and camera self-localization. First, all the objects of interest for the soccer game (field carpet, field and goal lines, goals, beacons, robot players, ball) are detected using color segmentation and some simple rules. No external objects, as for example, spectators or human referee's legs are detected. The detection operation allows also the identification of goals, beacons and the ball, because each of them has a defined form and color composition (see current definitions in [30]). The identification of field and goal lines is performed using the relative distance from the detected lines to the already identified beacons and goals The identification of the robot players is much more difficult, and is performed using local image descriptors as proposed in [28]. Real-time robot identification is achieved thanks to the computation of the local descriptors only in the detected robot regions, and because of this identification is carried out only at the beginning of each robot tracking sequence (see the feedback connection from *Tracking* to *Object Perception* in figure 1). The camera self-localization is computed using the pose of the landmarks (goals and beacons) and the lines, and the internal parameters of the camera.

The *Tracking* module is in charge of tracking the moving objects, i.e. the ball and the robot players. The implemented tracking system is built using the mean shift algorithm [29], applied over the original image (not the segmented one). The seeds of the tracking process are the detected ball and robot players. As in [31], a Kalman Filter is employed for maintaining an actualized feature model for mean shift. In addition, a fast and robust line's tracking system was implemented. Using this system it is not necessary the detection of the lines in each frame.

The *Motion & Action Analysis* module is in charge of analyzing the game dynamics and the actions performed by the players (e.g., kicking or passing), and detecting game relevant events (goal, ball out of the field, illegal defender, etc.). This analysis is carried out using information about static and moving detected objects, and the game rules, which are retrieved from the *Game Rules* database. The outputs of the *Motion & Action Analysis* module are refereeing decisions (e.g. goal was scored by team A), and game statistics (e.g. player 2 from team A score a goal) that are stored in the corresponding database.

3.2 Object Perception

(i) *Beacons, Goals, Carpet and Ball Detection and Identification.* Landmarks (beacons and goals), the field carpet and the ball are perceived using a standard color-based vision method that consists of three main stages: color classification (segmentation), blob formation and object detection. Color classification and blob formation are relatively common principles amongst RoboCup teams and standard solutions for these tasks already exist (for further information read for example [27]). In the RoboCup Four-Legged league all game objects has a distinctive form and a distinctive color composition, therefore they can be detected (identified) in the color space using simple rules (e.g. [26]). For instance, the ball is spherical and orange, and there is no other orange object in the field. Usually, false detections are filtered out using some geometrical considerations (e.g. to be near the image horizon). See [26] for details.

(ii) *Robots Detection and Identification.* Sony legged robots are non-rigid objects, and therefore their form is difficult to describe. However, they can be detected in the color segmented images using some simple heuristics, and the facts that (a) robots are over the field carpet, and (b) robots have a different color composition than the other field objects (landmarks and ball). Therefore, after detecting all other field objects, non-green patches over the field carpet are robot candidates. The color composition of these patches is checked. Robots are white (ERS7), while their uniforms are blue or red. Hence, each robot candidate patch should include two of these colors in a given amount to be considered a robot. This color information can also be used to classify the robot in one of two classes: blue team robot or red team robot. Certainly, other colors can also be present in the candidate patches, but in a much smaller amount (black because of the shadows, or other colors because of a wrong segmentation). After the robots' detection and team classification, the remaining task is the robot identification, that is, to know the robot identity or number. This identification is performed by calculating local descriptors (SIFT features [32]) in the candidate patch, and by comparing those descriptors against descriptors of robots' templates, already stored in a robot model database. For performing this task we use the same methodology proposed in [28]. It is important to remember that all robots and also the ball are tracked in the *Tracking* module. Therefore, during the time that the robots (and the ball) are tracked (several frames) they do not need to be detected and identified again.

(iii) *Lines Detection and Identification.* Images can be analyzed very quickly using a grid of vertical and horizontal scan virtual rays; lines perpendicular to these scan rays can be found easily, generating line keypoints, which can be used to achieve line detection. There are two ways for generating these line keypoints: (a) by detecting differences in the intensity channel along the scan line, and selecting those that exceeds a given threshold, or (b) by making a convolution between the scan line and a set of edge detection filters (e.g. [−1;+2;−1]) of different sizes (fast implementation using differences between pixel's values), and selecting those points where the filtering operation exceeds a given threshold and that correspond to a local maximum over the line and over the neighboring scales. We choose the second option because is more robust against motion blur, and line scale and orientation changes in the image. Candidate line keypoints that do not have green (carpet) pixels around are filtered out. Then, the remaining candidate line keypoints are summarized using a Hough Transform, which allows the determination of the lines' parameters. A last test for accepting the candidate lines consist on generating three parallel scan lines, where the middle scan line corresponds to the detected line. If the mean intensity value of the central scan line is much larger than the intensity value of the lateral scan lines, the line candidate is accepted. After line detection, the line classification is performed using the relative distance from the detected lines to the already detected beacons and goals, and some simple rules. For example, the yellow goal line is the nearest parallel line to the yellow goal.

(iv) *Camera Self-Localization.* This feature is not required in all applications, but in cases where it is desired to obtain an exact description of the game (e.g., automatic generation of training data for learning complex behaviors or real game data to be used in simulations). The camera self-localization is implemented using the standard procedures employed in the Four-Legged league, which makes use of the landmarks information, lines information, and intrinsic parameters of the camera (see for example [26]).

3.3 Tracking

The *Tracking* module is in charge of tracking all moving objects: the ball and all visible robot players (up to 8). The implemented tracking system is performed using the mean shift algorithm [29] applied over the original image (not the segmented one). We use RGB color histograms as feature vectors (model) for mean shift, with each channel quantized to 32 bins. The feature vector is weighted using an Epanechnikov kernel (see [29]). For each detected and identified moving object a certain mean shift model is initialized.

As in [31], a Kalman Filter is employed for maintaining an actualized feature model for mean shift. The state vector of the Kalman filter is the weighted feature vector (RGB histogram) of the object under tracking. Thus, in each tracking process a

Kalman filter estimates the state of the corresponding mean shift model. This allows that longer tracking episodes can be maintained. Each time that the tracking of a moving object is lost, the corresponding object is detected and identified again.

Considering that the camera can also be mobile, it is important to track the lines. The lines have a second-order dynamic model whose state corresponds to the line inclination and position, which permits future line position prediction. The lines are tracked in a simple and robust way by generating small lateral scan lines that are perpendicular to each line prediction. Then for each line, the maximum intensity value in each lateral scan line is determined. Maxima points that are positioned over one of the two extremes of the lateral scan line are discarded. The remaining maxima points for each line are clustered into line segments. Line segments with less than 8 maxima points or with no overlap (over the line axis) with the previous line detection are discarded. Then, a least-square analysis of the surviving maxima gives the new line position estimation. If no line segment has more than 8 maxima points, the sizes (scale) of the perpendicular scan lines are enlarged and the process is repeated. If the scan lines are larger than a threshold (12 pixels in our implementation), the line tracking process fails and the line model is destroyed.

Using the described tracking processes, we are able to track in real time (30 fps) all game moving objects and the lines.

3.4 Motion and Action Analysis

This module is in charge of analyzing the game dynamics, determining the actions performed by the players, and the game relevant events. This analysis is carried out using the information of the static and moving detected objects, and the game rules (defined in [30]).

Most of the game situations can be analyzed using information about the position in the field of the ball and the robot players, the exact localization of the field lines and goal lines, and the identity of the last robot that touches the ball. For instance, for detecting a goal event, a ball crossing one of the two goal lines should be detected. The identity of the scoring robot is the identity of the last robot that touches the ball. Thus, using simple rules the following situations (definitions taken from [30]), can be analyzed:

- *Goal:* "A goal is achieved when the entire ball (not only the center of the ball) goes over the field-side edge of the goal line, i. e. the ball is completely inside the goal area".
- *Robots kickoff positioning:* "In the ready state, the robots should walk to their legal kick-off positions. These positions are always located inside their own side of the field. Only two field players of the attacking team can walk to positions between the centerline and the middle of their side. They may put their leg(s) on the center circle line, but no leg may be inside the circle line. The other field players (one of attacking team, three of defending team) have to be located behind the middle of their side (none of their legs are allowed to be in front of the line connecting the centers of beacons), and must have at least two feet outside the penalty area. In contrast, the goal keeper must have at least two feet inside the penalty area", and "The robots have a maximum of 30 seconds to reach their positions".

- *Ball leaving the field:* "A ball is considered to have left the field when there is no part of the ball over the green field inside the boundary line".
- *Global Game Stuck:* "No robot touches the ball for 30 seconds".
- *Robot leaving the field:* "A robot that leaves the 4mx6m carpeted area will be removed for 30 seconds as per the Standard removal penalty".
- *Illegal defense:* "The vertical projection of the goalie to the goal line should not occupy more than 50 percent of the length of the goal mouth".

Moreover, in the case of the *Goal*, *Ball leaving the field* and *Robot leaving the field* situations, the identity of the protagonist robot (the scoring robot, the robot leaving the field or the robot that through out the ball) is determined (SIFT descriptors).

However, there are some other situations that are much harder to analyze, because it is required either to judge the intention of the players (e.g. robot obstruction) or to solve visual occlusions that difficult the determination of the relative position of the robot legs or the ball (e.g. ball holding). We are currently working towards a robust detection of these harder situations (see list bellow). For instance, using SIFT descriptors for determining which robots are looking to the ball (a similar idea was implemented in [28]), allows solving the *Robot obstruction* situation. The list of non-completely solved situations is (definitions taken from [30]):

- *Ball Holding:* "A robot which does not leave enough open space around the ball will be penalized as 'Ball Holding' if that situation continues more than 3 seconds or if the robot moves the ball over 50cm while continuing to hold the ball".
- *Goalie Pushing:* "When the goalie is in its own penalty area (2 feet on or inside line), no attacker may be in contact with the goalie for more than 3 seconds or be pushing the goalie indirectly through the ball for more than 3 seconds."
- *Field Player Pushing:* "Any robot pushing another robot for more than 3 seconds will be removed from play for 30 seconds as per the standard removal penalty", and "The closest robot (including the goal keeper) to the ball on each team, if it is within 2 dog-lengths of the ball, cannot be called for pushing"
- *Illegal defender:* "Having three legs inside the penalty area is the definition of being in the penalty area and that situation is not allowed for defending field players", but "This rule will be applied even if the goalie is outside of the penalty area, but not if an operational defender is pushed into the penalty area by an opponent."
- *Robot obstruction:* "When a robot that is not heading for the ball is actively, and intentionally, blocking the way to the ball for a robot of the other team, this is an obstruction"

Summarizing, for refereeing purposed we can automatically detect the following situations: correct *Robots kickoff positioning*, *Goal event*, *Ball leaving the field*, *Robot leaving the* field, *Global Game Stuck*, and *Illegal defense*. The game statistics that can be computed in our current implementation are: % of the time that the ball is in each half of the field, number of goals of each team and team player, number of direct kicks to the goal by each team and each team player, number of times that each team and each team player sent the ball out of the field, number of illegal defense events of each goalie, number of times that each team player leaves the field, number of global game stuck events, and time required for automatic kickoff positioning by each team.

4 Experimental Results

Experiments were carried out with the objective of characterizing the proposed system. Quantitative results are presented for the tracking of objects (lines, players and ball). Qualitative results are presented for the some events detection.

4.1 Object Tracking

Line tracking. For characterizing the line tracking subsystem, 19 video sequences were analyzed (30 fps, around 14.5 seconds each one, all together 8,313 frames). Lines to be tracked are initialized and the tracked. During the video analysis most of the lines are correctly tracked (see table 1). However, in some cases lines cannot be tracked, generally due to strong camera movements, and the line's model is destroyed. In other cases the tracking of some lines became confuse and other object (generally other lines) starts to be tracked. All these cases, including the situation when lines get out of the image as the camera moves are summarized in table 1.

Player tracking. For characterizing the player tracking subsystem, the same 19 video sequences used for line tracking analysis were employed. During the video analysis most of the players are correctly tracked (see table 1). However, in some cases players cannot be tracked, generally due to strong camera movements, and the player's model is destroyed. In other cases the tracking of some players became confuse and other object (generally other players) starts to be tracked. All these cases, including the situation when players move out of the image as the camera moves are summarized in table 1.

Ball tracking. Ball tracking is done using two sources of information: the mean shift information and (very fast) ball detection subsystem. These both information sources are fused using a Kalman. The same 19 video sequences were used for the ball tracking analysis In all the analyzed frames, the ball tracking has no lost tracking or incorrect tracking problems.

Table 1. Tracking statistics

	Line Object	Player Object
Total number of analyzed frames	8313	8313
Total number of initialized objects	83	38
Total number of objects which get out of the image	26	13
Total number of objects with lost tracking	9	3
Total number of objects with incorrect tracking	2	2

4.2 Event Detection

The event detection capabilities of the proposed system are still not well characterized, we just have qualitative results, which are very promissory. In figures 2 and 3 we show two exemplar video sequences where it can be seen the detections of a *Goal* event and a *Ball leaving the field* event, respectively. In these sequences the ball tracking window is shown in blue, while the players' tracking windows are shown in red. Every time a

Fig. 2. Selected frames from a robot scoring sequence. Robot/ball tracking window in red/blue. The goal detection event (frame 115) is shown by a red window out of the ball blue window.

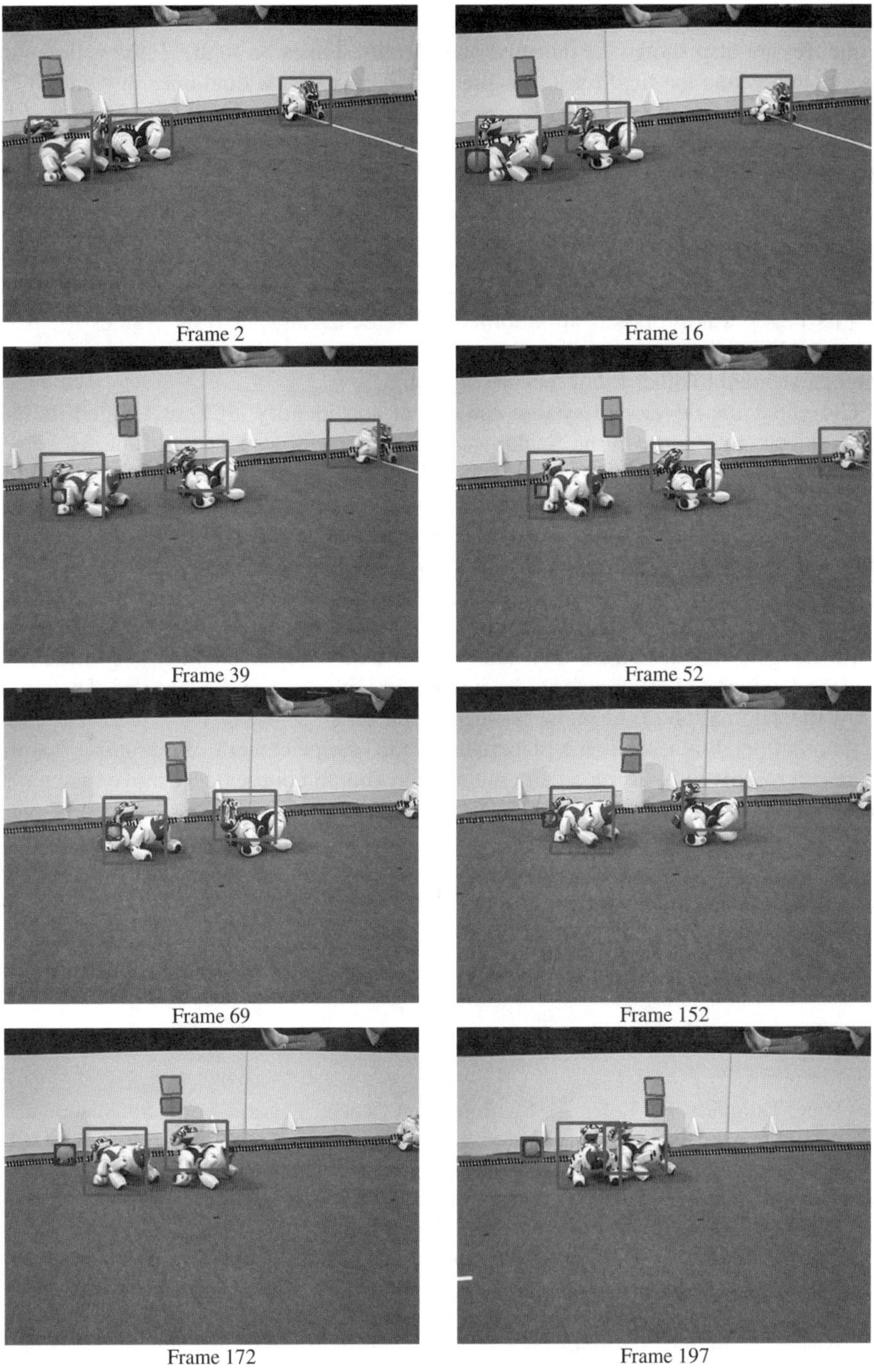

Fig. 3. Selected frames from a ball leaving the field sequence. Robot/ball tracking window in red/blue. The ball out of the field is marked by a red window out of the blue one (frame 197).

ball event occurs, a red window is displayed out of the ball's blue window. In the sequences are also shown the detected and identified lines; in figure 2, the yellow goal line in yellow and the field lines in white, and in figure 3 the field lines in white. It can also be seen the small lateral lines (perpendicular to each goal and field line) used for line tracking. Finally, the detected and identified landmarks are also indicated with boxes (yellow goal in figure 2 and pink-cyan beacon in figure 3).

5 Conclusions

In this paper was proposed an automated refereeing and analysis system for robot soccer. Although this system is originally developed for the Four-Legged league, it can be extended to other robot soccer contexts.

Currently, the developed system can detect and identify all Four-Legged defined field objects, and perform the tracking of the moving objects in real-time. However, not all defined situations can be robustly detected, mainly because the detection of some complex situations requires either to judge the intention of the players or to solve visual occlusions. In this sense the proposed system is still under development, but we think that the preliminary results show its potential.

We believe that developing a robust automated referee could be a joint initiative of the whole Four-Legged league. Using the framework here-proposed, other teams could develop and test their own analysis module. Moreover, in a near future the proposed automated referee could be integrated with the game controller. This could allow the automation of the game controlling process.

Another pending issue is the placement of the referee camera. We propose that this issue can be solved by using a cameraman who should continuously move the camera and focus it towards the game significant situations. In this way the whole game analysis and decision process could be automated, and humans will only need to move the referee camera o cameras, and to manipulate the robots in some specific situations (manual placement, robot pick-up, and removal of a penalized robot from the field of play). This is a necessary intermediate step towards having robot referees in a future.

References

1. Laue, T., Spiess, K., Röfer, T.: SimRobot - A General Physical Robot Simulator and Its Application in RoboCup. In: Bredenfeld, A., Jacoff, A., Noda, I., Takahashi, Y. (eds.) RoboCup 2005. LNCS (LNAI), vol. 4020, Springer, Heidelberg, 2006 (in press)
2. Zagal, J., Ruiz-del-Solar, J.: A Dynamically and Visually Realistic Simulator for the RoboCup Four Legged League. In: Nardi, D., Riedmiller, M., Sammut, C., Santos-Victor, J. (eds.) RoboCup 2004. LNCS (LNAI), vol. 3276, pp. 34–45. Springer, Heidelberg (2005)
3. Assfalg, J., Bertini, M., Colombo, C., Del Bimbo, A.: Semantic Annotation of Sports Videos. IEEE MultiMedia 9(2), 52–60 (2002)
4. Assfalg, J., Bertini, M., Colombo, C., Del Bimbo, A., Nunziati, W.: Semantic Annotation of soccer videos: automatic highlights identification. Computer Vision and Image Understanding 92(2-3), 285–305 (2003)

5. Babaguchi, N., Kawai, Y., Kitahashi, T.: Event Based Video Indexing by Intermodal Collaboration. IEEE Transactions on Multimedia 4(1), 68–75 (2002)
6. Bertini, M., Del Bimbo, A., Nunziati, W.: Soccer Video Highlight Prediction and Annotation in Real Time. In: Roli, F., Vitulano, S. (eds.) ICIAP 2005. LNCS, vol. 3617, pp. 638–644. Springer, Heidelberg (2005)
7. D'Orazio, T., Ancona, N., Cicirelli, G., Nitti, M.: A ball detection algorithm for real soccer image sequences. In: Proc. Int. Conf. Patt. Recog. – ICPR'02, Canada, (August 11-15, 2002)
8. Duan, L.Y., Xu, M., Chua, T.S., Tian, Q., Xu, C.S.: A mid-level representation framework for semantic sports video analysis. In: Proc. of ACM MM'03, Berkeley, USA, pp. 33–44. ACM Press, New York (November 2-8, 2003)
9. Ekin, A., Tekalp, A.M.: Automatic soccer video analysis and summarization. IEEE Trans. on Image Processing 12(7), 796–807 (2003)
10. Han, M., Hua, W., Xu, W., Gong, Y.H.: An integrated baseball digest system using maximum entropy method. In: Proc. of ACM MM'02, pp. 347–350. ACM Press, New York (2002)
11. Li, B., Sezan, M.I.: Event detection and summarization in American football broadcast video. In: Proc. SPIE Conf. on Storage and Retrieval for Media Databases, vol. 4676, pp. 202–213 (2002)
12. Nepal, S., Srinivasan, U., Reynolds, G.: Automatic detection of 'Goal' segments in basketball videos. In: Proc. ACM MM'01, Ottawa, Canada, pp. 261–269. ACM Press, New York (2001)
13. Pingali, G., Jean, Y., Carlbom, I.: Real time tracking for enhanced tennis broadcasts. In: Proc. IEEE Conf. Comp. Vision and Patt. Rec. – CVPR'98, pp. 260–265. IEEE Computer Society Press, Los Alamitos (1998)
14. Pingali, G., Opalach, A., Jean, Y.: Ball tracking and virtual replays for innovative tennis broadcasts. In: Proc. Int. Conf. Patt. Recog. – ICPR'00, Barcelona, Spain, pp. 4152–4146 (2000)
15. Rui, Y., Gupta, A., Acero, A.: Automatically extracting highlights for TV Baseball programs. In: Rui, Y., Gupta, A., Acero, A. (eds.) Proc. of ACM MM'00, pp. 105–115. ACM Press, New York (2000)
16. Sudhir, G., Lee, J., Jain, A.K.: Automatic classification of tennis video for high-level contentbased retrieval. In: Proc. of IEEE Int. Workshop on Content-based Access of Image and Video Database, pp. 81–90. IEEE Computer Society Press, Los Alamitos (1998)
17. Tovinkere, V., Qian, R.J.: Detecting semantic events in soccer games: Towards a complete solution. In: Proc. ICME'01, pp. 1040–1043 (2001)
18. Wan, K., Yan, X., Yu, X., Xu, C.S.: Real-time goalmouth detection in MPEG soccer video. In: Proc. of ACM MM'03, Berkeley, USA, pp. 311–314 (2003)
19. Wan, K., Yan, X., Yu, X., Xu, C.S.: Robust goalmouth detection for virtual content insertion. In: Proc. of ACM MM'03, Berkeley, USA, pp. 468–469. ACM Press, New York (2003)
20. Wang, J.R., Paramesh, N.: A scheme for archiving and browsing tennis video clips. In: Proc. of IEEE Pacific-Rim Conf. on Multimedia - PCM'03, Singapore, IEEE Computer Society Press, Los Alamitos (2003)
21. Wang, J.R., Parameswaran, N.: Survey of Sports Video Analysis: Research Issues and Applications. In: Conferences in Research and Practice in Information Technology VIP'03, Sidney, vol. 36, pp. 87–90 (2004)
22. Yow, D., Yeo, B.L., Yeung, M., Liu, B.: Analysis and presentation of soccer highlights from digital video. In: Proc. ACCV'95 (1995)

23. Yu, X., Xu, C.S., Leong, H.W., Tian, Q., Tang, Q., Wan, K.W.: Trajectory-based ball detection and tracking with applications to semantic analysis of broadcast soccer video. In: Proc. of ACM MM'03, Berkeley, USA, pp. 11–20. ACM Press, New York (2003)
24. http://www.hawkeyeinnovations.co.uk/
25. http://www.questec.com/
26. Röfer, T., et al.: German Team, Technical Report, RoboCup 2005, Four-legged league (February 2006) Available on http://www.germanteam.org/GT2005.pdf
27. Quinlan, M.J., et al.: The NUbots Team Report, RoboCup 2005, Four-legged league (February 2006), Available on http://www.robots .newcastle.edu.au/ publications/ NUbot FinalReport2005.pdf
28. Loncomilla, P., Ruiz-del-Solar, J.: Gaze Direction Determination of Opponents and Teammates in Robot Soccer. In: Bredenfeld, A., Jacoff, A., Noda, I., Takahashi, Y. (eds.) RoboCup 2005. LNCS (LNAI), vol. 4020, Springer, Heidelberg (2006)
29. Comaniciu, D., Ramesh, V., Meer, P.: Kernel-Based Object Tracking. IEEE Trans. on Pattern Anal. Machine Intell. 25(5), 564–575 (2003)
30. RoboCup 2006 Four-legged Leagues official rules: Available on February 2006 in http://www.tzi.de/4legged/pub/Website/Downloads/Rules2006.pdf
31. Peng, N.S., Yang, J., Liu, Z.: Mean shift blob tracking with kernel histogram filtering and hypothesis testing. Pattern Recognition Letters 26, 605–614 (2005)
32. Lowe, D.G.: Distinctive Image Features from Scale-Invariant Keypoints. Int. Journal of Computer Vision 60(2), 91–110 (2004)

Detecting Motion in the Environment with a Moving Quadruped Robot

Peggy Fidelman[1], Thayne Coffman[2], and Risto Miikkulainen[1]

[1] Department of Computer Sciences
The University of Texas at Austin
[2] Department of Electrical and Computer Engineering
The University of Texas at Austin

Abstract. For a robot in a dynamic environment, the ability to detect motion is crucial. Motion often indicates areas of the robot's surroundings that are changing, contain another agent, or are otherwise worthy of attention. Although legs are arguably the most versatile means of locomotion for a robot, and thus the best suited to an unknown or changing domain, existing methods for motion detection either require that the robot have wheels or that its walking be extremely slow and tightly constrained. This paper presents a method for detecting motion from a quadruped robot walking at its top speed. The method is based on a neural network that learns to predict optic flow caused by its walk, thus allowing environment motion to be detected as anomalies in the flow. The system is demonstrated to be capable of detecting motion in the robot's surroundings, forming a foundation for intelligently directed behavior in complex, changing environments.

Keywords: robot vision, image processing.

1 Introduction

The ability to detect motion is important to a robot in a novel or changing environment. Motion can potentially be a very significant clue about which parts of the environment are interesting or dangerous. For example, consider a consumer robot in the home, such as the commercially available Sony Aibo[1]. Motion can give it clues to where humans are located in its environment, which will help it interact with them more effectively. It can also be used to direct the robot's attention to potential danger, such as a stack of books sliding off of a desk or the family dog preparing to pounce on it. As another example, consider a surveillance robot [2]. Modern surveillance systems are primarily based on motion detection. If a robot must stop its own motion in order to detect motion in its surroundings, much of the advantage of using a robot instead of a simpler system (such as a set of stationary cameras) is lost.

If a robot is to be able to deal with truly novel environments, it is also an advantage for the robot to have legs rather than wheels. Legs allow traversal of highly uneven surfaces. A legged robot can step over obstacles or climb stairs,

G. Lakemeyer et al. (Eds.): RoboCup 2006, LNAI 4434, pp. 219–231, 2007.

whereas analogous feats are often impossible for a wheeled robot. However, the motion of a legged robot is not as smooth as that of a wheeled robot. This characteristic introduces additional challenges for detecting motion in the environment while the robot itself is moving.

The main contribution of this paper is a method for detecting motion from a quadruped robot while it is walking at its maximum speed (approximately 35cm/sec). The overall architecture is inspired by that of Lewis [3], but our method differs in several important ways: preprocessing of optic flow has been eliminated, a substantial postprocessing step has been added to the output of the neural network, and the details of most of the architecture's components have been redesigned. These innovations make our method effective in a less constrained environment than Lewis's method requires, and also allow use of a significantly faster-moving robot.

The paper is organized as follows. Section 2 gives a background on optic flow and describes related work that uses optic flow for navigation and obstacle detection on mobile robots. Section 3 introduces the method for detecting external motion. Section 4 describes the setup of the system used in the experiments, as well as the details of the experiments themselves. Section 5 presents the results of these experiments, and Section 6 discusses these results as well as possible directions for the future.

2 Background and Related Work

2.1 Optic Flow

Optic flow is a way of describing the apparent motion between two images of the same scene taken in quick succession. It is typically expressed as a vector field, with a two-dimensional vector for each pixel in the first image, representing vertical and horizontal displacement. These vectors give a complete description of where each pixel in the first image appears to have moved in the second image.

The apparent motion in two dimensions depends on the actual motion in three dimensions in a complex manner. Sometimes it is difficult to determine whether the motion is the result of the camera moving or objects in the scene moving. In other cases, it is clear that an object is moving and not the camera, but the actual direction of the object's motion cannot be determined because only part of the object is visible (this effect is known as the aperture problem).

For an intuitive understanding of some of these issues, consider a passenger looking out the right-hand side window of a car at an adjacent vehicle while s/he is waiting for a stoplight to turn green. Without prior knowledge, the scenes s/he perceives if the car appears to move to the left could be interpreted in a number of ways – the other car might be moving forward and the passenger's car might still be stationary, or the other car might be stopped and the passenger might be moving backward, or both cars might be moving forward or backward at different rates. Thus, while the optic flow field contains a wealth of knowledge, interpretation can be difficult.

Optic flow is formulated in terms of instantaneous (in space and time) image intensity gradients. The key formula defining the optic flow between two images in a sequence is

$$I_{t+dt}(x + u(x,y), y + v(x,y)) = I_t(x,y), \tag{1}$$

where $I_t(x,y)$ is the image intensity at each pixel at time t, $I_{t+dt}(x,y)$ is the image intensity at time $(t + dt)$, $u(x,y)$ is the horizontal flow at each pixel, and $v(x,y)$ is the vertical flow at each pixel. Thus the horizontal and vertical optic flows generate a "mapping" between corresponding pixels in the two images.

A perfect solution to the optic flow formula is rarely available because of effects such as noise and occlusion. Instead of trying to compute a total solution, most approaches attempt to minimize the error in equation (1) summed over all pixels in the image. Computation of optic flow is an underconstrained problem; therefore, in addition to minimizing the error in the pixel matching between images, a smoothness constraint is typically included. This constraint is justified because for images of real-world objects (which are, in general, smooth and connected) the optic flow field is likely to be smooth at almost every pixel. The objective function for optimization then becomes

$$\sum_{(x,y)} E^2(x,y) = \sum_{(x,y)} \left\{ (I_x u + I_y v + I_t)^2 + \lambda(u_x^2 + u_y^2 + v_x^2 + v_y^2) \right\},$$

with

$$f_x = \frac{\partial f}{\partial x} \quad , \quad f_y = \frac{\partial f}{\partial y} \quad , \quad f_t = \frac{\partial f}{\partial t}$$

where $f \in \{I, u, v\}$. The correctness and smoothness components to the optimization are clearly visible as the first and second group of terms in the objective function, respectively. As the optic flow field generates a more accurate matching between pixel intensities in the two images, the first group of terms will decrease towards zero. As the field becomes smoother (showing less variation between adjacent pixels), the second group of terms will decrease towards zero. The relative importance of these terms is regulated by the parameter λ. Because formulation of the optic flow field is based on local gradient information, its computation is often more accurate when images are only incrementally different.

Optic flow is often computed via iterative relaxation, by one variation or another of an approach developed by Horn and Schunk in the 1980s [4]. In this approach, the proposed solution is initialized, and on each iteration, the solution is refined by propagating information from each pixel to its local neighbors through a local averaging of the optic flow field. This process is guided by the equations

$$u^k(i,j) = \bar{u}^{k-1}(i,j) \; - \; I_x(i,j)\frac{P(i,j)}{D(i,j)},$$

$$v^k(i,j) = \bar{v}^{k-1}(i,j) \; - \; I_y(i,j)\frac{P(i,j)}{D(i,j)},$$

$$P = I_x\bar{u} + I_y\bar{v},$$

$$D = \lambda^2 + I_x^2 + I_y^2.$$

In these equations, $u^k(i,j)$ and $v^k(i,j)$ are the kth iteration's estimates of the horizontal and vertical flow at pixel (i,j), and $\bar{f}(x,y)$ is the local average of function $f(\cdot)$ near pixel (x,y). Iteration continues until a convergence condition is reached or a maximum number of iterations, $k_{\max}$, have occurred. Although many alternatives have been studied, this 25-year-old approach continues to be one of the most common and effective methods of optic flow computation.

2.2 Optic Flow in Navigation

Optic flow is useful in navigation and obstacle detection, and several groups have used it for these purposes in robots [5, 6]. In general, the robots are wheeled (or airborne) instead of legged, which means that their typical motion is smoother. This regularity of motion makes normal optic flow easier to characterize, so abnormalities such as the ones caused by obstacles can readily be detected.

A method for detecting obstacles in the walking path of a bipedal robot using optic flow information was recently developed by Lewis [3]. In this method, a neural network uses the robot's joint angles and gait phase to predict optic flow events. This prediction is compared to observed optic flow events, and any significant difference between the two indicates that an obstacle has been detected.

The success of Lewis's approach indicates that optic flow can be a source of useful information even on legged robots. It also suggests that neural networks are a promising tool for overcoming the difficulty of characterizing normal optic flow on legged robots, as discussed above. However, Lewis's experiments take place in a highly constrained environment. The robot is tethered so that it walks in a circle, and its camera is fixed at a slight downward angle so that it is always focusing on the ground slightly in front of it, which is the same at all points around the circle (with the exception of the obstacles it must detect during the testing phase). It also walks very slowly (2cm/sec).

In order to be generally useful, a motion detection method for a legged robot should be free of these constraints. If the robot cannot move freely through non-uniform environments and still detect motion, there is little benefit to using a legged robot at all – a wheeled robot or even a stationary camera could probably do the same task more reliably. In addition, a robot should ideally not have to slow down its own movements in order to accommodate the motion detection algorithm.

The method presented in this paper is effective on a freely moving robot walking at its top speed (35cm/sec). It also differs from Lewis's approach in that it operates on the raw optic flow field rather than on preprocessed data (optic flow "events"), and a substantial postprocessing step has been added as part of comparing the neural network's prediction to the actual observed optic flow. It is not clear to what extent the predictor neural network differs from that of Lewis, because details of his architecture are not available.

3 Motion Detection Method

The proposed method is depicted in Figure 1. First, the optic flow in the image is calculated. The resulting vector field is then given as input to a neural network along with information about the current position in the robot's walk cycle and readings from its three accelerometers. This neural network outputs a prediction of the optic flow to be seen in the next image. When the robot receives this next image, the optic flow is calculated and compared with the network's prediction. A postprocessing algorithm then looks for discontinuities in the difference between the calculated and predicted optic flow to determine where in the image the externally moving objects are likely to be found.

3.1 Optic Flow

Training and testing sequences of optic flow were obtained from the Aibo and transferred to a set of image files on a desktop PC's hard disk. Image sequences

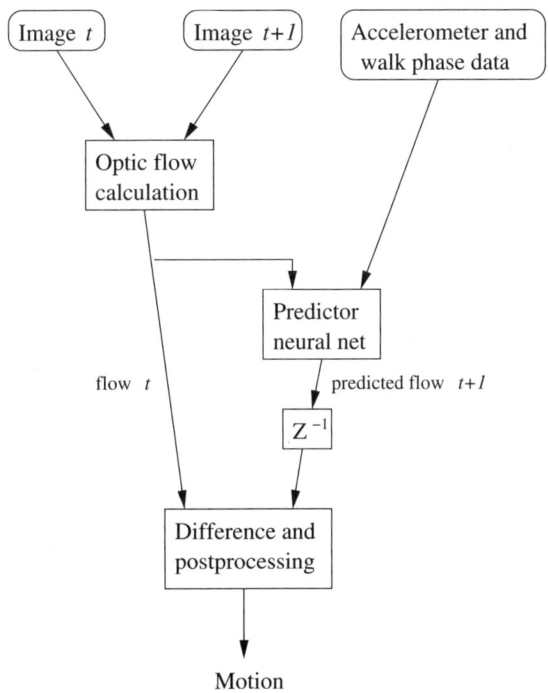

Fig. 1. An overview of the method for detecting moving objects in the robot's environment. (Z^{-1} is a one-step delay operator.) Optic flow, accelerometer readings, and information about the current phase of the robot's walk are given as inputs to a neural network, which then predicts the optic flow to be observed at the next timestep. The difference between the prediction and the observed optic flow is then used to detect any locations in the image that contain moving objects.

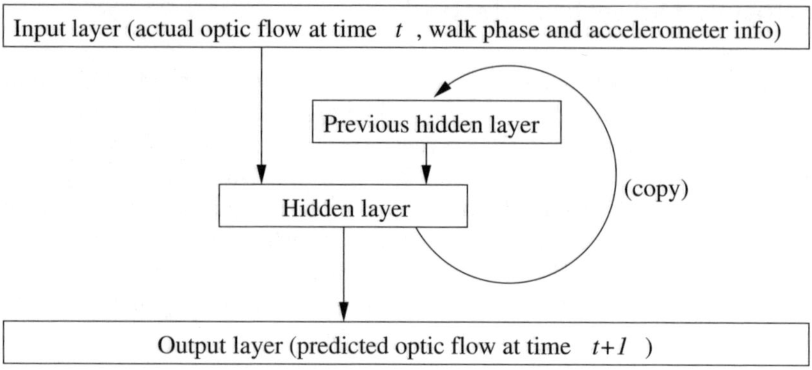

Fig. 2. The simple recurrent network used to predict the optic flow. All projections are full and feed-forward, except for the projection from the hidden layer to the previous hidden layer. This projection is a direct copy of the hidden layer activations into the previous hidden layer. In this way, some context is kept in the network, allowing it to retain information from previous inputs.

were then loaded off the disk, the optic flow calculations were run, and the resulting flow fields were stored back to the disk for later use in training or comparison against predicted flow. Although in principle these calculations could be done in real time on the Aibo, offline experiments test the method more efficiently.

The optic flow fields were computed by a Matlab implementation of Horn & Schunk's iterative relaxation algorithm described in Section 2.1. Identical parameters, $\lambda = 1.0$ and $k_{\max} = 1000$, were used to compute all training and testing data. These parameter choices resulted in very smooth fields. While generally smooth fields might be expected given the Aibo's known motion characteristics and environment, additional work will have to be performed in the future to understand whether using parameters that generate more rapidly-varying optic flow fields could allow more accurate motion discrimination.

3.2 Predictor Neural Network

The predictor neural network, shown in Figure 1, accepts three types of information as input. The first is an optic flow field. The second is a single number indicating the robot's position in the walk cycle at the time of the second image (of the two images used to calculate the optic flow field). The third is a set of accelerometer data corresponding to the same image. In the experiments described in this paper, a simple recurrent network architecture [7] was used [1] (Figure 2).

The network is trained on sequences of optic flow fields, which are generated from sequences of robot camera images in which there is no external motion. Thus, the network is effectively learning to produce what the next optic flow

[1] Informal experimentation indicated that a simple recurrent network performs better than a three layer feed-forward network. However, such a non-recurrent network is also capable of some success at this task.

field should look like *if there is no external motion*, given the last optic flow field observed and information about the robot's acceleration and position in the walk cycle. By comparing this prediction to the actual optic flow observed at the next timestep, it is possible to see which parts of the image exhibit unpredictable motion, indicating where objects moving relative to the world can be found in the image.

3.3 Postprocessing

Consider the vector field resulting from taking the (vector) difference of the predicted and actual optic flow fields. Taking the magnitude of each of the elements of this vector field results in a matrix of the same size, called the *difference field.*

The discontinuous motion of the robot's camera and the noisy nature of real-world sensor data make the optic flow prediction task quite difficult. Although the neural network typically does a reasonable job of predicting optic flow for image sequences containing no external motion, its prediction is occasionally entirely wrong. Therefore, the naive approach – simply taking the size of the difference field at each point to be the likelihood of external motion at that point – will not suffice.

However, these magnitudes do contain useful information. Because the neural network is trained with images from an environment in which nothing other than the robot itself is moving, all of its targets during training were optic flow fields with coherent motion. So, assuming the images contain no external motion, even when the network makes a wrong prediction that prediction will be more or less "equally wrong" at all points. If there is external motion in the scene, however, there will often be sharp discontinuities in the difference field, which can be discovered by running an edge detection algorithm on each difference field.

The edge detection used in the postprocessing step is rather unconventional. Many edge detection algorithms are designed to find the best edges in an image, even if that image has only poor candidates for edges. However, if there are no sharp edges in the difference field at some timestep, there is probably no motion in the image. Therefore, in this case the postprocessing algorithm should not find any edges. To this end, first a binary version of the difference field is obtained by finding its maximum element and replacing every element extremely far away from this maximum[2] with a zero, and replacing the rest with ones. Then a conventional edge detection algorithm (such as the Laplacian of Gaussian method) is run over this binary difference field.

4 Experimental Procedure

The images used in the experiments discussed here were acquired by a Sony Aibo ERS-7 walking across a standard 2004 RoboCup legged league field [8]. The Aibo

[2] In practice, to be "extremely far away from the maximum," an element must be very close to zero and the maximum over the difference field must be large. This constraint enforces that all edges found in the next step will correspond to very sharp edges in the original difference field.

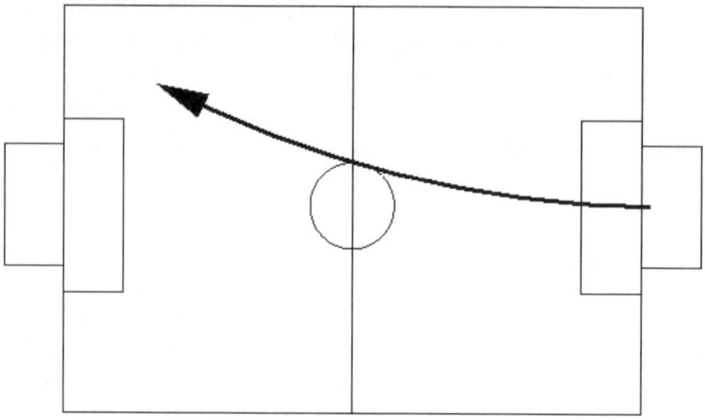

Fig. 3. A typical trajectory used in our experiments

has three degrees of freedom in each of its four legs as well as its head. It has a CMOS camera in the head, from which it is possible to capture approximately 25 images per second[3]. The robot always starts close to the center of the yellow goal facing outward toward the field center (although not always in the exact same location) and then walks most of the way across the field[4] using the fastest available forward walk (approximately 35cm/sec)[9]. Due to the Aibo's slight left-right weight asymmetry, the forward walk curves slightly to the right over long distances. Thus a typical trajectory looks like the one depicted in Figure 3.

The resolution of the Aibo images was first reduced by a factor of 6 in both the horizontal and vertical directions by averaging 3-by-3-pixel blocks of half-resolution Aibo images. Images were converted from full color to grayscale before the optic flow computation, but no other image preprocessing (histogram equalization, deblurring, etc.) was performed.

The resulting images and flow fields consisted of 35 columns by 27 rows. This lowered image resolution allowed a simpler neural network to be trained and decreased the runtime of optic flow computations. Exact runtime performance was not recorded, but computing the optic flow field for one frame pair required approximately 1s of CPU time on a 1.8GHz desktop machine. Real-time performance will require that the current implementation be optimized for speed and translated from Matlab to a language more suitable for embedded operation on the Aibo.

As discussed in Section 3.2, a simple recurrent architecture was used for the predictor neural network. This network had a 200-unit hidden layer. It was trained with backpropagation (using momentum) on data from six runs of the robot on an empty field, where each run consisted of approximately 150 sequential images. Training of this network took approximately 1050 epochs.

[3] The hardware is capable of capturing 30 frames per second, but software overhead reduces this number somewhat.

[4] The length of these trajectories is constrained by the amount of memory available on the Aibo.

5 Results

As was discussed in Section 1, motion detection on a robot is most useful for alerting the robot to potential anomalies in its environment, particularly those that suggest there are changes taking place. Thus, robots will typically want to react immediately as soon as any motion is detected; the value of such information decays very rapidly. However, this means false positive motion detections are particularly dangerous – each one is likely to distract the robot from its primary task unnecessarily. Thus, before incorporating this technology into any robotic system, it is important not only to demonstrate that it can provide useful information about motion, but also that it can do so while keeping false positives to a minimum.

The results shown here reflect these priorities. The parameters of the post-processing algorithm were set to make the system maximally resistant to false positives. Then, to judge the classification accuracy of the system, it was applied to four sets of images from trajectories of the sort shown in Figure 3. Two of these runs contained one moving robot, one contained three moving robots, and one contained no external motion. Each image in the four sets was divided into 9 sectors (Figure 4), and each sector was labeled by hand as containing motion or not. If less than 1/4 of the sector contained a moving object, the sector was labeled as not containing motion. This ground truth was compared to the motion detected by the system.

The system correctly labeled a significant portion of the image sectors containing motion, despite the postprocessing parameter values that virtually eliminated

Fig. 4. An example image showing division into sectors, for use in the quantitative evaluation of the motion detection method. Though both robots and the ball in this image are moving, the middle sector would not be labeled as containing motion, because less than 1/4 of it contains moving objects.

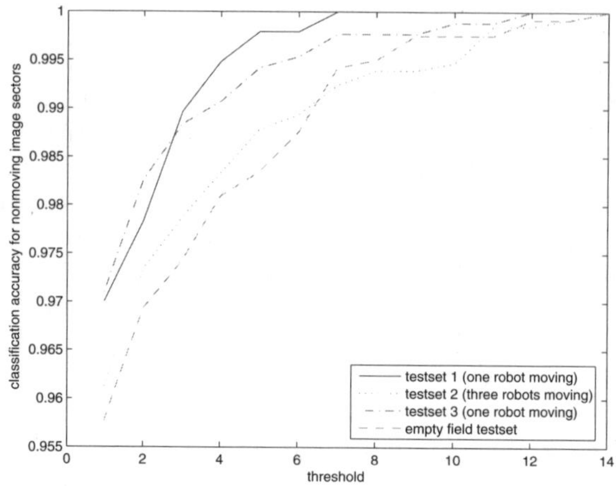

Fig. 5. Classification accuracy for image sectors not containing motion. The x-axis corresponds to the number of edge pixels in a sector required for motion to be detected in that sector, and the y-axis is the fraction of total image sectors not containing motion that were correctly labeled as not containing motion. As more edge pixels are required for a sector to be labeled as containing motion, accuracy in labeling nonmoving sectors increases (i.e., false positives decrease), though there is a tradeoff with accuracy in labeling moving sectors (see Figure 6). Note, however, that even when this threshold is set to 1, accuracy is greater than 95% for all testsets.

false positive motion detections (see Figures 5 and 6). Note that if these parameters are set to less extreme values, classification accuracy of sectors containing motion can be improved, so in this sense Figure 6 reflects the "worst-case" result for motion detection. However, the current settings have the considerable advantage that image sectors labeled as containing motion are virtually certain to actually contain motion.

The typical qualitative behavior of the system is shown in Figures 7 and 8. Figure 7 contains four sequential frames from one test run of the system. The robot in the foreground is moving at full speed; all other parts of the image are stationary relative to the world. For comparison, Figure 8 shows typical errors from a testing run with no external motion. In this testing run, over 95% of the sectors were correctly labeled as containing no motion (see Figure 5).

6 Discussion and Future Work

Based on informal observations, the system appears to detect motion more reliably when the moving object is closer to the robot and moving more rapidly. Because the robot's own motion is so rapid, it is understandable that slow-moving objects would be hard to detect. Near motion will appear more rapid in two dimensions; also, because of the downsampling, a moving object sufficiently far away will appear as a single pixel whose color is changing slightly.

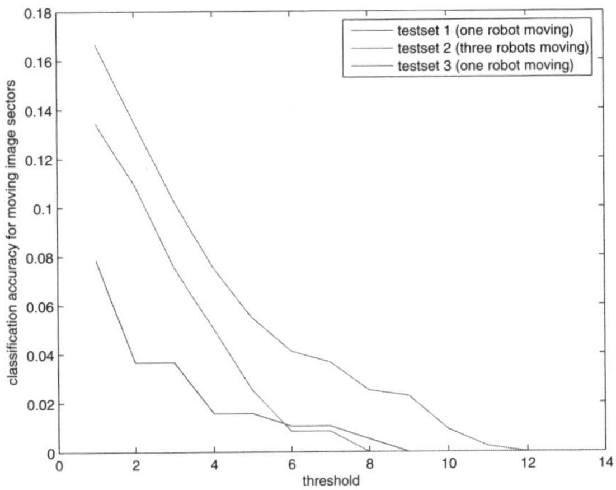

Fig. 6. Classification accuracy for image sectors containing motion. The system can detect a significant fraction of the motion in the robot's environment while maintaining the extremely low false positive rate shown in Figure 5. The x-axis corresponds to the number of edge pixels in a sector required for motion to be detected in that sector. As more edge pixels are required, fewer sectors containing motion are correctly labeled, leading to a tradeoff between accuracy in motion detection and elimination of false positives (Figure 5).

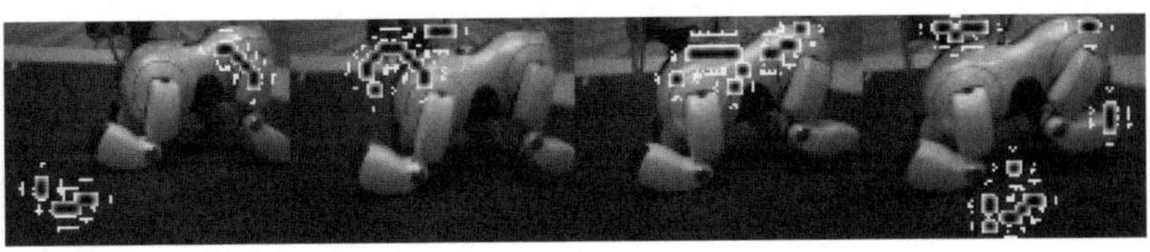

Fig. 7. Four sequential frames from a test run. The robot in the images is moving to the left at full speed. Overlaid black squares on the image indicate the discontinuities found by the postprocessing algorithm; when enough of these squares appear in a sector of the image, the system concludes that this sector contains motion.

Fig. 8. Some typical errors on a field with no motion. The overlaid squares only indicate discontinuities, and enough of these squares must appear in the same sector for motion to be detected there. In the run from which these images were taken, in which there was no motion on the field, 83% of the frames contained no discontinuities of this type, and over 95% of the image sectors were correctly labeled as containing no motion.

This observation suggests an important direction for future work: extension of the system to work with larger images. Although the predictor neural network will have to be larger, it should not make training intractable: training time for the current network only requires a few hours of CPU time. Moreover, optic flow calculation over a half-resolution image (as opposed to the 1/6 resolution images currently used) has already been verified to be tractable.

Another important direction for future work is to implement the system to run in real time on an Aibo robot. This extension is plausible given the size of the images. The Aibo has a 576 MHz 64 bit RISC processor that allows for significant amounts of onboard computation. Although the computation required to process the images would reduce the rate at which images can be captured, a slower gait could compensate for any challenges posed to the optic flow algorithm by the increased time between images.

7 Conclusion

A method was presented for detecting external motion from a quadruped robot while it is walking freely and quickly. The system is resistant to false positives and is sufficiently accurate on real-world sensor data, and it is able to process this data with a speed that suggests that future onboard implementation is possible. Thus, it provides a way for a legged robot to sense motion in its environment, allowing it to direct its attention more intelligently, and ultimately making it more able to negotiate novel or changing environments.

Acknowledgments

Thanks to Mohan Sridharan for his help in implementing edge detection and to Uli Grasemann for his help with image dimensions. Thanks also to the UT Austin Villa robot soccer team for their efforts in creating the Aibo code base that supports this project. This research was supported in part by NSF EIA-0303609, NSF CAREER award IIS-0237699, and ONR YIP award N00014-04-1-0545.

References

1. Sony: Aibo robot (2004), `http://www.sony.net/Products/aibo`
2. Rybski, P.E., Stoeter, S.A., Erickson, M.D., Gini, M., Hougen, D.F., Papanikolopoulos, N.: A team of robotic agents for surveillance. In: Proceedings of the Fourth International Conference on Autonomous Agents (2000)
3. Lewis, M.A.: Detecting surface features during locomotion using optic flow. In: Proceedings of the IEEE International Conference on Robotics and Automation, Washington, DC, IEEE Computer Society Press, Los Alamitos (2002)
4. Horn, B.K.P., Schunck, B.: Determining optical flow. Artificial Intelligence 17, 185–203 (1983)

5. Bhanu, B., Das, S., Roberts, B., Duncan, D.: A system for obstacle detection during rotorcraft low altitude flight. IEEE Transactions on Aerospace and Electronic Systems 32, 875–897 (1996)
6. Young, G.S., Hong, T.H., Herman, M., Yang, J.C.: Safe navigation for autonomous vehicles: a purposive and direct solution. In: Proceedings of SPIE: Intelligent Robots and Computer Vision XII: Active Vision and 3D Methods, vol. 2056, pp. 31–42 (1993)
7. Harris, C.L., Elman, J.L.: Representing variable information with simple recurrent networks. In: Proceedings of the 11th Annual Conference of the Cognitive Science Society, pp. 635–642 (1989)
8. RoboCup Technical Committee: Sony four legged robot football league rule book (2004), `http://www.tzi.de/~roefer/Rules2004/Rules2004.pdf`
9. Stone, P., Dresner, K., Fidelman, P., Jong, N.K., Kohl, N., Kuhlmann, G., Sridharan, M., Stronger, D.: The UT Austin Villa 2004 RoboCup four-legged team: Coming of age. Technical Report UT-AI-TR-04-313, The University of Texas at Austin, Department of Computer Sciences, AI Laboratory (2004)

Using Temporal Consistency to Improve Robot Localisation

David Billington, Vlad Estivill-Castro, René Hexel, and Andrew Rock

Griffith University, Australia

Abstract. Symbolic reasoning has rarely been applied to filter sensor information; and for data fusion, probabilistic models are favoured over reasoning with logic models. However, we show that in the fast dynamic environment of robotic soccer, Plausible Logic can be used effectively to deploy non-monotonic reasoning. We show this is also possible within the frame rate of vision in the (not so powerful) hardware of the AIBO ERS-7 used in the legged league. The non-monotonic reasoning with Plausible Logic not only has algorithmic completion guarantees but we show that it effectively filters the visual input for improved robot localisation. Moreover, we show that reasoning using Plausible Logic is not restricted to the traditional value domain of discerning about objects in one frame. We present a model to draw conclusions over consecutive frames and illustrate that adding temporal rules can further enhance the reliability of localisation.

1 Introduction

Despite a large body of research into symbolic reasoning in Artificial Intelligence, symbolic reasoning has rarely been applied to filter sensor information. For data fusion, probabilistic models are favoured over reasoning with logic models. For example, for combining information from several sources, their reliability is modelled using probabilities and "reasoning with uncertainty" is performed using general models that include applications to sensor fusion [8, and references].

However, reasoning has always been regarded as a fundamental capability of intelligent systems. Progress in reasoning is remarkable [9,11] and has become a classical aspect of intelligent systems technology (for example, 4 chapters are dedicated to uncertain knowledge and reasoning in [14], at present the most widely accepted textbook in Artificial Intelligence and 57th most cited computer science publication ever). Moreover, many examples of reasoning relate to robotic situations, as is the long-lived example of the blocks world where in order to stack objects a robot that can hold only one object must reason about which other objects to remove from above the object it needs to transport. Even in the context of agent systems [16], the agent is expected to do some reasoning after sensing the environment in order to select what action to carry out next. However, in realistic robotic environments, the domain is dynamic[1] and

[1] The environment is dynamic if it will evolve in the time gap between the sensors collecting information and the agent performing an action [16].

G. Lakemeyer et al. (Eds.): RoboCup 2006, LNAI 4434, pp. 232–244, 2007.

non-deterministic[2]. Robotic soccer is also inaccessible[3] and the agent needs to work in teams against teams of adversaries. While the robot may carry with it stable domain knowledge regarding the field and settings of RoboCup, some of the conclusions of its reasoning regarding a particular situation may need to be withdrawn in light of new evidence. This capacity to retract previous derivations is called *non-monotonic reasoning* (and we differentiate it from *belief revision* as we are not to update the domain knowledge).

Non-monotonic reasoning has long been considered too complex for real-time settings. Visual robot localisation in the 4-legged league places particular stringent demands. Video camera systems typically operate at a rate of 30 frames per second, which allows only about 30 ms to perform full image recognition, feature extraction, and consistency verification. Moreover, poor lighting conditions can make colour calibration extremely difficult. More often than not, vision systems make errors in object recognition (in particular, they may occasionally miss the landmarks for localisation or report non-existent objects as visible).

Dynamic, inaccessible and non-deterministic environments have prompted the use of reactive architectures and/or hybrid systems. The influential works by Brooks [4] have also lessened the interest in using symbolic/logic approaches. We argue here that a computable non-monotonic logic has a role to play in making sense of inputs and filtering sightings for localisation. Moreover, we present a method that uses time as an important factor when reasoning about the consistency of the perceived world. In this work we propose to use a combination of both the time and the value domain in a robot localisation system to make more solid decisions about consistent and inconsistent objects. We show how the incremental extensibility property of Plausible Logic can be used to increase the robustness against temporary inconsistencies. Section 2 examines the state of the art in robot localisation. Section 3 explains our approach and shows the rule based system we propose. Section 4 contains an experimental evaluation of our system. Section 5 concludes with a discussion on the impact of our findings.

2 Background and Related Work

Localisation based on a camera presents data quality issues over other positional sensors. In the 4-legged league, localisation is based on only one camera with restricted field of vision mounted on a head with fast movement. Such hardware (and the vision software) produces errors due to many causes.

For real-time localisation of robots, Kalman Filters (KF) pose some problems. Alternative techniques have since emerged to address these problems in robot localisation. These techniques use grid-based hidden Markov Models (MM) and Monte Carlo Localisation (MCL) [6,7,15]. Bayes' theorem is used in all three approaches [7,15] to integrate into the current belief about one's position in the world $Prob(\boldsymbol{x}_t)$ (1) the data $\boldsymbol{o}_t$ from a sensor, (2) the prediction $\boldsymbol{x}'_{t+1}$ by a motion

[2] An environment is non-deterministic if an action may not have the expected outcome [16], like a skid because the surface is smoother than anticipated.

[3] Information about the entire environment may not be possible to collect [16].

model, and (3) a previous pose belief $Prob(\boldsymbol{x}_{t-1})$. Conditional probabilities are used to represent prior and posterior knowledge about the state of the world. Every observation $\boldsymbol{o}_t$ and action $\boldsymbol{a}_t$ is then used to update the positional belief.

MCL has been shown to be superior to the Extended KF [7] and to MM [7]. They seem the most effective method for the 4-legged league [13]. This paper does not argue for the elimination of these techniques. However, these approaches have problems with inconsistencies. MCL is slow to converge relative to the accuracy of the sensors, and until recently little theoretical foundation existed for some of its fixes [15]. Because of these drawbacks, it is important that the observations from sensors be reliable (otherwise, convergence is too slow or the artifacts to handle the kidnap problem introduce other high modes in the representation of the distribution). For example, in soccer, the two goals are at opposite sides of the field. With a single, forward-looking camera, it is impossible to see both goals. The question is what to do if vision does report both goals at the same time. The challenge is to estimate $Prob(\text{visible scene}|\boldsymbol{pos})$, where the visible scene is formed of a description of all visible objects, and where $\boldsymbol{pos}$ is a vector for the current belief. To avoid the daunting task of building a table for the values $Prob(\text{visible scene}|\boldsymbol{pos})$ at all scenes and positions one would like to at least decompose this into separate probabilities (as per objects in the scene). In the simple case of two visible objects, one would like to treat the objects as independent and weight the current belief in proportion to the product $Prob(See\_front\_goal|\boldsymbol{pos}) \cdot Prob(See\_back\_goal|\boldsymbol{pos})$ (for example when seeing both goals). The difficulty becomes defining $Prob(X\_is\_visible|\boldsymbol{pos})$ for each land mark X taking into account that $\boldsymbol{pos}$ has significant noise regarding the orientation and pan of the head of the Sony AIBO. Therefore, $Prob(X\_is\_visible|\boldsymbol{pos})$ is unlikely to be set to zero in any case. As a result, the MM and MCL approaches create at least a new mode in the distribution modelling the location and KF suffers an enlargement of the covariance matrix.

Domain knowledge says that seeing both goals is impossible. Thus, for all postures $\boldsymbol{pos}$, $Prob(See\_front\_goal \cap See\_back\_goal|\boldsymbol{pos}) = 0$ is usually adopted. This leads back to a large sets of special cases and the impossibility of decomposing $Prob(\text{visible scene}|\boldsymbol{pos})$ into simpler functions. We conjecture that all teams at RoboCup have incomplete systems to deal with these special cases (but survive reasonably well because many cases are uncommon and the frame rate allows the robot to recuperate from these pathological cases[4]). The handling of inconsistencies within the localisation module becomes a series of logical checks. The code which filters observations that are considered inconsistent soon becomes a large piece of software that is hard to verify for correctness or completeness.

Our first thesis is that such a filter of inconsistent observations is better handled by some logic. The second thesis is that such a logic should not only be capable of ruling out observations, but allow reasoning about them to provide informative inputs for localisation. This treatment of cases could potentially be

[4] However, from the perspective of software quality and reliability for robots around humans, the system must be capable of acting properly if it faces seeing all 6 landmarks in one frame.

treated by a classifier system (a la machine learning). Namely, learning or making explicit a function P that maps a visible scene and a pose $\boldsymbol{pos}$ into a suitable value $Prob(\text{visible scene}|\boldsymbol{pos})$. The impossibility of a comprehensive training set rules out using a decision tree, a decision list, an artificial neural network, or even support vector machines to learn P.

Moreover, the design implicit in existing approaches deals only with the consistency problem in the value domain. That is, because of the large number of special cases, the system can not take advantage of other information for localisation, most notably time. For example, if vision reports both goals in one frame, the usual rule of thumb is to discard both observations (even if in the previous frame and the next frame show only one goal and the 3 consecutive frames have the front goal in common). We will show here how reasoning with Plausible Logic (PL) [3,12] about the previous frame can even take advantage of this. KF, MCL, and the MM approaches deal with this situation over time, but for those 3 frames they loose height in the peaks of their distributions.

3 Using a Model of Time to Improve Consistency

Plausible Logic. Non-monotonic reasoning allows to draw conclusions from a collection of facts and beliefs, but also withdraws those inferences in light of new evidence that challenges previous inferences. Plausible Logic [2] is a formalism for non-monotonic reasoning [1] that is implementable. Currently, the only available implementation, DPL, is in Haskell. Moreover, with respect to other non-monotonic logics, Plausible Logic (PL) is capable of identifying proofs that demand an infinite loop and as a result it can halt. One additional characteristic of PL is that it has several algorithms to establish a conclusion, each of them weighing differently the evidence in favour and against a proposition and allowing a conservative or risky approach to accepting the proposition. This is because PL differentiates between propositions derived using only factual information from those derived using plausible information. This is achieved because there are several proof algorithms each resulting in a certain degree of confidence on the validity of the proposition. PL reduces to classical propositional logic if only factual information is used. However, when determining the provability[5] of a formula, the algorithms in PL can deliver three values (that is, it is a three-valued logic). The proof algorithms terminate assigning the value $+1$ to the formulas that have been proved. It assigns the value 0 when the formula cannot be proved and attempting so will cause infinite recursive looping. It assigns the value -1 when the formula is not provable and does not generate a loop. Because PL uses different algorithms, it can handle a closed world assumption (where not telling a fact implies the fact is false) as well as the open world assumption by which not being told means nobody knows. The β algorithm for PL uses the closed world assumption while the π algorithm uses the open world assumption.

The information that constitutes a PL program is encoded in three types of rules. The first type are strict rules of the form $A \rightarrow l$ and the semantics

[5] Provability here means determining if the formula can be verified/proved.

$\{\} \rightarrow quail(Quin)$ /* Quin is a quail */
$quail(x) \rightarrow bird(x)$ /* Quails are birds */
$R_1 : bird(x) \Rightarrow fly(x)$ /* Birds usually fly */ (a)

$\{\} \rightarrow quail(Quin)$ /* Quin is a quail */
$quail(x) \rightarrow bird(x)$ /* Quails are birds */
$R_2 : quail(x) \Rightarrow \neg fly(x)$ /* Quails usually do not fly */ (b)

Fig. 1. Two knowledge bases. Are they inconsistent?

is that if all literals in A are proved, then we can conclude l (that is, this is ordinary implication). An example of a strict rule is $human(x) \rightarrow mammal(x)$ to represent that *Humans are mammals.* The second type of rule are plausible rules of the form $A \Rightarrow l$ and the semantics is that when there is no evidence against l, then A is sufficient evidence for concluding l. An example of a plausible rule is $bird(x) \Rightarrow fly(x)$ and we use this to represent that *Birds usually fly.* The third type of rules are defeater rules of the form $A \rightharpoonup \neg l$ with the intended meaning that if we cannot disprove A, then it is too risky to accept l. An example is $\{sick(x), bird(x)\} \rightharpoonup \neg fly(x)$ to encode that *Sick birds might not fly.* Plausible rules allow derivations although we may not be absolutely sure. Defeater rules prevent conclusions which would otherwise be too risky. This may happen in a long chain of conclusions from plausible rules.

PL provides a priority relation $R_1 > R_2$ between rules that represents that R_1 should be used over R_2. Later, we demonstrate the expressive power of this aspect; however, consider the example of the knowledge base in Fig. 1 (a)from which one would conclude that *Quin* usually flies. But, if one considers Fig. 1 (b), the correct conclusion is then that Quin usually does not fly. But what if both knowledge bases are correct; that is, both rules R_1 and R_2 are valid. We perhaps can say that R_2 is more informative as it is more specific and we add $R_2 > R_1$ to a knowledge base representing the beliefs of a robot who knows both. PL reaches the proper conclusion that Quin usually does not fly while if it finds another bird that is not a quail, the robot would accept that it flies.

Another example of the power of hierarchies between rules can be seen in the 3 Laws of Robotics. In fact, humans describe situations commonly in this way. That is, a human expert will usually define a general rule, and present the next rule as a refinement. Rules further down continue to polish the description. This style of development is not only natural, but allows incremental refinement. Indeed, the knowledge elicitation mechanism known as *Ripple Down Rules* [5] extracts knowledge from humans experts by refining a previous model by identifying the rule that needs to be expanded by detailing it more.

Temporal Model. We illustrate our approach with one simple model. A purely spatial DPL model for 0, 1, or 2 landmarks within one frame has already been shown to be feasible and also it has been extended to 3 and even 4 landmarks [10]. Moreover, the properties of PL make it possible to incrementally construct more complex world models by basing them on proven simpler models [10]. Thus, we base the temporal model on the same domain knowledge as the spatial model [10]. Fig. 2 shows the facts of the model for the 2005 4-legged league

```
% The things we might see
type GoalType = {FG, BG}.
type PostType = {LP, RP, RBP, LBP}.
type Landmark = GoalType + PostType.
% Some landmarks are diametrically opposite
type Opp(x <- Landmark, y <- Landmark - {x}).
default ~Opp(x, y).
Opp(FG, BG).   Opp(RP,  LBP).   Opp(RBP, LP).
Opp(BG, FG).   Opp(LBP, RP).    Opp(LP,  RBP).
```

```
% Landmarks closer together than opposite,
% have a real left-to-right order.
type LR(x <- Landmark, y <- Landmark - {x}).
default ~LR(x, y).
LR(LP, FG).    LR(LP, RP).    LR(FG,  RP).
LR(FG,  RBP).  LR(RP, RBP).   LR(RP, BG).
LR(RBP, BG).   LR(RBP, LBP).  LR(BG, LBP).
LR(BG, LP).    LR(LBP, LP).   LR(LBP, FG).
```

Fig. 2. The facts about the 4-legged soccer field

field[6]. Colour coding allows the identification of landmarks as Front Goal (FG), Back Goal (BG), Left Post (LP), Right Post (RP), Right Back Post (RBP) and Left Back Post (LBP). However, in a temporal model, the objects may have been visible in the previous frame or in the current frame and vision now reports sightings with respect to a frame (ie the predicate is now $See(x, f)$).

```
type Frame = {PF, CF}. type See(x <- Landmark, f <- Frame).
type SeeLtoR(x <- Landmark, y <- Landmark - {x}, f <- Frame).
```

Sightings may be transient (did not last across consecutive frames) or persistent (the object is in both frames).

```
type Tra(x <- Landmark).   R1: {} => ~See(x,f).   R2: {} => ~Tra(x).
R3: {See(x, PF), ~See(x, CF)} => Tra(x).  R3: {~See(x, PF), See(x, CF)} => Tra(x).  R3 > R2.
type Per(x <- Landmark). R4: {} => ~Per(x). R5: {See(x, PF), See(x, CF)} => Per(x). R5 > R4.
```

Nothing is consistent unless we get at least a transient or a persistent sighting.

```
type Cs(x <- Landmark). R6: {} => ~Cs(x).
R7: {Tra(x)} => Cs(x). R7 > R6. R8: {Per(x)} => Cs(x). R8 > R6.
```

Seeing two opposite landmarks is grounds for inconsistency (even persistently or transiently).

```
R9: {Opp(x, y), Per(x), Per(y)} => ~Cs(x). R9 > R7. R9 > R8.
R10: {Opp(x, y), Tra(x), Per(y)} => ~Cs(x).
R11: {Opp(x, y), Tra(x), Tra(y)} => ~Cs(x). R10,R11 > R7.
```

What it means for two objects to be in a transient left-to-right order? This happens if vision sees the objects in the previous frame in that order but does not see them in the current frame in such order, or they are seen in the current frame in that order but they were not seen in the previous frame in such order.

```
type TraLtoR(x <- Landmark, y <- Landmark - {x}).  R14: {} => ~TraLtoR(x, y).
R15: {SeeLtoR(x, PF, y, PF), ~SeeLtoR(x, CF, y, CF)} => TraLtoR(x, y).
R15: {~SeeLtoR(x, PF, y, PF), SeeLtoR(x, CF, y, CF)} => TraLtoR(x, y).  R15 > R14.
```

However, objects are persistently seen in a left-to-right order if the sighting of that relationship happened in the previous and current frame.

```
type PerLtoR(x <- Landmark, y <- Landmark - {x}).  R16: {} => ~PerLtoR(x, y).
R17: {SeeLtoR(x, PF, y, PF), SeeLtoR(x, CF, y, CF)} => PerLtoR(x, y).  R17 > R16.
```

Finally, seeing landmarks out of order is grounds for inconsistency. But we only overwrite those rules that may have suggested consistency.

```
R18: {LR(x, y), Per(x), Per(y), PerLtoR(y, x)} => ~Cs(x).
R18: {LR(x, y), Per(x), Per(y), PerLtoR(y, x)} => ~Cs(y).  R18 > R8.
R19: {LR(x, y), Tra(x), Per(y), TraLtoR(y, x)} => ~Cs(x).
R19: {LR(x, y), Per(x), Tra(y), TraLtoR(y, x)} => ~Cs(y).  R19 > R7.
R20: {LR(x, y), Tra(x), Tra(y), TraLtoR(y, x)} => ~Cs(x).
R20: {LR(x, y), Tra(x), Tra(y), TraLtoR(y, x)} => ~Cs(y).  R20 > R7.
```

[6] Previous versions of the field are very similar.

4 Implementation and Evaluation

Running PL off-line but evaluating on-line. For the 2005 RoboCup, it seemed rather difficult to re-implement DPL in C++ for the Sony AIBO. However, DPL not only provides the algorithms for obtaining proofs but provides an interface for presenting facts, and describing a theory [10]. With the interface, we could indicate which predicates are inputs and for what output predicates we needed proofs. The system summarises the proof results as a logic expression for each output predicate in terms of the relevant inputs. We then could request the logic expressions to be simplified (optimised by reducing the number of terms).

The model is executed in DPL (outside the robot) and as a result we obtain a header file in C++ with expressions for the requested outputs [10]. The robot executes a template method (named `Run()`) where all the C++ expressions (one for each landmark) are evaluated after each frame. PL is used on the Sony AIBO through a C++ template method. The template method is executed every time a new frame arrives and the vision module analyses an image.

```
void  Consistency::Run()
{INIT_ALL_FALSE(); UPDATE_ALL(); CHECK_NEW_LANDMARKS(); PLACE_CS_ALL(); COPY_ALL_BOOL(); }
```

These procedures are defined by computer generated glue code macros. The `INIT_ALL_FALSE()` macro creates the definitions of C++ Boolean variables for all landmarks and sets all Booleans that identify if a landmark is visible in the current frame to false. `UPDATE_ALL()` queries the reports of the vision module for the current frame. `CHECK_NEW_LANDMARKS()` makes sure that sightings are for the current frame and previous sighting also have the correct fame number relative to the current frame. `PLACE_CS_ALL()` will evaluate the output expressions for which we requested outputs and if a landmark evaluates to true, it will forward the sighting to the localisation module (or any other module that may benefit from it, like the action to kick when the front goal is visible). `PLACE_CS_ALL()` will have as many `if` statements as outputs requested. The expression of the `if` statement is defined in the pre-computed simplified proofs. For example, testing (evaluating) the `Cs_FG` macro, correspond to asking if we have a consistent sighting of the front goal. Finally `COPY_ALL_BOOL()` shifts the current Boolean values to the Boolean variables corresponding to previous frames (so the previous frame values are correctly set for the next execution of the template method `Run()`).

If the front goal was not seen in this frame, then `INIT_ALL_FALSE()` would have set the variable to false and `UPDATE_ALL()` would not have changed *FG*'s value, so no landmark sighting is forwarded to localisation. However, if the front goal was visible `UPDATE_ALL()` would have (1) set *FG* to true (previously initialised to false in `INIT_ALL_FALSE()`) and (2) provided a pointer to such an object so other attributes about the landmark can be evaluated (like its size or if it is to the left or right of another landmark). The `if` statement in `PLACE_CS_ALL()` would fire, resulting in localisation receiving the sighting information about the front goal if the model determines it is consistent with other objects sighted in this and other previous frames.

Simulator. The evaluation of temporal properties is particularly hard in an environment where sighting errors only occur sporadically at a frame rate higher than 25Hz. Thus, first we implemented a simulator that allows reproducible scenarios of what vision might report to the localisation system and used it to validate the correctness of the PL expressions resulting from the off-line execution. To minimise the risk of testing different algorithms in the simulator and the robot, we used the same C++ code that was generated from the PL rules in both the simulator and the robot. In order to provide a way to consistently set and evaluate a scene, the simulator wraps the C++ expressions in a graphical user interface (GUI) (refer to Fig. 3).

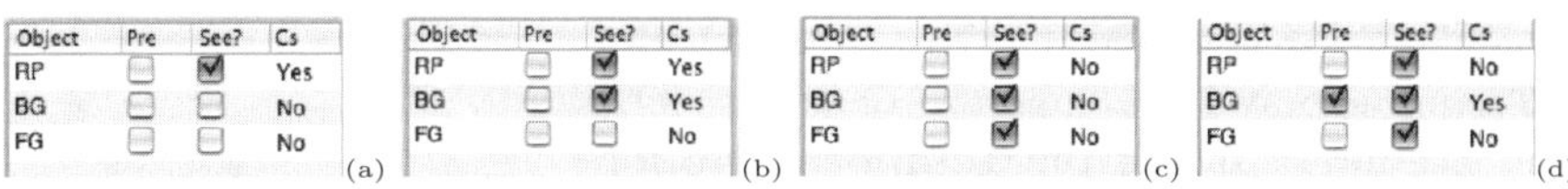

Fig. 3. (a) The simulator showing one particular object (RP) as visible in the current frame. (b) Two consistent objects in the current frame. (c) Three inconsistent sightings in the current frame. (d) The persistent back goal (BG) wins over the temporary sightings of the right post (RP) and the front goal (FG).

The user of the simulator can place landmarks on rows, and the succession of rows represents the order in which these objects are seen (with top to bottom representing left to right in the field of vision). The first column shows the name of the landmark. The next two columns allow the user to select what the visibility state is for the previous (`Pre`) and current (`See?`) frames. The rightmost column shows the output of the consistency module (`Cs`) after performing its reasoning about the state of the world. Furthermore, the GUI allows the dragging and dropping of objects to change the order, as well as addition (`Add`) and deletion (`Delete`) of landmarks.

Fig. 3 (b) shows two consistent sightings. Even though the two landmarks RP and BG were not visible in the previous frame, they are consistent with each other, allowing them to be forwarded on to the localisation module. In fact, this is the same result that a traditional value domain reasoning system would obtain. This is also true for Fig. 3 (c) where we can see three objects that are inconsistent with each other. Since all the objects only occur within one single frame, the only conclusion that can be drawn is that nothing is consistent in that scene. Once information varies over time, a richer belief about the environment can be formed. Fig. 3 (d) shows the same scenario as Fig. 3 (c), but this time the temporal properties of the visible objects vary. The back goal that was visible in the previous frame as well as the current frame is given precedence over the right post and the front goal that were only visible in a single frame.

Temporal test can be combined as in Fig. 4 (a). Objects that are consistent in either space or both space and time are ruled as being consistent in the world view of the system. Only inconsistencies that persist over both space and time will force the system to conclude that nothing is consistent (Fig. 4 (b)).

Evaluation on the robot. We have analysed the effectiveness of the approach in an Sony AIBO ERS-7 in the lab and in the actual RoboCup 2005 competition. In particular, evaluating the CPU costs on board is important since the file `roboconsistency.h` can be quite large. For the **Spatial Model** [10], this file has 4700 tokens for the 6 expressions that determine the consistency of the landmarks. For the **Temporal Model** (with 3 outputs) this file has 1,336 tokens. The **Complete Model for 3** [10], has 12,530 tokens. However, these expressions are in conjunctive normal form and the terms are literals of Boolean variables. The C++ compiler can optimise their evaluation, and in fact we profiled the execution times on the AIBO. Evaluating all 6 expressions in the **Spatial Model** required only 60±1 microseconds (this is with a 99% confidence interval), and evaluating all 6 expressions in the **Complete Model for 3** required only 110±3 microseconds (also 99% confidence interval). For the **Temporal Model**, we evaluated 3 expressions and this required 41±1 microseconds (this is also a 99% confidence interval). However, in all three models the constant overhead of initialising the Boolean values and retrieving the pointers from the vision module was 30 ± 1 microseconds. This means that the evaluation of the large expressions is comparable to the initialisation of a few Boolean variables and in fact orders of magnitude faster than the processing of a frame by the vision module or the processing of a localisation update (the revision of the position belief because of sensor inputs and action) in the localisation module.

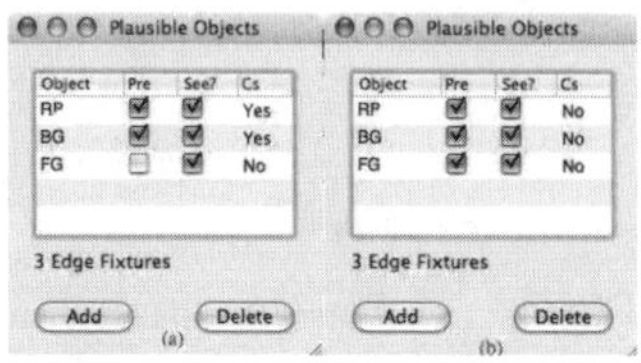

Fig. 4. (a) The two persistent landmarks (RP and BG)are consistent with each other, but inconsistent with the front goal (FG). (b) Nothing is consistent in this view.

Although all cases were tested, we illustrate only when both goals were reported by vision on the same frame. We evaluated during games (with a log on the robot's memory stick) and also in the lab with a log on a `telnet` connection that displays the ID of the objects reported by vision and the outcomes of the non-monotonic reasoning. Fig. 5(a)-5(c) shows images at the venue in Osaka where the consistency module filtered phantom objects for localisation. We have enlarged the captured image on board, then the blobs of the colour as the second largest and the objects reported by vision appear on three screens on the bottom right corner. The left most of these bottom images displays the sightings for goals. Fig. 5(a) shows that phantom sightings occur even with the regular colour-coded objects in the field. The ball has enough yellow pixels to be confused for a yellow goal against a blue goal. Fig. 5(b) shows that the blue match score and timer appears as a goal on a frame with the yellow goal. While our vision system has an analysis for filtering objects above the field of vision, the fact that the Sony AIBO has a head with three degrees of freedom and has legs that during pursuit of the ball make positions and angles of vision that cannot always rule this case out. Fig. 5(c) shows another case where natural lighting and

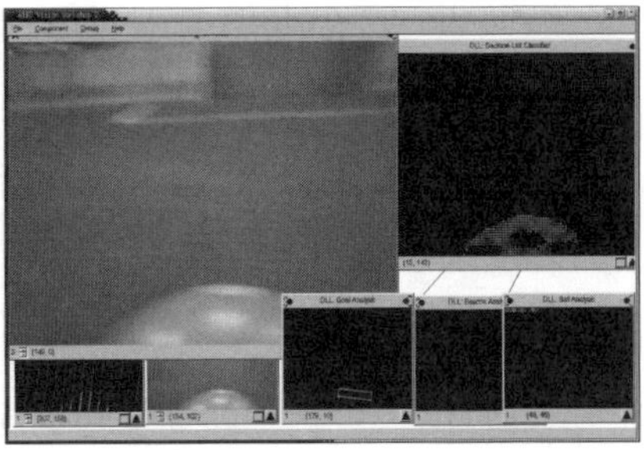

(a) The ball has enough yellow pixels to be confused for a yellow goal against a blue goal

(b) The blue timer appears as a goal with the yellow goal

(c) A window of the field is confused for a blue goal against a yellow goal

(d) The lab setting allows sequences of frames where two goals are reported by vision

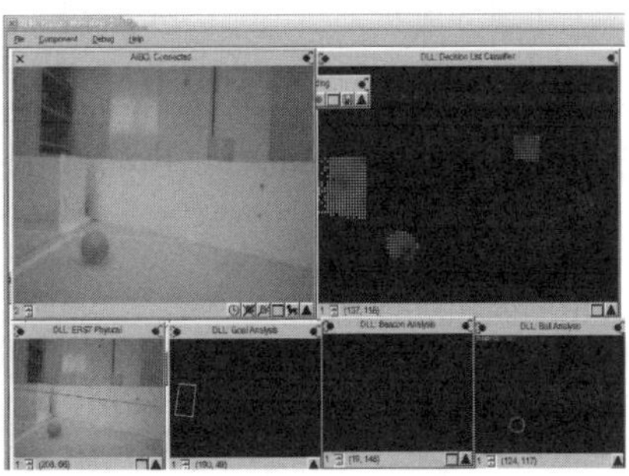

(e) The blue goal can be (dis) covered quickly

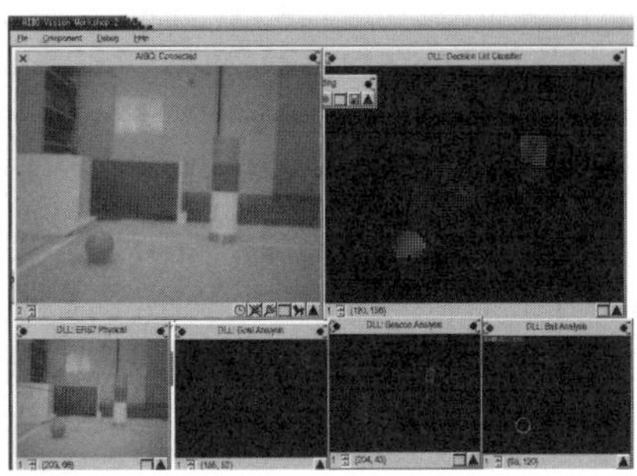

(f) The yellow goal can be (dis) covered quickly

Fig. 5. RoboCup 2005 examples and lab examples

off the field objects result in phantom sightings. In this case, a window registered enough blue pixels to be reported as a blue goal together with the yellow goal.

Fig. 5(d)-5(f) shows a lab setting where we can rapidly produce opposite goals in a frame and immediately after block one goal, or the other. In the log, we found sequences where the robot is seeing only the front goal and reports it as consistent. When the back goal appears as well, for that first frame, the front goal remains consistent and the back goal is labeled inconsistent (note that in the discussion of the KF, MM, and MCL localisers we indicated that the frame with both goals becomes not usable). If the back goal persists with the front goal for one more frame, then both goals are now labeled inconsistent (note that the model can easily be adjusted if a different effect is desired besides having two consecutive frames with both goals to rule them out). If the front goal drops out, then the back goal in the previous frame and the back goal in the current frame both become consistent. We have no space here to discuss more examples of the versatility of the modelling here, but using the RP also helps since the blue goal is in the right left-to-right order with respect to the post.

5 Discussion and Conclusions

We have experimented with modelling the RoboCup field for the 4-legged league without PL [10]. This results in large models that grow exponentially as more landmarks are introduced and, where little can be inferred. An attempt with other logics also results in very large models [10]. The challenge is then to consider what the always imperfect vision module may report. That is, in a single frame, the analysis of an image may actually perceive two blobs of yellow colour and one of blue that are rectangular enough for all of them to be considered as goals. Again, any software/logic that rules out two rectangular blobs of yellow, perhaps on the basis that one is larger than the other, or one is above the field of vision, or one is next to green, is performing some reasoning based on domain knowledge. And what we are arguing here is that if all those ways of ruling out sightings of landmarks are not concentrated in a single module represented in logic, then the software is very likely to have such rules in several modules, resulting in high coupling of several modules, and more seriously, in incomplete and inconsistent modelling of the reasons why some sightings are ruled out before they are used for localisation. As the robots move to more realistic environments more reasoning is needed (now the new field does not have a barrier of white around it, there are more phantom objects in the audience [7]).

We have no space here to discuss what other logics produce as the rules for when vision reports two objects which are landmarks that have a third landmark in between. What are the cases for when we have two objects which are landmarks that have a third landmark to the side. Then, we have to progress to when vision reports exactly 3 landmarks. Seeing exactly 4 landmarks results in 120 rules and so on. The point we are making is that even from the software engineering, software verification and validation point of view, we need a complete and correct logic theory of the consistency of the vision reports.

Because of this large number of cases, the knowledge elicitation approach (attempting to make P explicit with something like Ripple Down Rules) will also constitute a large model of deeply nested `if-then-else` rules. Since the hierarchy is basically a tree, there is the potential of replicating some subtrees. But more seriously, Ripple Down Rules would be a binary valued logic incapable of

[7] See Fig. 5 for examples of this in the RoboCup-05 setting.

reaching more generic conclusions and reasoning capabilities as PL(for example, switching between the closed world assumption and the open world assumption). Although the **Spatial Model** and the **Temporal Model** have relations for only two objects in the same frame, they correctly derive conclusions for most cases with three objects in a frame. In particular, these two models can correctly rule out the inconsistent object when the left post is correct, but the goal and right post are inverted. A case they miss is when the left post and the goal are in correct order, but the right post appears leftmost. This again reflects the power of modelling with PL as opposed to modelling without it. The PL is analysing the pairs within the triplet in sight. And while two of the pairs are consistent (those involving the left post), the pair involving the goal and the right post indicates both of these are inconsistent. The **Complete Model for 3 objects** gets this case correct and again resolves correctly almost all sightings of 4 objects.

We included reasoning in a hybrid architecture (symbolic and reactive) for useful tasks within the time bounds of a dynamic environment like robotic soccer. The next step is to implement PL in C++ so that it can run on board the Sony AIBO. In this way, the robot will not be required to carry on board proofs of all possible outputs of interest, but just execute reasoning as needed.

References

1. Antoniou, G.: Nonmonotonic Reasoning. MIT Press, Cambridge, MA (1997)
2. Billington, D.: The proof algorithms of plausible logic form a hierarchy. In: Zhang, S., Jarvis, R. (eds.) AI 2005. LNCS (LNAI), vol. 3809, pp. 796–799. Springer, Heidelberg (2005)
3. Billington, D., Rock, A.: Propositional plausible logic: Introduction and implementation. Studia Logica 67, 243–269 (2001)
4. Brooks, R.A.: Intelligence without reason. In: ICJAI-91. 12th Int Joint Conf on Artificial Intelligence, San Mateo, CA, pp. 569–595. Morgan Kaufmann, San Francisco (1991)
5. Compton, P., et al.: Ripple down rules: possibilities and limitations. In: 6th Banf AAAI Knowledge Acquisiiton for Knowledge Based Systems Workshop (1991)
6. Fox, D., Burgard, W., Thrun, S.: Markov localization for mobile robots in dynamic environments. Journal of Artificial Intelligent Research 11, 391–427 (1999)
7. Gutmann, J.-S., Fox, D.: An experimental comparison of localization methods continued. In: IEEE/RSJ Int. Conf. Intelligent Robots and Systems, Lausanne (2002)
8. Haemmi, R., Hartmann, S.: Modeling partially reliable information sources: A general approach based on Dempster-Shafer theory. Information Fusion (forthcoming)
9. Marek, V.W., Truszczynski, M.: Nonmonotonic Logic: Context-Dependent Reasoning. Springer, Heidelberg (1993)
10. Billington, D., Estivill-Castro, V., Hexel, R., Rock, A.: Non-monotonic Reasoning for Localisation in RoboCup. In: Sammut, C. (ed.) Australasian Conf. on Robotics and Automation. UNSW, Sydney (2005) CD-ROM
11. Rich, E., Knight, K.: Artificial Intelligence, 2nd edn. McGraw-Hill, New York (1991)
12. Rock, A., Billington, D.: An implementation of propositional plausible logic. In: Edwards, J. (ed.) 23rd Australasian Computer Science Conf. Australian Computer Science Comms, pp. 204–210. IEEE Computer Society, Los Alamitos (2000)

13. Röfer, T., Jüngel, M.: Fast and robust edge-based localization in the SONY four-legged robot league. In: Polani, D., Browning, B., Bonarini, A., Yoshida, K. (eds.) RoboCup 2003. LNCS (LNAI), vol. 3020, Springer, Heidelberg (2004)
14. Russell, S., Norvig, P.: Artificial Intelligence: A Modern Approach, 2nd edn. Prentice-Hall, Englewood Cliffs, NJ (2002)
15. Thrun, S., Fox, D., Burgard, W., Dellaert, F.: Robust Monte Carlo localization for mobile robots. Artificial Intelligence 128, 99–141 (2001)
16. Wooldridge, M.: An Introduction to MultiAgent Systems. Wiley, New York (2002)

Multi-cue Localization for Soccer Playing Humanoid Robots

Hauke Strasdat, Maren Bennewitz, and Sven Behnke

University of Freiburg, Computer Science Institute, D-79110 Freiburg, Germany
strasdat@gmx.de, {maren,behnke}@informatik.uni-freiburg.de,
http://www.NimbRo.net

Abstract. An essential capability of a soccer playing robot is to robustly and accurately estimate its pose on the field. Tracking the pose of a humanoid robot is, however, a complex problem. The main difficulties are that the robot has only a constrained field of view, which is additionally often affected by occlusions, that the roll angle of the camera changes continously and can only be roughly estimated, and that dead reckoning provides only noisy estimates. In this paper, we present a technique that uses field lines, the center circle, corner poles, and goals extracted out of the images of a low-cost wide-angle camera as well as motion commands and a compass to localize a humanoid robot on the soccer field. We present a new approach to robustly extract lines using detectors for oriented line pints and the Hough transform. Since we first estimate the orientation, the individual line points are localized well in the Hough domain. In addition, while matching observed lines and model lines, we do not only consider their Hough parameters. Our similarity measure also takes into account the positions and lengths of the lines. In this way, we obtain a much more reliable estimate how well two lines fit. We apply Monte-Carlo localization to estimate the pose of the robot. The observation model used to evaluate the individual particles considers the differences of expected and measured distances and angles of the other landmarks. As we demonstrate in real-world experiments, our technique is able to robustly and accurately track the position of a humanoid robot on a soccer field. We also present experiments to evaluate the utility of using the different cues for pose estimation.

1 Introduction

The knowledge about the poses of the robots during a soccer game is one of the prerequisite features for successful team play. Many approaches to localization on the RoboCup field or less structured environments have already been presented. Several approaches exist that use distance information provided by a proximity sensor, like a laser scanner [1], or images of an omnidirectional camera [2,3,4,5] for localizing a robot in the RoboCup environment. However, for some types of robots such sensors do not seem to be appropriate since they do not agree with their design principle. Humanoid robots, for example, which are constructed to resemble a human, are typically only equipped with directed cameras.

In this paper, we present an approach to vision-based localization of a humanoid robot that uses a single wide-angle camera. Many problems exist that make localization

G. Lakemeyer et al. (Eds.): RoboCup 2006, LNAI 4434, pp. 245–257, 2007.

for humanoid robots a complex problem. The robot can only observe a constrained part of the soccer field due to the directed camera and due to occlusions caused by moving objects. The camera images are distorted and often blurred because of camera motion. Furthermore, the roll angle of the camera changes continously and can usually only be roughly estimated. The camera perspective, however, has a significant impact on the measured distances and angles to landmarks. Finally, compared to a wheeled robot equipped with odometry sensors, we get a very noisy estimate by dead reckoning, i.e. , the prediction of the robot's pose based on executed motion commands.

We apply the well-known Monte-Carlo localization (MCL) technique [6] to estimate the robot's pose (x, y, ϕ), where (x, y) denotes the position on the field and ϕ is the orientation of the robot. MCL uses a set of random samples, also called particles, to represent the belief of the robot about its pose. We extract environment-specific landmarks out of the camera images. Features we use for localization are the field lines, the center circle, the corner poles, and the goals. Additionally, we consider compass data as well as the motion commands sent to the robot.

We present a new technique to robustly extract lines out of images. In contrast to previous approaches that find line transitions on scan lines in the image [2,7], simply detect edges based on certain color changes [3,8,4], or using classical kernels that detect edge gradients in greyscale images [9], we extract line segments out of an image by applying four different detectors that find oriented line points. The detectors guarantee that green is detected on both sides of a line. Since the detectors provide a reliable estimate of the orientations of the individual line points, they are localized well in the Hough domain and we can direct the search towards lines that match these orientations. Before applying the Hough transform to find lines in the two main orientations, we locate the center circle.

For a discrete set of possible poses of the robot, we compute the largest lines that are expected to be visible from that pose. During localization, these lines are then compared to the largest currently observed lines. To get a much more reliable estimate how well two lines fit, we do not only compare their Hough parameters. We also take into account a measure that estimates the positions and lengths of the lines. In the observation model of the particle filter, we additionally consider the differences between measured and expected distances and angles of the other landmarks as well as the compass data. As we demonstrate in practical experiments with a humanoid robot in the RoboCup environment, our technique is able to robustly and accurately track the pose of the robot. In order to assess the benefit of using the different types of features, we present experiments illustrating their influence on the localization result.

This paper is organized as follows. After discussing related work in the following section, we introduce our robot and the RoboCup environment. In Section 4 we describe the Monte-Carlo localization technique that is applied to estimate the robot's pose. In Section 5 we explain how to extract lines out of the images provided by our camera. Afterwards, we present the observation model used for MCL in Section 6. Finally, in Section 7, we show experimental results illustrating the robustness of our approach to estimate the robot's pose.

2 Related Work

Many systems that perform vision-based localization in the RoboCup domain have already been presented. None of them can directly be applied for localization using our humanoid robot as we will point out in the following.

Several localization systems make use of the field lines. De Jong et al. [8] identify edges and divide the image in subimages afterwards. They apply the Hough transform in each subimage assuming that the line segments (even the ones corresponding to the center circle) are mostly straight. They propose to estimate the position of the robot by considering the distance from the observed line segments to the closest expected lines for poses in a local area around its last pose. However, no pose tracking experiments are presented. Furthermore, it remains unclear how to choose an appropriate discretization of the image. Iocchi and Nardi [10] apply an extended Kalman filter for localization with a perspective camera. They match observed and model lines in the Hough domain. One assumption for this method to work is that odometry data yields a good estimate of the robot's pose. This cannot be guaranteed in our case. Furthermore, we have to deal with the problem of changing camera perspectives during walking. Marques and Lima [4] apply a similar approach of line matching in the Hough domain. They search color transitions on circles in omnidirectional images. Röfer et al. [7] distinguish between four different types of lines. In each image, they search for line transitions corresponding to these types and randomly draw three transitions for each. During MCL, the measured angles to the drawn points are then compared to the expected angles of the closest points of each type. Furthermore, they take into account other landmarks like goals and poles. Due to the unpredictable roll angle of the camera, it is in our case not enough to consider only a few points on the lines for localization. We have to use as much information about the lines as we can extract.

Lauer et al. [2] use line transitions extracted out of omnidirectional images and apply gradient decent to minimize an error term in order to locally find the best match between the lines in the current image and the field lines. It is not clear yet how this method performs under real conditions, especially in case of sensor noise. Schulenburg et al. [3] apply a Kalman filter and combine omnivision with laser data for localization. They search for color transitions in the omnidirectional image and find lines in the Cartesian space. Their approach associates each point to a unique line. In order to avoid the data association problem and in order to be able to deal with noisy data, we search for lines in the Hough domain. The approach presented by von Hundelshausen and Rojas [5] uses a technique to track the green regions and search for transitions on the boundaries of these tracked regions. Tracking regions is much easier in omnidirectional images since, due to occlusions during a dynamic soccer game, usually the small field of view changes rapidly when using a directed camera only.

Enderle et al. [11] rely only on particular landmarks during MCL and do not take into account the field lines. This approach may have problems in case of frequent occlusions as they typically occur during a soccer game. Note that Hoffmann et al. [12] propose to utilize negative information during localization. In case the robot did not detect a certain landmark, this information could be used to exclude some states. The idea has so far only been tested in simple setups and not during a soccer game where occlusions are likely to cause problems.

3 The Design of Our Robot and Environmental Setup

The left image of Figure 1 depicts one of our robots, which we used to carry out the experiments presented in this paper. The height of the robot is 60cm and it has 19 degrees of freedom, which are driven by servo motors. We use a Pocket PC, which is located in the chest, to control the robot. The Pocket PC is equipped with a 520MHz processor and 128MB RAM. The robot uses a compact flash color camera with a wide-angle lens to get information about the environment. The lens is located approximately at the position of the larynx. The large field of view ($150°$ horizontally and $112°$ vertically) allows the robot to see at the same time the ball at its feet and objects above the horizon (see right image of Figure 1). The camera delivers images at a rate of up to 5fps with a resolution of 320×240 pixels. Additionally, the robot is equipped with a compass, which is located in its head, to facilitate localization.

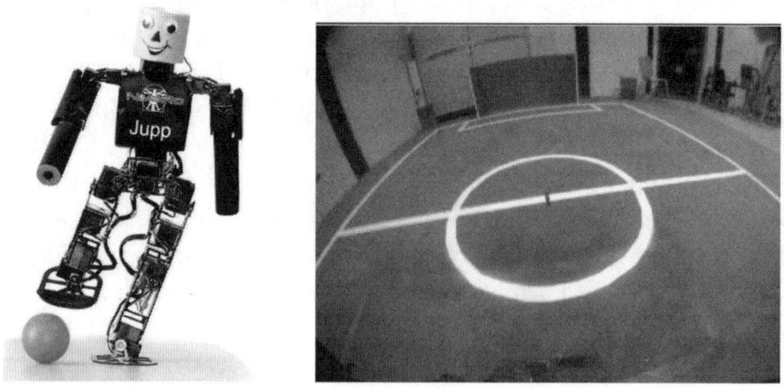

Fig. 1. The left image shows one of our humanoid robots. The robot is controlled by a Pocket PC and uses the images of a low-cost wide-angle camera for perceiving the relevant information about the environment. An image captured from the robot's perspective while it was walking can be seen on the right.

The KidSize soccer field in the Humanoid League has a size of $4.5\text{m} \times 3\text{m}$. Objects that can be used for localization are the two goals (colored blue and yellow), the field lines, the center circle, and four corner poles.

4 Monte Carlo Localization

To estimate the pose x_t of the robot at time t, we apply the well-known Monte-Carlo localization (MCL) technique [6], which is a variant of Markov localization. MCL recursively estimates the posterior about the robot's pose:

$$
\begin{aligned}
& p(x_t \mid z_{1:t}, u_{0:t-1}) \\
& = \eta \cdot p(z_t \mid x_t) \cdot \int_{x_{t-1}} p(x_t \mid x_{t-1}, u_{t-1}) \cdot p(x_{t-1} \mid z_{1:t-1}, u_{0:t-2})\, dx_{t-1} \qquad (1)
\end{aligned}
$$

Here, η is a normalization constant resulting from Bayes' rule, $u_{0:t-1}$ denotes the sequence of all motion commands executed by the robot up to time $t-1$, and $z_{1:t}$ is the sequence of all observations. The term $p(x_t \mid x_{t-1}, u_{t-1})$ is called motion model and denotes the probability that the robot ends up in state x_t given it executes the motion command u_{t-1} in state x_{t-1}. The observation model $p(z_t \mid x_t)$ denotes the likelihood of making the observation z_t given the robot's current pose is x_t.

MCL uses a set of random samples to represent the belief of the robot about its state at time t. Each sample consists of the state vector $x_t^{(i)}$ and a weighting factor $\omega_t^{(i)}$ that is proportional to the likelihood that the robot is in the corresponding state. The update of the belief is carried out according to the sampling importance resampling particle filter. First, the particle states are predicted according to the motion model. For each particle, a new pose is drawn given the executed motion command since the previous update. In the second step, new individual importance weights are assigned to the particles. Particle i is weighted according to the likelihood $p(z_t \mid x_t^{(i)})$. Finally, a new particle set is created by sampling from the old set according to the particle weights. Each particle survives with a probability proportional to its importance weight. This step is also called resampling.

In order to allow for global localization, e.g., in case of the "kidnapped robot problem", a small amount of the particles is replaced by particles with randomly drawn poses.

5 Finding Lines Using Detectors for Oriented Line Segments and the Hough Transform

In this section, we introduce our approach to robustly locate lines in the images of the wide-angle camera. Figure 2 illustrates the whole process of finding individual line points, extracting line segments, determining their position in egocentric coordinates and their transformation into the Hough space. The details are described below.

5.1 Extracting Oriented Line Segments

To reduce the influence of different lighting conditions, we color-classify pixels in the YUV color space using the pie-slice method [13]. In a multistage process, insignificant colored pixels are discarded so that only the colors of the relevant features remain. To find lines, we are only interested in the pixels that are classified as white or green.

Line points can be detected by processing the color-classified image as follows. We apply elongated Gaussian kernels to determine the likelihood of pixels being part of a line. In short, we search for elongated white regions that have green neighbors on two opposite sides. By taking into account the relative difference of the sum of the likelihoods of green and white pixels within a neighborhood, we estimate the likelihood that a pixel corresponds to a line point. We use Gaussian kernels of two different sizes to search for closer line points, which contain more white pixels and which are in the lower part of the image, and for line points that are farer away. In order to estimate the orientation of a line point, we consider four different orientations and apply individual kernels (see Figure 3) to compute the individual likelihoods. For each line point with a

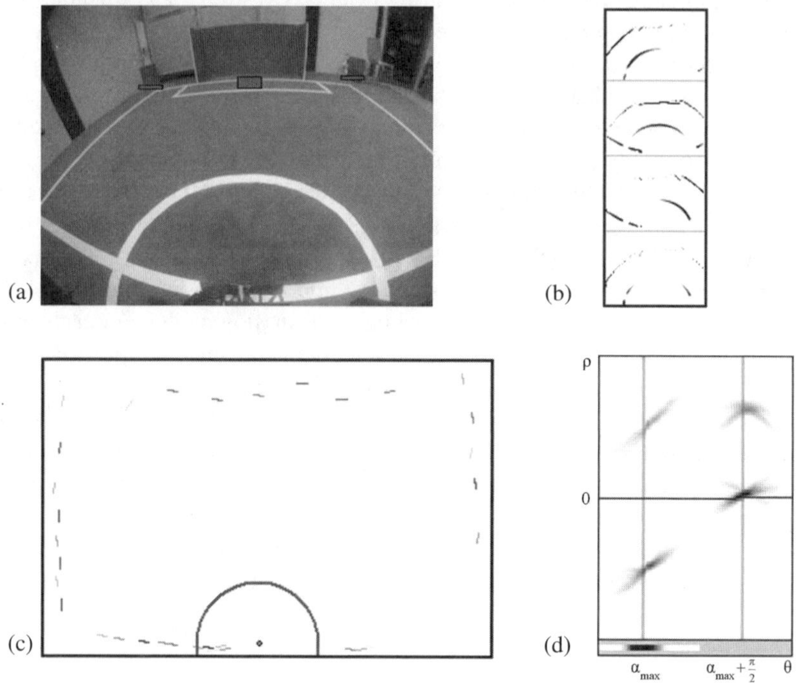

Fig. 2. Illustration of the extraction of lines: The robot's current field of view can be seen in image (a). The four small images (b) show responses of four oriented line detectors. Image (c) depicts the extracted line segments projected into the egocentric space. The center circle is detected and removed. Image (d) shows the representation of the line segments in the Hough space, where the major orientation α_{max} is estimated and the main lines are detected.

likelihood above a threshold, we store the orientation that yielded the maximum likelihood as well as the best neighbor orientation. Figure 2 (b) illustrates the responses of the line detectors for the example image shown in Figure 2 (a).

After identifying the line points in the image, we project them onto the soccer field (see Figure 2 (c)). To get the metric egocentric coordinate on the soccer field corresponding to an image coordinate, we first eliminate the radial distortion in the image and apply an affine transformation afterwards. The parameters of the projective transformation can be calculated from four pairs of points on the field and the corresponding points in the camera image. Here, we assume a fixed viewing angle of the camera onto the field. Note that after the undistortion process and the projection, the line segments have a rather high distance to each other.

In addition to the coordinates of the line points, the corresponding two orientations are also transformed from the image space into the metric egocentric space. We estimate the actual orientation of a line point by computing the weighted mean of these transformed orientations using their likelihoods. Thus, we have a set of weighted line points that are described by their positions and their orientations relative to the robot.

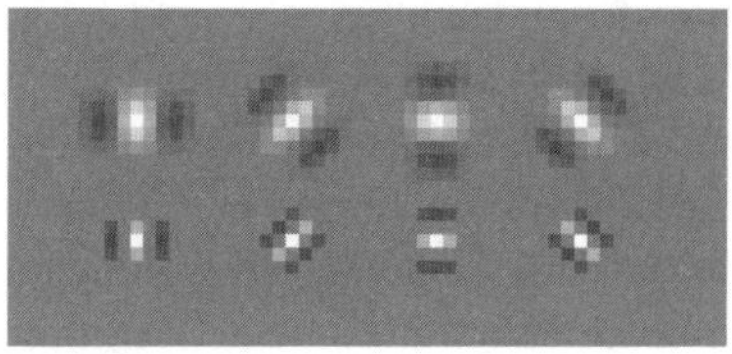

Fig. 3. The four kernels used for the detection of oriented line segments in two different sizes

To improve robustness, we do not transform individual line points into the Hough space (see next section). Instead, we merge line points that are close in the image space and have similar orientations to one longer line segment. This way, we can deal with false detections caused by noisy observations.

5.2 The Hough Transform for Line Extraction

The Hough transform is a robust method to find lines fitting a set of 2D points [14]. It relies on a transformation from the Cartesian plane in the Hough domain. The following curve in the Hough domain is associated with a point (x, y) in the Cartesian plane:

$$\rho = x \cdot \cos(\theta) + y \cdot \sin(\theta) \quad (2)$$

Here, ρ is the perpendicular distance from the origin and θ is the angle with the normal. This curve describes all the lines that go through (x, y), since each point in the Hough space corresponds to a line in the 2D Cartesian space.

By discretizing the Hough space into cells $h(\theta_i, \rho_j)$, one can search for local maxima. These local maxima in the Hough domain correspond to the best fitting lines for the input points. The Hough transform has the advantage that it is quite robust to sensor noise, since outliers do not affect the local maxima. Furthermore, since distances of points on the lines are not considered, it can deal with occlusions.

Before we apply the Hough transform to find the best fitting lines for the extracted oriented segments, we locate the center circle. Whenever large parts of the circle are visible in the image, this can impair the line detection. Parts of the circle can be misclassified as short lines and the circle can potentially affect the estimation of the main orientations. To avoid these problems, we first identify the center circle following the approach presented by de Jong et al. [8]. We consider the individual line segments and vote for the positions at a distance of the radius of the center circle, orthogonal to the orientation of the segment. By determining the largest cluster of points and identifying the segments that voted for it, we find the segments corresponding to the center circle. To avoid false positive detections, we only interpret a cluster as the center circle if the line segments that voted for it have a large range of orientations. The corresponding segments are eliminated, i.e. they are not transformed into the Hough space. Additionally, the knowledge about the position of the center circle is used in the observation model of the particle filter (see next section).

In the standard Hough transform, each line segment (x, y) votes for all bins $h(\theta, \rho)$ which fulfill Eq. (2). Since we have already estimated the orientation of the line segments, we only have to vote for a small subset of bins, which reduces the computational costs. In particular, we accumulate its likelihood in the bins $h(\theta \pm \epsilon, \rho)$ that correspond to the estimated orientation θ of a segment. Here, ϵ indicates that we consider a local neighborhood of θ whose bins are also incremented. In this way, we direct the search towards lines that fit the preestimated orientations. In our current implementation, we use a discretization of 2.5° and 6cm for the Hough space.

In general, to locate lines in the Hough space one has to search the entire space for local maxima. In the RoboCup domain, we only have two possible orthogonal orientations for the field lines. This allows us to use a robust method for finding lines that additionally reduces the computational costs: We can determine the two main orientations by adding the bins corresponding to α and $\alpha + \frac{\pi}{2}$, with $\alpha \in [0; \frac{\pi}{2}[$ and finding the maximum:

$$\alpha_{max} = \underset{\alpha=\theta_i \bmod \frac{\pi}{2}}{\operatorname{argmax}} \sum_{\rho_j} h(\theta_i, \rho_j) \tag{3}$$

Finally, we search for maxima in the bins of α_{max} and $\alpha_{max} + \frac{\pi}{2}$, respectively. In this manner, we extract from the Hough space four field lines, two for each main orientation, that are used in the observation model of the particle filter (see next section). Figure 2 (d) illustrates the representation of the segments in the Hough space (the darker the bins the higher the values) as well as the main orientation α_{max}.

6 Observation Model

In this section, we describe how to integrate the observations made by the robot in order to update the weights of the particles.

6.1 Corner Poles and Goals

The corner poles and goals can be detected by processing the color classified image. Using constraints about the number of pixels of certain colors in neighborhoods, these landmarks can be found quite robustly. The egocentric coordinates of the landmark's lower borders are obtained using the projective transformation. For each such landmark, we estimate the distance and angle to the individual particle poses. To compute the likelihood of a landmark observation, we evaluate the differences between expected and measured distances and angles of landmarks

$$p_{landmark} = \exp\Big(-\frac{\|d_e - d_o\|}{2 \cdot \sigma_1^2}\Big) \cdot \exp\Big(-\frac{\|\beta_e - \beta_o\|}{2 \cdot \sigma_2^2}\Big), \tag{4}$$

where σ_1 and σ_2 are the variances of the Gaussians, d_e and d_o are the expected and measured (observed) distance to the landmark, and β_e and β_o are the expected and measured angle of the landmark.

6.2 Center Circle and Lines

If the center circle can be detected, the position of its center is estimated as explained in Section 5.2. To compute the likelihood of this observation, we also evaluate the difference between the expected and the observed position

$$p_{circle} = \exp\Big(-\frac{\|c_e - c_o\|}{2 \cdot \sigma_3^2}\Big). \tag{5}$$

Here, c_e denotes the expected and c_o the measured position of the center circle, and σ_3 is the variance of the Gaussian evaluating the distance.

Line matching is a bit more complicated. First, it should be noted that transforming line segments into the Hough space has a serious shortcome. Hough parameters describe straight lines without starting and endpoints. Thus, significant information gets lost when line segments are transformed into the Hough space. To overcome this drawback, we additionally compute the value u_s that corresponds to the mean position of the line segments s on the straight line l as well as its standard deviation $\sigma(u)$.

In particular, we compute for each single line segment s belonging to l a value

$$u_s = x \cdot \sin(\theta) - y \cdot \cos(\theta), \tag{6}$$

where (x, y) is the the position of s in the Cartesian space and θ is the angular Hough parameter of l. The value u_s indicates, in the egocentric space, the displacement of s along l from the perpendicular of l that goes through the origin. Then, we determine for l the mean $\bar{u}$ and the standard derivation $\sigma(u)$ of the u_s values of its segments. A low value of $\sigma(u)$ corresponds to a short line. On the other hand, a long line results in a high value of $\sigma(u)$. Furthermore, $\bar{u}$ is an indication for the position of the line. These correlations in the Cartesian space are lost when applying the plain Hough transform. In order to more reliable estimate how well two lines fit, we do not only compare their Hough parameters but also their $\bar{u}$ and $\sigma(u)$ values. This is especially important in case the actual role angle of the camera is different from the predicted angle.

In the observation model, we do not take into account all observed lines. Instead, we only consider the largest ones. For a discrete set of possible poses on the soccer field, we precompute the six largest field lines (according to the number of line points) that are expected to be visible from that pose as well as their parameters θ, ρ, $\bar{u}$, and $\sigma(u)$. From the lines extracted out of the current image, we determine for each of the two main orientations two lines by finding local maxima in the Hough space. We assign each of these observed lines to one of the four largest lines that are expected to be visible from the particle pose. Here, we follow a nearest neighbor approach and use the Hough parameters for comparison.

To compute the likelihood of line observations, we evaluate the differences between expected and measured Hough coordinates $h = (\theta, \rho)$ of matched lines, as well as the differences between the expected and measured $\bar{u}$ and $\sigma(u)$. Note that we only consider the lines corresponding to the nearest neighbor matchings m

$$p_{lines} = \prod_m \exp\Big(-\frac{\|h_e^m - h_o^m\|}{2 \cdot \sigma_4^2} - \frac{\|\bar{u}_e^m - \bar{u}_o^m\|}{2 \cdot \sigma_5^2} - \frac{\|\sigma(u)_e^m - \sigma(u)_o^m\|}{2 \cdot \sigma_6^2}\Big). \tag{7}$$

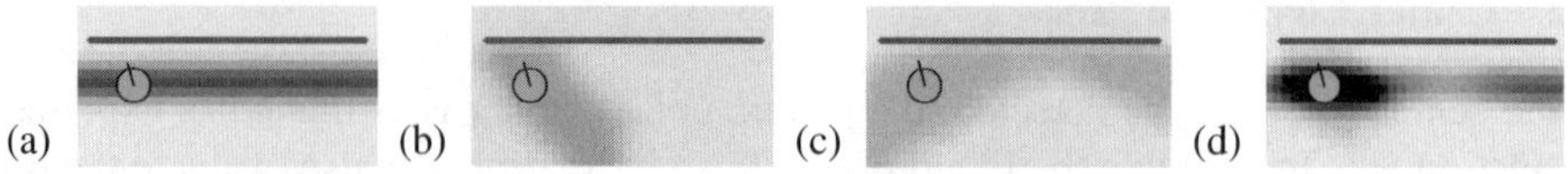

Fig. 4. Likelihood of positions given the observed line using (a) the Hough parameters only, (b) the $\bar{u}$ value, (c) the $\sigma(u)$ value and (d) all four parameters (the darker the more likely)

Here again, the σ's are the variances of the Gaussians, e and o denote the expected and measured values. Figure 4 shows the likelihood of positions given the line observation from the indicated pose.

6.3 Compass

To eliminate ambiguities, which can be caused by the symmetric field lines, the robot can additionally use its compass. Compass data is evaluated with an analogous function that is based on the difference between measured and expected values to yield the likelihood p_{com}. Since the compass data is not that precise, we use a Gaussian with a rather high variance to evaluate the difference.

6.4 Integration of All Observations

Finally, we can define for a particle with the pose $x_t^{(i)}$ the likelihood of making the observation $z_t = \{landmarks, circle, lines, compass\}$ as the product of the individual likelihoods

$$p(z_t \mid x_t^{(i)}) = \prod_{landmarks} p_{landmark} \cdot p_{circle} \cdot p_{lines} \cdot p_{com}. \tag{8}$$

Note that we use a confidence value c_j for all individual observations o_j. c_j indicates how sure we are that o_j was correctly observed. We ensure that no observation has a likelihood p_j smaller than $(1 - c_j)$:

$$p_j = (1 - c_j) + c_j \cdot p_j. \tag{9}$$

7 Experimental Results

In order to evaluate our approach to estimate the pose of a humanoid robot on the RoboCup soccer field, we carried out an experiment in which the robot was controlled via joystick to six different poses that were marked with tape on the field. We take these marked positions as "ground truth". Therefore, part of the errors in the localization results is due to the problem of joysticking the robot exactly onto the marked positions.

Since the robot does not possess any odometry sensors, we perform dead reckoning to estimate the pose of the robot based on motion commands sent to the robot's base. The control input consists of the robot's current gait vector that controls the lateral, sagittal, and the rotational speed of omnidirectional walking. The estimated velocities are integrated over time to determine the relative movement. However, due to slippage on the ground the dead reckoning estimate is highly unreliable. We use therefore a rather high noise in the motion model of the particle filter.

Figure 5 shows the true pose of the robot as well as the estimates at the six marked poses using the different cues. It should be noted that in the beginning, the robot globally localized itself on the field. We determined the position of the robot by clustering the particles and computing the weighted mean of the particles of the maximum cluster. In the figure, we only illustrate the pose estimate based on all available cues and the pose estimate resulting when ignoring the lines. As can bee seen, we obtain a much more robust and accurate estimate by considering the lines in the observation model. For example, shortly before the robot reaches position 6 it has a false positive detection of a landmark in the laboratory environment. This results in a false pose estimate when the lines are not used for localization. The average error in the x/y-position was $20cm$ and the average error in the orientation of the robot was $5°$ when the lines were used for localization. The analogous values were $66cm$ and $11°$ in the case that the lines were not used. We also performed experiments in which we ignored the corner poles or the compass. We observed that the lines are the most relevant feature.

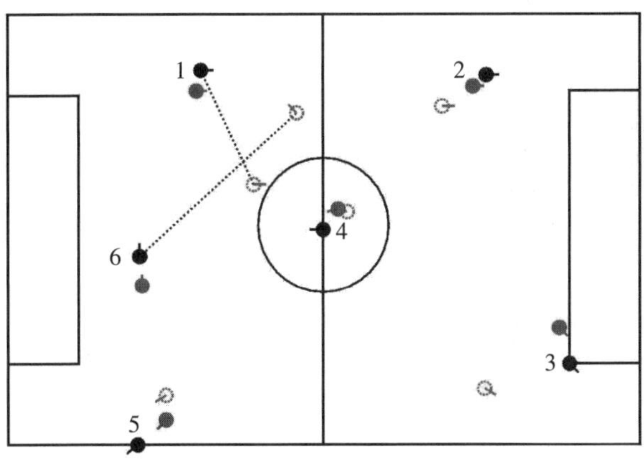

Fig. 5. Result of a localization experiment with our humanoid robot. The black (filled) circles correspond to the true pose, the red (gray) filled circles to the localization result using all cues, and the green (unfilled) circles to the estimate based on all cues except the lines. As can be seen, by considering all available cues the localization works robustly and accurately.

8 Conclusions

In this paper, we presented a robust approach to accurately localize a humanoid robot on the soccer field. This is a challenging task since data from an omnivision camera or a distance sensor is not available. The image of the perspective camera covers only a constrained part of the environment and, additionally, the roll angle of the camera can only be roughly estimated. However, it has a high influence on the measured positions of objects. A further problem is that the robot lacks odometry sensors and that dead reckoning provides extremely noisy pose estimates. For reliable localization, we use the the field lines, the center circle, the corner poles, the goals as well as compass data. We developed a new method to robustly extract lines out of noisy images. The main idea is that we apply robust detectors for oriented line points in the image and aggregate them locally to oriented line segments. In the Hough domain, we then can direct the search towards lines that fit these orientations. While comparing an observed line to a model line during localization, we do not only consider their Hough parameters but also an estimate of their positions and lengths. As we demonstrated using a humanoid robot on the soccer field, our system provides a robust and accurate pose estimate.

Acknowledgment

This project is supported by the German Research Foundation (DFG), grant BE2556/2-1.

References

1. Gutmann, J.S., Weigel, T., Nebel, B.: Fast, accurate, and robust self-localization in polygonal environments. In: Proc. of the IEEE/RSJ Int. Conf. on Intelligent Robots and Systems (IROS) (1999)
2. Lauer, S.L.M., Riedmille, M.: Calculating the perfect match: An efficient and accurate approach for robot self-localisation. In: Bredenfeld, A., Jacoff, A., Noda, I., Takahashi, Y. (eds.) RoboCup 2005. LNCS (LNAI), vol. 4020, Springer, Heidelberg (2006)
3. Schulenburg, E., Weigel, T., Kleiner, A.: Self-localization in dynamic environments based on laser and vision data. In: Proc. of the IEEE/RSJ Int. Conf. on Intelligent Robots and Systems (IROS) (2003)
4. Marques, C.F., Lima, P.U.: A localization method for a soccer robot using a vision-based omni-directional sensor. In: Stone, P., Balch, T., Kraetzschmar, G.K. (eds.) RoboCup 2000. LNCS (LNAI), vol. 2019, pp. 96–107. Springer, Heidelberg (2001)
5. von Hundelshausen, F., Rojas, P.: Localizing a robot by the marking lines on a soccer field. In: Proc. of the Computer Vision Winter Workshop (CVWW) (2003)
6. Dellaert, F., Fox, D., Burgard, W., Thrun, S.: Monte Carlo localization for mobile robots. In: Proc. of the IEEE Int. Conf. on Robotics & Automation (ICRA), IEEE Computer Society Press, Los Alamitos (1998)
7. Röfer, T., Laue, T., Thomas, D.: Particle-filter-based self-localization using landmarks and directed lines. In: Bredenfeld, A., Jacoff, A., Noda, I., Takahashi, Y. (eds.) RoboCup 2005. LNCS (LNAI), vol. 4020, Springer, Heidelberg (2006)
8. de Jong, F., Caarls, J., Bartelds, R., Jonker, P.: A two-tiered approach to self-localization. In: Birk, A., Coradeschi, S., Tadokoro, S. (eds.) RoboCup 2001. LNCS (LNAI), vol. 2377, pp. 405–410. Springer, Heidelberg (2002)
9. Roberts, L.: Machine perception of three-dimensional solids. In: Tippett, J., et al. (eds.) Optical and Electro-optical Information Processing, pp. 157–197. MIT Press, Cambridge (1965)
10. Iocchi, L., Nardi, D.: Hough localization for mobile robots in polygonal environments. Robotics & Autonomous Systems 40, 43–58 (2002)
11. Enderle, S., Ritter, M., Fox, D., Sablatnög, S., Kraetzschmar, G.K., Palm, G.: Vision-based localization in robocup environments. In: Stone, P., Balch, T., Kraetzschmar, G.K. (eds.) RoboCup 2000. LNCS (LNAI), vol. 2019, pp. 291–296. Springer, Heidelberg (2001)

12. Hoffmann, J., Spranger, M., Göhring, D., Jüngel, M.: Exploiting the unexpected: Negative evidence modeling and proprioceptive motion modeling for improved markov localization. In: Bredenfeld, A., Jacoff, A., Noda, I., Takahashi, Y. (eds.) RoboCup 2005. LNCS (LNAI), vol. 4020, Springer, Heidelberg, 2006 (to appear)
13. Thomas, P.J., Stonier, R.J., Wolfs, P.J.: Robustness of colour detection for robot soccer. In: Proc. of the 7th International Conference on Control, Automation, Robotics and Vision (ICARCV) (2002)
14. Hough, P.V.C.: Machine analysis of bubble chamber pictures. In: Proc. of the Int. Conf. on High Energy Accelerators and Instrumentation (CERN) (1959)

Proprioceptive Motion Modeling for Monte Carlo Localization

Jan Hoffmann

Institut für Informatik
Humboldt-Universität zu Berlin, Germany
jan@aiboteamhumboldt.com

Abstract. This paper explores how robot localization can be improved and made more reactive by using an adaptive motion model based on proprioception. The motion model of mobile robots is commonly assumed to be constant or a function of the robot speed. We extend this model by explicitly modeling possible states of locomotion caused by interactions of the robot with its environment, such as collisions. The motion model thus behaves according to which state the robot is in. State transitions are based on proprioception, which in our case describes how well the robot's limbs are able to follow their respective motor commands. The extended, adaptive motion model yields a better, more reactive model of the current robot belief, which is shown in experiments. The improvement is due to the fact that the motion noise no longer has to subsume any possible outcome of actions including failure. In contrast, a clear distinction between failure and normal, desired operation is possible, which is reflected in the motion model.

1 Introduction

Due to the dynamic nature of the RoboCup environment, collisions with other robots occur frequently. Collisions are particularly bad for the Sony Aibo (or for that matter: any legged robot), as two robots easily get one another's legs entangled, making directed locomotion impossible. On the one hand, this leads to impaired mobility, which the robot needs to recover from as quickly as possible. On the other hand, collisions leave the robot badly localized since its actions did not have the expected result.

Detecting collisions is difficult in the Sony Aibo as it does not have any dedicated touch sensors on its shell. Even with the Aibo's touch sensors in its paws it is difficult to detect if the robot's leg touches the ground. To cope with this lack of dedicated sensory input and to provide a degree of proprioception, the movement of the robot's legs are examined [10,5]. The robot's legs (and to some extent its head) are the parts of the robot that come into contact with other robots or the environment when collisions occur. It was found that not only the overall movement of the robot but the movement of individual limbs is impaired during a collision. Monitoring the deviation of intended motions (control) to actual motion of limbs (servo readings), collisions can be detected [5].

G. Lakemeyer et al. (Eds.): RoboCup 2006, LNAI 4434, pp. 258–269, 2007.

To understand the effect of collisions on localization, one has to consider the motion update of the Bayes filter employed [12]:

$$\text{bel}^-(x_t) \longleftarrow \int p(x_t|x_{t-1}, u_{t-1})\text{bel}(x_{t-1})dx_{t-1} \tag{1}$$

$$\text{bel}(x_t) \longleftarrow \eta p(z_t|x_t)\text{bel}^-(x_t) \tag{2}$$

In the motion update (Eqn. 2), the (prior) localization belief bel^- is updated using knowledge about the action currently performed by the robot, i.e., $p(z_t|x_t)$. For a moving robot, this intended action is given by the motor commands and the expected outcome. This *probabilistic motion model* is commonly obtained by having the robot move about in its environment and measuring the deviation between desired and actual motion. Collisions are usually not included explicitly in the motion model as they are difficult to detect and the effect on the outcome of motions is hard to model. Particle filters offer a certain robustness with respect to errors caused by such un-modeled phenomena by means of random search. In these cases, the motion error is generally assumed to be larger than the actual error observed during calibration to account for unforeseen events. This generally requires a large particle count and reduces the accuracy of the localization: neither does the particle distribution reflect the uncertainty brought about by collisions, nor does it properly reflect the relative certainty when the robot is moving freely.

In contrast, we will show how collision detection can be used to create an adaptive motion model by explicitly modeling different states of robot mobility. This establishes a dynamic level of trust in odometry. The amount and shape of noise associated with locomotion is modeled according to the state of mobility of the robot. The resulting belief representation better resembles the current state of the robot which allows it to recover more quickly from collisions as particles are more spread out. It also allows the robot to monitor the belief entropy and thus determine if it can go on with its task or if it is better to stop and re-localize before proceeding.

The importance of better modeling arises in part from the low particle count used due to limited computational resources. With high particle counts, areas of low probability are still represented by a small number of particles. If something unexpected happens, such particle filters can recover quickly as unlikely areas are not completely deserted. As the particle count is lowered, however, particle filters become more susceptible to errors caused by un-modeled events as unlikely regions of the state space may not be represented at all. Spreading out the particle distribution when disturbances occur helps cope with such events and yield quicker convergence when sensor data is acquired.

Related Work. Legged robots are generally believed to be able to deal with a wider range of environments and surfaces than wheeled robots. The many designs of legged robots vary mainly in the number of legs, ranging from insectoid or arachnoid with 6, 8 or more legs[2],4-legged, such as the Sony Aibo used in our research [4,7], to humanoid with 2 legs (biped) [1,9].

Apart from the current action failing, collisions (and subsequently being stuck) have severe impact on the robot's localization as odometry is used in the localization process [3,11]. As stated before, the Sony Aibo lacks dedicated sensors that would allow the robot to detect touching objects and collisions. This leaves two sensors to achieve proprioception and thus collision detection, the accelerometer and the directional sensors of the robot's servos:

We found the accelerometer data of the Aibo to be very noisy and used it only in situations where low pass filtering could be applied. One such application is to have the robot's head look parallel to the ground. The most important application, though, is to figure out if the robot has fallen over and then trigger a recovery action. In very static, well controlled environment, the accelerometer signal can be also be classified to differentiate between surfaces that the robot is walking on [13].

The Aibo uses servo motors in its leg joints. A servo is an electrical motor with an integrated position feedback device and controller. The position/direction measurements of these servos can be used to achieve proprioception. Quinlan et al. [10] show how this can be used to effectively monitor the traction of the robot, namely by comparing the current servo direction measurements of the leg joints to reference values gathered in a prior calibration run. It relies solely on the direction measurements and does not take into account the control commands. In contrast, our work uses the approach presented by [5], which is based on the assumption that there is, in fact, a similarity between intended and actual motion and the variance of the sensor signal is bounded for the entire period of the motion. Under these assumptions, training can be greatly simplified as the number of parameters required to describe unhindered motion can be reduced by orders of magnitude. The reference value is proportional to the control signal, which is calculated by the walking engine as the inverse kinematic calculation is performed [4].

While improving the sensing model has been the focus of many papers (see bibliographical remarks in [12]), modeling the motion has received little attention and quite crude models prevail in the context of robot localization [11]. In [8], the motion of the ball is modeled in detail, taking into account interactions with its environment using a Bayes net. Our approach takes this idea and applies is to model robot motion, using proprioception as evidence for what state the robot might find itself in.

Outline. We will first summarize the concept of probabilistic motion modeling and describe the motion model used for the Sony Aibo and how this model can be enhanced. We then illustrate the benefits of adaptive motion modeling in several experiments.

2 Probabilistic Motion Model

The motion model is a probabilistic model of the outcome of control action u_t performed by the robot. It is given by the conditional probability density

$$p(s_t|u_t, s_{t-1}). \tag{3}$$

It describes the motion update of the Bayes filter in Equations 1 and 2. There are two types of approaches to modeling the motion of the robot: one is called velocity motion modeling whereas the other is based on odometry [12]. In wheeled robots, an odometer is used to count the number of turns of the wheels as the robot moves. Given the diameter of the wheels, the distance traveled can be calculated. In contrast, the velocity motion model is based on the predicted outcome of control actions. Both velocity- and odometry-based motion modeling suffer from drift and slippage. However, the odometer constantly *measures* the turns of the wheels whereas discrepancies of actual motion and model are not compensated in the velocity motion model. As the Aibo has no wheels and deducing the distance traveled from the leg's movement is very difficult, we use a velocity-based motion model.

Motion Model of the Sony Aibo. Although the Aibo uses legs for locomotion and does not have an odometer, the term odometry is commonly used to describe how far the robot has traveled given a sequence of control command. A control command is also called a *motion request* and consists of desired robot speeds $\boldsymbol{m} = (\dot{x}, \dot{y}, \dot{\theta} = \omega)^{\mathrm{T}}$ with angular velocity ω. Given the motion request, the distance traveled of the robot can be calculated [11,12]. The true distance traveled, however, is never exactly the same in two subsequent runs. Odometry is subject to cumulative errors (drift) and by itself does not account for slipping, sliding, or skidding. We model the odometry error as a normally distributed random variable of finite variance, resulting in the effective speed $\boldsymbol{m}_{\text{eff}}$ and effective locomotion (distance traveled) $\Delta\boldsymbol{r}$:

$$\boldsymbol{m}_{\text{eff}} = \boldsymbol{m} + \boldsymbol{\epsilon}\,(\dot{x}, \dot{y}, \omega) = \begin{pmatrix} \dot{x} \\ \dot{y} \\ \omega \end{pmatrix} + \begin{pmatrix} \epsilon_{\dot{x}}(\dot{x}, \dot{y}, \omega) \\ \epsilon_{\dot{y}}(\dot{x}, \dot{y}, \omega) \\ \epsilon_{\omega}(\dot{x}, \dot{y}, \omega) \end{pmatrix} \tag{4}$$

$$\Delta\boldsymbol{r} = \Delta t \cdot \boldsymbol{m} \tag{5}$$

$$\Delta\boldsymbol{r}_{\text{eff}} = \Delta t \cdot \boldsymbol{m}_{\text{eff}} \tag{6}$$

The true velocity of the robot equals the command velocity plus an additive error ϵ of zero mean. Note that ϵ is a function of all three control inputs, i.e., drift may cause the robot to turn although $\omega = 0$. In a simple model, the noise is assumed gaussian and the standard deviation σ_i of the probability density function (PDF) of ϵ_i is assumed linearly proportional to the weighted sum of the control inputs, requiring a total of 9 parameters to describe the odometry error:

$$\sigma_i(\dot{x}, \dot{y}, \omega) = \alpha_{i,1}\ \dot{x} + \alpha_{i,2}\ \dot{y} + \alpha_{i,3}\ \omega \tag{7}$$

When a particle filter is employed, this means for each particle, an error is sampled from the error PDF associated with ϵ. This error is added to the motion request $\boldsymbol{m} + \boldsymbol{\epsilon} = \boldsymbol{m}'$, which is in turn used to calculate the new robot pose at time t.

In the actual Monte Carlo Localization (MCL) implementation used by the GermanTeam, each particle's pose is updated using the motion request $\boldsymbol{m}$. The odometry error $\Delta t \cdot \boldsymbol{\epsilon}$ is only added after this update. Furthermore, the standard deviation in Equation 7 is approximated in the following fashion:

$$\sigma_i(\dot{x}, \dot{y}, \omega) \approx \sigma_i(|v|, \omega) \tag{8}$$

with

$$|v| = \sqrt{\dot{x}^2 + \dot{y}^2}, \tag{9}$$

assuming that the effect of the translational error is the same in dimensions x and y. The robot pose of a sample is thus updated by:

$$\boldsymbol{r}_t \leftarrow \boldsymbol{r}_{t-1} + \Delta\boldsymbol{r} + \Delta t \begin{pmatrix} \beta_1 |v| \text{ rand}(-1,1) \\ \beta_2 |v| \text{ rand}(-1,1) \\ (\beta_3 |v| + \beta_4 \omega) \text{ rand}(-1,1) \end{pmatrix} \tag{10}$$

Where $\beta_1, ..., \beta_4$ are parameters describing the error model and rand$(-1,1)$ is a function that returns a random number in the range $[-1,1]$ (uniform distribution is used for computational speed). The values of parameters $\beta'_i = \Delta t \cdot \beta_i = 8\text{ms} \cdot \beta_i$ used in the implementation are: $\beta'_1 = 0.1\text{ms}$, $\beta'_2 = 0.02\text{ms}$, $\beta'_3 = 0.002(\text{rad/m})\text{ms}$, $\beta'_4 = 0.2\text{ms}$.

3 Adaptive Motion Model

Some preliminary work on integrating information about collisions into the motion model was presented in [6]. We extend this naive approach by more accurately modeling collisions and slip. The underlying idea is to separate the components of the error brought about by the inherent odometry error $\boldsymbol{\epsilon}_{\text{odo}}$ and the error caused by slippage and collisions $\boldsymbol{\epsilon}_{\text{col}}$. This allows to lower the noise added per step when the robot is thought to move freely (accurate odometry) and to only add large amounts of noise (uncertainty) if collisions are detected. The effective speed $\boldsymbol{m}_{\text{eff}}$ of the robot is thus modeled as:

$$\boldsymbol{m}_{\text{eff}} = \boldsymbol{m} + \boldsymbol{\epsilon}_{\text{odo}} + \boldsymbol{\epsilon}_{\text{col}} \tag{11}$$

The robot pose is thus given as (cf. Equation 10):

$$\boldsymbol{r}_t = \boldsymbol{r}_{t-1} + \Delta\boldsymbol{r} + \boldsymbol{\epsilon}_{\text{odo}}\Delta t + \boldsymbol{\epsilon}_{\text{col}}\Delta t \tag{12}$$

$$= \boldsymbol{r}_{t-1} + \Delta\boldsymbol{r} + \boldsymbol{\epsilon'}_{\text{odo}} + \boldsymbol{\epsilon'}_{\text{col}} \tag{13}$$

This separation improves localization accuracy when the robot moves about freely while at the same time enabling it to more quickly recognize collision events. The development was helped by the advent of means to better calibrate the walking engine [4], making odometry much more precise as long as movement is unhindered.

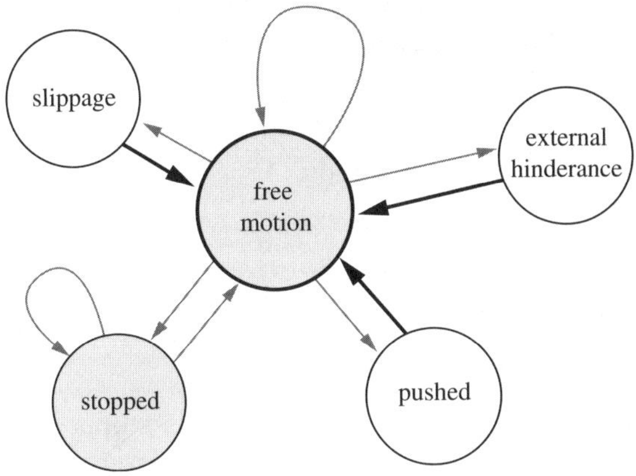

Fig. 1. Bayes net describing the mobility of the robot motion. State transitions happen when collisions occur; transition probabilities not shown.

3.1 Modeling States of Mobility

We will show how the state of mobility of the robot can be modeled. This is done much like the state of the ball is modeled in [8]. The Bayes net in Fig. 1 models the things that can happen to the robot with regards to its ability to move. State transitions away from the "free" state are triggered by collision events. The possible states and their effect on the robot's position and orientation are:

Free motion. The robot moves freely, no internal or external disturbances occur, locomotion error is determined by the odometry error only.

Slippage. This state subsumes motion disturbances like slipping and skidding that occur without external disturbance and that are often caused by abrupt changes of the motion request.

The effective translational speed of the robot is reduced and the robot orientation is subject to error:

$$v' = v \text{ rand}(\beta_1, 1.0) \tag{14}$$
$$\omega' = \omega + \beta_2 \text{ rand}(-1.0, +1.0) \tag{15}$$

with $0 \leq \beta_1 < 1$ and angle β_2.

External hindrance. The robot's motion cannot be executed as intended, e.g., because it is running into a wall. The effective motion is smaller than the motion request.

Effective translational speed and angular velocity are reduced:

$$v' = v \text{ rand}(\beta_3, 1.0) \tag{16}$$
$$\omega' = \omega \text{ rand}(\beta_4, 1.0) \tag{17}$$

with $0 \leq \beta_3 < 1$ and $0 \leq \beta_4 < 1$.

Pushed. The robot is being pushed by another robot; a force acts upon it resulting in the robot being turned and displaced.

$$\alpha = \text{rand}(-\pi, +\pi) \tag{18}$$
$$d = \beta_5 \text{ rand}(0, 1) \tag{19}$$
$$x' = x + d \cos(\alpha) \tag{20}$$
$$y' = y + d \sin(\alpha) \tag{21}$$
$$\theta' = \theta + \beta_6 \text{ rand}(-1.0, +1.0) \tag{22}$$

with displacement error β_5 and angle β_6.

Stopped. The robot runs into an obstacle and is stopped dead in its track. The translational speed of the robot becomes zero. However, it is often observed that robots turns when being stuck:

$$v' = 0 \tag{23}$$
$$\omega' = \beta_7 \text{ rand}(-1.0, +1.0) \tag{24}$$

with angle β_7.

Variables used: translational speed of the robot $|v| = \sqrt{\dot{x}^2 + \dot{y}^2}$, robot orientation θ, angular velocity $\omega = \dot{\theta}$, model parameters β_i, and rand(a, b) is a function that returns a random value in the interval $[a, b]$.

When the robot is in the "stopped" state, it remains in this state for some time until it has freed itself by a recovery action or the entanglement with another robot has somehow been resolved. Characteristically, when the robot is stopped, it will continuously bump into the obstacle, unable to free itself for a few moments. Only when it has freed itself will it no longer detect (frequent) collisions. It therefore remains in the stopped state as long as collisions are detected at high frequency.

When collisions occur, we observed that the robot tends to turn towards the cause of the collision, usually making matters worse. The motion of the robot's legs on the side where the collision occurs is hindered, slowing them down; this results in a difference in the forward component of the motion of the left and right legs, causing the robot to turn like a differential drive vehicle would. The maximum turning speed caused by the difference in speeds is given by $\omega = v_{\text{left}}/w$ (w = lateral distance between the robot's feet). The collision percept z_c contains information about the location of a collision. The angular velocity of the robot thus is changed in Equation 24 in the following fashion:

$$\omega' = \frac{v}{w}\, c(z_c)\, \beta_i \text{ rand}(0, +1.0) \tag{25}$$

with:

$$c(z_c) = \begin{cases} +1 & \text{if collision left} \\ -1 & \text{if collision right} \\ 0 & \text{if collision left and right} \end{cases} \tag{26}$$

This enables us to model a robot running into a wall, turning towards it until finally facing it. When facing it, collisions will be detected both left and right, marking the end of the turning motion.

4 Experimental Results

In the following experiments, a robot moves forward at constant speed and experiences one or more collisions. The belief represented by the particle distribution is generated without external perception, i.e., it is solely based on odometry, odometry error, and proprioception-based collision error. This is done to emphasize the effect of the proposed motion model. In an actual application where vision percepts are constantly integrated into the robot's belief, the adaptive motion model makes sure that the particle distribution is spread out enough to model the belief's uncertainty and to allow quicker re-localization after collision events.

The parameters used in our experiments were hand-tuned; automating the process of tuning poses a challenging machine learning task and remains future work.

4.1 No Collision

Fig. 2 a) shows the particle distributions of a robot moving forward without experiencing any collisions. Snapshots of the particle distribution are shown in two second intervals. The robot moves at a speed of 200mm/s and it thus takes the robot ten seconds to cross the distance of two meters as indicated. In this illustration and in the following ones, the particle distributions are made up of 100 particles. The initial particle distribution has a standard deviation of zero, i.e., all particles represent the same robot pose $\boldsymbol{r}_0 = (x_0, y_0, \theta_0)$.

The figure contrasts the particle distribution using the standard GermanTeam motion model and the distribution using the adaptive motion model. The latter adapts to the situation of moving freely by using a lower odometry error resulting in a more confined, low entropy distribution. Since the standard motion model is a function only of the robot speed, a trade off between accuracy and robustness is made: the particle distribution needs to spread out as collisions and hindrances may occur but at the same time the noise added needs to be limited for the distribution to remain stable and to not diverge too much. The standard model therefore has a higher entropy than the adaptive motion model.

4.2 Single Brief Collision

In the first collision experiment, illustrated in Fig. 2 b), a robot moves forward at $|v| = 200$mm/s, runs into another robot for about 2s and then continues to walk forward. The particle distribution of the robot integrating collision information is of typical sickle shape caused by angular disturbances; it is also more spread out after the collision and accounts for the various potential turning motions

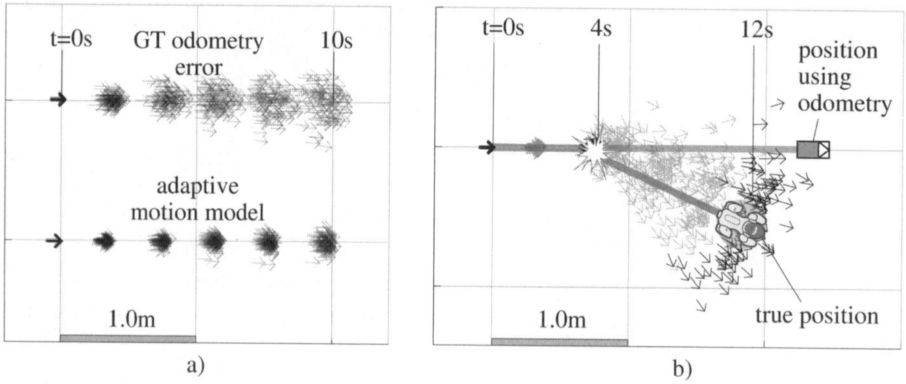

Fig. 2. a) Particle distribution representing a robot moving forward at constant speed experiencing no collisions. The *top* distribution uses the basic motion model used by the GermanTeam, the *bottom* distribution uses the adaptive motion model. Since the adaptive model "knows" that the robot is moving freely, the error of dead reckoning is smaller.
b) Robot running into an obstacle and turning in the process; it then continues to move freely. Note how the particle distribution has become much more spread out compared to the unhindered movement.

experienced by the robot. Information about the side on which the collision occurred is integrated and results in a non-symmetric PDF. Not taking into account collisions and only using odometry, the robot believes to have traveled in a straight line and also to have traveled further than it actually has.

As intended, the uncertainty associated with the belief increases after the collision. Fig. 3 (bottom) shows the relative "sum of squared deviations" [5] during the run, indicating when collisions are detected. Entropy is used to describe the uncertainty associated with the particle distribution [12]:

$$H_p(s) = -\sum p(s_i) \log_2 p(s_i) \tag{27}$$

The entropy is divided into orientation and position entropy. As an artifact of the way that entropy is calculated (grid based), the values of $H_p(s)$ start at zero and remain zero in early stages of each run. This is because at the very beginning, all particles fall within the center cell, resulting in $H = 1 \log_2 1 = 0$.

Before the collision, the entropy of the adaptive motion model is lower than that of the standard motion model. When the collision occurs, the amount of motion noise is increased, which can be noticed in both the position and the orientation entropy. After the collision, the position entropy of the adaptive model continues to rise caused by particles diverging due to the angular uncertainty in conjunction with the robot moving forward.

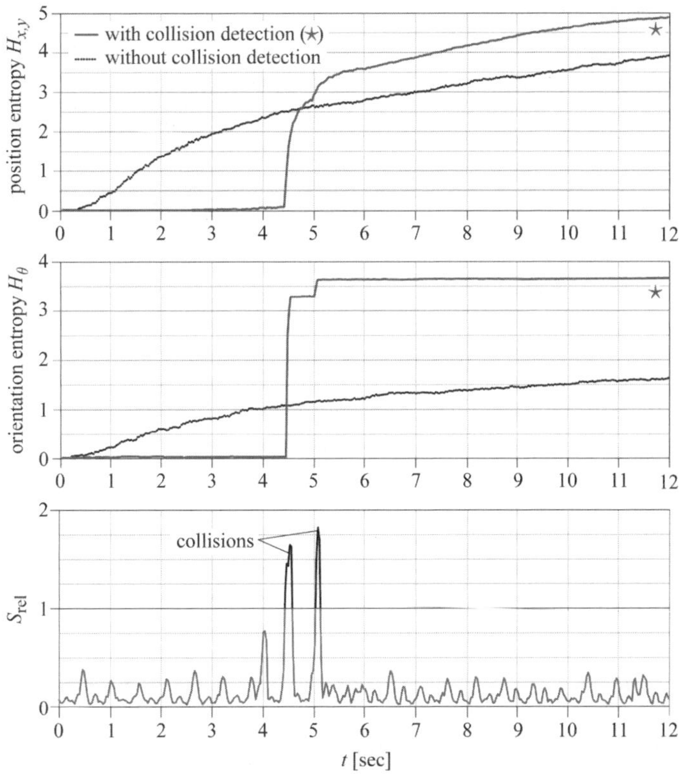

Fig. 3. Impact of collision on the entropy of the particle distribution. Values of S_{rel} greater than 1 are interpreted as collisions. The marked curves (*) represent the entropy of the particle distribution using the adaptive motion model. Without using collision information, the entropy continuously rises over time. Incorporating collision information, the entropy is lower *before* the collision and increases drastically as collisions are detected.

4.3 Two Subsequent Collisions

In this experiment, the robot moves forward but experiences two collisions in brief succession, one on its left and one on its right (Fig. 4). The two collisions somewhat compensate each other in terms of resulting robot orientation. However, the robot is slowed down by the collisions and the total distance traveled is reduced compared to the robot moving freely.

The particle distribution is spread out quite a bit, accounting for the two collisions and their probable outcomes. The impact on the particle distribution of the second collision is not as prominent as the distribution is already quite disturbed. Note that the distribution already is a little patchy in some areas, i.e., not all areas of the PDF are equally well described by the particle distribution due to the small number of particles employed.

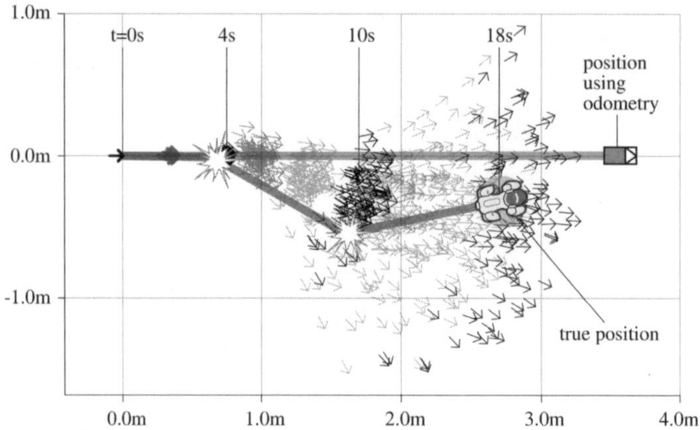

Fig. 4. Robot experiencing two collisions in brief succession. Note that the distance the robot has actually traveled is much shorter than the distance calculated using only odometry.

5 Conclusion

We were able to show how motion of a mobile robot can be modeled more closely to improve Monte Carlo Localization (MCL). The basic velocity-based motion model of a legged robot is enhanced by modeling various states describing the robot's ability to move and perform actions. As evidence for state transitions, proprioception is used, e.g., to detect if the robot has run into an obstacle. Whereas the simple velocity-based motion model tries to model all these states in a single PDF, the adaptive model offers a much more accurate description of robot motion and thus offers several advantages over the simple model: a) the particle count used in MCL can be lowered as the particle distribution better describes the belief PDF; b) when the robot is moving freely, the particle distribution remains relatively confined as it does not need to spread out to accommodate for potential action failures; c) when the robot does run into something, the error caused by this is more quickly reflected in the particle distribution.

The resulting particle distribution describing the robot's belief is more reactive and more accurately describes the situation that the robot is in. This has interesting applications to robot control and active vision when the current belief uncertainty is taken into account for control.

Acknowledgments

Work was funded by the German Research Foundation, Priority Project 1225. The software framework was developed by the GermanTeam; source code and documentation available for download at http://www.germanteam.org.

References

1. Behnke, S., Müller, J., Schreiber, M.: Toni: A soccer playing humanoid robot. In: Bredenfeld, A., Jacoff, A., Noda, I., Takahashi, Y. (eds.) RoboCup 2005. LNCS (LNAI), vol. 4020, Springer, Heidelberg (2006)
2. Clark, J.E., Cham, J.G., Bailey, S.A., Froehlich, E.M., Nahata, P.K., Full, R.J., Cutkosky, M.R.: Biomimetic design and fabrication of a hexapedal running robot. In: Proc. of the 2001 Intl. Conference Robotics and Automation (ICRA), IEEE Computer Society Press, Los Alamitos (2001)
3. Dellaert, F., Fox, D., Burgard, W., Thrun, S.: Monte carlo localization for mobile robots. In: Proc. of the 1999 IEEE Intl. Conference on Robotics and Automation (ICRA), vol. 2, IEEE Computer Society Press, Los Alamitos (1999)
4. Düffert, U., Hoffmann, J.: Reliable and precise gait modeling for a quadruped robot. In: Bredenfeld, A., Jacoff, A., Noda, I., Takahashi, Y. (eds.) RoboCup 2005. LNCS (LNAI), vol. 4020, Springer, Heidelberg (2006)
5. Hoffmann, J., Göhring, D.: Sensor-actuator-comparison as a basis for collision detection for a quadruped robot. In: Nardi, D., Riedmiller, M., Sammut, C., Santos-Victor, J. (eds.) RoboCup 2004. LNCS (LNAI), vol. 3276, Springer, Heidelberg (2005)
6. Hoffmann, J., Spranger, M., Göhring, D., Jüngel, M.: Exploiting the unexpected: Negative evidence modeling and proprioceptive motion modeling for improved markov localization. In: Bredenfeld, A., Jacoff, A., Noda, I., Takahashi, Y. (eds.) RoboCup 2005. LNCS (LNAI), vol. 4020, Springer, Heidelberg (2006)
7. Hornby, G., Takamura, S., Yokono, J., Hanagata, O., Yamamoto, T., Fujita, M.: Evolving robust gaits with Aibo. In: Proc. of the 2000 Intl. Conference on Robotics and Automation (ICRA), IEEE Computer Society Press, Los Alamitos (2000)
8. Kwok, C., Fox, D.: Map-based multiple model tracking of a moving object. In: Nardi, D., Riedmiller, M., Sammut, C., Santos-Victor, J. (eds.) RoboCup 2004. LNCS (LNAI), vol. 3276, Springer, Heidelberg (2005)
9. Dario, C.L.P., Guglielmelli, E.: Humanoids and personal robots: design and experiments. Journal of Robotic Systems 18(2) (2001)
10. Quinlan, M.J., Murch, C.L., Middleton, R.H., Chalup, S.K.: Traction monitoring for collision detection with legged robots. In: Polani, D., Browning, B., Bonarini, A., Yoshida, K. (eds.) RoboCup 2003. LNCS (LNAI), vol. 3020, Springer, Heidelberg (2004)
11. Röfer, T., Brunn, R., Dahm, I., Hebbel, M., Hoffmann, J., Jüngel, M., Laue, T., Lötzsch, M., Nistico, W., Spranger, M.: GermanTeam 2004: The German national RoboCup team. In: Nardi, D., Riedmiller, M., Sammut, C., Santos-Victor, J. (eds.) RoboCup 2004. LNCS (LNAI), vol. 3276, Springer, Heidelberg (2005)
12. Thrun, S., Burgard, W., Fox, D.: Probabilistic Robotics. MIT Press, Cambridge, MA, USA (2005)
13. Vail, D., Veloso, M.: Learning from accelerometer data on a legged robot. In: Proc. of the 5th IFAC Symp. on Intelligent Autonomous Vehicles (IAV-2004) (2004)

Autonomous Planned Color Learning on a Legged Robot

Mohan Sridharan[1] and Peter Stone[2]

[1] Electrical and Computer Engineering
smohan@ece.utexas.edu
[2] Department of Computer Sciences,
The University of Texas at Austin
pstone@cs.utexas.edu

Abstract. Our research focuses on automating the color-learning process on-board a legged robot with limited computational and memory resources. A key defining feature of our approach is that instead of using explicitly labeled training data it trains autonomously and incrementally, thereby making it robust to re-colorings in the environment. Prior results demonstrated the ability of the robot to learn a color map when given an executable motion sequence designed to present it with good color-learning opportunities based on the known structure of its environment. This paper extends these results by demonstrating that the robot can plan its own such motion sequence and perform just as well at color-learning. The knowledge acquired at each stage of the learning process is used as a bootstrap mechanism to aid the robot in planning its motion during subsequent stages.

Keywords: Robot Vision, Color Learning.

1 Introduction

The first step for most teams, upon arrival at RoboCup, in any of the real robot leagues, is *color calibration*: mapping raw camera pixels to color labels such as *white* or *pink*. Due to differences in lighting conditions and object colors between the teams' labs and the competition venue, pre-trained vision modules are unlikely to work "out of the box." Also, the time required for color calibration (more than an hour in the legged league) leads to multiple days of setup time before each competition, a costly proposition from the perspective of reserving the venue. But both soccer-playing and rescue robots must eventually be able to operate in natural, changing lighting conditions, as soon as possible after arriving on site. One way to dramatically reduce this time is to enable the robot to autonomously learn the desired colors from the environment.

The most common approach to color calibration is manual labeling of a small subset of the color space, which is used to label the values of nearby pixels and produce the color map. Instead, we specify the properties of objects in the robot's environment (locations, color labels, and sizes), but no information on the pixel values corresponding to the colors is given. The known locations and *structure* (color labels and sizes) of objects are used to seed the color-learning process and

G. Lakemeyer et al. (Eds.): RoboCup 2006, LNAI 4434, pp. 270–278, 2007.

plan the corresponding motion sequence. When illumination conditions change, assigning pixel-label biases could require human supervision each time, or fail altogether. Our method does not suffer from this problem since no information is needed regarding the pixel values that correspond to each color.

The problem of color segmentation takes as input the color-coded model of the world with a representation of the size, shape, position and color labels of objects of interest. A stream of input images are provided and the robot's initial position (and its joint angles over time) are known. The desired output is a *Color Map* that assigns a *color label* to each point in the color space. But the process is constrained to work within the limited memory and processing resources of the robot and it should be able to cope with the rapid motion of the limited-field-of-view camera, with the associated noise and image distortions.

Building on our previous work [9], where the robot learnt colors by moving through a pre-defined motion sequence that was generated manually, here we enable the robot to autonomously plan its motion sequence for any given configuration of objects, based on environmental knowledge and heuristic constraints on its motion sequence. Further, it simultaneously learns colors and localizes, and incrementally performs better at both these tasks.

2 Background Information

The SONY Aibo, *ERS-7*, is a four legged robot whose primary sensor is a CMOS camera with a field-of-view of 56.9^o (hor) and 45.2^o (ver), providing the robot with a limited view of its environment. The images have a resolution of 208×160 pixels and are captured in the *YCbCr* format at $30Hz$. The robot has 20 degrees-of-freedom (dof). It also has noisy touch sensors, IR sensors, and a wireless LAN card for inter-robot communication. The camera jerks around a lot due to the legged locomotion, and images possess common defects such as noise and distortion. Figure 1 shows the robot and the $4.4m \times 2.9m$ playing field.

Fig. 1. An Image of the Aibo and the field

On the robot, visual processing typically occurs in two stages: color segmentation and object recognition (see [6]). Color segmentation is a well-researched field in computer vision with several good algorithms [4,10]. But these involve computation that is infeasible to perform on autonomous robots with computational and memory constraints. In the RoboCup domain, the methods applied range from the baseline approach of creating mappings from the YCbCr values to the color labels [2], to the use of decision trees [11] and axis-parallel rectangles in the color space [3]. All of them involve an elaborate training process where the color map is generated by hand-labeling several (≈ 25) images over a period of at least an hour.

The color map is used to segment the image and construct connected constant-colored regions, which are used to detect useful objects (e.g. markers). The robot uses the markers to localize itself on the field and coordinates with its teammates to score goals on the opponent. All processing, for vision, localization, locomotion, and action-selection, is performed on board the robots, using a 576MHz processor. Though games are currently played under constant and reasonably uniform lighting conditions, a change in illumination over several days often forces teams to re-calibrate the vision system. Also, the overall goal of eventually playing against humans in natural lighting puts added emphasis on the ability to learn the color map in a very short period of time.

Attempts to automatically learn the color map on the Aibos have rarely been successful. In one approach, edges are detected and closed figures are constructed to find image regions corresponding to known environmental features [1]; color information from these regions is used to build the color classifiers. This is time consuming even with the use of offline processing and requires human supervision. In [7], a color map is learnt using three layers of color maps, with increasing precision levels. This is still not as accurate as the hand-labeled one and additional constraints are required to disambiguate the colors. Schulz and Fox [8] present another example where colors are estimated using a hierarchical Bayesian model with *Gaussian* priors.

Our approach does not need color priors. It enables the robot to *autonomously plan* its motion to *learn* the color map, using the knowledge of location and *structure* of the objects, in less than *five minutes*. It involves very little storage and the resultant color map is comparable in segmentation accuracy to the hand-labeled one that take more than an hour of human effort. Note that we provide a world model instead of a color map and/or the motion component. This removes the manual-intensive component and enables the robot to function in different environmental settings.

3 Problem Specification

As described in [9], to recognize objects and operate in a color-coded world, a robot typically needs to recognize a certain discrete number (N) of colors ($\omega \in [0, N-1]$). A complete mapping identifies a color label for each possible point in the color space:

$$\forall p, q, r \in [0, 255], \qquad \{C_{1,p}, C_{2,q}, C_{3,r}\} \mapsto \omega|_{\omega \in [0,N-1]} \tag{1}$$

where C_1, C_2, C_3 are the three color channels (e.g. YCbCr), with the corresponding values ranging from $0 - 255$.

We represent each color by a three-dimensional (3D) Gaussian model with mutually independent color channels, i.e. no correlation among the values along the three color channels. Though more expressive color representations, such as histograms, have been used extensively in the literature, and the independence assumption does not hold perfectly in practice, we determined, using empirical data and the statistical technique of bootstrapping [5], that a 3D Gaussian model with independent channels closely approximates reality. In addition to simplifying calculations the Gaussian has the advantage that the mean and variance are the only statistics that need to be stored for each color. This makes the learning process feasible to execute on mobile robots with constrained processing power.

Under the three-dimensional Gaussian model with independent channels, the *apriori* probability density functions (color $\omega \in [0, N-1]$) are given by:

$$p(c_1, c_2, c_3|\omega) \sim \frac{1}{\sqrt{2\pi}\prod_{i=1}^{3}\sigma_{C_i}} \cdot \exp -\frac{1}{2}\sum_{i=1}^{3}\left(\frac{c_i - \mu_{C_i}}{\sigma_{C_i}}\right)^2 \tag{2}$$

where, $c_i \in [C_{i_{min}} = 0, C_{i_{max}} = 255]$ represents the value at a pixel along a color channel C_i while μ_{C_i} and σ_{C_i} represent the corresponding means and variances.

Assuming equal priors, the *aposteriori* probabilities for each color are:

$$p(\omega|c_1, c_2, c_3) \propto p(c_1, c_2, c_3|\omega) \tag{3}$$

For each pixel, the color label corresponds to the color that has the maximum *aposteriori* probability.

4 Autonomous Color Learning

Our learning algorithm is summarized in Algorithm 1 and specific details are described below. The basic color learning component (lines $9-14$) was described in [9] while the rest of the algorithm deals with the motion sequence planning.

The robot starts off at a known position in its world model and the locations of various color coded objects are known. The robot has no initial color information (means and variances of all colors are zero) but it has the list of colors to be learnt ($Colors[]$). It also has an array of structures ($Regions[][]$) — a list for each color. Each structure corresponds to an object of a particular color and stores a set of properties for that region, such as its size (length and width) and its three-dimensional location (x,y,z) in the world model. Both the starting pose of the robot and the object locations can both be varied between trials, which causes the robot to also modify the list of candidate regions for each color.

Given the robot's limited field of view, it is essential to adjust its pose to focus on objects with the colors of interest. This can be extremely challenging in the initial stages due to the inherent inaccuracy of the motion model (due to slippage) and the initial lack of visual information. Geometric constraints on the position of the objects are essential to resolve conflicts. These heuristic constraints depend on the robot and the problem domain. In our case, they are:

- No two objects should occupy the same position in the world model — there should be a minimum distance (600mm) between two objects.
- No two objects of the same dimensions can be within 90° of each other (with respect to the corresponding robot position) if they each consist of only one unknown color.

Algorithm 1. Planned Autonomous Color Learning

```
Require: Known initial pose (but can be varied across trials).
Require: Color-coded model of the robot's world - objects at known positions that
   can change between trials.
Require: Empty Color Map; List of colors to be learnt - Colors[].
Require: Arrays of colored regions, rectangular shapes in 3D space; Regions[][]. A
   list for each color, consisting of the properties (size, shape) of the regions of that
   color.
Require: Ability to navigate (approximately) to a target pose (x, y, θ).
 1: i = 0, N = MaxColors
 2: Time_st = CurrTime, Time[] — the maximum time allowed to learn each color.
 3: while i < N do
 4:   Color = BestColorToLearn( i );
 5:   TargetPose = BestTargetPose( Color );
 6:   Motion = RequiredMotion( TargetPose )
 7:   Perform Motion {Monitored using visual input and localization}
 8:   if TargetRegionFound( Color ) then
 9:     LearnGaussParams( Color )
10:     Learn Mean and Variance of color from candidate image pixels
11:     UpdateColorMap()
12:     if !Valid( Color ) then
13:       RemoveFromMap( Color )
14:     end if
15:   else
16:     Rotate at target position.
17:   end if
18:   if CurrTime − Time_st ≥ Time[Color] or RotationAngle ≥ Ang_th then
19:     i = i + 1
20:     Time_st = CurrTime
21:   end if
22: end while
23: Write out the color statistics and the Color Map.
```

The order in which the colors are to be learnt is computed dynamically and greedily (*BestColorToLearn()* — line 4); it chooses the best color one at a time without actually planning ahead for where it will be after learning that color. This is based on:

1. The amount of motion (distance) that is required to place the robot in a location suitable to learn the color.
2. The existence of a region that can be used to learn that color without requiring the knowledge of any other (as of yet) unknown color.

The goal is to learn colors with minimal motion, so as to increase the chances of being well-localized. Once a color order is chosen, for the first color in the list, (*Color*), the robot determines the *best candidate region* to learn that color from.

Once the candidate is determined, the robot calculates the pose that would be best suited to recognize this candidate region – *BestTargetPose()* (line 5).

The robot then determines (*RequiredMotion()* — line 6) and executes the motion sequence to place it in the target position. The motion to the target position is monitored (*visual feedback*) using the current knowledge of colors to recognize objects and localize to the correct location. Once it gets close to the target location, the robot searches for candidate regions that satisfy the heuristic constraints of size and shape for the region that it is looking for. The actual world-model definitions in the structure *Regions[Color][best-candidate-region]* are dynamically modified by the robot, based on its pose and standard geometric principles, to arrive at suitable constraints.

The robot stops when either the candidate region is found or the target position is reached. If the candidate region is not found (*TargetRegionFound()*, line 8, is *false*), it is attributed to slippage and the robot turns in place, searching for the candidate region. The world model and heuristic constraints resolve any conflicts that arise. Once such a region is found, the robot stops, with the region at the center of its visual field. Then the robot proceeds to learns the color (*LearnGaussParams()* - line 9). Each pixel in candidate region is accepted as a member of the color class being learnt if it is sufficiently distant from the means of the other known color classes. The *mean* and *variance* of the accepted pixels define the color's 3D Gaussian. The learnt Gaussians are used to generate the $128 \times 128 \times 128$ color map (*UpdateColorMap()* - line 11) around once every five seconds. The updated color map, in addition to being used to segment subsequent images and validate the color parameters currently learnt (lines 12-14), helps the robot *localize* itself and move to suitable locations to learn the other colors. The learning algorithm *bootstraps*, with the knowledge available at any given instant being exploited to plan and execute the subsequent tasks efficiently.

If the robot has rotated in place for more than a threshold angle (Ang_{th}) and/or it has spent more than a threshold amount of time learning a particular color ($Time[Color]$), the robot transitions to the next color in the list. The process continues until the robot has attempted to learn all the colors.

Note that instead of providing a color map and/or the motion sequence each time the environment or the illumination conditions change, we just provide the positions of various objects in the robot's world and have it plan its motion sequence autonomously. This significantly reduces the amount of manual input required in our color learning approach [9]) while still learning colors much faster than the baseline approach of hand-labeling several images.

5 Experimental Results

Our previous work [9] demonstrated the ability of the robot to learn the colors when provided with an appropriate action sequence. Here, we show that the robot can succeed at this task while planning its owb action sequence.

To localize, the robot has to learn five colors - *white, green, yellow, blue, pink*, and we measure both its segmentation accuracy and localization accuracy (the robot uses this color map to move to a few positions on the field).

Table 1. Successful Planning and Localization Accuracy

Config	Success (%)	Localization Error		
		X (cm)	Y (cm)	θ (deg)
Worst	70	17	20	20
Best	100	3	5	0
avg	90.0 ± 10.7	8.6 ± 3.7	13.1 ± 5.3	9.0 ± 7.7

One challenge in experimental methodology was to measure the robot's planning capabilities in qualitatively *difficult* setups (configurations of the objects and robot initial position). We asked seven graduate students with experience working with the robots to pick a few test configurations which they thought would challenge the algorithm. For each configuration, we let the robot execute its color learning algorithm and measured the number of successful learning attempts: an attempt is deemed a success if all five colors are learnt.

In Table 1 we tabulate the performance of the robot in its planning task over these configurations. It also shows the localization accuracy of the robot using the learnt color map. The results in the table indicate the performance of the robot over 15 configurations, with 10 trials for each configuration. The robot is able to plan its color learning task and execute it successfully in most of the configurations (that were designed to be adversarial) and the localization accuracy is comparable to that obtained with the hand-labeled color map ($\approx 6cm, 8cm, 4deg$ in X, Y, and θ).

One configuration where the robot performs worst is shown in Figure 2. Here, the robot is forced to move a large distance to obtain its first color-learning opportunity (from position 1 to position 2). This motion sometimes leads the robot into positions that are quite far away from its target location (position 2) and it is then unable to find any candidate image region that satisfies the constraints for the yellow goal. Currently, failure in this initial stage strands the robot without any method for recovery: a suitable recovery mechanism using additional geometric constraints is an important area for future work. Note that the 30% failure rate in this case is entirely due to the unreliability of the robot's motion model: the color-learning plan generated by the robot is quite reasonable.

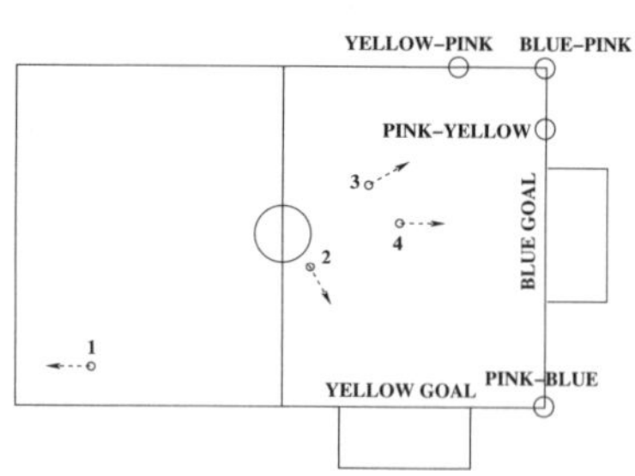

Fig. 2. Sample Configuration where robot performs worst

To test the segmentation accuracy of the learnt color map, we generated a color map by hand-labeling images [6]. We refer to this color map as *HLabel*. We compared the labeling provided by the two color maps (*HLabel* and *Learnt*) with the that provided by a human observer, the *Ground Truth* (*GTruth*). Only the colors of the objects on the field and/or below the horizon matter because other regions are automatically rejected in the object recognition phase. Also, the *correct* classification result is unknown for several background pixels in the image. So, the observer only labeled pixels that appear on or around the field and they were compared with the classification provided by the two color maps. On average, $\approx$ 6000 of the 33280 pixels in each image get labeled by the observer. The average classification accuracies for *HLabel* and *Learnt*, when compared with *GTruth*, are 99% and 96.7% respectively. We then tested the algorithm under different illumination conditions in addition to testing the algorithm's independence to color labels (labeling all *pink* objects as *blue* and vice versa does not pose any problems). This confirms our hypothesis that a *repainting* of the environment in any way, from just changing color shades, to scrambling colors entirely, does not disrupt our approach. Sample results for these experiments are available on-line: *www.cs.utexas.edu/users/AustinVilla/?p=research/auto_vis.*

6 Conclusions and Future Work

We have presented an approach that automatically plans a motion sequence to learn the desired colors on-board a legged robot with limited computational and storage resources. The corresponding segmentation and localization accuracies comparable to that obtained by the previous approach of having the robot learn the color map by executing a prespecified motion sequence [9]. The robot is able to *plan* its motion sequence dynamically in different world configurations based on heuristic constraints. The planned color learning can be repeated under different illumination conditions and object configurations, exploiting the inherent *structure* in the environment.

Our approach may apply to much more general environments, such as robots in homes or industrial settings. All that's needed is an environmental model, with the locations of distinctive features labeled. A major premise of this research is that generating such a model is significantly easier for a human than labeling pixels or generating a good motion path for color learning. This is reasonable, for example, whenever the configuration of objects in the world changes less frequently than the lighting conditions.

Currently, the color map is learnt from a known starting position without any prior knowledge of colors. We are working on learning colors from an *unknown* starting position on the field. Ultimately, we aim to develop efficient algorithms for a mobile robot to function autonomously under completely uncontrolled natural lighting conditions.

Acknowledgments

We would like to thank the members of the UT Austin Villa team.This work was supported in part by NSF CAREER award IIS-0237699, ONR YIP award N00014-04-1-0545, and DARPA grant HR0011-04-1-0035.

References

1. Cameron, D., Barnes, N.: Knowledge-based autonomous dynamic color calibration. In: Polani, D., Browning, B., Bonarini, A., Yoshida, K. (eds.) RoboCup 2003. LNCS (LNAI), vol. 3020, Springer, Heidelberg (2004)

2. Sony Legged League Team CM-Pack: In: Birk, A., Coradeschi, S., Tadokoro, S. (eds.) RoboCup 2001. LNCS (LNAI), vol. 2377, Springer, Heidelberg (2002)
3. Cohen, D., Ooi, Y.H., Vernaza, P., Lee, D.D.: RoboCup-2003: The Seventh RoboCup Competitions and Conferences. In: Polani, D., Browning, B., Bonarini, A., Yoshida, K. (eds.) RoboCup 2003. LNCS (LNAI), vol. 3020, Springer, Heidelberg (2004)
4. Comaniciu, D., Meer, P.: Mean shift: A robust approach toward feature space analysis. PAMI 24(5), 603–619 (2002)
5. Efron, B., Tibshirani, R.J.: An Introduction to Bootstrap. Chapman and Hall Publishers, Sydney (1993)
6. Stone, P., et al.: UT Austin Villa 2004: Coming of Age, AI TR 04-313. Technical report, Department of Computer Sciences, UT-Austin (October 2004)
7. Jungel, M.: Using layered color precision for a self-calibrating vision system. In: The Eighth International RoboCup Symposium, Lisbon, Portugal (2004)
8. Schulz, D., Fox, D.: Bayesian color estimation for adaptive vision-based robot localization. In: IROS (2004)
9. Sridharan, M., Stone, P.: Autonomous color learning on a mobile robot. In: The Twentieth National Conference on Artificial Intelligence (AAAI) (2005)
10. Sumengen, B., Manjunath, B.S., Kenney, C.: Image segmentation using multi-region stability and edge strength. In: ICIP (2003)
11. Sony Legged League Team UNSW: In: Birk, A., Coradeschi, S., Tadokoro, S. (eds.) RoboCup 2001. LNCS (LNAI), vol. 2377, Springer, Heidelberg (2002)

Sensor Modeling Using Visual Object Relation in Multi Robot Object Tracking

Daniel Göhring and Jan Hoffmann

Institut für Informatik
LFG Künstliche Intelligenz
Humboldt-Universität zu Berlin
Unter den Linden 6
10099 Berlin, Germany
http://www.aiboteamhumboldt.com

Abstract. In this paper we present a novel approach to estimating the position of objects tracked by a team of mobile robots. Modeling of moving objects is commonly done in a robo-centric coordinate frame because this information is sufficient for most low level robot control and it is independent of the quality of the current robot localization. For multiple robots to cooperate and share information, though, they need to agree on a global, allocentric frame of reference. When transforming the egocentric object model into a global one, it inherits the localization error of the robot in addition to the error associated with the egocentric model.

We propose using the relation of objects detected in camera images to other objects in the same camera image as a basis for estimating the position of the object in a global coordinate system. The spacial relation of objects with respect to stationary objects (e.g., landmarks) offers several advantages: a) Errors in feature detection are correlated and not assumed independent. Furthermore, the error of relative positions of objects within a single camera frame is comparably small. b) The information is independent of robot localization and odometry. c) As a consequence of the above, it provides a highly efficient method for communicating information about a tracked object and communication can be asynchronous.

We present experimental evidence that shows how two robots are able to infer the position of an object within a global frame of reference, even though they are not localized themselves.

1 Introduction

For a mobile robot to perform a task, it is important to model its environment, its own position within the environment, and the position of other robots and moving objects. In RoboCup, the most important object to track is, naturally, the ball. The task of estimating the position of an object is made more difficult by the fact that the environment is only partially observable to the robot.

G. Lakemeyer et al. (Eds.): RoboCup 2006, LNAI 4434, pp. 279–286, 2007.

In hybrid architectures [1], basic behaviors or skills, such as following a ball, are often based directly on sensor data, e.g., the ball percept. Maintaining an object model becomes important if sensing resources are limited and a short term memory is required to provide an estimate of the object's location in the absence of sensor readings.

Robots often use an egocentric model of objects relevant to the task at hand, thus making the robot more robust against global localization errors. A global model is used for communicating information to other robots [8], to commonly model a ball by many agents with Kalman filtering [2] or to model object-environment interactions [5]. In all cases, the global model inherits the localization error of the observer.

We address this problem by modeling objects in allocentric coordinates from the start. To achieve this, the sensing process needs to be examined more closely. In a typical camera image of a RoboCup environment, the image processing could, for example, extract the following percepts: *ball*, *opponent player*, and *goal*. Percepts are commonly considered to be independent of each other to simplify computation, even if they are used for the same purpose, such as localization [7].

When modeling objects in relative coordinates, using only the respective percept is often sufficient. However, information that could help localize the object within the environment is not utilized. That is, if the ball was detected in the image right next to a goal, this helpful information is not used to estimate its position in global coordinates.

We show how using the object relations derived from percepts that were extracted from the same image yields several advantages:

Sensing errors. As the object of interest and the reference object are detected in the same image, the sensing error caused by joint slackness, robot motion, etc. becomes irrelevant as only the relation of the objects within the camera image matters.

Global localization. The object can be localized directly within the environment, independent of the quality of current robot localization.

Communication. Using object relations offers an efficient way of communicating sensing information, which can then be used by other robots to update their belief by sensor fusion.

Outline. We will show how relations between objects in camera images can be used for estimating the object's position within a given map. We will present experimental results using a Monte-Carlo Particle Filter to track the ball. Furthermore, we will show how communication between agents can be used to combine incomplete knowledge from individual agents about object positions, allowing the robot to infer the object's position from this combined data.

Our experiments were conducted on the color coded field of the *Sony Four Legged League* using the Sony Aibo ERS-7.

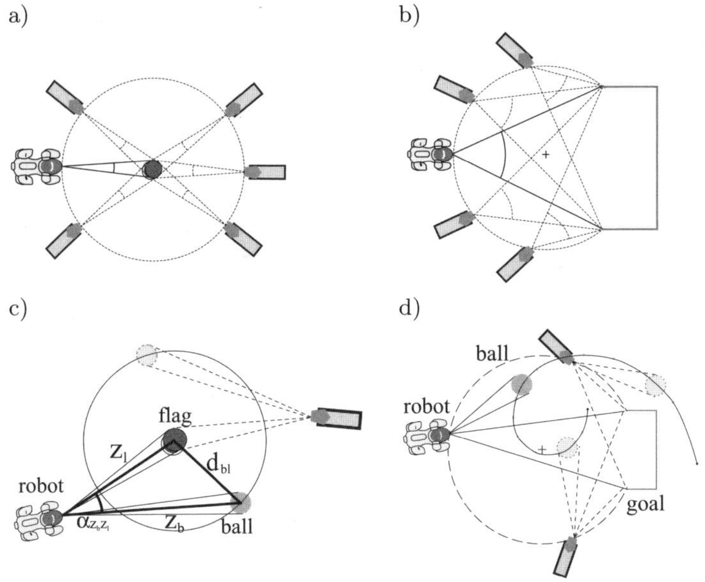

Fig. 1. Single percept: a, b) When a flag or a goal is seen, the robot can calculate its distance to it, but not its exact position, a circle remains for all possible robot positions. Two percepts in one image c, d) a flag/goal and a ball let the robot determine the ball's position relative to the flag/goal; all possible positions of the ball relative to the flag/goal form a circle/spiral arc.

2 Object Relation Information

In a RoboCup game, the robots permanently scan their environment for landmarks as there are flags, goals, and the ball. The following section presents the information gained by each perception.

2.1 Information Gained by a Single Percept

If the robot sees a two colored flag, it actually perceives the left and the right border of this flag and thus the angle between those two borders. Because the original size of landmarks is known, the robot is able to calculate its own distance to the flag and its respective bearing (Fig. 1 a). In the given approach we don't need that sensor data for self localization, but for calculating the distance from other objects as the ball to the flag.

If a goal is detected, the robot can measure the angle between the left and the right goal-post. For a given goal-post angle the robot can calculate its distance and angle to a hypothetical circle center (Fig. 1 b).

If a ball is perceived, the distance to the ball and its direction relative to the robot can be calculated. Lines or line crossings can also be used as reference marks, but the sensor model for lines is more complex than for a goal or a flag as there are many equally looking line segments on the field. For simplicity reasons we didn't use line information in the given approach.

2.2 Information Gained by Two Percepts Within the Same Image

If the localization object is visible together with another landmark, e.g., a flag or a goal, the robot does not only get information about distances to both objects but also information about the angle between them. With the *law of the cosine* the distance from the ball to a flag can be calculated (Fig. 1 c).

When a goal and a ball were seen, a similar determination of the position can be done for the ball, but the set of possible solutions leads to a spiral curve (Fig. 1 d). But one landmark and one ball alone are not sufficient to exactly determine the ball's position. One possibility to overcome this limitation would be to scan for other landmarks and take this information into account, but this could be time consuming. Another approach would be to let the robots communicate and interchange the necessary information for an accurate object localization. This has two advantages:

1. Apart from communication time which takes about two or three tenth of a second, information transfer between robots is cheap in resources, as only few data needs to be transferred.
2. Many robots can gather more information than a single robot, because many robots can see more than one robot.

Now we want to describe a possible implementation of this approach. As the sensor data of our Aibo ERS-7 robot are not very accurate, we have to cope with a lot of sensor noise. Furthermore, the probabilistic distribution is not always unimodal, e.g., in cases where the observations lead to more than one solution for possible ball positions. This is why a simple Kalman filter would not be sufficient [5]. We chose an implementation using a Monte-Carlo Particle Filter because of its ability to model multimodal distributions and its robustness to sensor noise. Other approaches as Multi Hypothesis Tracking or Grid Based algorithms might work also [4].

3 Monte-Carlo Filter for Multi Agent Object Localization

Markov localization methods, in particular Monte-Carlo Localization (MCL), have proven their power in numerous robot navigation tasks, e.g., in office environments [3], in the museum tour guide Minerva [9] and in the highly dynamic RoboCup environment [6]. MCL is widely used in RoboCup for object and self localization [7] because of its ability to model arbitrary distributions and its robustness towards noisy input data. It uses Bayes law and Markov assumption to estimate an object's position. The probability distribution is represented by a set of samples, called particle set. Each particle represents a pose hypothesis. The current belief of the object's position is modeled by the particle density, i.e., by knowing the particle distribution the robot can approximate its belief about the object state. The ball position is modeled relative to the field, which makes it independent from robot motions. The a-priori belief is updated by sensor data z_t, therefore called update step. Our update information is information about object relations as described in section 2. Therefore a sensor model is needed, telling the filter how accurate the sensor data are. The localization is being ititialized with $Bel(s_0)$ at $t = t_0$. The particles from the particle set are distributed arbitrarily across the field. Every ball position is equally uncertain. If sensor data is gained, the particle set will be updated and after a few steps converge to a certain area.

3.1 Monte-Carlo Localization, Implementation

Our hypotheses space has two dimensions for the position q on the field. Each particle s^i can be described as a state vector $\overrightarrow{s}^i$

$$\overrightarrow{s}^i = \begin{pmatrix} q_x^i \\ q_y^i \end{pmatrix} \tag{1}$$

and its likelihood p^i.

The likelihood of a particle p^i can be seen as the product of all likelihoods of all gathered evidences [7], which means in our case that for all landmark-ball pairs a likelihood is being calculated. From every given sensor data, e.g., a landmark l and a ball (with its distances and angles relative to the robot) we calculate the resulting possible ball positions relative to the landmark l, as described in section 2.2. The resulting arc will be denoted as ξ^l. We showed in 2.2 that ξ^l has a circular form, when l is a flag and a spiral form, when l is a goal. The shortest distance δ^l from each particle $\overrightarrow{s}^i$ to ξ^l is our argument for a Gaussian likelihood function $\mathcal{N}(\delta, \mu, \sigma)$, where $\mu = 0$ and with a standard deviation σ, which is determined as described in the next section. In fact, the sensor model is more complex than a Gaussian, but assuming it to be Gaussian showed to be a good approximation. The likelihood is being calculated for all seen landmarks l and then multiplied:

$$p^i = \prod_{l \in L'} \mathcal{N}(\delta^l, 0, \sigma) \tag{2}$$

In cases without new evidence all particles get the same likelihood. After likelihood calculation, particles are resampled.

Multi Agent Modeling. To incorporate the information from other robots, percept relations are communicated to other robots. The receiving robot uses the communicated percepts for likelihood calculation of each particle the same way as if it was its own sensor data. This is advantageous compared to other approaches:

- Some approaches communicate the particle distributions. But when, as in our examples, two robots only know the arcs or the circular function on which the ball could be found, this would increase position entropy rather than

Table 1. Object distance and angle standard deviations

Object		Standard Deviation σ	
	Distance in mm	σ_{Dst} in mm	σ_{Ang} in Rad
Ball	1500	170	0.015
Flag	2000	273	0.019
Goal	2000	25	0.021
Flag- Ball-Diff.	500	196	0.008
Goal- Ball-Diff.	500	175	0.0054

decreasing it. Communicating whole particle sets can also be very expensive in resources.
- By communicating percept relations rather than particles, every robot can incorporate the communicated sensor data to calculate the likelihood of its particle set.

Because of this, we decided to let every robot communicate every percept relation (e.g., flag, ball) it has gathered to other robots.

Sensor Model. For the sensor model, we measured the standard deviation σ^l by letting a robot take multiple images of certain scenes: a ball, a flag, a goal and combinations of it. The standard deviation of distance differences and respectively angle differences of objects in the image relative to each other were measured as well. The robot was walking the whole time on the spot to get more realistic, noisy images. The experiment results are shown in table 1.

It can be seen that the standard deviation for the distance from the ball to the flag (or goal) is smaller than the sum of the distance errors given a ball and a flag (or goal). The same can be said for the angle standard deviation. This gives evidence that the sensor error for percepts in the same image is correlated, due to walking motions and head swings.

4 Experimental Results

The Aibo ERS-7 robot serves as a test platform for our work. In our experiment, two robots try to localize and to model the ball in an egocentric model. As a result each robot maintains a particle distribution for possible ball positions, resulting from self localization belief and the locally modeled ball positions. In the next step the two robots communicate their particle distribution to each other (or a part of it). After communication each robot creates a new particle cloud as a combination of its own belief (the own particle distribution) and the communicated belief (communicated particle distribution). We want to check how this algorithm performs in contrast to our presented algorithm in situations, where self localization is not

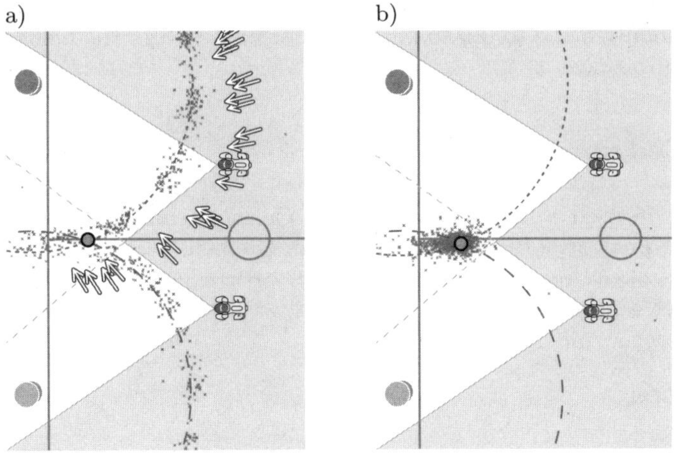

Fig. 2. Experiment with two flags: a) no percept relations communicated, the robots are self localizing (arrows show SL-particles of the upper robot schematically), the ball positions (cloud of dots) are modeled egocentric, transformed into global coordinates, then communicated to the other robot and merged with its ball particle distribution. b) No self localization needed, percept relations used as described, two robots communicating object relations for calculating the particle distribution; the small circle at the center line marks the real ball position in the given experiment.

possible, e.g., when every robot can only see one landmark and the ball. We placed both robots in front of a different landmarks with partially overlapping fields of view, such that both robots could see the ball (Fig. 2).

One can see from the experiments that there is almost no convergence to a confined area for the case in which the two robots are communicating their particle distributions to each other. In case of percept communication, the particle distribution converges nicely to a confined area.

5 Conclusion

Object relations in robot images can be used to localize objects in global coordinates. Without having to be localized at all, it can accurately estimate the position of an object within a map of its environment using nothing but object relations. Furthermore, we were able to show how the process of object localization can be sped up by communicating object relations to other robots. Two non-localized robots are thus able to both localize an object using their sensory input in conjunction with communicated object relations.

Future Work. Future work will investigate the use of other landmarks (e.g., field lines) for object localization. An active vision control is currently being developed to gain more images containing object relations, e.g., looking at the ball and landmarks at once if possible. Furthermore, we will investigate how data about commonly modeled objects in field coordinates, e.g., the ball can be used for self localization.

Acknowledgments

Program code used was developed by the GermanTeam, a joint effort of the Humboldt University of Berlin, University of Bremen, University of Dortmund, and the Technical University of Darmstadt. Source code is available for download at http://www.germanteam.org.

References

1. Arkin, R.: Behavior-Based Robotics. MIT Press, Cambridge, MA, USA (1998)
2. Dietl, M., Gutmann, J., Nebel, B.: Cooperative sensing in dynamic environments. In: IEEE/RSJ International Conference on Intelligent Robots and Systems (IROS'01), Maui, Hawaii (2001)
3. Fox, D., Burgard, W., Dellart, F., Thrun, S.: Monte carlo localization: Efficient position estimation for mobile robots. In: Proceedings of the Sixteenth National Conference on Artificial Intelligence and Eleventh Conference on Innovative Applications of Artificial Intelligence (AAAI), pp. 343–349. AAAI Press/MIT Press, Cambridge (1999)
4. Gutmann, J.-S., Fox, D.: An experimental comparison of localization methods continued. In: Proceedings of the 2002 IEEE/RSJ International Conference on Intelligent Robots and Systems (IROS), IEEE Computer Society Press, Los Alamitos (2002)
5. Kwok, C., Fox, D.: Map-based multiple model tracking of a moving object. In: Nardi, D., Riedmiller, M., Sammut, C., Santos-Victor, J. (eds.) RoboCup 2004. LNCS (LNAI), vol. 3276, pp. 18–33. Springer, Heidelberg (2005)
6. Lenser, S., Bruce, J., Veloso, M.: CMPack: A complete software system for autonomous legged soccer robots. In: AGENTS '01: Proceedings of the fifth international conference on Autonomous agents, pp. 204–211. ACM Press, New York (2001)
7. Röfer, T., Jüngel, M.: Vision-based fast and reactive monte-carlo localization. In: Polani, D., Bonarini, A., Browning, B., Yoshida, K. (eds.) Proceedings of the 2003 IEEE International Conference on Robotics and Automation (ICRA), pp. 856–861. IEEE Computer Society Press, Los Alamitos (2003)
8. Schmitt, T., Hanek, R., Beetz, M., Buck, S., Radig, B.: Cooperative probabilistic state estimation for vision-based autonomous mobile robots. IEEE Transactions on Robotics and Automation 18(5), 670–684 (2002)
9. Thrun, S., Fox, D., Burgard, W.: Monte carlo localization with mixture proposal distribution. In: Proceedings of the National Conference on Artificial Intelligence (AAAI), pp. 859–865 (2000)

Robust Color Segmentation Through Adaptive Color Distribution Transformation

Luca Iocchi

Dipartimento di Informatica e Sistemistica
Università di Roma "La Sapienza"
Via Salaria 113 00198 Rome Italy
iocchi@dis.uniroma1.it

Abstract. Color segmentation is typically the first step of vision processing for a robot operating in a color-coded environment, such as RoboCup soccer, and many object recognition modules rely on that.

Although many approaches to color segmentation have been proposed, in the official games of the RoboCup Four Legged League manual calibration is still preferred by most of the teams. In this paper we present a method for color segmentation that is based on an adaptive transformation of the color distribution of the image: the transformation is dynamically computed depending on the current image (i.e., it adapts to condition changes) and then it is used for color segmentation with static thresholds. The method requires the setting of only a few parameters and has been proved to be very robust to noise and light variations, allowing for setting parameters only once when arriving at a competition site.

The approach has been implemented on AIBO robots, extensively tested in our laboratory, and successfully experimented in the some of the games of the Four Legged League in RoboCup 2005.

1 Introduction

RoboCup soccer is a color-coded environment, where colors are used to define principal objects needed for the robots to perform their robot tasks. Recognition and positioning of colored beacons and goals in the field are used for self-localization and reactive behaviors, while the recognition of the orange ball feeds behaviors and coordination tasks.

Consequently, color segmentation is typically the first step of the vision system of a robot playing RoboCup soccer. Since good color segmentation allows for easy implementation of object recognition and localization, most of the robot vision systems are based on fast and accurate implementation of such process. Conversely, it is also possible to recognize and locate objects from a rough segmentation (e.g., [3]), applying more sophisticated recognition techniques (e.g., region growing) at a later stage. However, this second approach may be less reliable or require more computational resources.

Many approaches to color segmentation have been proposed in the RoboCup soccer scenario. Some of these approaches can be implemented in real-time only on robots with adequate computational resources (e.g., Middle-size robots [1]).

G. Lakemeyer et al. (Eds.): RoboCup 2006, LNAI 4434, pp. 287–295, 2007.

An effective implementation of on-line color segmentation on AIBO robots has been reported in [2]. The paper presents an adaptive non-parametric method that computes color classes by representing them as cuboids in the YUV color space with different precision layers. Experimental results of this approach show good computational performance (about 5 fps). However, manual calibration is still preferred during the games [4], since it provides for fast and accurate results, accepting the drawback of time consuming manual setting that is often repeated several time (e.g., just before each game) and for each robot (since color cameras have different response on different robots).

In this paper we present an approach to color segmentation that has been implemented on AIBO robots and used for RoboCup soccer in the Four Legged League. The method performs an adaptive transformation of the color distribution of the image, that is dynamically computed during robot task and used for segmentation. The approach integrates the following advantages: 1) it computes a dynamic transformation providing for robustness to noise and adaptivity to variable light conditions; 2) it uses static thresholds for fast segmentation; 3) it does not require time consuming manual calibration (only a few parameters must be set when arriving in a new location).

An effective implementation can be obtained by computing the transformation function periodically, for example every 25 frames (about 1 second), and when the robot is not in a critical phase of the game (e.g., about to approach the ball). In such steps results of color classification are stored in a lookup table (or color table), that is used for classification of the subsequent frames. In this way we reach the maximum performance of the image processing module (20-24 fps) for most of the frames (when such transformation is not computed), and periodically a lower performance (currently, slightly less than 100 ms) when this transformation is computed.

Experimental results show the effectiveness of the proposed method: a large data set of labeled images has been used to evaluate the approach in different locations and conditions. Furthermore, we have experimented the method during some of the games in RoboCup 2005. In that case we have set parameters once when we arrived at the competition site and then we used these parameters for some of the games and for the variable light challenge (with the very same code and settings), noticing no difference in the overall behavior of the team with respect to matches in which we have used manual calibration.

2 Color Distribution Transformation

The approach to color segmentation proposed in this paper is based on an adaptive transformation of the color distribution of an image and subsequent static thresholding. The aim of this approach is to integrate the robustness to noise and light changes typical of dynamic methods with efficiency of static ones.

Let us denote an image as $\mathcal{I}$, and each pixel of an image as $i \in \mathcal{I}$. A color space $\mathcal{C}$ is a color representation of an image as captured by a color camera (e.g., RGB, YUV, HSV). We will use the notation $c(i)$ to represent the color of pixel i in a given color space.

Color classes can be seen as a partition of the color space $\mathcal{C}$, $\mathcal{CC} = \{\mathcal{C}_1, ..., \mathcal{C}_n\}$, such that $\mathcal{C}_k \subset \mathcal{C}$ and $k_1 \neq k_2 \Rightarrow \mathcal{C}_{k_1} \cap \mathcal{C}_{k_2} = \emptyset$. It is convenient to consider a complete coverage for the color classes, i.e. $\mathcal{C} = \cup_{k=1..n} \mathcal{C}_k$, possibly defining a special color class including undefined (or uninteresting) colors. We also define the color distribution of an image $\mathcal{I}$ in the color space $\mathcal{C}$ by $D_{\mathcal{I}}(\gamma)$ as the number of pixels with color $\gamma \in \mathcal{C}$.

For a formal definition of our method we define a transformation function $\tau : \mathcal{C} \rightarrow \mathcal{C}$, and we call *$\tau$-distribution* the new color distribution obtained by applying the function τ to the color distribution of an image, and *τ-transformation* the operation of computing the *τ-distribution* of an image. Having τ, color classification is obtained by assigning to each pixel in the image one class within the predefined set of color classes

$$f_{\tau,\mathcal{CC}}(i) = k \;\; \texttt{such that} \;\; \tau(c(i)) \in \mathcal{C}_k \tag{1}$$

The objective of the transformation $\tau(\cdot)$ is to modify the color distribution of an image in such a way that the subsequent use of fixed thresholds can be effective for color segmentation, i.e., both robust to noise and light variations and efficient. In particular, we would like that the new color distribution do not vary too much in presence of light variations, specially in proximity of the thresholds. To obtain an adaptive method, τ is periodically computed from the current images the robot acquires.

In this paper we refer to the color component H of the color space HSV, because it is a mono-dimensional space that allows for distinguishing most of the colors of interest in RoboCup. The transformation $\tau(\cdot)$ is computed by the following algorithm, where the color space H if discretized to 360 integer values (i.e., $H = [0, 360)$).

Algorithm 1. Color distribution transformation

Input: Image $\mathcal{I}$, window size δ
Output: Function $\tau(h)$, $h \in H$ (explicit representation though a vector $\tau[0:359]$)

for each $\gamma \in [0:359]$ set $\tau(\gamma) = \gamma$
compute color distribution $\mathcal{D}_{\mathcal{I}}(\gamma)$
for $\gamma = \delta/2$ to $360 - \delta/2$ **do**
  $\tau[\gamma] = \mathbf{argmax}_{\lambda \in [\gamma-\delta/2, \gamma+\delta/2)} \{\mathcal{D}_{\mathcal{I}}(\lambda)\}$
end for
while $\tau[\gamma] \neq \gamma \wedge \tau[\gamma] \neq \tau[\tau[\gamma]]$ **do**
  $\tau[\gamma] = \tau[\tau[\gamma]]$
end while

The function τ is first initialized to the identity function, i.e. $\tau(\gamma) = \gamma$, then two steps are performed: the first step assigns to $\tau[\gamma]$ the value $\gamma^* \in [\gamma - \delta/2, \gamma + \delta/2]$ for which $\mathcal{D}_{\mathcal{I}}(\gamma^*)$ is maximum in such interval; the second step performs a transitive closure of the τ function. The value δ is used to limit the interval in which the maximum value is searched. This value is set in such a way to process a significant interval of values in the color space: smaller values of δ may fail to accumulate colors over a single component (thus producing more peaks), larger values may collapse different colors in a single peak. By empirical tests, we have determined that a value around 10 provides good results in our implementation.

An example is shown in Figure 1. Figure 1b) shows the original distribution (lighter color), the τ-distribution (darker color) in the H color space and three values for the orange-yellow threshold (vertical lines), respectively 51, 57, and 64. The transformation τ is instead given in Figure 1c). The images in Figure 1d) are the results of segmentation with such different values for the orange-yellow threshold, by using τ-distribution (left side) and original distribution (right side). It is possible to see that the segmentation obtained with τ-distribution is more robust since it essentially returns the same correct segmentation for all the values of the threshold, while the segmentation obtained with the original distribution produces some bad classification (orange is classified as yellow when the threshold is lower, yellow is classified as orange when the threshold is higher). This example shows that segmentation on the τ-distribution is more robust. In fact, any value for the orange-yellow threshold between 51 and 64 are good.

The approach presented in this section can also be extended to automatically generate a color table from a set of images, for example by selecting for each value in the color space the color class that has been chosen more often during the sequence. This is useful for off-line automatic calibration or for generation of a first color table to be manually refined.

3 Experiments

In order to evaluate the approach presented here we have performed several experiments. In this section we first discuss about performance metrics of color segmentation algorithms, and then present the results of our experiments.

Performance metrics. Defining performance metrics for color segmentation can be relatively easy. For example, we may manually label a set of images by correctly classifying each pixel and then compute classification rate of a segmentation method as the number of correctly classified pixels. Manual labeling can be performed only on a subset of pixels relative to important objects in the scene (see for example [5]). This is a good metric for evaluating color segmentation methods, but it requires a lot of time to generate the ground truth (since several pixels on each image must be manually classified). Moreover, typically object recognition processes that follow the segmentation phase are somewhat robust to small errors in classification, making this measure perhaps too much demanding. In fact, to our knowledge it does not exist a large data set of images labeled in this way, that can be used as a standard benchmark for segmentation algorithms (the one reported in [5] is limited to only a few dozens of images).

In this paper we propose an alternative metric for evaluation of a segmentation algorithm. This metric can be considered as an approximation of the above one, with the advantage of being much easier to be used to produce ground truth,

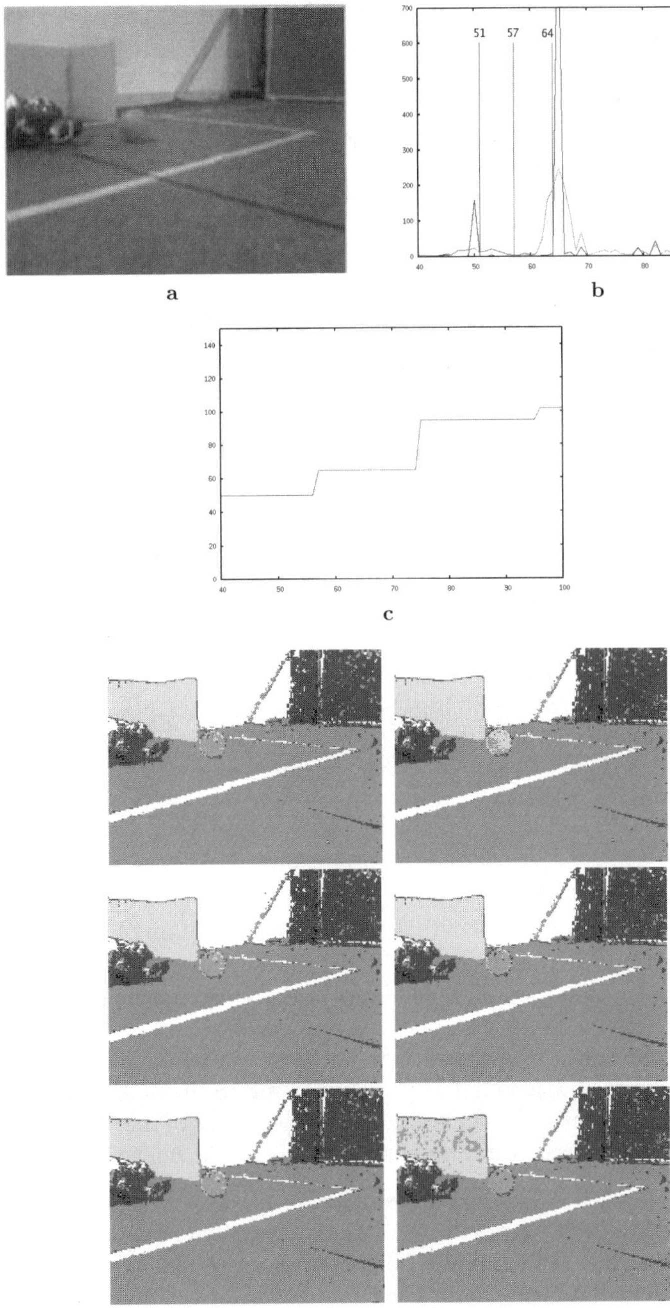

Fig. 1. a) Original image; b) Color distributions; c) A portion of $\tau(\cdot)$ function for image a); d) Color segmentation

thus allowing for easily generating larger data sets to be used for comparing different methods.

Instead of labeling each pixel on an image, we define a 4-sided polygon around any object that must be recognized in the environment. For the RoboCup soccer environment we chose to consider ball, goals and beacons. When objects are partially occluded or in case of non-polygonal shapes (e.g., a goal seen closely), we anyway define the best fitting 4-sided polygon around the object. The ball is instead modeled with an ellipse.

From a labeled image and the result of color segmentation on that image, we can measure the percentage of correctly classified pixels within such polygons. Let us denote with $\eta_o(\mathcal{I}_t)$ the percentage of correctly classified pixels of object o at frame $\mathcal{I}_t$. We want to measure the performance of the method over all the frames, i.e. over the entire data set $\{\mathcal{I}_t\}$. A simple choice would be to compute average and standard deviation of this value over t. However, by plotting typical responses of $\eta_o(\mathcal{I}_t)$ over t we see that the outcome is usually multi-modal, thus average and standard deviation seem not to be a good measure for the overall sequence. We found, for example, that in presence of motion blurring the percentage $\eta_o(\mathcal{I}_t)$ for the beacons may be very low. In order to summarize the behavior of the system over time, we propose to compute a distribution over η_o, counting how many times the pixels within the object o are correctly classified with a percentage rate of at least η_o. More specifically, we want to define $V_o(\lambda)$ as the percentage of times in which at least a percentage λ of pixels within the object o have been correctly classified. For example, $V_{ball}(0.5)$ expresses the percentage of frames in which at least half of the pixels of the ball have been correctly classified. $V_o(\lambda)$ ($\lambda \in [0,1]$) is thus defined as

$$V_o(\lambda) = \frac{|\{\mathcal{I}_t | \eta_o(\mathcal{I}_t) \geq \lambda\}|}{|\{\mathcal{I}_t | \mathcal{I}_t \text{ contains object } o\}|}$$

This value can be directly related to the capabilities of an object recognition module. For instance, we have empirically determined that current implementation of our object recognition module allows for correct recognition of objects if at least about 60 % of the pixels are correctly classified.

To evaluate the rate of false positives we have computed the ratio between the number of pixels classified with colors belonging to the objects (i.e., orange, yellow, sky blue and pink) that are outside of the polygons of the objects provided by the groud truth and the total number of pixels in the image.

Experimental evaluation. Evaluation of the proposed method has been performed by computing the functions $V_{ball}(\lambda)$, $V_{goal}(\lambda)$, and $V_{beacon}(\lambda)$ over a number of different data sets, comparing the results of the approach presented in this paper and segmentation obtained by manual definition of a color table. The data sets include at 208x160 pixels resolution and other internal information

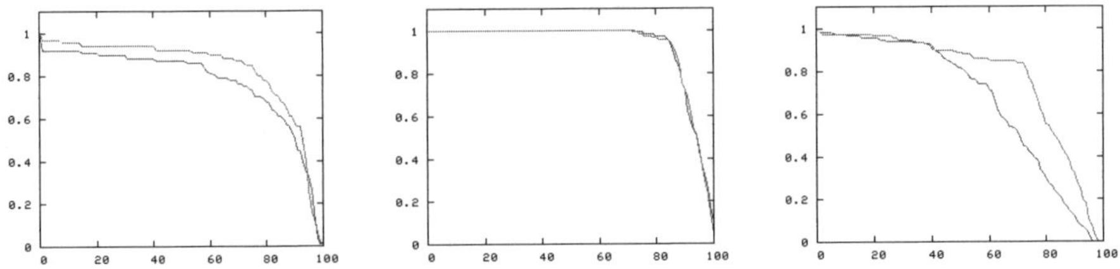

Fig. 2. First data set: $V_{ball}(\lambda)$, $V_{goal}(\lambda)$, $V_{beacon}(\lambda)$

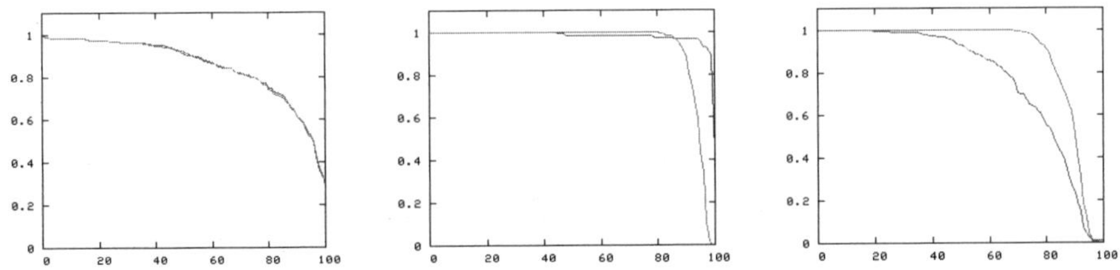

Fig. 3. Second data set: $V_{ball}(\lambda)$, $V_{goal}(\lambda)$, $V_{beacon}(\lambda)$

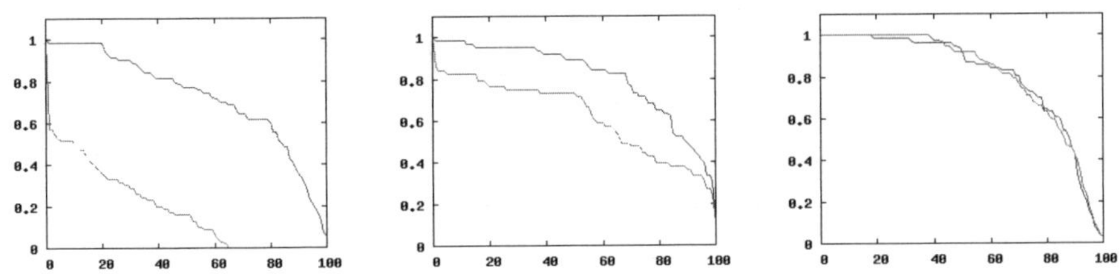

Fig. 4. Osaka data set: $V_{ball}(\lambda)$, $V_{goal}(\lambda)$, $V_{beacon}(\lambda)$

about the robot, they have been been taken from different locations (including Paderborn 2005, Osaka 2005 and our lab) and manually labeled as described above[1]. The first two data sets (containing 395 and 593 frames) used in the experiments reported here are taken from two different locations of our laboratory, using the same elements for the field and the same kind of illumination devices, but in different external light conditions. The third data set (193 frames) has been taken during RoboCup 2005 in Osaka.

For the first data set we have manually created a color table for static segmentation and tuned the parameters of the method proposed in this paper. We then use the same configuration for all the three data sets. For each data set,

[1] The data sets have been used also for evaluating object recognition and localization tasks and are available from `http://www.dis.uniroma1.it/~spqr`.

Table 1. False positives average and maximum rates

Data Set	Our method avg/max	Static segmentation avg/max
1	0.006 / 0.044	0.010 / 0.057
2	0.013 / 0.062	0.033 / 0.094
3	0.070 / 0.242	0.065 / 0.225

we have thus evaluated the functions $V_o(\lambda)$ for each object (ball, goal, beacon) and the false positive error rates, obtained with the method presented here and with static segmentation.

The results are reported in Figures 2, 3, 4, where red (darker) plot indicates our method, while the green (lighter) plot is the result of static segmentation, and in Table 1. As expected, static segmentation outperforms the dynamic method presented here in the first data sets, where it has been calibrated. However, the graphs also show that our method is very competitive. Also the rate of false positives are similar. In the second data set, that is similar to the first one but in different external conditions, we still have comparable results. However, while there is not a clear distinction in the functions $V_o(\lambda)$ (see Fig. 3), the false positive error rates are larger when using static segmentation. In the third data set, that is quite different from the first one, instead there is a clear difference in the performance (see Fig. 4). Our method provides for higher robustness to different lighting conditions, and to different environments.

These experiments show that the method proposed here is competitive with static segmentation obtained by manual calibration and robust to different light conditions and different environments.

4 Conclusions

In this paper we presented a dynamic color segmentation approach based on adaptive transformation of the color distribution of an image, that is suitable for implementation on robots with low computational resources and effective in presence of noise and illumination changes. The approach has been successfully implemented on AIBOs and extensively experimented in laboratory as well as during official RoboCup games. A new evaluation approach has been proposed and a large data set of labeled images have been created to evaluate and compare different methods.

The present method requires setting only a few parameters usually once arrived at a new location, and then it is robust to different light conditions over a typical competition period. As future work we intend to exploit machine learning techniques for learning the few parameters the method require for realizing a complete non-parametric method.

References

1. Anzani, F., Bosisio, D., Matteucci, M., Sorrenti, D.G.: On-line color calibration in non-stationary environments. In: Bredenfeld, A., Jacoff, A., Noda, I., Takahashi, Y. (eds.) RoboCup 2005. LNCS (LNAI), vol. 4020, Springer, Heidelberg (2006)
2. Jüngel, M.: Using layered color precision for a self-calibrating vision system. In: Nardi, D., Riedmiller, M., Sammut, C., Santos-Victor, J. (eds.) RoboCup 2004. LNCS (LNAI), vol. 3276, Springer, Heidelberg (2005)
3. Lovell, N.: Illumination independent object recognition. In: Bredenfeld, A., Jacoff, A., Noda, I., Takahashi, Y. (eds.) RoboCup 2005. LNCS (LNAI), vol. 4020, Springer, Heidelberg (2006)
4. Nisticò, W., Röfer, T.: Improving percept reliability in the sony four legged league. In: Bredenfeld, A., Jacoff, A., Noda, I., Takahashi, Y. (eds.) RoboCup 2005. LNCS (LNAI), vol. 4020, Springer, Heidelberg (2006)
5. Sridharan, M., Stone, P.: Autonomous color learning on a mobile robot. In: Proc. of AAAI (2005)

H_∞ Filtering for a Mobile Robot Tracking a Free Rolling Ball

Xiang Li and Andreas Zell

Wilhelm-Schickard-Institute, Department of Computer Architecture,
University of Tübingen, Sand 1, 72076 Tübingen, Germany
{xiang.li,andreas.zell}@uni-tuebingen.de

Abstract. This paper focuses on the problem of tracking and predicting the location and velocity of a rolling ball in the RoboCup environment, when the ball is pushed consecutively by a middle-size omnidirectional robot to follow a given path around obstacles. A robust algorithm based on the H_∞ filter is presented to accurately estimate the ball's real-time location and velocity. The performance of this tracking strategy was also evaluated by real-world experiments and comparisons with the Kalman filter.

1 Introduction

In many mobile robots applications, the robots are required not only to adapt themselves to the external situation, but also have the ability to interact with the environment. Estimating and predicting the motions of moving objects are the foundation for the interaction tasks. For example, when robots play the football, it is very important to detect and predict the ball's position and velocity, so that the robot can catch the ball, push it through obstacles, and shoot it in the goal. In this paper we focus on tracking and predicting the location and velocity of a rolling ball in the RoboCup domain with a middle-size omnidirectional robot, under the condition that the ball is consecutively pushed by the robot.

Kalman filters ([1], [2], [3], [9]) have been used in many ball tracking problems. They provide efficient and convenient minimum-mean-square-error solutions for the state estimation problem, considering that both the process and the measurement noises of the target system are assumed as Gaussian with known statistical properties ([8]). Besides that, multiple model filters based on Kalman filters reveal much better performance than the single model filter in some applications. As one of the multiple model filters, the interacting multiple model (IMM) algorithm, used for the object tracking in the RoboCup ([5]), utilizes a Kalman filter for each mode of the target movement. However, in practical situations, the uncertainties of the target system and the measurements normally do not satisfy the Gaussian assumption, and the noise statistics is usually not available.

To avoid thinking about these uncertainties, a method to build a predictive model of the ball's movement is used in the estimation of the ball's position and velocity ([7]). It models a free rolling ball's movement as the linear movement

G. Lakemeyer et al. (Eds.): RoboCup 2006, LNAI 4434, pp. 296–303, 2007.

and estimates the model parameters using ridge regression. By comparing the observed and predicted ball's positions, the method can also recognize the change points of the ball's movement. Due to the requirement of a buffer to store the observations of the ball's movement and the estimation of model parameters, the memory occupancy and computational complexity of this method are highly increased.

In this paper, we present a robust algorithm based on the H_∞ filter for an omnidirectional robot to track a rolling ball in the RoboCup domain. The H_∞ filter does not require priori knowledge of the noise statistics, only assuming that the noise signals have finite energy. Unlike the Kalman filter providing the minimum variance of the estimation error, the H_∞ filter provides the minimal effect of the worst noise on the estimation error. Experiments with a real omnidirectional robot show that this approach is efficient and yields highly robust estimations of the ball's location and velocity.

2 Problem Formulation

The ball tracking problems in the RoboCup domain are challenged by the interactions between the robots and the ball. These frequent interactions usually result in a highly non-linear movement of the ball, and it is very difficult to precisely estimate the uncertainty distribution of the interactions. Moreover, the measurement accuracy of the ball's position is also limited by the sensors and the corresponding signal processing algorithms. This paper focuses on tracking a rolling ball when it is consecutively pushed by an omnidirectional robot to follow a given path. Considering the uncertainty of the interactions between the robot and the ball, we utilize a new approach based on the H_∞ filter to estimate the ball's location and velocity.

The discrete representation of the ball's dynamics is described by the following equations:

$$\dot{p}_{k+1} = \dot{p}_k + \ddot{p}_k T \tag{1}$$

$$p_{k+1} = p_k + \dot{p}_k T + \frac{1}{2}\ddot{p}_k T^2 \ , \tag{2}$$

where p is the position of the ball, $\dot{p}$ and $\ddot{p}$ are respectively the velocity and acceleration of the ball. T is the sampling interval and k is the index of the sampling interval. We define a state vector consisting of the position and velocity as $x_k = [p_k, \dot{p}_k]^T$. Knowing the measurement value is the ball's position, we build the system model of the ball as follows:

$$x_{k+1} = \begin{bmatrix} 1 & T \\ 0 & 1 \end{bmatrix} x_k + \begin{bmatrix} T^2/2 \\ T \end{bmatrix} u_k \tag{3}$$

$$y_k = \begin{bmatrix} 1 & 0 \end{bmatrix} x_k \ , \tag{4}$$

where u denotes the system input and is equal to the acceleration $\ddot{p}$ which is completely determined by the friction of the ground and the pushing operation from the robot. But in the practical situation, the previous equation (3) can not give the precise state values because of the noise due to the rugged carpet ground and other unfortunate realities, and the precise output values can not be obtained from the equation (4), since measurement noise decreases the reliability of the measurement data. So a more precise mode is given as

$$x_{k+1} = \begin{bmatrix} 1 & T \\ 0 & 1 \end{bmatrix} x_k + \begin{bmatrix} T^2/2 \\ T \end{bmatrix} u_k + w_k \tag{5}$$

$$y_k = \begin{bmatrix} 1 & 0 \end{bmatrix} x_k + v_k \ , \tag{6}$$

where w is called process noise and v is called measurement noise.

As we do not know exactly the friction of the ground, the moment when the robot collides the ball, and the corresponding effect of the collision on the ball's movement, the system input u is not available. But we can consider u as additional process noise and unify u with the process noise w. Then a more realistic system model is

$$x_{k+1} = \begin{bmatrix} 1 & T \\ 0 & 1 \end{bmatrix} x_k + \begin{bmatrix} T^2/2 \\ T \end{bmatrix} w_k \tag{7}$$

$$y_k = \begin{bmatrix} 1 & 0 \end{bmatrix} x_k + v_k \ . \tag{8}$$

3 Robust H_∞ Filtering

As mentioned earlier, the Kalman filter requires the priori knowledge of statistical properties of the system and measurement noises, which are really hard to obtain practically. The ball filter with predictive model, described in [7], could bring a higher computational cost and memory occupancy, so this filter is not very efficient for the fast tracking problem. As a robust filter strategy, the minimax H_∞ strategy has the same efficient computation as that of the Kalman filter, and does not depend on the known noise statistics, but on the assumption of a finite disturbance energy. Consider the following linear system:

$$x_{k+1} = A_k x_k + B_k w_k \tag{9}$$

$$y_k = C_k x_k + v_k \ , \tag{10}$$

where $x_k \in \Re^n, w_k \in \Re^m, y_k \in \Re^p, v_k \in \Re^p$. A_k, B_k and C_k are matrices with appropriate dimension, (A_k, B_k) is controllable and (C_k, A_k) is detectable. Unlike the Kalman filter, which is interested in the estimation of the system state x_k, the H_∞ filter concerns the linear combination of x_k

$$z_k = L_k x_k \ . \tag{11}$$

The output matrix L_k is selected by the user according to the different applications. In our problem, we care about the ball's location and velocity, which just constitute the system state, so here L_k is specified as an identity matrix. The

H_∞ filter computes the estimated state $\hat{z}_k$ based on the measurement Y_k , where $Y_k = \{y_k, 0 \le k \le N-1\}$, and evaluates the estimation error by a performance measure, which can be regarded as an energy gain:

$$J = \frac{\sum\limits_{k=0}^{N-1} \|z_k - \hat{z}_k\|_{Q_k}^2}{\|x_0 - \hat{x}_0\|_{p_0^{-1}}^2 + \sum\limits_{k=0}^{N-1} \left(\|w_k\|_{W_k^{-1}}^2 + \|v_k\|_{V_k^{-1}}^2 \right)} \tag{12}$$

where N is the size of the measurement history, Q_k, p_0, W_k, V_k are the weighting matrices for the estimation error, the initial conditions, the process noise and the measurement noise. Moreover, $Q_k \ge 0, p_0^{-1} > 0, W_k > 0, V_k > 0$ and $((x_0 - \hat{x}_0), w_k, v_k) \neq 0$. The notation $\|x_k\|_{Q_k}^2$ is defined as $\|x_k\|_{Q_k}^2 = x_k^T Q_k x_k$. The denominator of J can be considered as the energy of the unknown disturbances, and the numerator is the energy of the estimation error. The H_∞ filter aims to provide an uniformly small estimation error $e_k = z_k - \hat{z}_k$ for any $w_k, v_k \in l_2$ and $x_0 \in R^n$, such that the energy gain J is bounded by a prescribed value:

$$\sup J < 1/\gamma \tag{13}$$

where sup denotes the supremum and $1/\gamma$ is the noise attenuation level. This condition keeps the robustness of the H_∞ filter, because the estimation energy gain is limited by $1/\gamma$ no matter what the bounded energy disturbances are.

To solve this optimal estimation $\hat{z}$ due to the bounded energy gain J, the H_∞ filter can be interpreted as a *minimax* problem ([10])

$$\min_{\hat{z}_k} \max_{(w_k, v_k, x_0)} J = -\frac{1}{2\gamma} \|x_0 - \hat{x}_0\|_{p_0^{-1}}^2 +$$

$$\frac{1}{2} \sum_{k=0}^{N-1} \left[\|z_k - \hat{z}_k\|_{Q_k}^2 - \frac{1}{\gamma} \left(\|w_k\|_{W_k^{-1}}^2 + \|v_k\|_{V_k^{-1}}^2 \right) \right] \tag{14}$$

where the estimation value $\hat{z}_k$ plays against the bounded energy disturbances w_k and v_k. Many strategies have been proposed for solving this *minimax* problem ([4]). We adopt a linear quadratic game approach ([10]), which does not require checking the positive definiteness and inertia of the Riccati difference equations for every step, but is implemented through recursive updating the filter gain H_k, the solution P_k of the Riccati difference equation, and the state estimation $\hat{x}_k$. The updating equations are given as follows:

$$\bar{Q}_k = L_k^T Q_k L_k \tag{15}$$

$$S_k = \left(I - \gamma \bar{Q}_k P_k + C_k^T V_k^{-1} C_k P_k \right)^{-1} \tag{16}$$

$$P_{k+1} = A_k P_k S_k A_k^T + B_k W_k B_k^T \tag{17}$$

$$H_k = A_k P_k S_k C_k^T V_k^{-1} \tag{18}$$

$$\hat{x}_{k+1} = A_k \hat{x}_k + H_k \left(y_k - C_k \hat{x}_k \right) \ , \tag{19}$$

where $P_0 = p_0$ and $P_k > 0$. I is the identity matrix.

Fig. 1. The omnidirectional robot equipped with a digital color camera and a hyperbolic mirror on the top

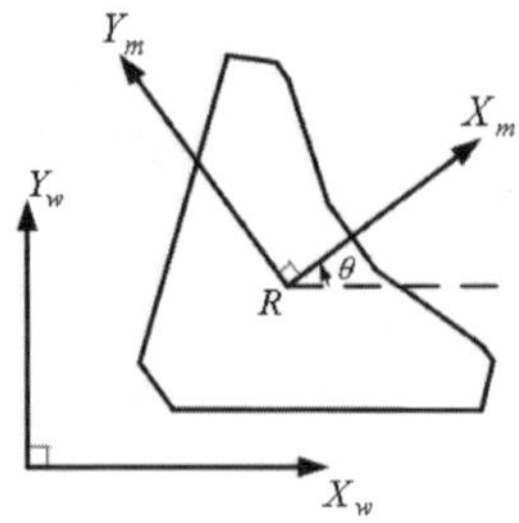

Fig. 2. Coordinate systems: $[X_w, Y_w]$ is the world coordinate system, $[X_m, Y_m]$ is the robot coordinate system

Apparently, these recursive equations have a similar form as the classic Kalman filter. Although we need not to know the statistics of noises w_k and v_k in the H_∞ filter, we should tune the weight matrices Q_k, p_0, W_k, V_k carefully, because these values determine the estimation error in the performance criterion (14). The weight matrices W_k, V_k can be chosen according to the experience about the noise. For example, if we know that the noises w is smaller than v, W_k should be smaller than V_k. p_0 is based on the initial estimation error. If we are highly confident about our initial estimation $\hat{Z}_0$, p_0 should be small. Similarly, if we care more about the precise estimations of some elements in the state, or some elements having bigger magnitude in their physical definition, the corresponding elements in the matrix Q_k can be set larger than others. As the performance criterion, γ can not be very large, because otherwise some eigenvalues of the matrix P may have magnitudes more than one. These eigenvalues prevent a proper derivation of the H_∞ filter equations, so that the H_∞ filter problem has no solution.

4 Experiments

The ball's observation values come from our omnidirectional view system and object detection process. Our omniderectional view system consists of a AVT Marlin F-046C color camera with a resolution of 780×580, which outputs signals up 50 times per second. In order to achieve a complete surrounding map of the robot, the camera is assembled pointing up towards a hyperbolic mirror which is mounted on the top of our omnidirectional robot, as shown in Fig.1. After obtaining the image from the camera, the other two processes, color calibration and distance calibration, map the colors to different classes based on the colors of objects and landmarks in the RoboCup domain, and the pixels in the image to the real world coordinates, respectively. At last, a fast object detection algorithm is used to get the ball's real world position, as described in [6].

While the camera image from the robot always displays the ball's relative position to the robot's position and orientation, the ball's relative position and velocity with respect to the robot coordinate system can be estimated directly by using the ball's observation values. When the ball's absolute position and velocity is required, the ball's observation values can be transformed into the world coordinate system, which is fixed in the robot playing field, by utilizing the robot's observation values. To prove the feasibility and the robustness of the H_∞ filter in estimating the ball's position and velocity with noisy observation values, we use the robot's odometer-based observation values in the experiments. The world coordinate system and the robot coordinate system are described in Fig.2.

All experiments were made in our robot laboratory having a half-field of the RoboCup-Middle size league. The H_∞ filter described in section 3 has been applied to tracking a rolling ball in the RoboCup domain, when the ball is pushed by a mobile robot to follow a linear path and a sinusoidal path with the constant desired translation velocity 0.3m/s and 0.5m/s respectively. The ball did not slide away from the robot during the whole pushing process because of the consecutive collisions with the robot. At every sampling time, the H_∞ filter estimated the ball's absolute position and velocity with respect to the world coordinate system, and the ball's relative position and velocity with respect to the robot coordinate frame. The noise attenuation level and weight matrices for estimating the x and y components were chosen as follows:

$$\gamma^x = 2.0,\ P_0^x = \begin{bmatrix} 30 & 0.004 \\ 30 & 2 \end{bmatrix},\ Q_k^x = \begin{bmatrix} 0.01 & 0 \\ 0 & 0.01 \end{bmatrix},\ W_k^x = 1,\ V_k^x = 10\ ;$$

$$\gamma^y = 1.5,\ P_0^y = \begin{bmatrix} 10 & 0.05 \\ 30 & 0.8 \end{bmatrix},\ Q_k^y = \begin{bmatrix} 0.1 & 0 \\ 0 & 0.1 \end{bmatrix},\ W_k^y = 10,\ V_k^y = 1\ .$$

To evaluate the performance of the H_∞ filter, a Kalman filter with assumed noise variance was also used to estimated the ball's position and velocity with the same observation values. The initial estimate error covariance matrices P_o and the probability distributions of process noise and measurement noise are chosen as follows:

$$p_0^x = \begin{bmatrix} 0.01 & 0.0001 \\ 0.003 & 0.005 \end{bmatrix},\ p(w^x) \sim N(0, 0.01),\ p(v^x) \sim N(0, 0.0001)\ ;$$

$$p_0^y = \begin{bmatrix} 0.01 & 0.0001 \\ 0.01 & 0.005 \end{bmatrix},\ p(w^y) \sim N(0, 1),\ p(v^y) \sim N(0, 0.0001)\ .$$

From the results shown in figures 3-8, we can see the H_∞ filter eliminated the high frequency components of the measurement and estimated the ball's position values sufficiently. Figures 5-8 show that the estimated positions from the H_∞ filter are slightly better than those from the Kalman filter. Figures 9-10 illustrate that the ball's velocity is effectively estimated and the H_∞ filter is better than the Kalman filter, while the estimated x-velocities from the H_∞ filter approach to the ideal robot's x-velocity 0.3m/s with less time and are more smooth than those from the Kalman filter.

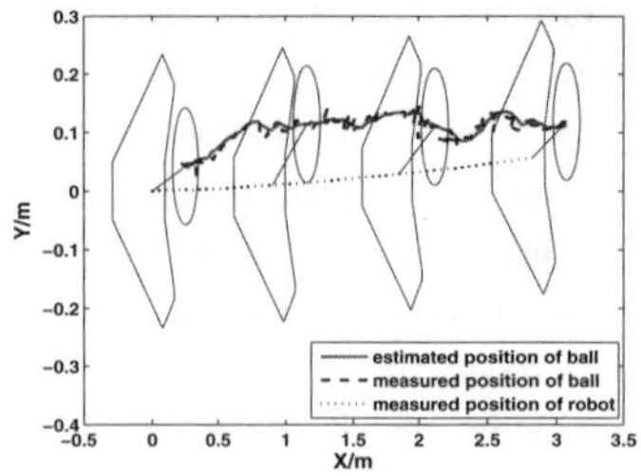

Fig. 3. Absolute positions of robot and ball along the linear path

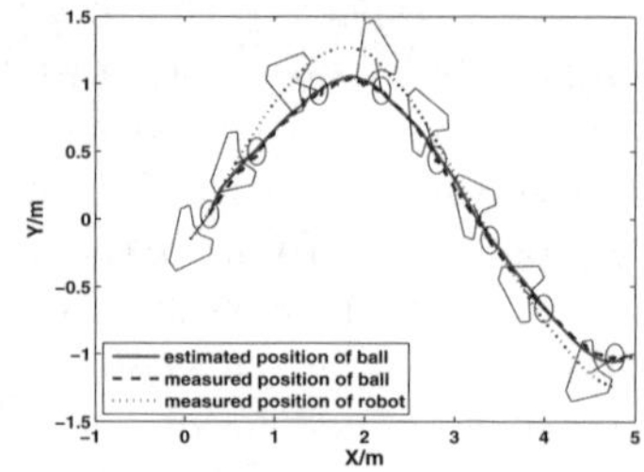

Fig. 4. Absolute positions of robot and ball along the sinusoidal path

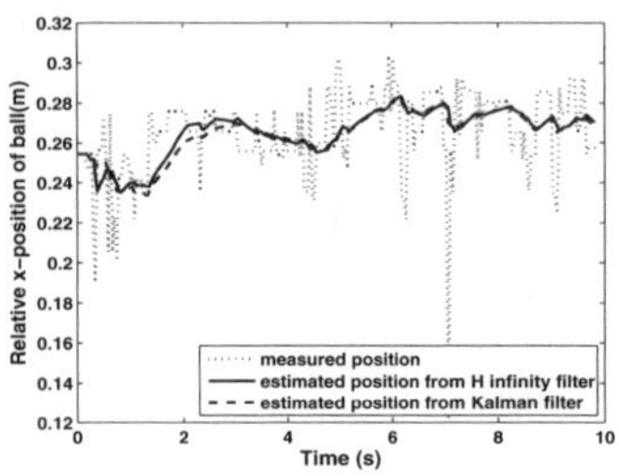

Fig. 5. Relative x-positions of ball along the linear path

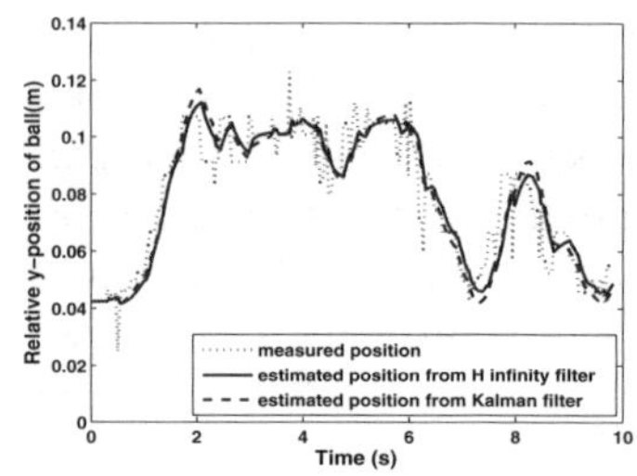

Fig. 6. Relative y-positions of ball along the linear path

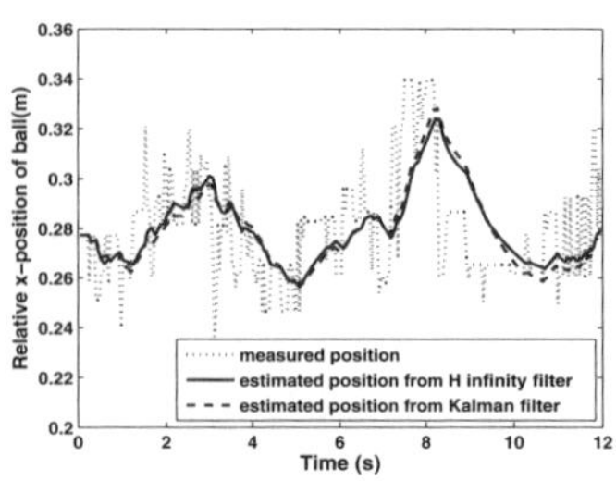

Fig. 7. Relative x-positions of ball along the sinusoidal path

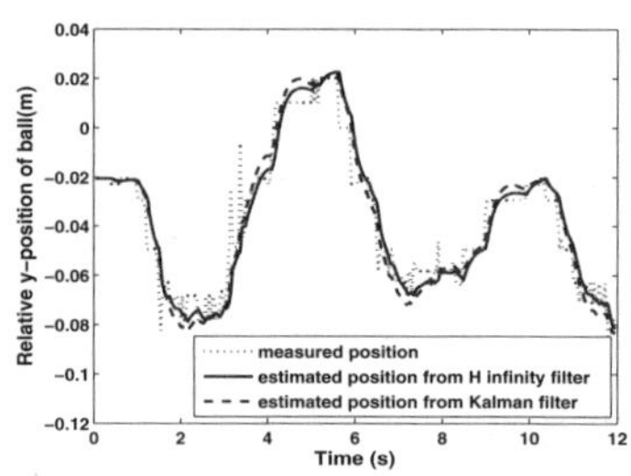

Fig. 8. Relative y-positions of ball along the sinusoidal path

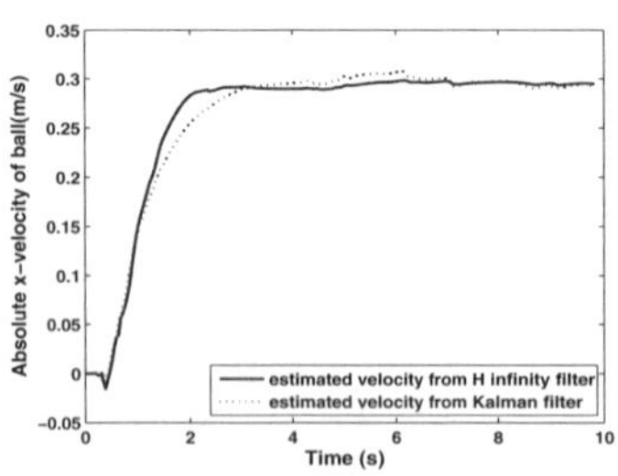

Fig. 9. Absolute x-velocities of ball along the linear path

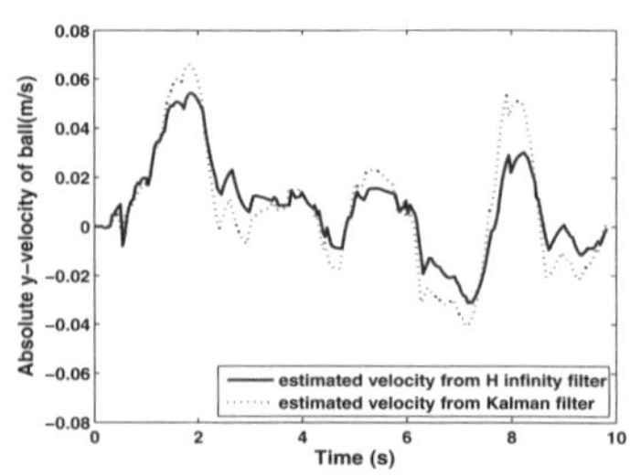

Fig. 10. Absolute y-velocities of ball along the linear path

5 Conclusion

In this paper we introduce a robust H_∞ filter, which does not require a priori knowledge about the statistical properties of the system and measurement noise, but only depends on the assumption of finite noise power. The recursive equations of the H_∞ filter are very similar to those of the Kalman filter, so the H_∞ has relatively low computation cost in the implementation and adapts to the real time estimation problem. With the real-world experiments, where the ball was following the given paths pushed consecutively by an omnidirectional robot, the performance of the H_∞ filter was evaluated by comparing the estimation values with those from the Kalman filter. The results of the estimated ball's position and velocity show that the H_∞ filter eliminates the high frequency noise components of the measurements and estimates the ball's position and velocity robustly in the pushing process. Moreover, the H_∞ filter in this application is shown to be superior to a Kalman filter, which requires manual tuning of the noise parameters.

References

1. Bar-Shalom, Y., Li, X.R., Kirubarajan, T.: Estimation with Applications to Tracking and Navigation. John Wiley, Chichester (2001)
2. Freeston, L.: Applications of the kalman filter algorithm to robot localisation and world modelling. Technical report, University of Newcastle, NSW, Australia (2002)
3. Gelb, A.: Applied Optimal Estimation. Cambridge (1974)
4. Hassibi, B., Sayed, A., Kailath, T.: Indefinite Quadratic Estimation and Control - A Unified Approach to H2 and H Infinity Theories. Society for Industial and Applied Mathematics, Philadelphia (1999)
5. Heinemann, P., Plagge, M., Treptow, A., Zell, A.: Tracking dynamic objects in a robocup environment in a robocup environment - the attempto tübingen robot soccer team. In: Polani, D., Browning, B., Bonarini, A., Yoshida, K. (eds.) RoboCup 2003. LNCS (LNAI), vol. 3020, Springer, Heidelberg (2004)
6. Heinemann, P., Rückstieß, T., Zell, A.: Fast and accurate environment moddelling using omnidirectional vision. In: Ilg, H.U., Bülthoff, H. (eds.) Dynamic Perception, Infix (2004)
7. Lauer, M., Lange, S., Riedmiller, M.: Modeling moving objects in a dynamically changing robot application. In: Furbach, U. (ed.) Advances in Artificial Intelligence, pp. 219–303. Springer, Heidelberg (2005)
8. Petrov, S.: Computer vision, sensorfusion und verhaltenssteuerung für fussballroboter. Master's thesis, Freie Universität Berlin, Institute für Informatik (2004)
9. Ruizdel-Solar, J., Vallejos, P.: Motion detection and tracking for an aibo robot using camera motion compensation and kalman filtering. In: Nardi, D., Riedmiller, M., Sammut, C., Santos-Victor, J. (eds.) RoboCup 2004. LNCS (LNAI), vol. 3276, Springer, Heidelberg (2005)
10. Shen, X., Deng, L.: Game theory approach to discrete H_∞ filter design. IEEE Transcation on Signal Processing 45(4) (1997)

Balancing Gains, Risks, Costs, and Real-Time Constraints in the Ball Passing Algorithm for the Robotic Soccer

Vadim Kyrylov

Simon Fraser University – Surrey
Surrey, British Columbia V3T 2W1 Canada
vkyrylov@sfu.ca

Abstract. We are looking for a generic solution for the optimized ball passing problem in the robotic soccer which is applicable to many digital simulated sports games with ball. In doing so, we show that previously published ball passing methods do not properly address the necessary balance between the anticipated rewards, costs, and risks. The multi-criteria nature of this optimization problem requires using the Pareto optimality approach. We propose a scalable and robust solution for decision making, as its quality degrades in a graceful way once the real time constrains are kicking in.

1 Introduction

1.1 Ball Passing Algorithms: State of the Art

Passing the ball to a teammate is a critically important player skill in many sports games with ball. Early RoboCup scholars have developed two reasonably good algorithms for the simulated soccer [1, 2]. In both the soccer agent chooses values of the direction of the kick and its force. In [1] the anticipated outcome of passing ball is evaluated with two heuristic indicators: (1) the tactical value of the end point and (2) the likelihood of that the receiving teammate will intercept the ball. This algorithm is searching for both direct and leading passes including passes to self. The tactical value is the only criterion for selecting the best option; the likelihood of success is used as a constraint. Although this method has proved to be rather good, it neglects risks such as the possible proximity of other opponents to the anticipated interception point. One more shortcoming is the requirement that the ball should be always intercepted by the receiver in the minimal time. Indeed, this may result in lost opportunities in executing leading passes when the ball is sent to the point of the field still reachable by the teammate and having better tactical value.

The algorithm implemented in [2] appears to be more sophisticated, as it is taking into account the opponent player congestion in the vicinity of the ball destination. Also are considered ball travel distance, opponent goal scoring opportunity if the pass is successful, and the possible outcomes if the ball would not be intercepted as intended. The decision is made by deliberating on 5 options for each receiving teammate: direct pass, leading pass, pass to the expected location of the teammate,

G. Lakemeyer et al. (Eds.): RoboCup 2006, LNAI 4434, pp. 304–313, 2007.

pass to a point near teammate having low congestion, and pass along a low congestion line. Each alternative is evaluated using 9 performance indicators. With the purpose of making a choice, these indicators are analyzed using a decision tree.

Even more advanced ball passing algorithm with player collaborating using aural messages was recently reported in a short paper [3]. As this algorithm is all based on a decision tree, it is possible that some indeed good ball passing options could be overlooked. This is a general shortcoming of decision trees; in what follows, we discuss this in more detail.

1.2 Unresolved Issues and Research Objectives

In this paper, we address three issues.

1. No benchmark. The existing algorithms are collections of sophisticated heuristics; it is still unknown to what extent they could be improved and what the benchmark solution is.
2. Smoothly balancing rewards and risk. We believe that implementing a continuous spectrum of risk-taking vs. risk aversive strategies by the soccer agent is highly desirable. However, in the existing methods this balancing does not render itself as a controlled feature.
3. Avoiding possible conflicts with the real-time constraints. Reduction of required computations in existing algorithms can normally be done by removing some branches in the decision tree. That may result in abrupt loss in the quality of decisions.

We resolve these issues using the multi-criteria decision analysis (MCDA). In doing so, we are pursuing the following objectives.

- Developing a theoretical framework for a totally optimal ball passing algorithm that could serve as a benchmark. We want this solution to be generic and thus reusable. This intention is standing in a concert with other RoboCup scholars looking for generic solutions [4].
- Fully identifying rewards, risks, and costs involved in passing the ball and demonstrating how they could be balanced in the proposed framework. We wish to offer a way to implementing a continuous spectrum of risk-taking and risk-aversive attitudes by the soccer player.
- Addressing the real-time constraints. We want to propose a truly scalable solution with just one parameter which determines the amount of the required computations. We also want to design a robust ball passing algorithm that would be resulting only in a gradual loss of the decision quality if we are forced to bypass some computations.

2 Rewards, Risks, and Costs in Ball Passing

Prior to developing the optimal decision making algorithm, we identify the presumably complete list performance criteria that govern the decision to pass the ball.

In doing so, we slightly modify the ball passing problem formulation as compared to [1, 2]. In our case, player with the ball considers all possible points (x, y) in the field and must decide to which point he should send the ball now and determine the ball speed in the end point. This end speed affects the probability of the successful interception by the receiving teammate; it also determines the ball travel time and thus the incurred risk. The decision is made by comparing performance indicators calculated for different ball passing options.

Once the passing player has made his choice of the point and of the ball end speed, he is able to determine the kicking force and direction, which are the actual decision variables. If the required kicking force exceeds the available limit, the point is just removed from the consideration. Likewise, points are eliminated if the perceived risks are prohibitively high.

Each remaining potential destination point for pass is assigned a vector criterion having continuous values of its m components, which are the performance indicators. So there is a two-dimensional decision space (*kicking_force, direction*) and an m-dimensional criterion space. For the analysis, we make two modifications. First, we replace the decision space by a three-dimensional one (x, y, *end_speed*) with only two coordinates being independent; this space is much easier to visualize. Second, in order to make our algorithm scalable when the real-time constraints are present, we replace the continuous decision space by a discrete one.

We split the decision criteria in three categories: (1) gains, (2) risks, and (3) costs.

Gains. We see two gains, or rewards, from passing the ball; we wish to maximize both.

Both are similar to the indicators used in [1] and [3], which served their purpose very well. The first is the tactical value of the point (x, y) where the ball will be sent to. This function encourages sending the ball close to the opponent goal and discourages destinations near own goal. The second reflects the chance to score the opponent goal from the ball destination. Its value depends on the anticipated number of opponent players between the opponent goal and the destination point (x, y) of the pass.

Risks. As proposed in [1], the risks involved in ball passing all are defined as soft constraints. We further improve this idea by dropping the requirement that the receiving player is intercepting the ball in minimal time. Rather, we assume that he must be chasing the ball if necessary. Hence we have more risk factors than our predecessors.

1. Opponent may reach (x,y) before the teammate. The risk function $r_1(x,y)$ is the time difference between the arrivals of the fastest teammate and the fastest opponent to this point.
2. Ball can be intercepted by the opponent on its way to (x,y). The risk function $r_2(x,y)$ is the time difference between the intended arrival time of the ball in (x,y) and the earliest time when it can be stolen by the opponent.
3. Teammate may be too late in point (x,y) after the ball rolls by. So the risk function $r_3(x,y)$ is the time difference between the arrivals of the teammate and the ball. However, this risk increases if the ball is moving in (x,y) too fast which is making it difficult to intercept.
4. Too many opponents may get close by. The risk function $r_4(x,y)$ is the time difference between the arrivals of the ball and the second fastest opponent in (x,y).
5. If the teammate fails to intercept the ball, it may cross the field boundary. The risk function $r_5(x,y)$ is minus the time remained until the ball crosses this line after bypassing (x, y).

6. The receiving player may have too low stamina to chase the ball. The risk function $r_6(x,y)$ is time when the receiving teammate reported low stamina less current time.
7. Ball may not reach the destination point at all, as (x, y) is too far away for given initial ball speed. As the ball movement is distorted by noise, the actual maximal ball traveling distance may differ from the calculated theoretical one, D_{max}. A soft constraint $r_7(x,y)$ is used to reflect this risk.

We want to minimize each of these seven risk factors. For convenience, they could be scaled so that they all are taking values in, say (0, 10).

Costs. The cost factor is the time required for obtaining the anticipated rewards, which we want to minimize. Taking this in consideration makes sense because the precision of the situation prediction substantially decreases with the forecast time. This criterion would be discouraging too long passes if, given all the rest conditions equal, shorter ones exist.

Concept demonstration. With the sole purpose of the concept demonstration used throughout this paper, we have designed an example with three simplifications. (1) Decision space is further reduced to determining the pass direction only; end speed in the destination point is a fixed parameter of the algorithm. (2) Only the tactical value of the end point is used as the reward. (3) Risk and costs merged in just one parameter by applying heuristic rules.

This allowed using two-dimensional displays for the visualization. The full-scale algorithm is treating all criteria separately.

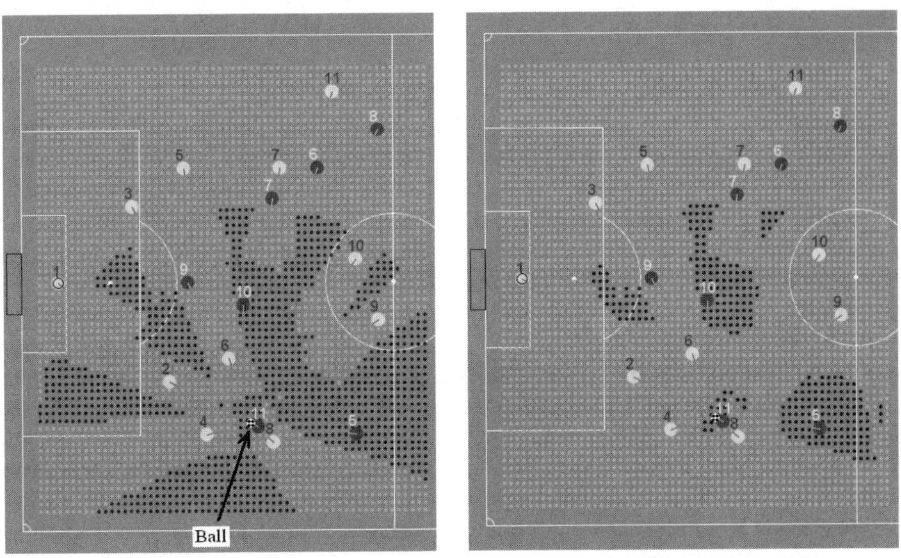

(a) applied risk factors 2 and 7 only (b) all seven risk factors applied

Fig. 1. Screenshots of the software tool for analyzing the soccer player tactics. Of the original 3600 points, most have been eliminated because the anticipated risk is inappropriately high.

Fig.1 shows a grid of 3600 points considered as candidates for passing the ball by player 11 from the right-hand team. Most are eliminated because the risk is too high. These points are shown in light gray; the darker points are the remaining alternatives. Player 11 must select the best one based on the vector of performance indicators available for each point.

We wish this decision to be optimal in some sense. This sense is the Pareto optimality.

3 Applying the Pareto Optimality Principle to Ball Passing

Pareto optimality, first originated in economics, is now a standard principle for solving vector optimization problems with conflicting criteria [5]. In what follows, we will replace the reward function with the negation thereof; thus we want all our criteria to be minimized simultaneously. In the general case, though, simultaneous minimization cannot be achieved. The Pareto optimality principle only offers a method for substantially reducing the set of decision alternatives by identifying among them the set of so-called non-dominated alternatives; altogether they are making the Pareto set, or the Pareto frontier.

By definition, the criteria vector $\mathbf{v}_i = \{v_{i1},\ldots,v_{im}\}$ is **dominated** by vector $\mathbf{v}_j = \{v_{j1},\ldots,v_{jm}\}$ if the following condition holds:

$$\forall k\{v_{ik} > v_{jk}\}. \tag{1}$$

This means that $\mathbf{v}_j$ is located inside the cone in $\mathbb{R}^n$ with the vertex $\mathbf{v}_i$, the sides of this cone being parallel to the coordinate subspaces $\mathbb{R}^{n-1}$. By definition, the Pareto set is the subset of non-dominated alternatives, i.e. whose cones do not contain other alternatives. The Pareto set is not necessarily convex, nor is it in the general case even connected. The computational complexity of determining the Pareto subset in the finite set with N elements is $O(N^2)$.

The meaning of a non-dominated alternative $\mathbf{v}_j$ is that outside the Pareto set there is no another alternative that outperforms $\mathbf{v}_j$ simultaneously by all criteria; at least one criterion value is worse, anyway. From this follows that the optimal decision should be sought within the Pareto set; all the rest alternatives could be eliminated as they are all inferior.

In Section 4, we will be also using a weakened version of the dominance relation, which is called ε-domination [5, 6]. The set of non-ε-dominated points is referred to as ε-Pareto set. Elements lying outside this set are having at least one criterion that is worse by more than ε.

Noteworthy that, eliminating ball passing alternatives before identifying the Pareto set, as it has been done in the existing algorithms, may result that some of the Pareto optimal points would be apparently removed without even evaluating thereof. This is exactly what may happen in decision trees. Unless the decision conditions are designed so carefully that any eliminations do not affect the Pareto set, there is no guarantee that the decision tree yields optimal solution to the problem in all cases. However, the trouble is in that such a decision tree is difficult to design, and for each new applied problem this must be done over and over again. On the other hand, the Pareto optimality principle offers a general solution.

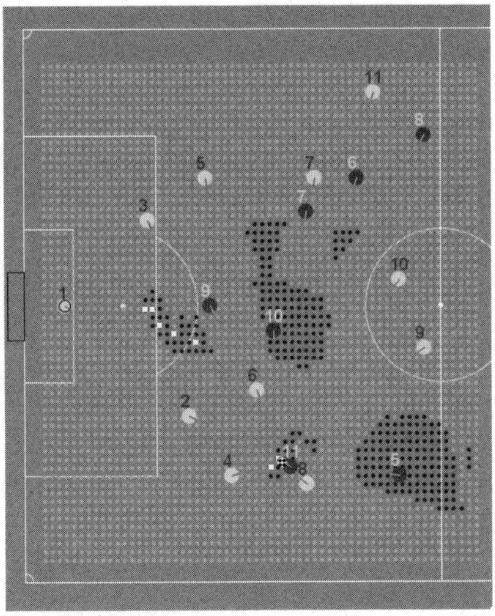

Fig. 2. Situation in Fig.1(b) with the points making the Pareto set shown in white

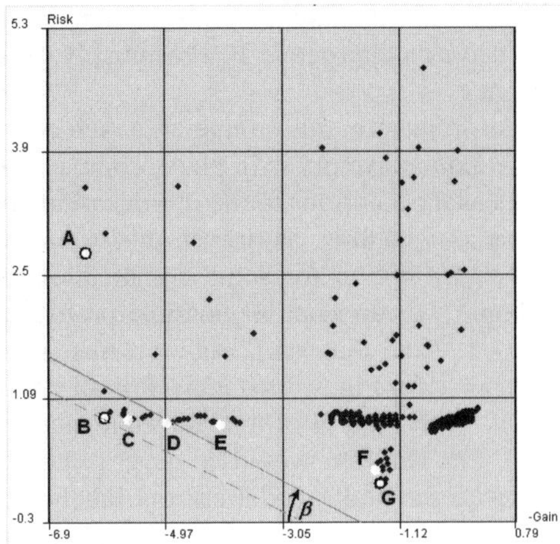

Fig. 3. The ball passing alternatives in the criterion space. Points in the Pareto set shown in white.

The Pareto set of the alternatives that player 11 in Fig.1(b) should be indeed choosing from is shown in Fig 2. This example suggests that either the leading pass to teammate 9 should be executed (five slightly different options), or player 11 should leave the ball for himself, i.e. execute so-called fast dribbling (two options). Passes to teammate 10 are not in the Pareto set. Note that some points near player 9 cannot be reached in minimal time; yet they are better for scoring the goal. Fig. 3 shows the situation as it occurs in the criterion space.

The MCDA theory leaves the final choice of the single alternative from the Pareto set up to the decision maker. In our case, however, it is the algorithm developer who must formalize the player preferences which could be used for searching the Pareto set. This search is exactly about *balancing* the rewards, risks, and costs; in what follows, we explain this idea.

A naïve approach suggests merging all criteria in just one and applying commonly known single-criterion optimization techniques. For example, one can use the utility function *U* of the decision variables (*x*,*y*), which is the weighted sum of risk *Risk* and gain *Gain*:

$$U(x,y) = -wGain(x,y) + (1-w)Risk(x,y), \tag{2}$$

where w is the positive weight, $0 \le w \le 1$; it reflects the importance of *Gain* for the decision maker, as compared to *Risk* whose weight is thus $1-w$. (Note the minus sign before *Gain*).

To find the solution, function (2) must be minimized. Equation $U(x,y) = c$, where c is some constant, in the criterion space represents the slant strait line shown in Fig.3. Search for the optimal solution in this case would be moving this line towards the origin by decrementing c until the line (shown in the dashed style) intersects with just one decision alternative **B**. Presumably, this would be the optimal, balanced solution sought.

Unfortunately, this simple approach does work only when the Pareto set is convex [6]. If non-convexity is in place, some elements of the Pareto set would be never rendered as the solutions to the optimization problem, no matter what values the decision maker assigns to w. However, this is counter intuitive, as each point in the Pareto set is the best option for some combination of the decision maker preferences. In our example we can scan all possible preferences by varying the weight w in the range $0 \le w \le 1$. Note that, as $\beta = \tan(w)$, this parameter determines the angle β of the line $U(x,y) = c$ (in Fig.3, w=0.335). For all possible weights, this would render only three points, **A**, **B**, and **G** of total seven available in the Pareto set (marked with black circles). The rest four would be never returned as solutions. This just illustrates the fact that with the multi-modal criteria functions which we are dealing with in the robotic soccer, a different way to finding the balanced optimal solution should be taken.

4 Searching the Optimal Ball Passing Decision in the Pareto Set

The different way is applying more sophisticated methods for searching the Pareto set that can work with non-convex problems. As there is a plethora of such methods, we will demonstrate just one, developed by the author of this paper. The method is called '*the randomized sequential elimination of the poorest alternatives*'. Because it does not rely on any information about the criteria functions, it is applicable to any MCDA problem with a finite Pareto set. This nicety, comes at rather low cost: with the total of K elements in the Pareto set, the computational complexity of this algorithm is $O(K^2)$. (Note that $K \ll N$, where N is the number of points in the set of the alternatives before any eliminations.)

The key assumption is that each criterion has its relative non-negative weight $w_1,\ldots,w_n$ whose sum is 1. They are reflecting the preferences of the developer of the decision making algorithm. In what follows, the set of weights is regarded as a probability distribution.

The algorithm has K-1 iterations, eliminating at a time one element from the Pareto having the worst value of j-th criterion. The criterion index j is randomly selected with probability w_j. Therefore, more important criteria tend to be chosen more frequently than the less important ones. The process ends when only one element remains in the working copy of the Pareto set. This is the approximation of the balanced, optimal solution to the problem. With K increasing, this approximation converges to the precise optimum.

The scarce discrete subset of the real infinite Pareto set like shown in Fig.3 yields too rough approximation. (Note this is what has remained from the original 3600 points.) Because further increasing the total number of points is not an option, we are using the ε-dominance relation instead of the strict one. This concession can be justified by that the criteria values are calculated with some errors, anyway. As we can guesstimate the standard deviation of these errors, we can choose ε of the same order of magnitude. As the ε-Pareto set will include near-optimal alternatives, it will be much denser. The application of the random elimination in this case would result in much smaller volatility of the solution.

Fig.4 and 5 give the idea of what happens to the situation in Fig.2 and 3 once ε-dominance is applied; the player indeed gets much more options to chose from. The cost for this is a slight deviation from the strict Pareto optimality and a longer, yet not prohibitive, computation time. The benefit is the better robustness of the solution search algorithm.

Assigning weights to the criteria in the proposed framework has a transparent meaning. Unlike using weights to sum up criteria similar to (2), in our method there is no way for that a higher value of one criterion apparently compensates for the insufficient value of the other. So the proposed technique allows easily modeling the continuous spectrum of risk taking and risk aversive attitudes of the decision makers. This is made possible by changing weights.

So far we have been using the example with the weight of *Gain* 0.335, i.e. *Risk* had about twice as much higher weight. This results in the risk-aversive decision shown in Fig. 4 and 5. Player 11 prefers to pass the ball to teammate 10 rather than taking the chance of sending the ball to teammate 9 whose position is much better. By changing the weight in favor of risk taking, it is indeed possible to persuade the player to pass the ball to player 9 (see Fig. 6, 7).

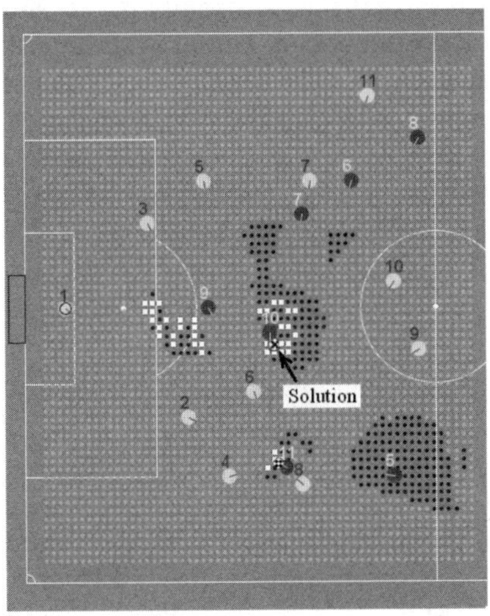

Fig. 4. Situation with the ε-Pareto set. Risky passes are avoided.

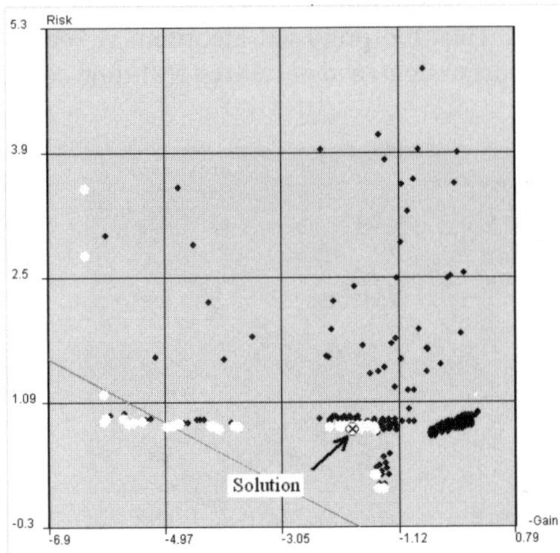

Fig. 5. The ball passing alternatives in the criterion space. The *Gain* weight is 0.335.

5 Addressing the Real-Time Constraints

As described so far, the optimal ball passing decision making algorithm in terms of computations appears to be even more demanding than the algorithms proposed in[1, 2, and 3]. In the first experiments in 2003 with our simulated soccer team *SFUnleashed* we have indeed found that the quality of decisions made by players while passing the ball non-monotonically depended on the total number of points *N*. Starting with small number of point, quality was noticeably increasing with N. Then, with greater N, we observed significantly decreased performance. Indeed, with large N the player process could not complete all required computations during one simulation cycle.

As it should be expected for a real-time system like robotic soccer, attempts to utilize all the player potential by using sophisticated optimization may be counterproductive because of the prohibitive computation time. Still we decided to find a way out so that the real-time constraints were not so restrictive. Our solution comprises two ways for the time reduction.

The first way is further reducing the number of alternatives that wittingly are not in the Pareto set; this can be done by replacing the equidistant grid (Fig.1-5) with randomly scattered points in the vicinity of each teammate (Fig.6, 7).

The second way is automatically adjusting the number of generated points N during run time with respect to the actually available time in the simulation cycle. As we know that the complexity of the whole method is $O(N^2)$, it is always possible to estimate affordable N in advance in the current simulation cycle and thus to prevent real-time constraints from kicking in. Reducing N would result in only gradually increase of the random deviations from the theoretically optimal solutions, without any abrupt losses in the quality of decisions on the average. This behavior is quite different from that of the algorithms based on decision trees whose real-time scalability is very limited. Thus the proposed algorithm is robust by design and is indeed scalable with respect to tightened or relaxed real-time constraints.

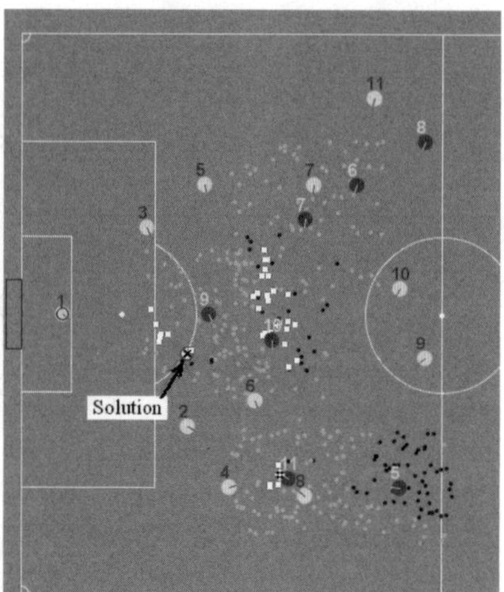

Fig. 6. Situation with 400 points randomly generated about the teammates

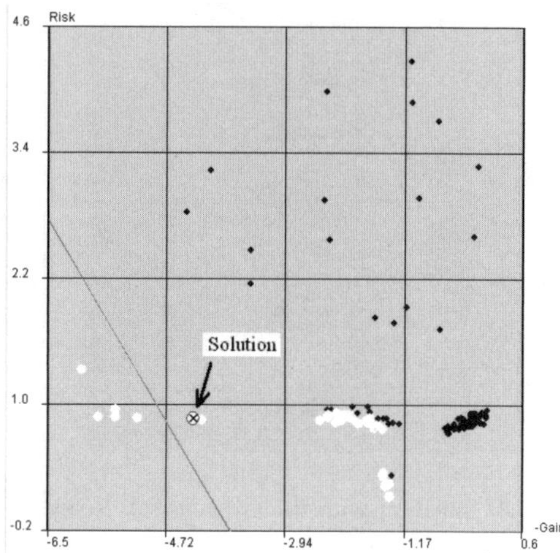

Fig. 7. The ball passing alternatives in the criterion space. The *Gain* weight is 0.614.

The full-blown algorithm is just a straightforward generalization of the simplified method illustrated in the above examples. The only difference is that instead of the two criteria function we are using all ten. The algorithms for computing these criteria have been described in Section 2; some of them are similar to that can be found in the RoboCup literature.

With the exception of particular performance criteria, the proposed optimal decision making framework is general enough to be applicable to a wide range of digital sports games with ball including all RoboCup leagues.

References

1. Stone, P., McAllester, D.: An Architecture for Action Selection in Robotic Soccer. In: Proceedings AGENTS'01, 5th International Conference on Autonomous Agents, May 28-June 1, 2001, Montreal, Quebec, Canada, pp. 316–323 (2001)
2. Reis, L.P., Lau, N.: FC Portugal Team Description: RoboCup 2000 Simulation League Champion. In: Stone, P., Balch, T., Kraetzschmar, G.K. (eds.) RoboCup 2000. LNCS (LNAI), vol. 2019, pp. 29–40. Springer, Heidelberg (2001)
3. Wang, C., Chen, X., Zhao, X., Ju, S.: Design and Implementation of a General Decision-making Model in RoboCup Simulation. International Journal of Advanced Robotic Systems 1(3), 207–212 (2004)
4. Dylla, F., Ferrein, A., Lakemeyer, G., Murray, J., Obst, O., Rofer, T., Stolzenburg, F., Visser, U., Wagner, T.: Towards a League-Independent Qualitative Soccer Theory for RoboCup. In: Nardi, D., Riedmiller, M., Sammut, C., Santos-Victor, J. (eds.) RoboCup 2004. LNCS (LNAI), vol. 3276, pp. 29–40. Springer, Heidelberg (2005)
5. Ehrgott, M.: Multicriteria Optimization, 2nd edn. Springer, New York (2005)
6. Miettinen, K.: Nonlinear Multiobjective Optimization. Kluwer Academic Publishers, New York (1998)

Learning in a High Dimensional Space: Fast Omnidirectional Quadrupedal Locomotion

Matthias Hebbel, Walter Nistico, and Denis Fisseler

Robotics Research Institute
Information Technology Section
Universität Dortmund
Otto-Hahn-Str.8
44227 Dortmund
forename.surname@uni-dortmund.de

Abstract. This paper presents an efficient way to learn fast omnidirectional quadrupedal walking gaits. We show that the common approaches to control the legs can be further improved by allowing more degrees of freedom in the trajectory generation for the legs. To achieve good omnidirectional movements, we suggest to use different parameters for different walk requests and interpolate between them. The approach has been implemented for the Sony Aibo and used by the GermanTeam in the Four-Legged-League in 2005. A standard learning strategy has been adopted, so that the optimization process of a parameter set can be done within one hour, without human intervention. The resulting walk achieved remarkable speeds, both in pure forward walking and in omnidirectional movements.

1 Introduction

Legged robots are advantageous over wheeled robots when the terrain, in which the robot operates, is jagged or uneven. On the other hand, to control the legs of a robot is a highly complex and challenging task because of the many degrees of freedom in moving a leg and the required properties like stability and achievable speed of the walk.

Our research is based on quadruped walking robots, namely the Sony Aibos. In robot soccer the speed and maneuverability of the robots play an important role. Being faster than the opponent gives a team an invaluable advantage because it will in general be faster at the ball and can control it first.

This paper is structured as follows. Before explaining the commonly used approaches to walk, we will briefly introduce the properties of the Sony Aibo ERS-7 robot. Then, in Sect. 2 we will suggest some improvements for the established and commonly used walking model and will subsequently propose in Sect. 3 a learning strategy which can cope with the problem to find optimal walking parameters in the resulting higher dimensional search space. Section 4 reports on the achieved results of both the extended walking model and the experiences made with the learning approach. Section 5 will finally conclude the paper.

G. Lakemeyer et al. (Eds.): RoboCup 2006, LNAI 4434, pp. 314–321, 2007.

1.1 The Experiment Platform

As a robot platform for the presented research, we used the commercially available Sony Aibo ERS-7. The Sony Aibo is a quadruped robot which comes equipped with a CMOS-camera as the most important sensor. Its legs have each 3 degrees of freedom, i.e. a hip abduction, a hip flexion and a knee flexion joint. The Sony Aibo is a truly autonomous robot since all the computation can be done on the on board MIPS IV processor with 576 MHz. For wireless communication the robot is equipped with a WLAN 802.11 compliant ethernet card.

All coordinates mentioned in this text are in the robots' coordinate system and are aligned as follows: the x-axis points to the forward direction, the y-axis points to the left side of the robot and the z-axis points up.

1.2 Related Work

Since the release of the first Sony Aibo model, a lot of research has been made on the walking style of this robot. In 2001 the so called wheel model has been introduced[1], which allows omnidirectional locomotion of the robot by treating the legs as wheels. Any kind of instantaneous movement of a robot can be described by a rotation about a certain point, the so-called *instantaneous center of rotation* (ICR). For walking straight forward without a rotational component, the ICR is located at infinity. The wheel model assumes that the steps of each foot perform a tangential movement on the circle around the ICR. The speed of the step can be calculated in the same way as for a wheel of a differential drive robot.

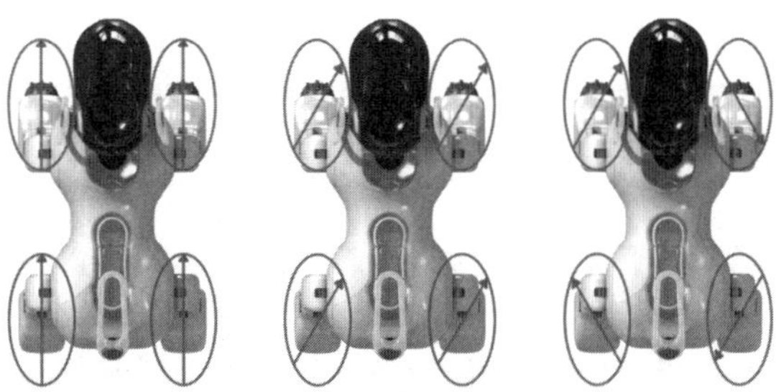

(a) Forward (b) Diagonal (c) Rotation

Fig. 1. The rotation of the leg locus for different walk requests

Like in a trot gait, the two diagonal opposite legs are always moved at the same time, e.g., two legs are always in the air while the other two legs remain in contact with the ground. Due to the duty factor of 0.5 of this walk, it is only dynamically stable, but it has turned out to be stable enough, even in a robot soccer match with a lot of pushing of other robots.

Further, the wheel model assumes, that the feet of the robot move on a certain locus. The according joint angles needed to control the feet on these loci are calculated by means of inverse kinematics. Figure 1 shows the rotation of these loci about the z-axis for different walk requests. The locus for the feet to move on, is described by several parameters[2,3,4]. In 2004 the team from UT Austin Villa used 17 parameters [5], the German Team 14 parameters [4] and the team from the University of Pennsylvania used 19 parameters[6] to define the walk.

Several approaches have been made to achieve faster walks with different loci, the most commonly used ones are rectangular, half-elliptical and trapezoidal loci. To optimize the parameters which describe the form of the loci, many learning approaches have been used[7,8].

These approaches have in common that only a single parameter set has been used for walking, and such set has only been tuned for fast forward walking. But a parameter set for fast forward walking is not automatically useful for fast backwards or sidewards walking.

2 Enhancements to the Walking Model

All known walking engines for the Sony Aibo make only use of static inverse kinematics. This means that the calculation of the desired joint angles to reach a specified position with the foot is only based on geometry and does not take into account physical properties like friction, moments of inertia or forces. While in industrial robotics very complex dynamic models are considered to calculate a trajectory[9], in mobile legged robotics this is not feasible due to the lack of computational power and adequate dynamic models.

2.1 Controlled and Real Walking Trajectories

The dashed paths in Fig. 2 show the controlled loci in the xz-plane for forward walking for the fore and hind legs; the loci in Fig. 2(a) and Fig. 2(b) are controlled on a rectangular path while the loci in Fig. 2(c) and 2(d) are controlled on a half-elliptical path. All other parameters like timing, step lengths etc. do not differ between the rectangular and half-elliptical loci of the same foot. The according real trajectories for the controlled trajectories are presented as solid curves in the Figures 2(a) – 2(d). Especially the locus of the front feet differs remarkably between the rectangular (Fig. 2(a)) and the half-elliptical control (Fig. 2(c)). A reason for this might be a slightly different angle of the robots' body while walking or the slipping of the feet on the ground. After having a look on the real trajectory of the feet in the yz-plane, we found out that especially the paths of the fast walks are bent while we would expect them to be a straight line.

2.2 Parameters Defining the Gait

Due to the observations on the real loci and the fact that especially fast walks differed the most from the controlled loci, we reasoned that more flexibility in the control of the feet could give better results in terms of faster walks. We decided to introduce three dimensional polygons instead of the common "flat" two dimensional shapes.

(a) Fore leg rectangular
(b) Fore leg half-elliptical
(c) Hind leg rectangular
(d) Hind leg half-elliptical

Fig. 2. The controlled locus (dashed) and real locus (solid) of the fore and hind legs for different trajectories

As mentioned in Sect. 1.2, the most common walking approaches use comparatively few parameters to describe the walk; this because hand tuning of walks is not feasible and the time to optimize the parameters with learning approaches also profits from a smaller search space due to the long time to evaluate the resulting speed of a parameter set.

However, we decided to use 3 dimensional polygons with n vertices $P_1, \ldots, P_n$. Additionally n timing parameters are needed to specify the amount of time needed for the foot to travel from vertex P_i to P_{i+1}, for $1 \leq i < n$, and from P_n to P_1. Further we restricted the walking gait to a trot gait, i.e. two diagonal opposite legs are at the same time in the air, while the other pair remains on the ground. The amount of overall parameters for a single leg is $3n + n$. Therefore, the number of parameters describing the walk for a 4 legged robot results to $16n$. With the reasonable constraint, that the pair of fore and the pair of hind legs are moving on mirror-symmetrical loci (but countercyclical in time), the number of parameters is reduced to $8n$.

To be able to find an optimum in the large search space, both a determined learning strategy and a fast and reliable measurement of the robots' speed is advantageous.

2.3 Different Parameter Sets for an Omnidirectional Walk

The ability to reach a desired point in any orientation is called omnidirectionality. In robot soccer, this ability gives the robots the advantage to be able to move quickly in any direction without having to rotate in advance, also it makes the control of the robot on the field much easier. The wheel model described in Sect. 1.2 allows omnidirectional movement by rotating the foot trajectories about the z-axis and scaling the legs' speed according to the radius of the circle around the ICR. Unfortunately a parameter set resulting in a fast forward speed (for example) is not necessarily useful for walking sideways or rotating. The fastest walk which was published on the Aibo ERS-210 in [8] was only useful for straight forward walking with only very small rotational components. Due to this fact the GermanTeam has used this parameter set in 2004 only for straight sprints to the ball[4], while in the "normal" game play they switched back to an omnidirectional parameter set.

Since "hard" switching of parameters while walking normally causes stumbling of the robot which can make him fall over or rotate unintentionally, we decided to implement a "soft" interpolation between parameter sets. In this case, we can use optimized parameters for certain walk requests without having the negative effects of stumbling or unwanted directional changes when changing the walk request. A detailed description of the interpolation is given in [10].

3 Parameter Optimization

As described in Sect. 2.2, the gait depends on a set of parameters $\mathbf{x} := (x_1, x_2, \ldots)$ with $x_i \in \mathbb{R}$. The parameters consist of 3 coordinates per vertex for the n vertices of the polygon and the n timing parameters per leg. The major goal for each parameter set is to reach the highest possible robot speed in the desired direction, i.e. to optimize the speed $\mathbf{v}$ of the robot with respect to the parameters $\mathbf{x} := (x_1, x_2, \ldots)$. Since the speed depends on the parameter set, the maximum speed $\mathbf{v}_k$ is a function of the parameter set $\mathbf{x}_k$, i.e. $\mathbf{v}_k := F(\mathbf{x}_k)$.

3.1 The Learning Strategy

For this challenging optimization process, we used the biologically inspired state-of-the-art $(\mu/\rho \overset{+}{,} \lambda)$ evolution strategy with self-adaption[11]. The evolution strategy operates on populations of individuals $\mathbf{a}$ and is based on the paradigm *survival of the fittest*. A parent population $\mathcal{P}_p^{(t)}$ is creating an offspring population $\mathcal{P}_o^{(t)}$ by making use of the operators *replication*, *mutation* and *recombination*. Due to the *mutation* and *recombination* operators the offspring individuals "differ" from their parents, e.g. they have different properties. The *selection* then decides which individuals will form the new parent generation $\mathcal{P}_p^{(t+1)}$, all other individuals will die out. When the *selection* operation is based on the mentioned paradigm *survival of the fittest*, in the course of the evolution process the properties of the parent generation will be optimized with respect to the *fitness* criteria of the selection operator. For more information about the chosen strategy please refer to [10].

3.2 Learning to Walk

As explained in Sect. 2.2, one requirement for the evolution in this high dimensional search space, is to be able to measure the speed of the robot quickly and precisely without constraining the robots' walk, like e.g. in [8]. For this reason, we developed a ceiling camera system which is mounted above the robot soccer field. The camera is attached to a server which is processing the camera images with a frame rate of 25 Hz. After detecting the robot in the image, the server broadcasts the position of the detected robot into the wireless network.

To let the walk evolution be as autonomous as possible, we developed a behavior which lets the robot walk on the field, always in the observation range of the ceiling cam. To determine the fitness F_k of an individual $\mathbf{a}_k$, we let the robot walk with the appropriate parameter set $\mathbf{x}_k$ and measure the speed. To keep the measurement error small, we allow 2 seconds walking, before starting to measure the speed to be sure, not to take acceleration effects into account. After this "warm up" phase, we take the starting position and after another 2 seconds the achieved position of the robot. By dividing the difference between the two taken points by 2, we get the average speed of the last 2 seconds' walk.

4 Results

4.1 Adaption of the Evolution Strategy

The evolution strategy described in Sect. 3.1 offers a lot of parameters to influence its behavior. In our first learning approaches, we used 8 vertices for each polygon and tried different values for the population sizes. In none of the tests, the speed started to converge and the fastest walk found for straight forward walking achieved only 33 cm/s which was still slower than most of the RoboCup teams were walking on the world championship in 2004.

In the next approach, we decided to reduce the search space by allowing only 4 vertices per polygon, i.e. reducing the search space to 32 dimensions. With this configuration, after a few generations a consistent speed improvement and a convergence of the speed was observable. As long as the population size μ and λ was big enough, the size did not have a big effect on the convergence speed; we chose $\mu = 6$ and $\lambda = 24$. This population size was big enough to explore the search space, so that the chance to get stuck in a local optimum was minimized. With these settings the speed converged already after 30 to 40 generations. The curve in Fig. 3 shows the run of the fitness of all offspring individuals during the evolution process to optimize the walking parameters for forward walking. The individuals in the first parent generation were all equal and hand generated. They all resulted in a forward speed of 280 mm/s. The fastest walk has been found after only 29 generations after less than one hour of training.

4.2 Achieved Speeds

The found parameter sets for all walking directions result in a faster speed than the walks presented on the RoboCup 2004. The maximum reached speeds are shown in Fig. 4.

A special parameter set has been found during the evolution, where we wanted to see, if the walking of the Aibo on the "elbows", like all of the RoboCup teams do, is really the most beneficial walk. For this experiment, we created parameter sets for the initial population, which let the Aibo walk with stretched legs, like a real dog. The resulting walk with these very uncommon starting parameters achieved a speed of 510 mm/s, which is certainly by far the fastest forward walk ever found on a Sony Aibo. But besides its fast speed, the walk had some unfortunate properties like unstableness and a lot of vibrations. These vibrations during the walk result in blurred camera images, thus we only used this walk when we wanted to sprint over a longer distance. The unwanted accelerations of the robot body which result in a shaking of the camera while walking of the 510 mm/s walk, our so called boost, are shown in Fig. 5 in comparison with the very few unwanted accelerations of the normal walk.

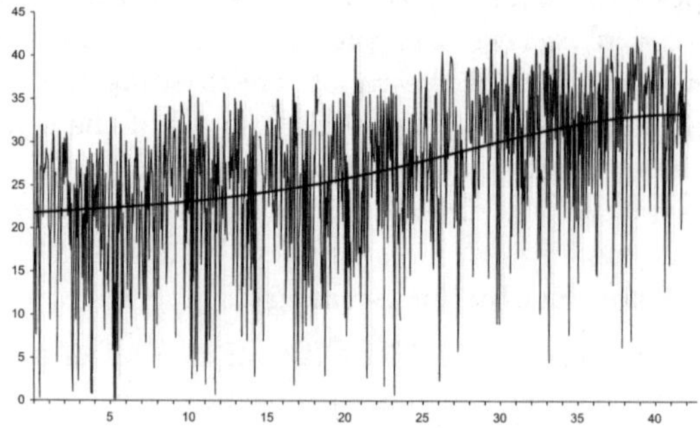

Fig. 3. The fitness of all offspring individuals during an evolution run

forward	backwards	sideways	diagonal	rotation
451 (510)	405	344	421	200

Fig. 4. Achieved speeds in mm/s, respectively $^\circ/s$ for different walk requests

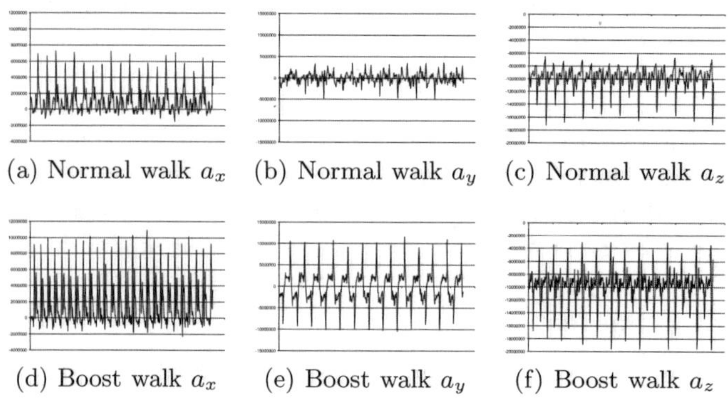

Fig. 5. Accelerations during walking straight forward for a normal walk and the "boost" walk

5 Conclusion

In this paper, we presented meaningful enhancements to existing and commonly used walking models for the Sony Aibo. Due to the fact that the enhanced model has more degrees of freedom to define a walk, a learning approach with external measurement of the fitness has been suggested. The learned walks with the described approach were in all directions more than 25% faster than existing walks, e.g. the walk of the German Team from 2004. The here described walk was also one of the reasons why the GermanTeam has won the RoboCup and was one of the fastest teams on the RoboCup championship 2005 in Japan.

Acknowledgment

The authors would like to thank the members of the GermanTeam for the fruitful cooperation and Microsoft MSDNAA for their financial support which made it possible for our team to participate with several team members in the RoboCup world championship in 2005 in Japan.

References

1. Hengst, B., Ibbotson, D., Pham, S.B., Sammut, C.: Omnidirectional locomotion for quadruped robots. In: Birk, A., Coradeschi, S., Tadokoro, S. (eds.) RoboCup 2001. LNCS (LNAI), vol. 2377, pp. 368–373. Springer, Heidelberg (2002)
2. Düffert, U., Hoffmann, J.: Reliable and precise gait modeling for a quadruped robot. In: Bredenfeld, A., Jacoff, A., Noda, I., Takahashi, Y. (eds.) RoboCup 2005. LNCS (LNAI), vol. 4020, Springer, Heidelberg (2006)
3. Kohl, N., Stone, P.: Policy gradient reinforcement learning for fast quadrupedal locomotion. In: Proceedings of the IEEE International Conference on Robotics and Automation (ICRA), IEEE Computer Society Press, Los Alamitos (2004)
4. Röfer, T., Laue, T., Burkhard, H., Hoffmann, J., Jüngel, M., Göhring, D., Lötzsch, M., Spranger, M., Altmeyer, B., Goetzke, V., von Stryk, O., Brunn, R., Dassler, M., Kunz, M., Risler, M., Stelzer, M., Thomas, D., Uhrig, S., Schwiegelshohn, U., Dahm, I., Hebbel, M., Nistico, W., Schumann, C., Wachter, M.: German Team Report 2004. Technical report, HU Berlin, TU Bremen, TU Darmstadt and University of Dortmund (2004)
5. Stone, P., Dresner, K., Fidelman, P., Jong, N.K., Kohl, N., Kuhlmann, G., Sridharan, M., Stronger, D.: The UT Austin Villa 2004 RoboCup four-legged team: Coming of age. Technical Report UT-AI-TR-04-313, The University of Texas at Austin, Department of Computer Sciences, AI Laboratory (2004)
6. Cohen, D., Ooi, Y.H., Vernaza, P., Lee, D.D.: Robocup 2004 Legged Soccer Team. Technical report, University of Pennsylvania (2004)
7. Kohl, N., Stone, P.: Machine learning for fast quadrupedal locomotion. In: The Nineteenth National Conference on Artificial Intelligence, pp. 611–616 (2004)
8. Röfer, T.: Evolutionary gait-optimization using a fitness function based on proprioception. In: Nardi, D., Riedmiller, M., Sammut, C., Santos-Victor, J. (eds.) RoboCup 2004. LNCS (LNAI), vol. 3276, Springer, Heidelberg (2005)
9. Craig, J.J.: Introduction to Robotics, Mechanics and Control. ch. 6, 7 Addison Wesley, Reading (1986)
10. Dahm, I., Fisseler, D., Hebbel, M., Nistico, W.: Learning fast walking patterns with reliable odometry information for four-legged robots. In: ISMCR'05 (2005)
11. Beyer, H.G., Schwefel, H.P.: Evolution strategies – A comprehensive introduction. Natural Computing 1(1), 3–52 (2002)

A Novel Approach to Efficient Monte-Carlo Localization in RoboCup

Patrick Heinemann, Jürgen Haase, and Andreas Zell

Wilhelm-Schickard-Institute, Department of Computer Architecture,
University of Tübingen, Sand 1, 72076 Tübingen, Germany
{heinemann,jhaase,zell}@informatik.uni-tuebingen.de

Abstract. Recently, efficient self-localization methods have been developed, among which probabilistic Monte-Carlo localization (MCL) is one of the most popular. However, standard MCL algorithms need at least 100 samples to compute an acceptable position estimation. This paper presents a novel approach to MCL that uses an adaptive number of samples that drops down to a single sample if the pose estimation is sufficiently accurate. Experiments show that the method remains in this efficient single sample *tracking mode* for more than 90% of the cycles.

1 Introduction

Self-localization has been a major research task in mobile robotics for several years. Especially in the Middle-Size-League of RoboCup where robots are expected to carry out cooperative tasks, it is crucial to know the robot's position in a common global coordinate system. Many different approaches to the localization task in RoboCup environments have been investigated over the last years. Nearly all of these methods are based on the relative distance and angle of features to the robot in two-dimensional field coordinates.

When walls still surrounded the RoboCup field, some teams used line segments extracted from the distance data of a laser range finder for a very fast localization ([4]). Today, the idea of extracting lines from distance data found its way into several other approaches, now using distances to field markings generated by camera systems. Iocchi *et al.* [6] as well as Marques *et al.* [7], and Jong *et al.* [1] presented algorithms where the lines are extracted using the Hough Transform. Instead of matching the lines to a model, Utz *et al.* [11] computed a distance to the model lines at given positions, to exploit the advantages of Monte-Carlo localization (MCL) [3]. Although it is obvious to extract lines as landmarks in a mostly polygonal environment like RoboCup, algorithms were developed, that were able to handle the raw distance data from the sensors. Probabilistic approaches like Markov localization use sensor data to assess given position estimates. As this is much easier than generating a position hypothesis through feature extraction and model matching, all algorithms based on raw distance data used Markov localization or more precise MCL, which became one of the most popular localization methods. Enderle *et al.* [2] presented an approach using the distance to walls extracted from camera images, while Hundelshausen

G. Lakemeyer et al. (Eds.): RoboCup 2006, LNAI 4434, pp. 322–329, 2007.

et al. [12], Röfer *et al.* [9,10] and Menegatti *et al.* [8] used the distance to the field markings. These algorithms mainly differ in the efficiency of the assessment of position estimates and the number of samples needed for the localization.

The self-localization algorithm presented in this paper is based on line points of the field's line markings. The fitness evaluation of different position estimates is based on a two-dimensional look-up table containing the distance to the next marking line for every position on the field [12,9,10]. To maintain a high accuracy a local search is used to iteratively improve the position estimation, as presented by Hundelshausen *et al.* [12]. In contrast to Hundelshausen *et al.*, however, who use their ideas only for dead-reckoning after an initial global localization, we combine the advantages of MCL with the iterative improvement of the position estimation. This algorithm uses an adaptive number of samples that is reduced to a single sample, if the position estimation of the previous cycle was sufficiently accurate. Experiments show that our algorithm remains in this efficient single sample *tracking mode* for more than 90% of the cycles. Nevertheless, it is still able to cope with the kidnapped robot problem.

The remainder of this paper is organized as follows: Section 2 is a detailed description of the proposed algorithm. Results concerning efficiency and accuracy of the algorithm are presented in section 3 and section 4 concludes the paper with an outlook on the future work.

2 Improved Monte-Carlo Localization

The major steps of the proposed algorithm are shown in Fig. 1 compared to a typical Monte-Carlo localization algorithm.

2.1 Initialization

When the algorithm starts the maximum number of samples N_{max} is generated and randomly distributed over the state space, which in RoboCup consists of the whole playable area of the field. If there is previous knowledge of the robot's pose, this knowledge can be represented by a different non-random initialization.

2.2 Application of the Motion Model

In the motion model the odometry information from the robot is incorporated. First, the samples of the set S are translated and rotated according to the observed motion a. Then a random gaussian noise is added proportional to the motion. In the tracking mode, i.e. only a single sample is used, this step only consists of translating and rotating the pose of the single sample.

2.3 Evaluation of the Sensor Model

The proposed algorithm uses the marking lines on a RoboCup soccer field as features for the self-localization. Several pixels in an omnidirectional camera image are identified as marking line points as shown by Heinemann *et al.* [5].

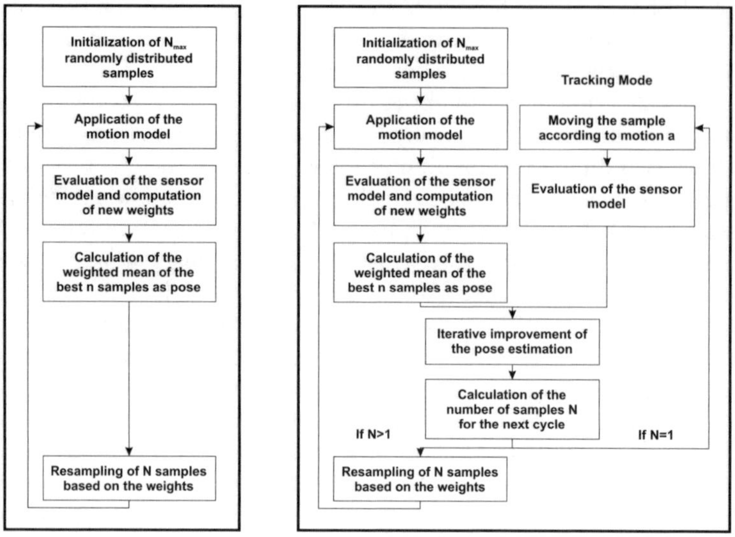

Fig. 1. The major steps of a typical Monte-Carlo localization algorithm (left) and our improved Monte-Carlo localization algorithm (right)

These pixels transformed into robot centered coordinates (x_j, y_j) serve as input for the sensor model introduced in this section.

For an efficient evaluation of the sensor model Röfer *et al.* [9,10] and Hundelshausen *et al.* [12] presented an idea of transforming the line points to the pose $l_i = (x_i, y_i, \theta_i)$ of the sample s_i, such that the coordinate system of the line points is located at position (x_i, y_i) and oriented according to θ_i. Denoting the new location of the line points as $(x_{i,j}, y_{i,j})$ and the vector from these points to their nearest model line in a model of the field as

$$f_{i,j} = (x_{m,j} - x_{i,j}, y_{m,j} - y_{i,j}), \tag{1}$$

an overall distance $D_{L,i}$ per sample can be calculated by summing over the squared distances to the model as

$$D_{L,i} = \frac{1}{j} \sum_j ||f_{i,j}||^2. \tag{2}$$

As these distances only depend on the position on the field they can be precomputed on a discrete grid and easily stored in a two-dimensional look-up table (*distance matrix*). In contrast to the original methods the proposed algorithm uses the squared distances to let line points with a higher distance to the next model line have an even greater influence than line points that are almost perfectly matched. As $D_{L,i}$ is only based on the symmetric marking lines on a RoboCup soccer field, it would be the same for at least two poses in each cycle. Thus, to resolve the symmetry the angle to the two differently coloured goals was introduced as an extra feature. From the omnidirectional image the angles $\widehat{\phi}_{1,i}$ and $\widehat{\phi}_{2,i}$ to the two goals are extracted. Comparing these angles with the expected angles at the pose of a sample $\phi_{1,i}$ and $\phi_{2,i}$ results in a goal distance

$$D_{G,i} = \left(\left\| \widehat{\phi}_1 - \phi_{1,i} \right\| + \left\| \widehat{\phi}_2 - \phi_{2,i} \right\| \right)^2 , \tag{3}$$

where $\|\cdot\|$ is the absolute value of the smaller angle difference accounting for the 2π period of angles. Again this distance is squared to let higher angular differences have a greater influence. The total distance value is computed as

$$D_i = (1 - \lambda) D_{L,i} + \lambda D_{G,i}, \tag{4}$$

with $\lambda \in [0, 1]$ representing the balance of the two distance terms, and finally, the weights are updated as

$$w_{i,t} = \alpha \frac{1}{D_i} , \tag{5}$$

with α such that $\sum_i w_i = 1$. In the tracking mode the distances are only computed for the calculation of the number of samples used in the next step N_{t+1}.

2.4 Iterative Improvement of the Pose Estimation

A preliminary pose estimation is calculated as weighted mean over all samples

$$\widehat{p} = (x, y, \theta) = \sum_n w_n l_n, \tag{6}$$

and is used as starting pose for the iterative improvement. In the tracking mode $\widehat{p}$ is the pose of the single sample.

In addition to the *distance matrix* Hundelshausen *et al.* [12] proposed a dead-reckoning approach for self-localization. By applying forces exerted on the transformed line points by the model lines an estimated position (x, y) is iteratively improved in both directions. Using the same forces a torque is computed which iteratively improves the orientation θ. Again, these forces can be precomputed and stored in a look-up table (*force matrix*). Here we use the force matrix to improve the pose estimation $\widehat{p}$ from the MCL in a number of iterations k. It contains the two-dimensional vectors $f_{i,j}$ from equation (1) that can be interpreted as a force exerted by the nearest model line proportional to the distance. A mean force acting on the pose estimation $\widehat{p}$ can be computed as

$$F = \frac{1}{j} \sum_j f_{i,j}. \tag{7}$$

A fraction of this force can be added to the pose estimation $\widehat{p}$ in each iteration to improve it regarding the position. In contrast to Hundelshausen *et al.* we compute a mean torque according to the estimated pose to improve the orientation estimation. It is computed over all line points as

$$M = \frac{1}{j} \sum_j (x_{i,j}, y_{i,j}) \times f_{i,j}. \tag{8}$$

Thus, in each iteration k a new pose estimation $\widehat{p}_k = (x_k, y_k, \theta_k)$ is generated by

$$(x_k, y_k) = (x_{k-1}, y_{k-1}) + \mu F \tag{9}$$
$$\theta_k = \theta_{k-1} + \nu M, \tag{10}$$

starting with the preliminary estimation

$$(x_0, y_0, \theta_0) = \widehat{p}. \tag{11}$$

The iterations can be seen as a local search that minimizes F and M and thus stabilizes the pose estimation by removing the noise from the weighted mean when $N_t > 1$ and reducing the tracking errors when $N_t = 1$. The search continues until a maximum number of iterations k_{max} is reached or the improvement between the iterations was too low. The final pose estimation p is the pose resulting from the last iteration. Please note that apart from inserting the improved pose estimation p into the sample set S_{t+1} the stochastic process of the Monte-Carlo Localization is not influenced by the iterative improvement.

2.5 Calculation of the Number of Samples

The number of samples needed for the next cycle is calculated depending on the distance D of the final pose estimation p according to equation (4) as

$$N_{t+1} = \begin{cases} N_{\max} & : \text{ if } \xi D \geq N_{\max} \\ \xi D & : \text{ if } 1 < \xi D < N_{\max} \\ 1 & : \text{ if } \xi D \leq 1 \end{cases}, \tag{12}$$

where ξ is a factor that controls how fast the number of samples n is reduced.

2.6 Resampling

If $N_t > 1$ the cycle ends with an importance resampling from the set of samples S with probability $w_{i,t}$ for resampling an old sample $s_{i,t}$. The sampling continues until the number of samples N_{t+1} for the next cycle was reached. To represent the improved pose information p in the sample set, this pose is inserted as new sample into the sample set S_{t+1} for the next cycle.

3 Results

This section presents results obtained by two experiments made in our robot lab on a half field of $7m$ width and $4m$ length. Throughout this section positions and orientations are given in meters and radians, respectively. In all experiments presented in this section we used the parameters given in table 1. The maximum number of samples N_{max} used was chosen such that it is comparable to a standard MCL approach with a fixed number of samples.

In a first experiment we compared the localization algorithm with a fixed number of samples $N = 200$, $N = 100$, $N = 50$ and the proposed method. As a

Table 1. Parameter set used for the experiments

Nmax	λ	ξ	μ	ν	k_{max}
200	0.1	2500	0.001	0.0003	20

database for the comparison we located the robot at a pose $p_1 = (1.04, 1.07, 2.1)$ and stored the detected line points of 98 images from the omnidirectional camera system. Afterwards, the robot was relocated to pose $p_2 = (1.64, 2.68, 0.0)$ where the line points from another 98 images were stored. The line points were used as input to 196 cycles of the localization algorithm without any odometry information, resulting in a kidnapped robot problem. Fig. 2 shows the estimation error of the algorithms. Independent of N the algorithms compute a very good pose estimation after at most 15 cycles, where the number of samples in the adaptive method drops to $N = 1$ in only 6 cycles. From cycle 20 to 90 the estimation error and the number of samples stay at the same level. In cycle 99 where the relocation of the robot happened the proposed algorithm immediately generates $N = N_{\text{max}}$ samples, reacting to the high estimation error. Apart from the algorithm with $N = 50$ samples all methods regenerate a good pose estimation after at most 10 cycles, whereas the number of samples in the adaptive method returns to $N = 1$ after 8 cycles, thus using only a single sample in 92.87% of the cycles. A fixed number of $N = 50$ samples without the iterative improvement is not able to handle the kidnapped robot problem in this case, as the estimation error does not recover after the relocation in cycle 98. Although the other three algorithms show comparable results concerning the estimation error, the proposed algorithm performes much better if the computation time is considered. Table 2 lists the mean computation time and estimation error.

With the second experiment we show that the method is also able to correctly track the pose of a moving robot. As a ground truth we used a laser scanner to record the true position of the robot. The robot was then manually controlled around the field. In the first run the mean speed was at 1 m/s, in the second run we raised the speed to 2 m/s. To show that the algorithm works for both differential drive and omnidirectional systems we first controlled the robot like a

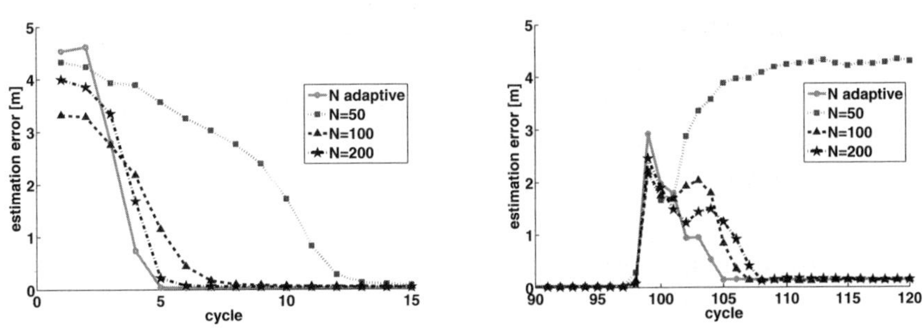

Fig. 2. Comparison of our algorithm to MCL with fixed numbers of samples

Table 2. Results of 196 cycles using different numbers of particles N

	N adaptive	$N = 50$	$N = 100$	$N = 200$
mean time	1.7632ms	3.6426ms	6.8223ms	14.2508ms
mean error	0.1936m	2.2515m	0.2075m	0.2057m

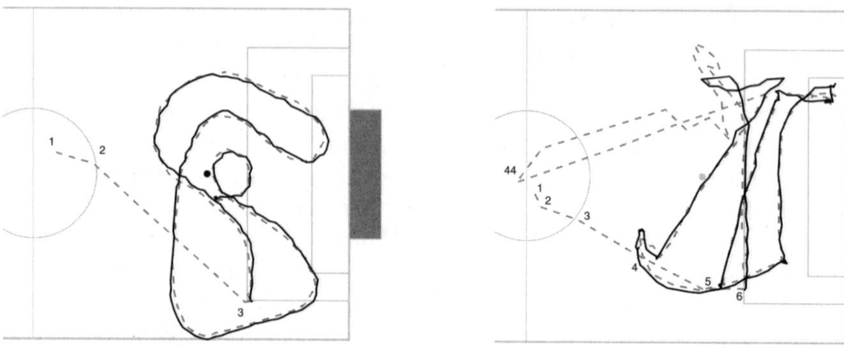

Fig. 3. This figure compares the position estimates of the proposed algorithm (dashed grey line) to the ground truth (solid black line). The algorithm needs up to 6 cycles to find the initial pose and then correctly tracks the robot. In the second run, the algorithm fails to track the position once, but relocalizes only a few cycles later.

differential drive and then the orientation was nearly fix in the second run. Fig. 3 (left) shows the results of the first run. In the beginning the weighted mean over the rndomly distributed samples results in a pose near the center of the field. After 3 cycles the starting pose of the robot is correctly estimated. Throughout the rest of the 212 cycles the estimated pose follows the path on the field with a mean accuracy of 9.89cm, using only a single sample in 97.17% of the cycles. The mean computation time for a cycle of the algorithm in this experiment was 1.5731ms on an Athlon XP 1800+ system. In the second run (Fig. 3, right) the algorithm needed 6 cycles to correctly estimate the starting position. The estimated position then follows the real position until the robot changes its direction very quickly two times in a row. Here the algorithm temporarily looses the track of the robot and distributes a higher number of samples (cycle 44). Thus, the mean accuracy without the initialization cycles was 23.97cm and the mean computation time increased to 4.32ms as only 90.64% of the cycles used the tracking mode.

4 Conclusion and Future Work

This paper presents an efficient combination of global Monte-Carlo localization with an adaptive number of samples and local position tracking. With a fast estimation of the samples' fitness and a local search for iterative improvement of the estimated pose the number of samples was reduced to a single sample resulting in a smooth transition between global localization and local pose tracking. We showed that the algorithm was able to handle the kidnapped robot problem and to track a moving robot. In all experiments the mean cycle time of the algorithm was leaving enough time for other important tasks like object detection and planning to be done in real-time.

References

1. de Jong, F., Caarls, J., Bartelds, R., Jonker, P.: A Two-Tiered Approach to Self-Localization. In: Birk, A., Coradeschi, S., Tadokoro, S. (eds.) RoboCup 2001. LNCS (LNAI), vol. 2377, pp. 405–410. Springer, Heidelberg (2002)
2. Enderle, S., Ritter, M., Fox, D., Sablatnög, S., Kraetzschmar, G., Palm, G.: Vision-based Localization in RoboCup Environments. In: Stone, P., Balch, T., Kraetzschmar, G.K. (eds.) RoboCup 2000. LNCS (LNAI), vol. 2019, pp. 291–296. Springer, Heidelberg (2001)
3. Fox, D., Burgard, W., Dellaert, F., Thrun, S.: Monte Carlo Localization: Efficient Position Estimation for Mobile Robots. In: Proceedings of the National Conference on Artificial Intelligence, pp. 343–349 (1999)
4. Gutmann, J., Weigel, T., Nebel, B.: Fast, Accurate, and Robust Self-Localization in Polygonal Environments. In: Proceedings of the IEEE/RSJ International Conference on Intelligent Robots and Systems (IROS '99) (1999)
5. Heinemann, P., Rückstieß, T., Zell, A.: Fast and Accurate Environment Modelling using Omnidirectional Vision. In: Dynamic Perception 2004, Infix (2004)
6. Iocchi, L., Nardi, D.: Self-Localization in the RoboCup Environment. In: Veloso, M.M., Pagello, E., Kitano, H. (eds.) RoboCup-99: Robot Soccer World Cup III. LNCS (LNAI), vol. 1856, pp. 318–330. Springer, Heidelberg (2000)
7. Marques, C., Lima, P.: A Localization Method for a Soccer Robot Using a Vision-Based Omni-Directional Sensor. In: Proceedings of EuRoboCup Workshop 2000 (2000)
8. Menegatti, E., Pretto, A., Pagello, E.: A New Omnidirectional Vision Sensor for Monte-Carlo Localization. In: Nardi, D., Riedmiller, M., Sammut, C., Santos-Victor, J. (eds.) RoboCup 2004. LNCS (LNAI), vol. 3276, pp. 97–109. Springer, Heidelberg (2005)
9. Röfer, T., Jüngel, M.: Vision-Based Fast and Reactive Monte-Carlo Localization. In: Proceedings of the 2003 IEEE International Conference on Robotics & Automation, pp. 856–861. IEEE Computer Society Press, Los Alamitos (2003)
10. Röfer, T., Jüngel, M.: Fast and Robust Edge-Based Localization in the Sony Four-Legged Robot League. In: RoboCup-2003: Robot Soccer World Cup VII. LNCS, vol. 3020, pp. 262–273. Springer, Heidelberg (2004)
11. Utz, H., Neubeck, A., Mayer, G., Kraetzschmar, G.: Improving Vision-Based Self-localization. In: Kaminka, G.A., Lima, P.U., Rojas, R. (eds.) RoboCup 2002. LNCS (LNAI), vol. 2752, pp. 25–40. Springer, Heidelberg (2003)
12. von Hundelshausen, F., Schreiber, M., Wiesel, F., Liers, A., Rojas, R.: MATRIX: A force field pattern matching method for mobile robots. Technical Report B-08-03, Free University of Berlin (2003)

Representing Spatial Activities by Spatially Contextualised Motion Patterns

Björn Gottfried and Jörn Witte

Artificial Intelligence Group, TZI, University of Bremen, Germany
{bg,jwitte}@tzi.de

Abstract. The interpretation of spatial activities plays a fundamental role in several areas, ranging from the analysis of animal behaviour to location-based assistance applications. One important aspect when interpreting spatial activities consists in relating them to their environment. A problem arises insofar propositional representations lack an appropriate attention mechanism to comprehend the spatiotemporal development of spatial activities. Therefore, we propose a diagrammatic formalism which allows spatial activities to get classified depending on their spatial context and provide a link to propositional formalisms. It shows that RoboCup soccer is particularly suitable for investigating these issues. In fact, alone the spatial activity of the ball teaches us to a considerable degree much about a game.

1 The Representational Problem of Spatial Activities

Making explicit spatial activities of animals and human beings, activity patterns play a fundamental role in several areas: patterns of animal movements are investigated providing a detailed picture of seasonal variability in the scale and patterns of movements [2]; location-aware shopping guides help people to find an efficient way through an unfamiliar shopping mall [1]. In these examples, the spatial activities of man and beast are analysed and interpreted.

Several approaches allow spatial activities to be represented, such as the calculus of events [4] and the situation calculus [6]. The intuition behind such approaches is that the world can be described in terms of situations and that the world changes from one situation to another one by performing specific actions. It is the axiomatic specification of situations and actions what characterises these approaches. Thus, their strengths consist in providing sound logical foundations for reasoning about actions and time. However, such logical languages fail to adequately represent how location related activities unfold themselves over time. The problem of representing activities in which objects change their location, so that their spatiotemporal development becomes explicit, is referred to as the *representational problem of spatial activities*. Solving this problem allows the spatial activities of objects to be valued, e.g. to distinguish whether someone moves around purposeful or purposeless and to determine the similarity of movements.

In this paper, we devote our attention to the representational problem of spatial activities. In section 2 a formalism is provided which complements logical languages about action and time in that it allows the spatial realisation of

G. Lakemeyer et al. (Eds.): RoboCup 2006, LNAI 4434, pp. 330–337, 2007.

activities to be dealt with more directly. Section 3 introduces a diagrammatic formalism using which frequency distributions about spatial activities can be derived. In order to investigate the representational problem of spatial activities it is necessary to clarify problems in a well defined testing environment, as it is provided by RoboCup soccer. In this way, it shows how RoboCup sticks to its principles to foster AI and intelligent robotics research, namely by illustrating representational issues on spatial activities. Section 4 shows how our diagrammatic formalism acts in concert with a logical language, allowing conclusions about matches to be derived on the basis of the spatial activity of the ball.

2 Contextualising Motion Patterns

A soccer game can be conceived of as a sequence of things happening: players dribble, they pass each other the ball, somebody else tries to stop the ball which is flying towards the goal. What is important for any soccer event is the spatial change of objects (players and ball). That is to say, for the purpose of characterising the behaviour of objects the only change that matters concerns spatial changes which correspond to patterns of changing positions. Connecting those positions we obtain trajectories of the objects. Parts of them correspond to specific events, and each event is described by a trajectory.

Furthermore, the pattern of changing positions which arises from the intention of a soccer player is called an activity pattern. More general, any pattern of some intentional or unintentional event is called a motion pattern:

Definition 1 (Motion pattern). *$\boldsymbol{T}$ denotes the infinite set of realisable trajectories, and $\boldsymbol{M}$ is a set of representatives of a partition of $\boldsymbol{T}$. Then, each sequence $M = m_1, ..., m_k$ with $m_i \in \boldsymbol{M}$ is called a motion pattern.*

For the most precise set of motion patterns it holds that there exists a bijective mapping between $\boldsymbol{T}$ and $\boldsymbol{M}$. However, such a set of motion patterns requires the consideration of infinite many cases. But we shall learn below that a small set of coarse motion patterns, consisting only of a few distinguishable cases, suffice for the representation in some cases. What will be our running example shows even the simplest case: it is only distinguished whether an object is motionless or whether it moves. Then, there are two equivalence classes: one is represented by the null-trajectory, referred to as $\mathsf{motionlessness}$, the other one, called motion, contains all other trajectories, i.e. $\boldsymbol{M} = \{\mathsf{motion}, \mathsf{motionlessness}\}$. A possible motion pattern is $M = (\mathsf{motion}, \mathsf{motionlessness}, \mathsf{motion}, \mathsf{motion})$, denoting any case in which an object moves somehow, keeps still, and moves again two times (the latter being equal to a single motion event). Eventually, the simplest motion patterns consist only of single elements of $\boldsymbol{M}$.

Much effort has been put into the development of methods for describing motion patterns, e.g. [7]. Though, it is frequently not the motion pattern itself that determines its meaning alone. Rather, changing the context in which a motion pattern occurs causes a semantic change. Taking into account its spatial context means to consider the motion pattern's environment. For instance, at the level of topology its spatial context extends the meaning of a motion pattern as follows:

Definition 2 (Topologically contextualised motion pattern). *$\boldsymbol{M}$ denotes a finite set of motion patterns, $\boldsymbol{R}$ denotes a finite set of regions, and $\boldsymbol{S}_t$ denotes a relation $\boldsymbol{S}_t \subseteq \boldsymbol{R} \times \boldsymbol{M}$. Then, each sequence $S = s_1, s_2, ..., s_n$ with $s_i \in \boldsymbol{S}_t$ is called a topologically contextualised motion pattern.*

$\boldsymbol{S}_t$ allows dependencies between motion patterns and regions to be represented, that is, the space where a motion pattern occurs is taken into account. For example, $\boldsymbol{R} = \{$leftHalf, rightHalf, lPenalty, rPenalty$\}$ and $\boldsymbol{M} = \{$motion, motionlessness$\}$. Then, $\boldsymbol{S}_t = \{$(leftHalf, motion), (rightHalf, motion), (lPenalty, motion), ... $\}$ and for the activity pattern on the left hand side of Fig. 1 it holds that $S =$ ((leftHalf, motion), (lPenalty, motion), (leftHalf, motion)).

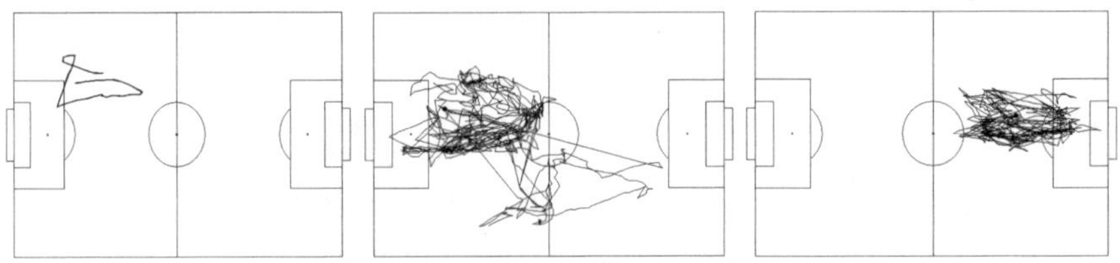

Fig. 1. Some trajectories, defenders of Apollo (middle) and AT-Humboldt (right)

As with the topological case, other spatial relations can also be considered, such as cardinal directions or distances between objects. We define accordingly and more general:

Definition 3 (Spatially contextualised motion pattern). *$\boldsymbol{M}$ denotes a finite set of motion patterns, $\boldsymbol{O}$ denotes a finite set of objects, and $\boldsymbol{S}$ denotes a relation $\boldsymbol{S} \subseteq \boldsymbol{O} \times \boldsymbol{M}$. Then, each sequence $S = s_1, s_2, ..., s_n$ with $s_i \in \boldsymbol{S}$ is called a spatially contextualised motion pattern.*

$\boldsymbol{O}$ denotes any spatial concept, such as cardinal directions or a specific set of landmarks. Any combination can be represented by $\boldsymbol{O}$, too, for instance, landmarks which are in a specific distance away from a motion pattern. However, here we shall exemplify the idea behind spatial contextualisations for motion patterns at the level of topology and confine the discussion on whether a motion pattern is (partly or completely) contained in a specific region.

The part of the trajectory t corresponding to any motion pattern s_k can be determined in different ways, amongst others, by the length of specific time intervals or by the spatial contextualisation itself. An example for the latter is the partition of t into parts at the boundaries of regions, i.e. whenever t crosses the boundary of a region a new part of the motion pattern is defined. Taking the example on the left hand side of Fig. 1, t is divided up into three parts since the boundary of the left penalty area is crossed twice. However, in the simplest case t would not be divided up into parts, and there would exist only a single spatially contextualised motion pattern. In our example $\boldsymbol{M}$ is defined as before and we will simply distinguish whether an object crosses specific regions or not.

3 A Diagrammatic Representation of Motion Patterns

After a conceptual framework for describing spatial activities has been introduced, we are able to solve the representational problem of spatial activities. The problem consists in representing the location of an object and its change in location, so as to make explicit its spatial activity. Solving this problem for topologically contextualised motion patterns amounts to project the trajectory of the object's spatial activity on its spatial environment. In this way, the topological context of the object's activity is made explicit. Such a *diagrammatic representation* allows model-based deductions, namely those which can be obtained by inspection processes [5]. For instance, it can be read off the diagram, to which we refer to as $\mathcal{D}$, which regions are crossed by the trajectory. The left hand side of Fig. 1 gives an example.

Definition 4 (Spatially contextualised diagram). *Let $t \in \boldsymbol{T}$ denote a trajectory and R the arrangement of a set of regions $\boldsymbol{R}$. Then, a spatially contextualised diagram $\mathcal{D}$ of motion patterns is obtained by projecting t onto R.*

$\mathcal{D}$ explicitly depicts the topologically contextualised spatial activity of the object, i.e. such a diagrammatic representation is only capable of representing a specific model. On the other hand, only valid models can be constructed, whereas propositional languages would allow invalid mappings to be considered, such as impossible changes of the object's positions. This is one of the most prominent advantages of diagrammatic representations (cf. [5]).

Accordingly to [3] we shall make use of graphical constructions for the purpose of model-based deductions. The graphical constructions we use are closely related to Euler diagrams which use topological properties such as enclosure and exclusion to illustrate set-theoretic notions of containment and disjointness, respectively. The advantages of such diagrams derive from their *built-in logics.* With these diagrams, we shall deduce the extent to which an object occupies specific regions in order to determine the object's activities regarding its environment. This is done by spatial templates which represent specific regions, r_k. Such a template is mapped onto $\mathcal{D}$ and an and-operation determines whether t crosses r_k. Especially, this operation can be extended in order to obtain the amount with which the object occupies r_k during its activities. By this means, a frequency distribution of the object's motion pattern is obtained, showing those regions which are most frequently visited by the object.

Definition 5 (Spatial template). *A spatial template $\mathcal{T}$ is a binary image depicting a specific, not necessarily connected region.*

For the purpose of applying $\mathcal{T}$ to $\mathcal{D}$, which is denoted by $\mathcal{T} \otimes \mathcal{D}$, both diagrams have either to be equal in size or it has to be defined how $\mathcal{T}$ is to be mapped onto $\mathcal{D}$. However, at any position a logical and-operation is applied which results into true if both diagrams are marked at that same position; note that a position in $\mathcal{D}$ can be marked k times depending on how often the trajectory crosses that position. Counting these positions it is determined to which extent t covers the region represented by $\mathcal{T}$. The resulting diagram marks those positions and can recursively be applied to further diagrams.

Definition 6 (Set-theoretic diagrammatic operations). *Let $\mathcal{R}$, $\mathcal{S}$, and $\mathcal{T}$ denote diagrams which are equal in size. Then, the following operations are defined in accordance to the set-theoretic operations:*

1. $\mathcal{R} = \mathcal{S} \otimes \mathcal{T}$*, $\mathcal{R}$ is marked wherever $\mathcal{S}$ and $\mathcal{T}$ are both marked (intersection);*
2. $\mathcal{R} = \mathcal{S} \oplus \mathcal{T}$*, $\mathcal{R}$ is marked where either $\mathcal{S}$ or $\mathcal{T}$ is marked (union);*
3. $\overline{\mathcal{R}}$*, $\mathcal{R}$ is marked wherever it was not marked before (complement).*

While the diagram explicitly represents information about topological and other geometric relations among the players and the pitch, in a sentential representation, such as in the situation calculus, these information is available only implicitly. Clearly, the diagrammatic representation is informationally equivalent to logical representations, but there are advantages concerning the computational efficiency. A trajectory mapped onto the regions of the pitch allows several questions about the spatial activity of a player directly to be answered: does he keep to his area, or where does he move? The example in the middle of Fig. 1 shows how a defender allows himself his position to left, while the defender of the other team (right same Fig.) does not allow himself to do this.

The computational advantage is due to the diagrammatic representation of the topologically contextualised activity patterns of the players. Such a representation guides attention, allowing the *player's area* as well as deviations from it to get inferred. By contrast, a pure logical representation requires to thoroughly search through all positions in order to determine the player's area and to recognise any deviations from it, lacking an appropriate attention mechanism. By means of predefined spatial templates, attention processes of visual inspection routines are simulated to comprehend the activities of players. It is the attention mechanism of any representation which determines its search strategy and which frequently makes diagrammatic representations computationally more efficient than sentential representations [5].

4 Interpreting Motion Patterns

In this section we will show how to derive qualitative information from spatial activities. Especially, we analysed games of the RoboCup 2005 2d-simulation league in Osaka. Exemplarily, we analyse the trajectory of the ball mapped onto the regions of a pitch (topological contextualisation), in order to generate answers concerning game and team behaviour. For this purpose, we combine logical with

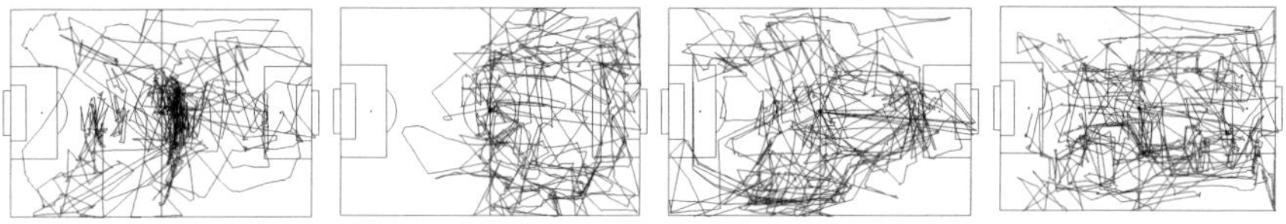

Fig. 2. Three most left: The trajectory of the ball during some games in the preliminary round; Right: The trajectory of the ball during the game *RoboSina vs. TsinghuAeolus*

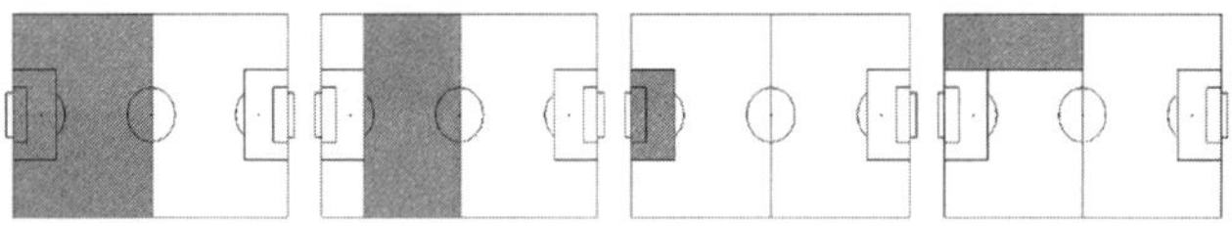

Fig. 3. Spatial templates used in our example (symmetrical regions are also used)

diagrammatic reasoning methods. Fig. 2, for example, shows a game which is quite centred, another one which is asymmetrical, and a third one showing a good wing play of one of the teams while the other team has a good defence.

In detail, we define a set of predicates (e.g. LEFTTEAM and MOTION, etc.) which can be expressed in first order predicate logic to allow spatial activities of soccer games to be formalised. Based on the particular predicate MOTION we provide a link to diagrammatic reasoning algorithms. Thus, the domain of this predicate will be defined in reference to the given spatial templates representing specific regions of a pitch (cf. Definition 5). While a spatial template determines to which degree a trajectory crosses a specific region, the MOTION predicate only distinguishes whether the region is significantly crossed by the trajectory or not. The significance is determined by the domain expert who, for instance, defines that MOTION holds as soon as the region is covered by the trajectory at least to a degree of 20%. Our example is based on a set of spatial templates which are shown in Fig. 3. These templates have to be defined by the domain expert, too.

Given terminological knowledge in terms of inference rules $\varepsilon_1, ..., \varepsilon_3$ which are based on a structure $\mathcal{A} = (U_A, I_A)$:

U_A = $\boldsymbol{A} \cup \boldsymbol{B} \cup \boldsymbol{R}$, a universe
$\boldsymbol{A}$ = {`ball`}, available trajectories
$\boldsymbol{R}$ = {`full, lPenalty, rPenalty, leftHalf, left18Yard, rightHalf, right18Yard, lTopCorner, lBotCorner, rTopCorner, rBotCorner`}, a finite set of regions
$\boldsymbol{B}$ = {`RoboSina, TsinghuAeolus`}, a set of teams

I_A(MOTION) $= \{(m, n) | m \in \boldsymbol{A}$ und $n \in \boldsymbol{R}\}$
I_A(LEFTTEAM) $= \{m | m \in \boldsymbol{B}\}$
...

A set of inference rules $\varepsilon_1, ..., \varepsilon_3$:

ε_1 : $\textsc{Good\_Defence}(team) \Rightarrow$
$\quad \textsc{LeftTeam}(team) \wedge \textsc{Motion}(\texttt{ball}, \texttt{left18Yard}) \wedge \neg\textsc{Motion}(\texttt{ball}, \texttt{lPenalty})$

ε_2 : $\textsc{Defending\_Deep}(team) \Rightarrow \textsc{LeftTeam}(team)$
$\quad \wedge(\textsc{Motion}(\texttt{ball}, \texttt{leftHalf}) \vee \textsc{Motion}(\texttt{ball}, \texttt{left18Yard}))$
$\quad \wedge(\neg\textsc{Motion}(\texttt{ball}, \texttt{rightHalf}) \vee \neg\textsc{Motion}(\texttt{ball}, \texttt{right18Yard}))$

ε_3 : $\textsc{Balanced\_Wing\_Playing}(team) \Rightarrow$
$\quad \textsc{LeftTeam}(team) \wedge\textsc{Motion}(\texttt{ball}, \texttt{rTopCorner}) \wedge\textsc{Motion}(\texttt{ball}, \texttt{rBotCorner})$

Using this logical representation we can use an abductive reasoning approach to make inferences on the basis of observed motion patterns, and consequently, to find possible reasons for the team behaviour.

Firstly, if we want to receive an affirmative answer for a specific question such as *'Does the RoboSina team has a good defence?'*, then it will be necessary to analyse all inference rules containing the $\textsc{Good\_Defence}(\texttt{RoboSina})$ predicate within the head of the rule, and try to prove these rules with the use of a backward-chaining algorithm. This process results in some atomic sentences which are to be proved within a diagrammatic inference algorithm (e.g. $\textsc{Motion}(\texttt{ball}, \texttt{left18Yard})$ and $\neg\textsc{Motion}(\texttt{ball}, \texttt{lPenalty})$).

Secondly, sentences given as the result of a diagrammatic reasoning process allow a game to be interpreted in several ways. For example, we take a look at the game *RoboSina vs. TsinghuAeolus* (see Fig. 4). In this case, a diagrammatic inference algorithm derives the following assertions:

$\neg\textsc{Motion}(\texttt{ball}, \texttt{full})$, $\neg\textsc{Motion}(\texttt{ball}, \texttt{leftHalf})$, $\textsc{Motion}(\texttt{ball}, \texttt{rightHalf})$,
$\textsc{Motion}(\texttt{ball}, \texttt{left18Yard})$, $\textsc{Motion}(\texttt{ball}, \texttt{right18Yard})$,
$\neg\textsc{Motion}(\texttt{ball}, \texttt{lTopCorner})$, $\textsc{Motion}(\texttt{ball}, \texttt{lBotCorner})$,
$\textsc{Motion}(\texttt{ball}, \texttt{rTopCorner})$, $\textsc{Motion}(\texttt{ball}, \texttt{rBotCorner})$,
$\neg\textsc{Motion}(\texttt{ball}, \texttt{lPenalty})$, $\textsc{Motion}(\texttt{ball}, \texttt{rPenalty})$

On the basis of these assertions, our inference rules, and additional knowledge such as $\textsc{LeftTeam}(\texttt{RoboSina})$ we are able to infer $\textsc{Good\_Defence}(\texttt{RoboSina})$ which means that the defending team is under great pressure but yet able to keep the ball out of the penalty area.

Having analysed all games of the RoboCup 2005 2d-simulation league, it shows that the trajectory of the ball provides indeed significant information about a game, identifying the proposed technique as a useful means for automatically evaluating games. Taking alone the predicate $\textsc{Balanced\_Wing\_Playing}$ shows the usefulness of spatially contextualised motion patterns: in about 57% of all the games (excluding those games where the wing play of both teams is equally well) that team with the better wing play wins (Fig. 4 show some examples). This indicates that the trajectory of the ball should indeed be taken into account for the

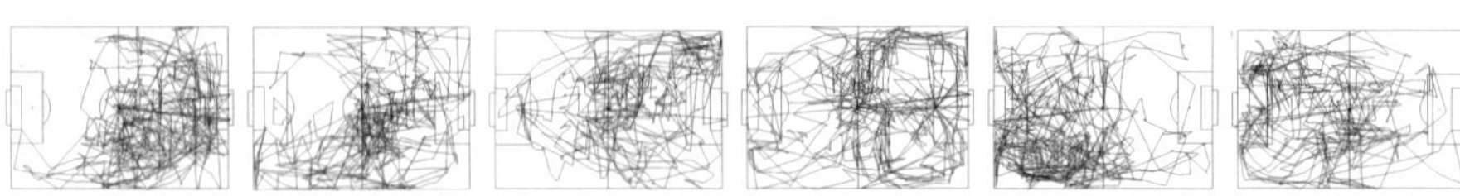

Fig. 4. Games which have been won by those teams with a good wing play

purpose of automatically deriving conclusions about a game. Clearly, combining appropriately a number of predicates more sophisticated inferences can be made.

5 In Sum

When observations about the spatial activity of objects are to be interpreted, we are in need of methods in order to understand their spatial activities. In particular, in soccer we want to comprehend the run of a play and we want to learn for the future. In our examples we analysed the ball's trajectory and all our conclusions are based on its spatial activties. For the purpose of analysing more precisely a game, the spatial activities of single players, their trajectories, and further spatially contextualised motion patterns are to be taken into account. For example, relations among players would tell us something about dummy runs, circulations, and corner kicks. Especially, the proposed formalism can be applied to compare those activities among different leagues. However, our primary aim here consisted in clarifying issues about spatial activities and to provide the framework for implementing interpretation systems on spatial activities. At the same time it should be clear that issues concerning the choice of spatial contextualisations and their interpretations are left to the domain expert.

Acknowledgements

We should like to thank Sebastian Hübner for valuable comments on this paper.

References

1. Bohnenberger, T., Jacobs, O., Jameson, A., Aslan, I.: Decision-Theoretic Planning Meets User Requirements: Enhancements and Studies of an Intelligent Shopping Guide. In: Gellersen, H.-W., Want, R., Schmidt, A. (eds.) PERVASIVE 2005. LNCS, vol. 3468, pp. 279–296. Springer, Heidelberg (2005)
2. Cushman, S.A., Chase, M., Griffin, C.: Elephants in space and time. OIKOS 109, 331–341 (2005)
3. Furnas, G.W., Qu, Y., Shrivastava, S., Peters, G.: The use of inter. graphic. constr. in problem solving with dynamic, pixel-level diagrams. In: Anderson, M., Cheng, P., Haarslev, V. (eds.) Diagrams 2000. LNCS (LNAI), vol. 1889, pp. 314–329. Springer, Heidelberg (2000)
4. Kowalski, R., Sergot, M.: A logic-based calculus of events. New Generation Computing 4, 67–95 (1986)
5. Larkin, J.H., Simon, H.A.: Why a Diagram is (Sometimes) Worth Ten Thousand Words. Cognitive Science 11, 65–99 (1987)
6. McCarthy, J.: Programs with common sense. In: Minsky, M. (ed.) Semantic Information Processing, ch. 7, pp. 403–418. MIT Press, Cambridge (1968)
7. Musto, A., Stein, K., Eisenkolb, A., Röfer, T., Brauer, W., Schill, K.: From motion observation to qualitative motion representation. In: Habel, C., Brauer, W., Freksa, C., Wender, K.F. (eds.) Spatial Cognition II. LNCS (LNAI), vol. 1849, pp. 115–126. Springer, Heidelberg (2000)

Mobile Robots for an E-Mail Interface for People Who Are Blind

V. Estivill-Castro and S. Seymon

School of ICT, Griffith University,
Nathan QLD 4111, Brisbane, Australia

Abstract. The availability of inexpensive robotic hardware has brought to realization the dream of having autonomous mobile robots around us. As such, the research community has recently manifested more interest in assisting robotic technology (see proceedings of the last two IEEE RO-MAN conferences, the emergence of the RoboCup@Home challenge at RoboCup and the first annual Human Computer Interaction Conference jointly sponsored by IEEE and ACM). Robots provide to the blind what was lost as textual interfaces were replaced by GUIs. This paper describes the design, implementation and testing of a first prototype of a multi-modal Human-Robot Interface for people with Vision Impairment. The robot used is the commercially available four legged SONY Aibo.

1 Introduction

It is now accepted that domestic robots will provide entertainment, care, and perform household chores. Inexpensive hardware has brought the realization that robots are moving out of industrial settings and are sharing the environments once considered only for humans. While robots may still be a long way from having generic capabilities, they can perform many useful, socially meaningful and productive tasks (like rescue robots or land-mine finding robots). Among the contributions that mobile robotics can make to the well being of humans are the entertainment, assistive and supportive roles for elderly humans, peoples with disabilities [19] and tutors for human [17]. However, for robots to constitute a useful interface between humans and intelligent systems or ubiquitous computing, many usability and Robot-Human interaction issues must be resolved.

For access by the blind, the conversion for textual output was considered trivial because of Text-To-Speech Technology or Braille output devices [8] while input could be handled by training on a keyboard. The introduction of GUIs and the mouse requires spatial display and more visual interaction resulting in challenging barriers for computer usage by the blind [18]. Multi-modal interaction can significantly facilitate computer usage by the blind [2]. While there has been significant interest in Robot-Human Interaction [3] and in assistive robotics [16], the suggestion of using robots as the multi-modal interface between the blind and ubiquitous computing has received very little attention [1,14]. *Ubiquitous Computing* is considered the next phase of ICT [10]. In this *Ambient Intelligence* [12],

G. Lakemeyer et al. (Eds.): RoboCup 2006, LNAI 4434, pp. 338–346, 2007.

buildings, domestic appliances, cars and many other devices are to provide hardware for intelligent agents that jointly provide Intelligent Environments. Such environments will be homes and offices where domestic robots would be part of the human-environment interface. Most approaches of robotic applications focus on assistive technologies [4] and orientation and navigation [6,7]. However, our approach is to consider the robot as a multi-modal replacement for the computer mouse and visual display. Robots as multi-modal interfaces have recently attracted interest [5]. We show that while general principles of user interface design are applicable to robots as interfaces, some new considerations emerge. In particular, users are not so prepared to accept robots that appear spontaneous or surprising while the user is attempting to complete a task.

The amalgamation of capabilities by computer technology in TVs, desktops, digital cameras and mobile devices makes possible what seemed impossible. Mobiles phones seemed useless to the deaf, but now, they are a flexible tool trough SMS and actions (like vibration) for incoming calls. These devices are becoming multi-modal interfaces and robotic capabilities would enable even more channels of communication through gestures and embodiment. Our thesis is that robots are to become an effective multi-modal interface for people who are blind. This thesis is illustrated with a prototype that uses the Sony AIBO legged robot as a multi-modal interface to applications on a PC with a 802.11b wireless card and Internet access. The Sony AIBO ERS-210 has a wireless card for an ad-hoc wireless connection to the PC. The PC provides the connectivity node and configuration facilities to enable tasks such as browsing the Internet, receiving and sending e-mail, and playing and recording audio files. The modes of communication for input include, physical manipulation, speech recognition, gesture recognition, and even posture. Output are also gestures of the robot, but the main output is sound since the intended users are people who are blind. Physical manipulation in this context is the touching of buttons, and movement of extremities, including the head and tail. A camera on the "nose" of the robot enables gesture recognition. An on-board speaker allows the system (through synthesised speech) to communicate available menus and options and request input. Sensors on all the joints allows for the monitoring of their positions. The system tracks when a joint is moved beyond a threshold from its current target position (because of human user shifting the limb), it triggers a signal that we can marshal and send a specific message to the PC. The system implements a state machine in the PC with state transitions that are the result of the joint moved and the current state. For instance, the moving of the tail to the left sets the system to select the next track, if it was in the audio player state.

We found [1] the ERS-210 provided enough spatial information and its design provided sufficient physical clues that made the project feasible. Fundamental operational issues were resolved. Namely, the robot could have its battery replaced and charged, and it would be strong and reliable enough to be operated manually by a blind person (for the ERS-7 the' outcome was unsatisfactory). Although there were some significant shortcoming in relation to tactile recognition of buttons, we added different textures to the three buttons [1].

Using the User Centred Design approach, we had several iterations to revise the specification of some aspects of the interface and the protocol. Two blind University students interacted 3 times each to evaluate and provide feedback on the usability of the system. Their input influenced the interface design.

2 Evaluation

We conducted two types of evaluations to draw conclusions regarding human-robot interaction. These were both usability assessments. Firstly, we video two full sessions with two blind university students. This process not only validated the User Centred Design that lead to a deployment where usability issues had been resolved, it also confirmed people who were blind from birth could operate it. The second technique was required due to the low numbers of subjects of the first evaluation. Sessions were now conducted with 15 full sighted adults. This subjects were not familiar with the Sony AIBO at all and wore a blindfold before the robot was presented and throughout the entire duration of the test.

In both case, each session lasted for 20 to 25 minutes. The actual exercise lasted between 15 and 20 minutes whilst a questionnaire was completed in the remaining time. The first part of the experiment involves the participant becoming familiar with the robot (mostly touching it, and experiencing the feedback of joints as well as its sounds). Participants were requested to pick the robot up, and explore it themselves. Once having completed this, the instructor would then direct their fingers to points of interest such as the tail, the textured buttons on the back and head, and the on/off button on the chest. Participant were asked to turn on the Sony AIBO and three applications were used to demonstrate the capability of the robotic interface. We refer to them as the *Audio Player*, the *Voice Recorder*, and the *E-mailer*. Participants were guided through the steps of each application once. Then, they were required to accomplish a simple task on each. For example, with the *Voice Recorder*, they were asked to to record a short message, and then review it. They are also given the opportunity to hear the messages in the out-box. The participant were taken through the process of composing an e-mail, send it and then verify that they could retrieve an e-mail.

Table 1. (a) Best default setting for mnemonic commands for the *Audio Player*. (b) Best default settings for mnemonic universal commands.

Audio Player mnemonic commands

Command	User's Action
Play	Head down
Stop	Head up
Pause	Tail up
Next Track	Tail left
Previous Track	Tail right

(a)

Universal mnemonic commands

Command	User's Action
Back	Button on the back
Stop Application	Close mouth

(b)

The Sony AIBO takes some time to initialise. The *Audio Player* is easiest and has basic controls of a media player. These are PLAY, STOP, PAUSE, NEXT TRACK, and PREVIOUS TRACK. Our usual mapping of these controls to the actual actions on the robot is listed in Table 1. While this mapping between actions (from the user on the robot) to effects of the *Audio Player* can be re-configured, we have found that the 4-legged shape of the robot allows for the user actions to be mnemonic. For example, the head movement down is a gesture of confirmation, acceptance and even obedience in dogs and many cultures.

The architecture uses the PC to manage the actual media. For example, a CD with musical content exist on the local disk of the PC. Typically, one can configure the track numbering by copying the audio-files to a directory. When music is selected, the sound is directed to the speakers of the PC (rather than the Sony AIBO) or another set of speakers in the environment. This achieves two issues. At a minimum, the music (or other audio, like iPod radio programs) is played through the speakers on the PC with higher clarity, and avoids high volume traffic on wireless networks. But more importantly, the intention of the *Audio Player* is not to reproduce media content on the Sony AIBO but in the environment in the sense of *Ambient Intelligence*. Thus, reproducing audio on the environment allows us to use the Sony AIBO clearly as an interface between the human and the surrounding environment.

Within each application there can be several different contexts [13]. A context can be seen as a step in a process. For instance, in the *E-mailer*, when users first enters the application, they need to specify whether they want to send an

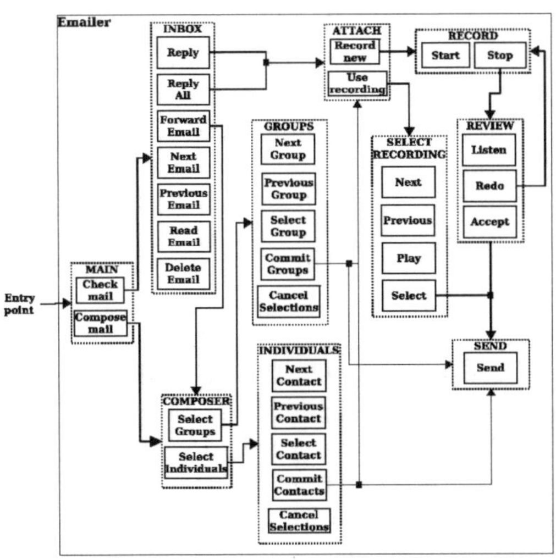

(a) The contexts for the *E-mailer* application.

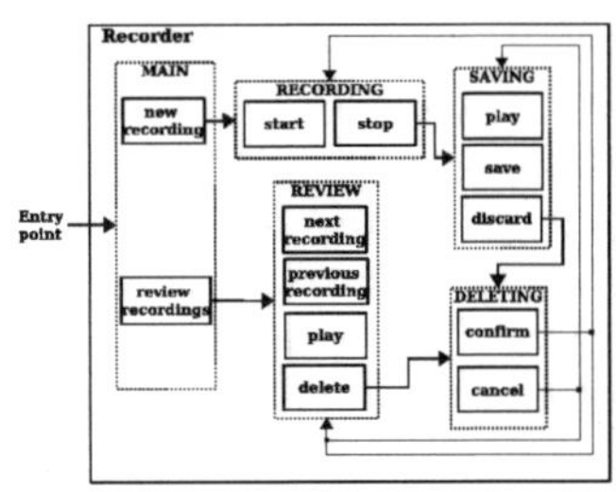

(b) The contexts for the *Voice Recorder*

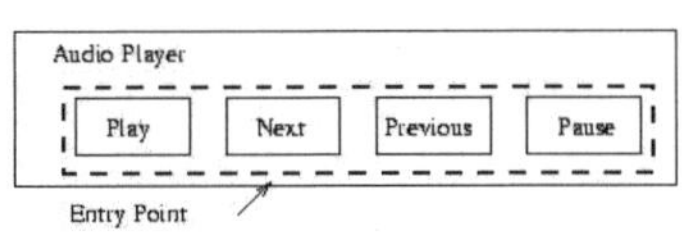

(c) The diagram of contexts for the *Audio Player*.

Fig. 1. The context diagrams reflect the complexity of the application email or review the in-box. This decision is the context. As there are only two things to do here, there will be only two actions mapped. Each context has its own button configuration, complementing the universal commands (refer to Table 1 (b)). Moving between contexts is the result of actions on the robot. For example, by selecting to review the in-box, the user moves to a different context where different options are possible (among the options in the new context we find listening to an e-mail in the in-box). We can measure the complexity of an application. The *Audio Player* is simple because it comprises of just one context. The universal commands listed in Table 1 (b) apply to every context. Touching the back of the Sony AIBO results in users moved to the previous context. That is, they are taken *back* to the previous menu (provided this makes sense). In some instances it is not possible to go back a step in the process as the action is permanent. For example, sending an email can not be undone. The back button effectively behaves as an an *undo* function. Whichever application is running, closing the mouth will stop the application and return the user to the root context of the system. The mnemonic for this is based upon the recommendation that to break and stop a dog fight, one shall grab the jaw of the dog.

Voice Recorder has two functions, recording and reviewing voice messages. The participants voice is captured on the stereo microphones on the Sony AIBO's head. The sound is streamed to the PC as an audio file. Logically, this constitutes an out-box of pre-recorded messages. It is not possible to use the on-board storage of the Sony AIBO as memory sticks are unsuitable for recording large amounts of audio.

We performed the developed process with an understanding of user interface design, but not adhering to any specific methodology, in order to allow more freedom that the prescriptive approaches and guidelines that the literature recommends for human computer interaction [11,13,15]. Once we completed the deployment, we compared with the guidelines suggested in the literature. These guidelines are in agreement with our final product. For illustration, we present a discussion regarding Shneiderman's "Eight Golden Rules of Interface Design" [13]. Rule 1 says "Strive for consistency". We found this rule necessary and applicable as we consistently trough contexts we used the tail to allow users to scroll forwards or backwards through lists. Rule 3 indicates "Offer informative feedback", we found necessary for the applications to respond to some commands by synthesised phrases. Like when users entered the *E-mailer* the robot would respond with the phrase "Emailer enabled". We also provided visual feedback trough the leds in the face of the Sony AIBO. While these are clearly of no value to blind users, they are helpful to trainers as well as for maintenance, and even set up by sighting people. Rule 4 mandates to "Design dialog to yield closure"; that is, to organise actions into groups with a beginning, a middle and an end. This is precisely what contexts achieve. Our *Back* button achieves Rule 5, namely, "Offer simple error handling" and Rule 6 "Permit easy reversal of actions". However, to stress the point in Rule 5 and prevent serious errors we believe it is intuitive to require, before an irreversible action is performed, confirmation where it is required to move the head of the robot down. We found that users were rapidly in control of the robot. They also felt in control. Thus, we believe we concur with Rule 7 that dictates one should "Support internal locus of control". The applications are simple; however, we found it necessary to provide users with a universal command that could repeat the options available and in this way comply with Rule 8 "reduce short-term memory load."

3 Discussion

Assistive technologies overcome some problems people who are blind face when using a GUI. These include Screen Readers and zoom displays. However, these are merely add-ons onto a system that has been designed for people whose main channel of interaction is sight. One would expect that commercial products would have been available to assist the blind in using e-mail and to interact with robots. But, we were rather surprised that we could only find very limited options. For example, we evaluated the Sony AIBO messenger software. While promoted as a complete e-mail application, it is closer to an e-mail reader. It has no functionality to compose and send e-mails. Moreover, it has not been designed for people who are blind. Therefore, our system is the first system to

1. provide a mobile robotic interface for *Ambient Intelligence*,
2. enable mnemonic commands because of the embodiment in a 4-legged dog looking robot, and
3. allow rapid learning and use by blind adults.

There were several lessons learned. With the blind participants, any kind of unexpected movement by the robot is distressing. Thus, in our application, the robot remains essentially motionless when in use. This is a very distinct aspect from current trends in GUIs and the World Wide Web. Current GUIs allow or enable push content on Internet navigation. Applications allow enabling the sudden appearance of wizards while most applications now have update wizards. These unexpected appearances on the visual display are in general far more easily tolerated by users than sudden movement, gestures, actions or sounds by the robotic interface. While it seems reasonable to prompt suddenly with a wizard on a visual display in order for some software update, robots as interfaces need more delicate consideration of this issue. At least, use of sudden actions should be restricted to limited situations (the ones we foresee are alarms or regular warnings, like updating a virus signature file or battery in need of charge).

Two other lessons learned are as follows. First, gaining familiarity with the use of a robot as an interface can be a rather quick process. It must certainly be an incremental process that builds and introduces functionality on top of already familiar functionality. If this principle is followed, then the learning of the interface and the modules is fast. The incremental process can be regulated in terms of the complexity of the functionality as follows. The depth of menus or the size of the sets of options per menu item, or the number of context should all be small for the first applications and then increase as the user progresses to more complex contexts. Note that our design found very useful reuse modules and reproduce sequences in the *Audio Player* for subtasks in the *Voice Recorder*. The recording and playing of messages in the *Voice Recorder* is part of the *E-mailer*. Second, to preserve the context and the familiarity gained, it is also important that the robot starts in a default state the user is significantly familiar with (thus, the simplest is the first application, and is always the starting point).

We use anthropomorphic robots to design mnemonic commands. For example, pulling the head back is a common practice to halt a dog or other quadrupeds. Since the user will have the robot standing laterally with the head close to the right hand (to operate the head with this hand) and the tail on the left hand, a push of the tail away from the user constitutes a move to go forward to the next track, while pulling the tail back for the user brings the application one track back (as we indicated, a right-handed versus a left handed user can easily reverse the position of the robot and the mapping of commands).

A multimodal system allows confirmation of a command by simultaneous production in two channels. However, the system should be configurable to allow unimodal use. The literature suggests that 95% to 100% of users preferred to interact in a multimodal manner when given the choice to use either speech, a pen or both [9]. We do see speech as becoming the core in the system. However, speech recognition demands that the context be limited in options and still seems very useful to preserve the "push the back" action to withdraw a command misunderstood or incorrectly indicated. Also, touching commands are a strong resource when speech is not viable, as in noisy environments.

The questionnaires revealed that the majority found tasks easy to complete (65%). The hardest aspect was remembering how to accomplish a task. However, observation with the blind adults shows that in fact, after one very short session, the steps are easy to remember and they can even describe the process to others. All participants described their experience as very enjoyable and only one person did not find the Sony AIBO easy to manipulate. Unfortunately, in two occasions the tail fell off. This caused some concern to the users at the time. We also notice that a few participants commented on being "worried about breaking it". However, once they grasped the amount of force required to generate a command, they found the setting satisfactory. All participants felt that this software was very useful in assisting vision impaired people. Even 8 sighted people thought that it would assist them in day to day computing. Moreover, these sighting people indicated they would use it if available at home. Note that sighting people would not be using it at home blindfolded. All of the participants were regular users of computers. Only 2 people would prefer it over traditional interfaces, because they were of the opinion that performing tasks would be more time consuming that on a GUI interface with a desk computer.

Currently the robot does very little computation. It indeed behaves almost as a "dumb" interface. However, in the same way as computing power increased to transform "dumb terminals" into machines capable of run significant graphics layouts and client software, we expect the power on inexpensive domestic robots to increase. This power needs to be put to use for good, reliable and comfortable human-computer interfaces. Naturally, one would expect that the improvements would be along the lines of speech recognition and gesture recognition. In particular, we expect these tasks will migrate from the desktop PC across to the robot. It also remains to study how configurable the interface and its settings shall be. The availability of configurable menus allows far more flexibility and users may tailor menus and settings to their needs. However, rapidly the user can be in a position that there is too much inconsistency between sequences to achieve tasks and the tool is too personalised. It seems that, in the same way any user can access a computer in a public library, robotic interfaces may need to develop standard interfaces, so some of them can be deployed for generic, rather than personalised use.

In summary, robots as interfaces to Ambient Intelligence allow anthropomorphic mnemonics. Multi-modality reduces the situations where a command is misunderstood. But, the willingness to accept a robot moving by itself in unexpected ways is low.

References

1. Bartlett, B., Estivill-Castro, V., Seymon, S., Tourky, A.: Robots for pre-orientation and interaction of toddlers and preschoolers who are blind. In: Roberts, J., Wyeth, G., (eds.) Proceedings of the 2003 Australasian Conference on Robotics and Automation, ACRA. CSIRO's (QCAT), Brisbane, Australia (December 1-3, 2003)
2. Blenkhorn, P., Evans, D.: Using speech and touch to enable blind people to access schematic diagrams. Journal of Network and Computer Applications 21, 17–29 (1988)
3. Ghidary, S.S., Nakata, Y., Saito, H., Hattori, M., Takamori, T.: Multi-modal interaction of human and home robot in the context of room map generation. Autonomous Robots Journal 13(2), 169–184 (2002)
4. Graf, B., Hans, M., Schraft, R.D.: Care-O-bot II — development of a next generation robotic home assistant. Auton. Robots 16(2), 193–205 (2004)
5. Hoshino, A., Kato, K., Takeuchi, J.: A chat information service system using a humanoid robot. In: Hoshino, A., Kato, K., Takeuchi, J. (eds.) 14th IEEE International Workshop on Robot and Human Interactive Communication, IEEE Computer Society Press, Los Alamitos (2005)
6. Kulyukin, V., Gharpure, C.: Ergonomics-for-one in a robotic shopping cart for the blind. In: Proceedings of the 2006 ACM Conference on Human-Robot Interaction (HRI 2006), Salt Lake City, Utah, ACM Press (to appear)
7. Kulyukin, V., Gharpure, C., Nicholson, J., Pavithran, S.: RFID in robot-assisted indoor navigation for the visually impaired. In: IEEE/RSJ International Conference on Intelligent Robots and Systems (IROS 2004), Sendai, Japan (2004)
8. Leventhal, J., Uslan, M., Schreier, E.: A review of Braille printers. Journal of Visual Impairment and Blindness 85, 364–350 (1991)
9. Oviatt, S.: Ten myths of multimodal interaction. Communications of the ACM 42(11), 74–81 (1999)
10. Raisinghan, M.S., Benoit, A., Ding, J., Gomez, M., Gupta, K., Gusila, V., Power, D., Schmedding, O.: Ambient Intelligence: Changing forms of human-computer interaction and their social implications. Journal of Digital Information 5(4), 8–24 (2004)

11. Raskin, J.: The Humane Interface. New Directions for Designing Interactive Systems. Addison-Wesley Publishing Co, Reading, MA (2000)
12. Remagnino, P., Foresti, G.L., Ellis, T. (eds.): Ambient Intelligence. Springer, Berlin (2005)
13. Shneiderman, B.: Designing the User Interface, 3rd edn. Addison-Wesley Publishing Co, Reading, MA (1998)
14. Steinfeld, A.: Interface lessons for fully and semi-autonomous mobile robots (2004)
15. Sutcliffe, A.G.: Human-Computer Interface Design, 2nd edn. MacMillan, London (1998)
16. Takahashi, Y., Jones, J., Koyama, H., Komeda, T.: Development of the assistive mobile robot system for coexisting and cooperating with human beings. Advanced Robotics 18, 473–496 (2004)
17. Tanaka, F., Fortenberry, B., Aisaka, K., Movellan, J.R.: Developing dance interaction between QRIO and toddlers in a classroom environment: Plans for the first steps. In: 2005 IEEE International Workshop on Robot and Human Interactive Communication (RO-MAN 2005), Nashville USA, pp. 223–228. IEEE Computer Society Press, Los Alamitos (2005)
18. Vanderheiden, G.C.: Nonvisual alternative displays techniques for output from graphical-based computers. Journal of Visual Impairment and Blindness 83(8), 383–390 (1989)
19. Werry, I., Dautenhahn, K., Odgen, B., Harwin, W.: Can social interactionskills be taught by a social agent? the role of a robotic mediator in autism therapy. In: Beynon, M., Nehaniv, C.L., Dautenhahn, K. (eds.) CT 2001. LNCS (LNAI), vol. 2117, pp. 57–74. Springer, Heidelberg (2001)

Robust and Efficient Field Features Detection for Localization

D. Herrero-Pérez and H. Martínez-Barberá

Dept. Information and Communication Engineering,
University of Murcia, 30100 Murcia, Spain
dherrero@dif.um.es, humberto@um.es

Abstract. In some Robocup leagues, specially in the four-legged league, robots make use of coloured landmarks for localisation. Because these landmarks have no correlation with real soccer, it seems a natural approach to remove them. But for this to be a reality, there are some difficulties that need to be solved, mainly an efficient and robust field features detection and an efficient localisation technique to manage such type of information. In this paper we deal with an approach for field features detection based on finding intersections between field lines which runs at frame rate in the AIBO robots. We also present some experimental results of the vision system and a comparison of the traditional coloured landmark localisation and the field features only localisation, both using a fuzzy-Markov localisation technique.

Keywords: Autonomous robots, fuzzy logic, image processing, localisation, state estimation.

1 Introduction

The Sony Four-Legged Robot (SFLR) League is one of the official leagues in Robocup, in which a standardised robot platform is used, the Sony AIBO. The main exteroceptive sensor is a camera, which can detect objects on the field. Objects are colour coded: there are four uniquely coloured landmarks, two goal nets of different colour, the ball is orange, and the robots wear coloured uniforms. However, in a real soccer field there are not characteristic coloured cues. The rules of RoboCup are gradually changed year after year in order to push progress towards the final goal. Removal of the artificial coloured beacons will be the next step in this direction.

Moreover, coloured landmarks and nets are not frequently perceived in game's conditions, because robots are constantly looking to field to find the ball. Therefore, natural landmarks over the field are constantly perceived, that is field lines, which can be used to update robot localisation more frequently. For these field lines to be successfully used, there are two problems that must be addressed: robust field features detection in real time, and robust localisation able to manage such information.

G. Lakemeyer et al. (Eds.): RoboCup 2006, LNAI 4434, pp. 347–354, 2007.

Preliminary work has been done by some teams in this league to allow the robot to self-localise without using the artificial beacons. For instance, the German Team and other teams use a sub-sampling technique to detect pixels that belong to the field lines [6], distinguishing between field lines along the field and field lines across the field to improve the goalkeepers localisation. These pixels are used in a Monte-Carlo localisation (MCL) schema [5]. Some teams use a kalman filter as their localisation approach, although it can not handle more than one position hypothesis. In order to overcome this problem, rUNSWift uses a Multiple Hypothesis Tracking approach called the multi-hypothesis localisation (MHL) [4]. In our case, we are using a fuzzy-Markov self-localisation technique, in which the robot location is modelled as a belief distribution on a $2\frac{1}{2}$D possibility grid [1]. This formalism allows us to represent and track multiple possible positions where the robot might be. Moreover, it only requires an approximate model of the sensor system and a qualitative estimate of the robot's displacement.

In this paper we propose an alternative solution to using field line detection for the localisation process, which is based on field line intersections detection. The proposed method is described in section 2. These detected features can be introduced in any localisation filter, as those mentioned above. Section 3 describes briefly how we introduce these perceptions in a fuzzy-Markov localisation filter. Section 4 presents some experimental results of the proposed method by way of the evaluation of the localisation accuracy. Finally, conclusions are presented.

2 Perception

The AIBO robots use a CCD camera as the main exteroceptive sensor. The perception process is in charge of extracting convenient features of the environment from the images provided by the camera. As the robot will localise relying on the extracted features, both the amount of features detected and their quality will clearly affect the process. Because of the league rules, all the processing must be done on board and for practical reasons it has to be performed in real time, which prevents us from using time consuming algorithms.

A typical approach for detecting straight lines in digital images is the Hough Transform and its numerous variants. The various variants have been developed to try to overcome the major drawbacks of the standard method, namely, its high time complexity and large memory requirements. Instead of using the field lines as references for the self-localisation, we use the corners produced by the intersection of the field lines (which are white). The two main reasons for using corners is that they can be labelled (depending on the type of intersection) and they can be tracked more appropriately given the small field of view of the camera.

2.1 Vision System Description

The vision system flowchart that we use on the AIBOs is depicted in Fig. 1. The source is the YUV images from the camera and the result is a set of features

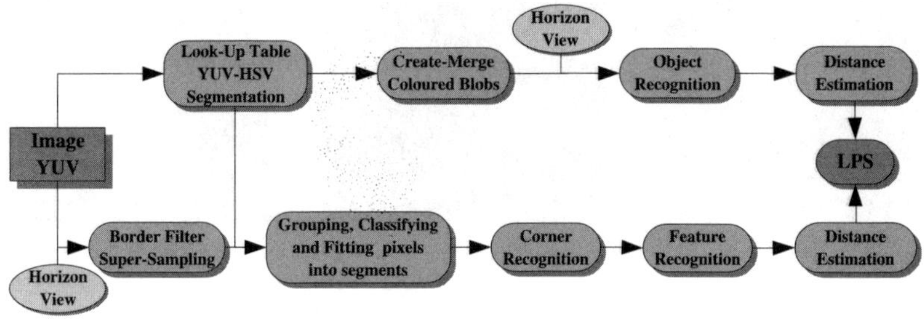

Fig. 1. Vision system flowchart

of the environment in robot-centric coordinates that is called *Local Perceptual Space* (LPS). There are two different paths: the detection of colour-coded object (ball, nets, robots, and landmarks) and the detection of field features (field line intersections). We will describe the second one.

The **first step** after taking a camera image is computing the horizon line, based on the pose of the robot and its head. The idea is looking for field lines only on the horizontal plane, thus saving computing time and avoiding false positives.

The **second step** is applying a border extraction filter to the brightness channel of the YUV image, in particular a Sobel filter is used (Figs. 2b and 3b). By sub-sampling the border-filtered image from the bottom up to the horizon line (Figs. 2a and 3a), transitions of the form non-border (black) → border (white) → non-border (black) are stored and considered for further processing.

The **third step**, which is shared with the coloured objects recognition path, is converting the YUV image to the HSV colour space (Figs. 2c and 3c). As this is a time consuming process, we perform the conversion using a look-up table with pre-computed values.

The **fourth step** is filtering out non-field line transitions from the candidate list. These candidate transitions are checked with the HSV colour segmented image to detect which ones have been produced by the field lines, in particular we consider the following transition types:

- **Carpet-to-line**. Transition from carpet (green) pixels to line (white) pixels.
- **Line-to-carpet**. Transition from line (white) pixels to carpet (green) pixels.
- **Carpet-to-net**. Transition from carpet (green) pixels to net pixels (cyan or yellow).

These labelled transitions are grouped together into sets of transitions that belong to the same straight segment. These segments are obtained using the *Recursive Iterative End Point Fit Algorithm* (RIEPFA) [2]. RIEPFA groups a set of points into segments by evaluating the distance of the points to candidate end segment points, which are recursively divided into smaller segments until the fit criteria is hold (point to segment distance). Segments are further evaluated to meet a minimum number of points and maximum point-to-point distance

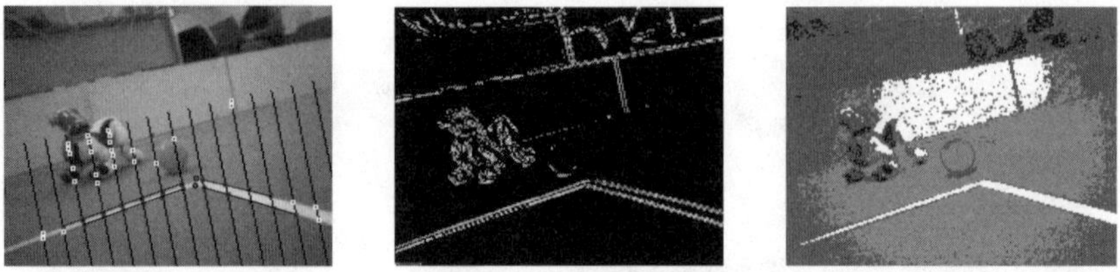

(a) (b) (c)

Fig. 2. Image from the goalkeeper position (a) Sub-sampling and intersections. (b) Sobel filter of the YUV brightness channel. (c) Colour segmentation of the HSV image.

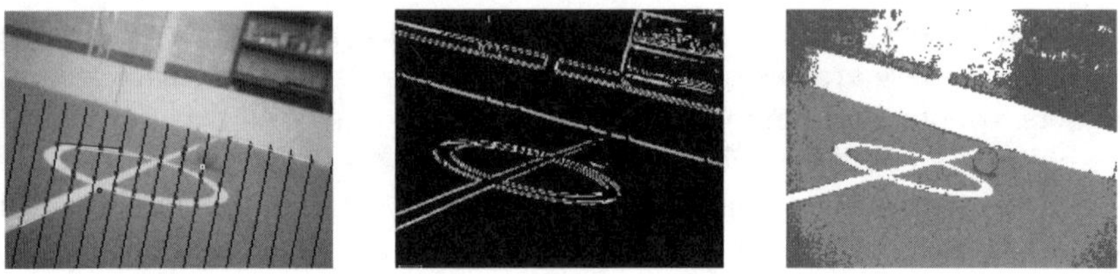

(a) (b) (c)

Fig. 3. Image from the defender position (a) Sub-sampling and intersections. (b) Sobel filter of the YUV brightness channel. (c) Colour segmentation of the HSV image.

criteria. The segments that are not rejected by the previous criteria are then labelled according to the transition pixels that originated the segment.

The **fifth step** is finding intersections between the labelled segments to produce field lines corners (Figs. 2a and 3a). The intersections are evaluated based on the segments labelling, segments orientation, and end points proximity. Valid intersections are then labelled as following:

- **Closed-intersection.** Intersection between *carpet-to-line* or *line-to-carpet* segments, with an angle less than 180 degrees.
- **Open-intersection.** Intersection between *carpet-to-line* or *line-to-carpet* segments, with an angle greater than 180 degrees.
- **Net-intersection.** Intersection between a *carpet-to-line* or *line-to-carpet* segment and a *carpet-to-net* segment. The angle is not taken into account.

The **sixth step** is grouping segment intersections into labelled field features, rejecting those that do not lie into any of the following categories:

- **Type C.** An *open-intersection* nearby of a *closed-intersection.* This feature can be found in the corners of the goal keeper area.
- **Type T-field.** Two *closed-intersections.* This feature can be found in the intersection of the field lines perimeter and any other type of line.
- **Type T-net.** A *closed-intersection* nearby of a *net-intersection.* This feature can be found in the intersection between the goal lines and the corresponding net.

The **seventh step** is the last one in the pipeline. It consists on computing the distance and orientation from the robot to each detected field feature. The pose of the camera is computed using the joint angles of the legs and the head. Then the pixel from the image that represents a field feature is projected onto the field, subject to the constraint that the feature is on the horizontal ground plane. Then the projection point is used to compute the corresponding distance and orientation.

2.2 Vision System Analysis

In this section we will discuss two important aspects of the field features detection procedure presented above: robustness and performance. Because the detected features will be used for locating the robot on the field, one key point is false positives. Reducing the possibility of false positives reduces the possibility of wrong position estimation. The goal of the presented detection procedure is the identification of field lines intersection, and the underlying idea is the filtering of candidates that might be false positives.

Several works detects transitions in segmented images [5] or extracts corners from grey-scale images [3]. But these approaches can not distinguish pixels belonging to the lines or to white robots. Although, these techniques can not distinguish pixels belonging to straight-lines or curve-lines. Furthermore, in the case the carpet perimeter is placed over a plain floor or there is a white hurdle, these pixels are labelled as lines. Thus, typical false positive cases are produced by the white robots, non-straight lines and non-normalised transition in the field perimeter.

All these cases can produce false positives what are very difficult to avoid without evaluate if the pixel belong to a straight line or not. Other technique to get straight line segments is the *Hough Transform*, but it has as main disadvantage the high computational cost.

3 Localisation

Once we are able to obtain a series of percept, be them coloured landmarks, nets or field features, we need a way to combine such information in order to estimate the robots pose on the field: the localisation filter. Although other techniques might be used, without loss of generality we describe how we model the uncertainty associated to the perceived field features using a fuzzy-Markov technique [1]. See [3] for more details.

4 Experimental Results

In order to validate the perception process described in the paper, two localisation experiments have been performed. In both cases the robot has been placed in the goal area, facing more or less to the opposite net. This is the typical goalkeeper position. For a goalkeeper localisation is critical because many behaviour depend on the absolute position. If localisation fails, the robot would start wandering around, possibly leaving the net clear to the other team. In both cases, the localisation process is initialised with a belief distributed along the whole field, that is it does not know its own location. Then the robot starts scanning its surroundings by moving its head from left to right. As soon as a feature is perceived, it is incorporated into the localisation process. The first experiment corresponds to the standard RoboCup scenario, with coloured landmarks and nets as the only perceptual source for localisation. The second experiment corresponds to a possible future RoboCup scenario, in which coloured landmarks have been completely removed, being the perceptual sources the nets and the field lines.

In the experiments we compare the estimated position with the real position. In order to measure the real robot position (actually tracking many robots and the ball) we use an external vision system. This is composed of an overhead camera with a wide angle lens mounted on a aluminium structure at 2.5 metres over the floor. The robot wears a coloured mark which allows computing both the position and orientation of the robot. Because of the high distortion of the lens, the accuracy of the position is of ±2 centimetres and ±5 degrees.

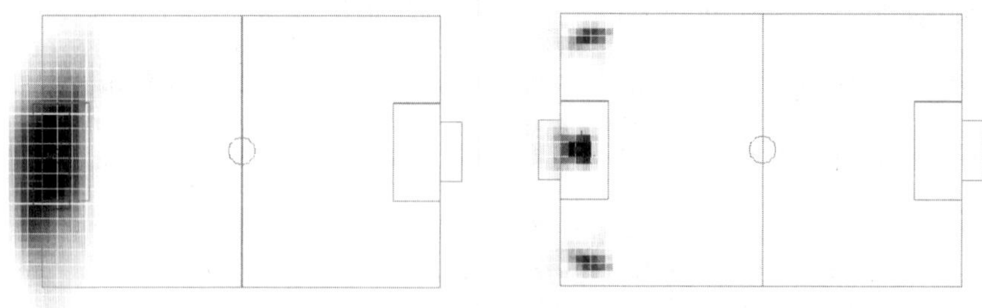

(a) (b)

Fig. 4. Resulting beliefs (a) Coloured landmarks and nets (b) Field lines and nets

The experiments start with belief distributed along the whole field and the robots are left for some seconds scanning the surroundings. The resulting beliefs of both experiments are shown in Fig. 4. The red box corresponds to the estimate of the uncertainty of the robot's position (obtained by finding the bounding box of the highest possibility area) and the red arrow corresponds to the estimate of the robot's position (obtained by defuzzification of the fuzzy belief with the centre of gravity). The comparison of the belief distributions shows that the uncertainty obtained in the standard RoboCup experiment (Fig. 4a) is larger than the one obtained with field lines (Fig.4b). This is because the coloured landmarks are further from the robot than the goal area lines are, thus having a larger uncertainty in the distance. Moreover, this uncertainty in the distance also depends on the colour calibration (the more pixels are segmented the closer the landmark will be perceived), while the field features are less sensible to this effect.

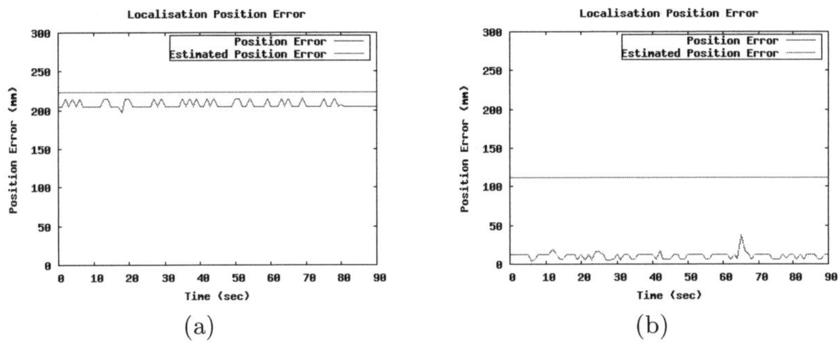

Fig. 5. Position error (a) using coloured landmarks and nets and (d) using field lines and nets

The reduction of the uncertainty conditions very much the accuracy of the localisation process, that is, less uncertainty usually leads to better accuracy. Fig. 5 shows the absolute position error over time, both for the coloured landmarks and field lines experiments. The absolute error is computed taking the difference between the estimated position (on-board) and the measured position (overhead camera). The estimated absolute error is computed from the uncertainty of the belief distribution and the estimated position (on-board). Because the belief distribution is a grid, the minimum estimated absolute error corresponds to the tessellation size, which corresponds to 100 mm in these experiments.

5 Conclusions

This paper presents a novel technique for recognising field features for localisation in the RoboCup Four-Legged League (FLL), although the results can be used in other leagues and scenarios. There is a push towards the removal of coloured landmarks in RoboCup leagues to make the fields more *soccer-like*. The typical approach in this context is the recognition of field lines. The described technique is based on the recognition of field line intersections, keeping into main the constraints of the FLL: on-board processing and the AIBO's camera. The computational burden of the process is low, and it can be set to run at frame rate on the AIBOs. In addition, most of the efforts of the technique are in the direction of avoiding false positives, which can lead to wrong localisation.

In order to validate the usability of the proposed vision process, the detected features are incorporated in a localisation filter. Without loss of generality, we use a fuzzy-markov grid in which we have modelled the perception and its associated uncertainty of different types of field line intersections. This localisation filter provides an effective solution to the problem of localisation of a legged robot in the RoboCup domain [1].

We have presented a series of experiments to show the performance of the localisation process in the standard RoboCup field (with uniquely coloured landmarks and nets) and the more soccer-like scenario (with field lines and uniquely coloured nets). We have compared the estimated robot position with the real one in both scenarios and have shown the feasibility of the removal of the landmarks. Moreover, in certain circumstances the field lines localisation can be more accurate than using landmarks (at least from the goalkeeper's position).

The standard RoboCup scenario provides unique coloured landmarks (that is, there is no ambiguity in the perception process), while the field lines scenario needs unique nets to cope with the natural symmetry of the field. Currently there are no plans to move from unique nets to equally coloured-nets (at least in the FLL), and thus this is not a drawback of the proposed method. Once the use of field lines and unique nets localisation is common in the leagues, there will be necessary any other means to break with the ambiguity, which are left as future work.

Acknowledgements

This work has been supported by CICYT project DPI2004-07993-C03-02.

References

1. Buschka, P., Saffiotti, A., Wasik, Z.: Fuzzy landmark-based localization for a legged robot. In: Intelligent Robots and Systems (IROS), Takamatsu, Japan, pp. 1205–1210 (2000)
2. Duda, R., Hart, R.y.: Classification and scene analysis. John Wiley and Sons, Chichester (1973)
3. Herrero-Pérez, D., Martínez-Barberá, H., Saffiotti, A.: Fuzzy self-localization using natural features in the four-legged league. In: RoboCup 2003, pp. 110–121 (2004)
4. Jensfelt, P., Kristensen, S.: Active global localization for a mobile robot using multiplehypothesis tracking. IEEE Transactions on Robotics and Automation 17(5), 748–760 (2001)
5. Röfer, T., Jüngel, M.: Fast and robust edge-based localization in the sony four-legged robot league. In: RoboCup 2003, Padova, Italy (2004)
6. Röfer, T., Laue, T., Thomas, D.: Particle-filter-based self-localization using landmarks and directed lines. In: Bredenfeld, A., Jacoff, A., Noda, I., Takahashi, Y. (eds.) RoboCup 2005. LNCS (LNAI), vol. 4020, Springer, Heidelberg (2006)

Coordination Without Negotiation in Teams of Heterogeneous Robots

Michael Isik[1], Freek Stulp[1], Gerd Mayer[2], and Hans Utz[2]

[1] Technische Universität München, Boltzmannstr. 3, D-85747 München, Germany
[2] University of Ulm, James-Franck-Ring, D-89069 Ulm, Germany
{isik,stulp}@cs.tum.de, {gerd.mayer,hans.utz}@informatik.uni-ulm.de

Abstract. A key feature of human cooperation is that we can coordinate well without communication or negotiation. We achieve this by anticipating the intentions and actions of others, and adapting our own actions to them accordingly. In contrast, most multi-robot systems rely on extensive communication to exchange their intentions.

This paper describes the joint approach of our two research groups to enable a heterogeneous team of robots to coordinate *implicitly*, without negotiation. We apply implicit coordination to a typical coordination task from robotic soccer: regaining ball possession. We discuss the benefits and drawbacks of implicit coordination, and evaluate it by conducting several experiments with our robotic soccer teams.

1 Introduction

Coordination of actions is essential to solve multi-agent tasks effectively. A striking aspect of human coordination is that we can achieve it with little or no communication. Consider two people assembling a bookcase. With apparent ease, actions are *anticipated* and coordinated: if I see you grab a screwdriver, I will take one of the shelves, and hold it in place, and so forth. Instead of communicating, humans achieve this by inferring the intentions of others. Once the beliefs and desires of the cooperating party are known, we simply imagine what we would do in that situation. This is called the Intentional Stance [1].

In previous research, we have achieved negotiation-free coordination, also called *implicit* coordination, within the AGILO RoboCuppers group [2]. Here, we extend and integrate this with another line of research, which is the formation of a mixed team [3]. Due to scientific as well as pragmatic reasons, there is a growing interest in the robotics field to join the efforts of different labs to form mixed teams of autonomous mobile robots. For many tasks, a group of heterogeneous robots with diverse capabilities and strengths is likely to perform better than one system that tries to encapsulate them all. Also, for many groups, the increasing cost of acquiring and maintaining autonomous mobile robots keeps them from forming a mixed team themselves. Furthermore, to allow all mid-size teams to participate at RoboCup 2006, many of them are required to form a mixed team. Our two groups, who have individually taken part in the RoboCup

G. Lakemeyer et al. (Eds.): RoboCup 2006, LNAI 4434, pp. 355–362, 2007.

mid-size league since 1998, have formed a mixed team with robots from the different labs. As almost all robots in this league, our robots are custom built research platforms with unique sensors, actuators, and software architectures. Therefore, forming a heterogeneous cooperative team presents an exciting challenge. One of these challenges is achieving robust coordination.

The standard solution in robotic teams is not to anticipate the actions of others, as humans do, but instead to extensively communicate utilities or intentions in a negotiation scheme. Previous work on robot coordination seems to have focussed almost exclusively on explicit coordination, as an overview paper on the key architectures shows [4].

In this paper, we discuss the benefits of implicit coordination, and implement it for our heterogeneous team. We apply implicit coordination to a typical coordination task from robotic soccer: regaining ball possession. Acquiring ball possession is a goal for the team as a whole, but only one of the field players is needed to achieve it. Of course, the robots must agree upon which robot will approach the ball. The intuitive underlying rule is that only the robot who is quickest to the ball should approach it. To infer the intentions of others, the agents first learn utility prediction models from observed experience. For the ball approach task, the utility measure is time, so the robots learn to predict how long it will take to approach the ball. During task execution, the robots locally predict the utilities for all robots, and globally coordinate accordingly.

The main contributions of this paper are: 1) learning temporal prediction models that take technical differences between the robot platforms into account 2) using these models to enable implicit coordination within a heterogeneous team of robots 3) demonstrating that coordination based on belief states is more robust than explicit coordination.

The rest of this paper is organized as follows. In the next section we describe how implicit coordination was implemented in our teams. Experimental results are presented in Section 3. In this section, we also discuss the benefits, as well as some drawbacks, of implicit coordination. Related work is presented in Section 4, and we conclude with Section 5.

2 Applying Implicit Coordination

In [2], we introduced a computational model for implicit coordination, that specifies three components are necessary for implicit coordination: 1) utility prediction models 2) knowledge of the states of others 3) the robots should have a shared performance model for joint actions. In our scenario, the latter component is a locker-room agreement [5] that only the quickest robot should approach the ball. Here, we apply this computational model to a team of heterogeneous robots. After presenting the two teams, we discuss the first two components of the computational model in more detail.

The Ulm Sparrows [6] are custom built robots, with infrared based near range finders and a directed camera. The available actuators are a differential drive, a pneumatic kicking device and a pan unit to rotate the camera horizontally

(270°). Each robot acts upon an egocentric belief state, using the camera as its main sensor. The AGILO RoboCuppers [7] are customized Pioneer I robots, with differential drive and a fixed forward facing color CCD camera. They act upon an allocentric belief state, which is acquired by cooperative state estimation. For the experiments we will present later, it is important that the variables are controllable and reproducible. Therefore, we have used our ground truth system to determine the positions of the robots with even more accuracy. This system uses three ceiling cameras to detect colored markers on top of the robots.

Utility Prediction Models. All robots must be able to predict their own ball approach time, as well as that of others. Therefore, they learn temporal prediction models from observed experience. Examples are gathered by navigating to random targets on the field, thereby measuring the time it took to approach the target. These measurements were acquired through the ground truth system. A model tree is then trained with these examples. Model trees are functions that map continuous or nominal features to a continuous value. They recursively partition the data, and fit linear models to the data in each partition. In previous research [2], we have shown that executing 300 navigation tasks yields a sufficient amount of training examples to learn an accurate prediction model. This takes about half an hour, so this is well within the continuous operational range of the robots. The mean absolute error of the models on a separate test set was 0.25s for the Ulm Sparrows, and 0.18s for the AGILO RoboCuppers. For more information about model trees, and how they can be used to learn action models of navigation tasks, we refer to [2].

Knowledge of the states of others. Predicting utilities for others, called perspective taking, can only be done if the robots have estimates of the other's states, which can be difficult if the robots only have local sensors. For instance, due to the limited view of our cameras, it is often not possible to see all the teammates. Therefore, our robots communicate their belief states to each other to achieve more coherent and complete beliefs about the world [3], which they use to determine their (joint) actions. This might seem contrary to the paradigm that we want to achieve coordination without communication. However, there are some important differences between communicating *intentions* and communicating *beliefs*, as we shall discuss in Section 3.2.

3 Experimental Evaluation

To evaluate if the learned prediction models and shared representations are sufficiently accurate for implicit coordination, we have conducted three experiments, one in a dynamic, one in a static environment, and one in simulation. For each experiment, we used one Sparrow and one AGILO robot. Each robot has a temporal prediction model for both robot types.

Dynamic environment experiment. In this experiment, the robots continuously navigated to random targets on the field, for about half an hour. The paths were generated such that interference between the robots was excluded.

At 10Hz, each robot records its own position and orientation, as well as that of its teammate and the ball. Each robot also logs the predicted approach time for both robots, and based on these times, which robot should approach the ball, in their view. Note that the robots never actually approach the ball.

Static environment experiment. In the previous experiment, it is impossible to measure if the temporal predictions were actually correct, and if potential inaccuracies caused the robots' estimate of who is quickest to be incorrect. Therefore a second experiment was conducted. The experimental set-up was as follows: Both robots navigate to different random positions and wait there. During the experiment, the target to approach is fixed and the same for both robots. Then, the robots are requested to record their own state, as well as that of their team mate. The robots compute the predicted approach times, and add them to the log-file. Then, one after the other, the robots are requested to drive to the goal position, and the actual approach duration is recorded. The log-files so acquired are almost identical to the ones in the dynamic experiment. The only difference is that they also contain the actual observed time for the robot. This static environment is less realistic, but allows us to compare the predicted time with the actually measured time for each robot.

Simulated experiment. Here, the experimental set-up is identical to the dynamic experiment. The simulator allows us to vary two variables that most strongly influence the success of implicit coordination. The first is communication quality. At random times, and for random durations, communication is switched off in both directions. By controlling the length of the intervals, we can vary between perfect (100%) and no (0%) communication. The second is the field of view of the robot. We can set the view angle of the robot's forward facing camera between 0 (blind) and 360 (omni-directional vision) degrees. The other robot and the ball are only perceived when in the field of view. Gaussian noise with a standard deviation of 9, 25 and 22 cm is added to the robot's estimates of the position of itself, the teammate and the ball respectively. These correspond to the errors we have observed on the real robots.

3.1 Results

Do the robots agree upon who should approach the ball? To answer this question, we simply determined how often the two robots agreed on which robot should approach the ball in the dynamic experiment, which was 96%.

Do the robots choose the quickest one? We would also like to know if the chosen robot is actually the quickest one to approach the ball. Of course, this could only be determined in the static experiment, in which the actual times it took each robot to approach the ball are known. A robot's decision to coordinate is deemed correct, if the robot that was the quickest was indeed predicted to be the quickest. The robots' choice was correct 92% of the time.

Are temporal prediction models necessary, or would a more simple value such as distance suffice? Using only distance as a rough estimate of the approach time, would save us the trouble of learning models. Although time is certainly strongly correlated with distance, using distance alone leads to significantly more incorrect coordinations. Agreement is still very good (95%), but the robot that is really the quickest is chosen only 68% of the time. So, when using distance, the robots are still very sure about who should approach it, but they are also wrong about it much more often.

When does implicit coordination fail? In the dynamic experiment, coordination succeeds 96% of the time. In the log-file, we labeled all examples in which exactly one robot decided to approach the ball with 'Success', and others with 'Fail'. A decision tree was then trained to predict this value. The learned tree is represented graphically in Figure 1. The main rule is that if the difference in predicted times between two robots is small, coordination is likely to fail, and if it is large, it is likely to succeed. This is intuitive, because if the difference between the times is large, it is less likely that estimation errors will invert which time is the smallest. Note that in between these two limits, there is a 'gray' area, in which some other rules were learned. They only accounted for a small number of example, so for clarity, we will not discuss them here.

In sports like soccer or volleyball, it is sometimes not completely clear who should go for the ball. Humans solve this problem by communicating their intention through an exclamation: "Mine!", or "Leave it!". The decision tree essentially provides the robots with similar awareness, as they predict when implicit coordination failure is likely. So, they could be used for instance to determine when robots should resort to explicit coordination.

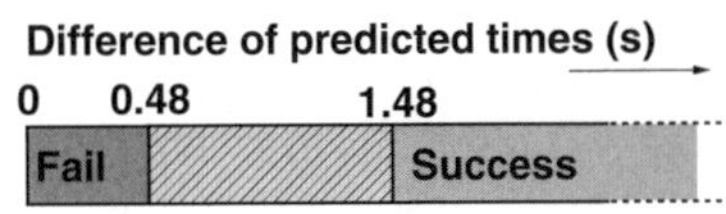

Fig. 1. Visualization of the decision tree that predicts coordination failure

How do communication quality and state estimation accuracy influence coordination? The results of the simulation experiment, which show how the performance of different coordination strategies depends on the quality of communication and the field of view, are depicted in Figure 2. Communication quality is the percentage of packets that arrive, and field of view is in degrees. The z-axis depicts coordination success, which is the percentage that only one robot intended to approach the ball.

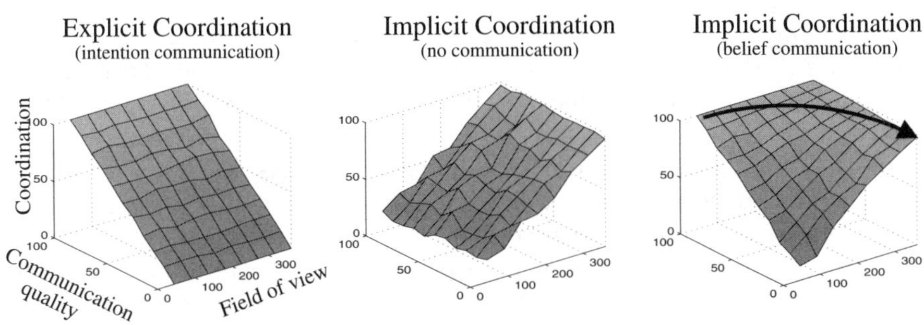

Fig. 2. Results of the simulation experiment, which show how the performance of coordination strategies depends on the quality of communication and the field of view

Since explicit coordination is based completely on communication, it is not surprising that it perfectly correlates with the quality of the communication, but is independent of the size of the field of view. No communications means no coordination, and perfect communication means perfect coordination. For implicit coordination without communication, the relation is converse. If a robot is able to estimate the states of others better, it is able to coordinate better. The third graph shows implicit coordination with belief state exchange (as used on our real robots). If the robot has another in its field of view, it determines the other's state through state estimation, otherwise it uses communication (if possible) to exchange beliefs. These states are then used to predict the utilities of others, independent if they were perceived or communicated. The graph clearly shows that this approach combines the benefits of both.

3.2 Discussion

There are several important benefits that implicit coordination without communication has over explicit coordination. First of all, protocols and arbitration mechanisms must be adopted between communicating entities to enable intention communication, which adds complexities and can degrade the system. It is generally argued that communication can add unacceptable delays in information gathering and should be kept minimal [8]. Furthermore, rescue robotics and autonomous vehicles operating in traffic are examples of domains in which robust communication is not guaranteed, but where correct coordination and action anticipation is a matter of life and death. Finally, human-robot interaction, a current research focus in for instance space exploration or rescue robotics, it cannot be expected of humans to continuously communicate their intentions. Instead, the robot must be able to anticipate a human's intentions, based on predictive models of human behavior. We consider implicit coordination to be essential for natural interaction between robots and humans, so adhering to explicit coordination will prevent robots from making a break-through into these application domains.

The most difficult aspect of implicit coordination is estimating the states of others. Especially for robots with a limited field of view, such as ours, this is problematic. Therefore, we resorted to the communication of beliefs to acquire a shared representation. This might seem contrary to our communication-free paradigm, but there is an important difference between communicating intentions and beliefs. We believe that improvements in sensor technology and state estimation methods will allow robots to autonomously acquire a increasingly complete and accurate estimation of the states of others. In RoboCup for instance, almost all mid-size teams have resorted to omni-directional vision to achieve exactly that. So, beliefs needed to infer the intentions of others are becoming more complete and accurate, independent of communication. The arrow in the third graph in Figure 2 depicts this trend. More accurate state estimation can essentially replace communication. This is certainly not the case for explicit coordination, which will always fully rely on communication.

Furthermore, the third graph in Figure 2 clearly shows that implicit coordination with belief exchange achieves better performance with communication loss than explicit coordination alone. Instead of complete coordination failure in case of communication loss, there is a graceful decay, because a second system based on state estimation can still be used to estimate the intentions of others.

Summarizing, improvements in sensor and state estimation will allow implicit coordination to depend less and less on belief communication. This is necessary to simplify communication schemes, increase coordination robustness, and enable human-robot cooperation. This work proposes a step in this direction.

4 Related Work

The idea of cross team cooperation has some tradition within the RoboCup leagues. The most similar mixed team cooperation effort was the Azzurra Robot Team, a mid-size team from various Italian universities. Their focus was on explicit role assignment and communication-based coordination strategies among the field players [9].

Previous research on cooperation has focussed almost exclusively on explicit coordination [4]. On the other hand, work on implicit coordination usually assumes that all agents have access to a central and global representation of the world, which is enabled by simulation, as in [10], or global perception, as in the RoboCup small-size league [8,11]. In all these papers, teammates are not reasoned about explicitly, but are considered to be mere environment entities, that influence behavior in similar ways to obstacles or opponents.

In [5] the issue of low band-width communication in the simulation league is dealt with by *locker-room agreements*, in which players agree on assigning identification labels to certain formations. During the game, only these labels, instead of complete formations, must be communicated.

Most similar to our work is [12], in which robots in the legged-league also coordinate through implicit coordination which is based on representations which are completed through the communication of belief states. Communication is essential, and assumed to be flawless. It is not investigated how communication loss influences coordination. The utility measure is a sum of heuristic functions, which are represented as potential fields. Whereas our utility models are grounded in observed experience, and have a well-defined meaning (e.g. execution duration in seconds), these heuristic functions have no clear semantics. Therefore, customizing these functions to individual robots is difficult, as the semantics of and interactions between them are not fully understood. However, this customization is essential for achieving efficient coordination in a heterogeneous team with robots with different dynamics and capabilities.

5 Conclusion

In this paper, we have discussed the necessity for implicit coordination in domains in which communication is unreliable or impossible. Relying on intention communication will prevent multi-robot systems from being applied in these domains. We have presented a system that achieves implicit coordination by predicting the utility of itself and others, and adapt its actions to the predicted intentions of others. Knowing the states of others is essential, so belief states are communicated. We have shown that this approach is more robust than communicating intentions, and have argued that improvements in sensors and state estimation will allow implicit coordination to become increasingly independent of communication. We have applied the system to a ball approach task from robotic soccer, and demonstrated its performance in several experiments.

Our current work aims at learning temporal models that take opponent robots into account. Because the state space of this problem is much larger, more training examples are needed. We will also learn more complex models that take into account the player's roles, as well as strategic considerations.

Acknowledgments. This work was partially funded by the Deutsche Forschungsgemeinschaft (German Research Foundation), in the SPP-1125, "Cooperating Teams of Mobile Robots in Dynamic Environments".

References

1. Dennett, D.: The Intentional Stance. MIT Press, Cambridge (1987)
2. Stulp, F., Isik, M., Beetz, M.: Implicit coordination in robotic teams using learned prediction models. In: Accepted for the IEEE Intl. Conf. on Robotics and Automation (ICRA), IEEE Computer Society Press, Los Alamitos (2006)
3. Utz, H., Stulp, F., Mühlenfeld, A.: Sharing belief in teams of heterogeneous robots. In: Proceedings of the International RoboCup Symposium (2004)
4. Gerkey, B.P., Matarić, M.J.: Multi-robot task allocation: Analyzing the complexity and optimality of key architectures. In: Proc. of the IEEE Intl. Conf. on Robotics and Automation (ICRA), IEEE Computer Society Press, Los Alamitos (2003)
5. Stone, P., Veloso, M.: Task decomposition, dynamic role assignment, and low-bandwidth communication for real-time strategic teamwork. Artificial Intelligence 110, 241–273 (1999)
6. Kraetzschmar, Mayer, Utz, et al.: The Ulm Sparrows 2003. In: Proc. of the Intl. RoboCup Symposium (2004)
7. Beetz, M., Schmitt, T., Hanek, R., Buck, S., Stulp, F., Schröter, D., Radig, B.: The AGILO robot soccer team experience-based learning and probabilistic reasoning in autonomous robot control. Autonomous Robots 17, 55–77 (2004)
8. Tews, A., Wyeth, G.: Thinking as one: Coordination of multiple mobile robots by shared representations. In: Intl. Conf. on Robotics and Systems (IROS) (2000)
9. Castelpietra, C., et al.: Coordination among heterogenous robotic soccer players. In: Proc. of the Intl. Conference on Intelligent Robots and Systems (IROS) (2000)
10. Sen, S., Sekaran, M., Hale, J.: Learning to coordinate without sharing information. In: Proceedings of the Twelfth National Conference on Artificial Intelligence (1994)
11. Veloso, M., Stone, P., Bowlin, M.: Anticipation as a key for collaboration in a team of agents: A case study in robotic soccer. In: Proceedings of SPIE Sensor Fusion and Decentralized Control in Robotic Systems II (1999)
12. Vail, D., Veloso, M.: Multi-robot dynamic role assignment and coordination through shared potential fields. In: Multi-Robot Systems, Kluwer, Dordrecht (2003)

Towards a Calibration-Free Robot: The ACT Algorithm for Automatic Online Color Training

Patrick Heinemann, Frank Sehnke, Felix Streichert, and Andreas Zell

Wilhelm-Schickard-Institute, Department of Computer Architecture,
University of Tübingen, Sand 1, 72076 Tübingen, Germany
{heinemann,sehnke,streiche,zell}@informatik.uni-tuebingen.de

Abstract. Many approaches for object detection based on color coding were published in the RoboCup domain. They are tuned to the typical RoboCup scenario of constant lighting using a static subdivision of the color space. However, such algorithms will soon be of limited use, when playing under changing and finally natural lighting. This paper presents an algorithm for automatic color training, which is able to robustly adapt to different lighting situations online. Using the ACT algorithm a robot is able to play a RoboCup match while the illumination of the field varies.

1 Introduction

The extraction of landmarks and objects from a camera image is a crucial task for robots. In RoboCup the color of such features is sufficient for their extraction. Although there were attempts to detect objects - mainly the ball - using form or texture [4],[12], the majority of the RoboCup teams uses algorithms exploiting the special colors to reduce the computational load. Most of these algorithms utilize a predefined subdivision of the three-dimensional color space into several color classes. In the Middle-Size-League (MSL) there are 6 color classes corresponding to different objects. If the image pixels are transformed into these color classes the extraction of the different objects can be very efficient [6]. The approaches mainly differ concerning the subdivision of the color space. Bruce *et al.* use a rectangular subdivision with a minimum and maximum threshold for each color class [2]. Others, such as Bandlow *et al.* use an arbitrary shape in the two color dimensions of the YUV color space and two thresholds in the intensity channel in order to achieve some independency of the lighting conditions [1].

However, with changing lighting, the colors change their representation in all three dimensions of the color space (cf. [10]). Therefore, our team uses a color look-up table that maps each color to its corresponding class. With such a table it is possible to model any subdivision of the color space. Nevertheless, even this approach will fail if the amount and the speed of changes in lighting conditions rises when playing at natural light. In addition, the time needed to manually train a color look-up table was still up to 5 minutes per robot which is too much if we want to be able to extremely reduce the setup time for the teams. Thus, there is clearly a need for a fast and automatic training of a look-up table that dynamically maps the colors to different color classes with changing lighting.

G. Lakemeyer et al. (Eds.): RoboCup 2006, LNAI 4434, pp. 363–370, 2007.

Related work on this topic includes the semi-automatic self-calibration system for MSL robots presented in [9], that subdivides the color space into circular clusters. However, the mapping of the clusters to their corresponding color class has to be decided by a human supervisor. Other methods from the Sony four-legged league require special movements of the camera to train the color look-up table [3] or simply train a color mapping for only three different illumination scenarios [11]. A very promising method is presented in [8]. This method retrains a look-up table without an a-priori known subdivision of the color space. However, this method would require too much computation time when applied to a 580×580 pixel image that is used on our MSL robots.

In this paper we present a new algorithm to automatically train such a table for a RoboCup robot, using only knowledge of the field geometry. By incorporating the pose of the robot computed by our self-localization [5], this algorithm is able to constantly retrain the mapping in conditions where the lighting changes. By keeping the amount of training per cycle as low as possible, the algorithm can be processed 50 times a second on our RoboCup MSL robots, while being capable of adapting to sudden changes in illumination in only a few cycles.

The remainder of the paper is organized as follows: the proposed algorithm is presented in detail in the following section. Section 3 emphasizes the robustness of the algorithm concerning changes in lighting by presenting experimental results, while the last section concludes this paper.

2 The Automatic Color Training Algorithm

The main idea of the automatic color training (ACT) algorithm is to automatically train a color look-up table, using the pose of the robot from the self-localization and a model of its environment to compute the expected color class for the image pixels. For this, the a priory known field parameters are used and a mapping from pixel to two-dimensional world coordinates is trained, using the field markings and a predefined pose of the robot [7].

However, the use of self-localization for the automatic color training results in a mutual dependency. Two features are used to overcome this mutual dependency. First, the image that is used for training the mapping from pixel to world coordinates is used for the training of an initial color look-up table. Second, the extraction of the green and the white color class is robust enough to cope with a sudden change in illumination as shown in section 2.1. This enables the self-localization to keep track of the robot until the other classes are adapted.

With the color values of the pixels and the expected color class a look-up table is trained. As there will obviously be errors in the expected color class, ACT tracks clusters of the color classes in the color space with a mean value and standard deviation, to filter out such errors. Only colors of pixels that fit into a sphere centered in the mean value of a color class with radius equal to a multiple of the standard deviation, are added to the look-up table, while colors of pixels that correspond to coordinates outside of the playable field are removed.

2.1 Computation of the Expected Color Class

Given the pose of the robot and the mapping from pixel to world coordinates, the algorithm can easily compute, which part of the field should correspond to a given pixel. Every pixel that corresponds to coordinates inside of the field is either classified as white field line or green floor, according to the field model. As green is the predominant color in the image in RoboCup, green pixels are usually classified as green, even if the pose estimation from the self-localization is not very accurate. White, however, is very rare in the image but has the main influence on the landmark-based self-localization. Therefore, a special treatment is used for pixels that are mapped to the white class. Only those pixels that have a higher intensity than their surrounding are ultimately used to train white. Given the intensity of a pixel $I(p_{x,y})$ at position (x, y) this filter is defined as

$$I(p_{x,y}) > \frac{1}{25} \sum_{i=x-2}^{x+2} \sum_{j=y-2}^{y+2} I(p_{i,j}). \tag{1}$$

Objects that extend into the third dimension cannot be mapped correctly. However, the region of the image that displays the goal can be defined, depending on the camera system. Only pixels inside this area are mapped to yellow or blue. To train black, the algorithm uses pixels that correspond to the chassis of the own robot. The ball color, though, is a problem for a calibration-free algorithm, as the ball is not static and there is no way of training the ball color without some previous knowledge about the color or position. All pixels that are mapped to a position outside of the field are assigned to the special color class *unknown*.

2.2 Adaptation of the Cluster for Each Color Class

For each color class $k = 1 \ldots 6$, ACT tracks a cluster in the color space with a mean value μ_k and a standard deviation σ_k resulting from the color values of the previous cycles. For color values $c = (u, v, w) \in \{0, C_{max}\}^3$, the parameters of the different clusters are initialized as

$$\mu_{k,0} = \frac{1}{2} (C_{max}, C_{max}, C_{max}) \tag{2}$$

$$\sigma_{k,0} = \frac{\sqrt{3}}{2} C_{max}. \tag{3}$$

Given the set of colors $X_{k,t} = c_1, \ldots, c_m$ of all pixels expected to belong to color class k at cycle t, these parameters are updated as

$$\mu_{k,t} = \frac{1}{\eta + 1} \left(\eta \, \mu_{k,t-1} + \frac{1}{m} \sum_{i=1}^{m} c_i \right) \tag{4}$$

$$\sigma_{k,t} = \frac{1}{\eta + 1} \left(\eta \, \sigma_{k,t-1} + \sqrt{\frac{1}{m-1} \sum_{i=1}^{m} (c_i - \mu_{k,t})^2} \right). \tag{5}$$

To save computation time, the algorithm uses only every 400th pixel, starting at a random pixel. The choice of η determines the responsiveness of the color look-up table update. A value of $\eta = 4$ was empirically determined as optimal, enabling the algorithm to extremely reduce the number of examined pixels. In order to avoid the cluster from collapsing, a lower bound of the standard deviation σ_{min} is introduced. As it is possible that the cluster is too small to include the new color values after a change of illumination, the standard deviation is then doubled to increase the size of the cluster until it includes these color values.

2.3 Add Colors to the Color Look-Up Table

To find out which colors are finally mapped to the classes, again a subset of every 400th pixel is selected. After the calculation of the expected color class, each color value is compared to the mean value of the corresponding color class. Given a color value c that is computed to belong to color class k, the mapping from c to k is only added into the look-up table if $\|\mu_k - c\| < \zeta\,\sigma_k$, with $\zeta > 1$, and $\|\cdot\|$ being the Euclidian norm and ζ being a threshold controlling the ratio between higher adaptability of the color look-up table and a higher false positive rate. The influence of ζ is investigated through experiments in section 3.1.

2.4 Remove Colors from the Color Look-Up Table

After the addition of mappings to the look-up table, colors that are mapped to the special *unknown* class are removed from the table. To process a large number of pixels outside of the field every 20th pixel is used to completely remove unwanted color mappings. A color value c that is expected to belong to the *unknown* class in this cycle but that was previously mapped to color class k is removed, if $\|\mu_k - c\| > \xi\,\sigma_k$, with $\xi > 1$, and ξ being a threshold controlling the ratio between a lower false positive rate and lower true positive rate. The influence of this threshold is also investigated in section 3.1.

3 Results

3.1 Influence of the Thresholds

To analyse the influence of the thresholds $\zeta\,\sigma_k$ and $\xi\,\sigma_k$ for adding and removing color mappings from the color look-up tables the algorithm was tested on three images with different brightness. For each image, several runs of the algorithm were started with different values for ζ using the appropriate pose estimation. After a few cycles the color look-up table converged to a stable state in each run and the quality or fitness of the resulting color look-up table was computed as the sum of the fitness of the k color classes

$$f = \sum_k \left(1 - \frac{TP_k}{TP_k + FN_k}\right), \tag{6}$$

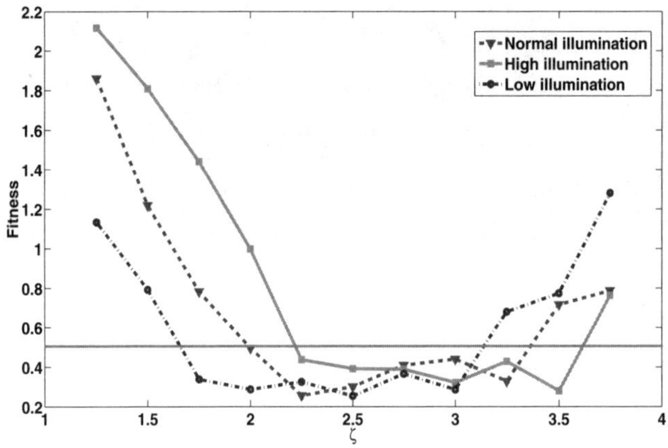

Fig. 1. Fitness of the color mapping of the resulting color look-up tables when ACT is applied to the test images

where TP_k is the number of true positives for class k and FN_k is the number of false negatives for class k. The fitness values for all runs are shown in figure 1. For the dark image, the algorithm converges to the best fitness for values $\zeta = [1.75, 3.0]$, while the interval of best fitness is $\zeta = [2.25, 3.25]$ and $\zeta = [2.25, 3.5]$ for the normal and the brightened image, respectively. On the one hand, lower values of ζ force the color classes to converge to a very small part of the color space that might not include all color values needed to correctly classify all pixels. If the image is very dark, however, the differences between the color classes and the deviation of the colors from the common mean per class are very low. Therefore good results can be achieved with lower values of ζ in the dark image. On the other hand, higher values of ζ enable the color class to spread out in the color space, resulting in a high standard deviation. This includes many colors that should not be mapped to this color class. Fortunately, there is a broad range of values $\zeta = [2.25, 3.0]$ for which all three images are classified with a very high fitness. For all subsequent experiments, a value of $\zeta = 2.5$ was used. The experiments done with the three test images to test the influence of ξ show that different values of ξ result in similar classification results for all three images. In fact, the selection of ξ for a good quality of the algorithm is depending far more on the selection of ζ. With $\zeta = 2.5$ the best results were achieved with a value of $\xi = [1.2, 1.5]$. For lower values of ξ, too many colors are removed from the table, while for higher values of ξ, too many colors from outside of the field remain in the table. For all subsequent experiments, a value of $\xi = 1.35$ was used.

3.2 Automatic Color Training on a Static Robot

This experiment demonstrates that a robot using the ACT algorithm is able to cope with sudden changes of lighting. First, the image on the left of figure 2 is used for training a color look-up table. The classified version of this image using

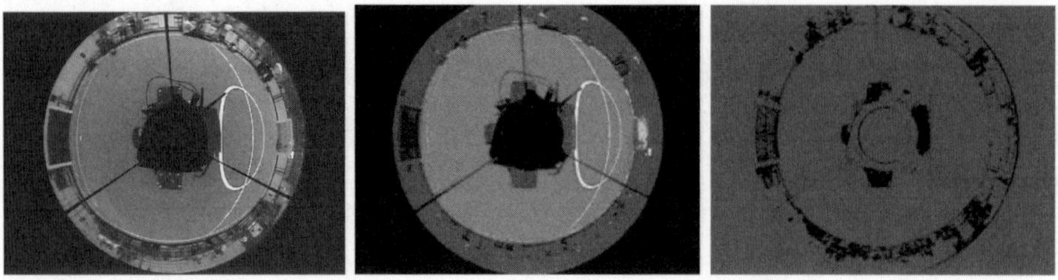

Fig. 2. The image in the middle shows the classification results using the look-up table trained by ACT on the left image. Using the same table to classify a darkened version of this image results in the classification shown on the right. Clearly, a robot with a static color mapping would have no chance of playing using such a classification.

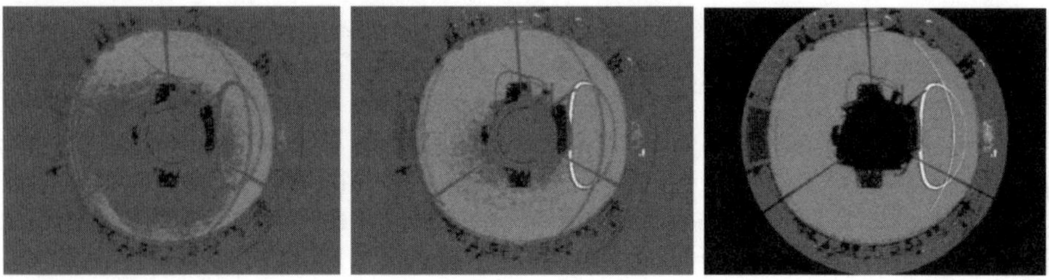

Fig. 3. The darkened image classified after 1, 2, and 8 steps of retraining the color look-up table of figure 2 using the ACT algorithm

the table trained by ACT is shown in the middle in figure 2. Then, to simulate such a sudden change in lighting, the same look-up table was used to classify a darkened version of this image generated by reducing the brightness and contrast to 50%. resulting in the classifaction shown on the right side in figure 2. Clearly, a robot with a static color lok-up table would have no chance of playing using such a classification. With the ACT algorithm, however, the color look-up table is adapted to a stable optimum in only 12 cycles, which would result in only 240ms without color classification. In addition, the look-up table is already very close to the optimum after 2 cycles, at least for the important color classes green and white. Figure 3 shows the classified image after 1, 2, and 8 cycles of adaption.

3.3 Automatic Online Color Training on a Moving Robot

As the presentation of results from a moving robot is very difficult, one experiment was carried out to show that the algorithm embedded in the rest of the robot control system including the self-localization is able to handle a completely wrong pose estimation. For that, three different images were consecutively fed into the ACT algorithm, two from a known pose of the robot and one from a pose where the robot was located 2.5m away from the old pose and rotated by

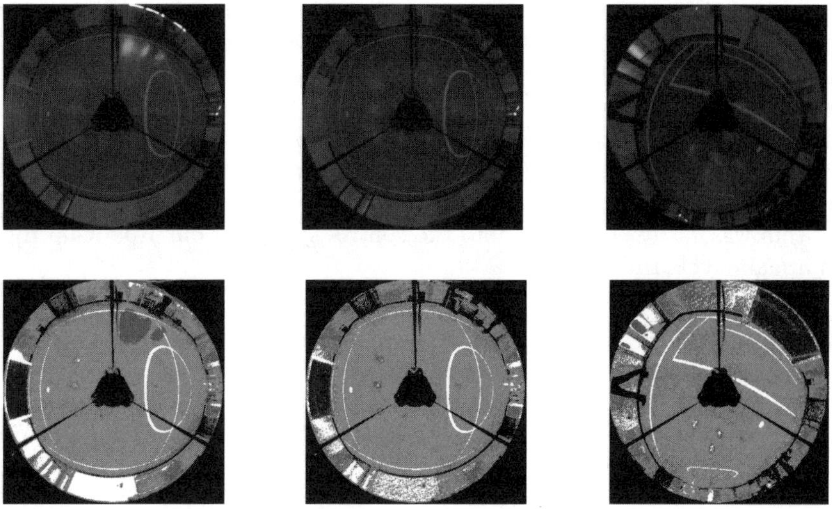

Fig. 4. After a few cycles of adaptation the ACT algorithm is able to train a very good color look-up table in all three situations, even with a completely wrong pose estimation (right)

165 degrees (cf. figure 4, top row). While the self-localization uses the trained color look-up table for the pose estimation, the ACT algorithm always used the known pose as estimation. The classification results are shown in figure 4, bottom row. ACT is able to keep a very good color look-up table, even if the pose used for retraining the table is completely wrong. As the quality of the look-up table adapted by ACT is hardly depending on the quality of the pose estimation, the mutal dependency of the self-localization and the color training is no problem when using the ACT algorithm.

In order to use the ACT algorithm for online training on a moving, soccer playing robot running other processes like self-localization, the time needed to compute one cycle of the ACT algorithm has to be very low. Fortunately, in all the experiments presented in this paper, the computation time was below 4ms for one cycle on an Athlon XP 2400+ with 2GHz. This fits into the main cycle time on our RoboCup MSL robots that are capable of running all processes needed to control the robot in a 20ms cycle on their Pentium-M 2GHz computer.

4 Conclusions

This paper presents an algorithm for automatic online training of a look-up table that maps the colors of a three-dimensional color space onto different color classes used for the detection of objects and landmarks in camera images. For that, the ACT algorithm incorporates knowledge about its environment to compute which colors correspond to which color class. ACT consecutively adapts the look-up table to changing lighting situations resulting in a robust classification of the image pixels. The presented results show that the main parameters of the algorithm can be chosen in a way to produce good results over a large variety of lighting scenarios. Finally, the algorithm was implemented on a RoboCup MSL robot for online training of the color look-up table. Here, the mutual dependency of the color training and the self-localization is shown to have very little impact on the robustness of the algorithm, as the color training is very stable, even for a completely wrong pose estimation. With a cycle time of only 4ms the ACT algorithm was easily embedded into the control system of our RoboCup robots with a main cycle time of 20ms.

References

1. Bandlow, T., Klupsch, M., Hanek, R., Schmitt, T.: Fast Image Segmentation, Object Recognition and Localization in a RoboCup Scenario. In: 3. RoboCup Workshop, IJCAI'99 (1999)
2. Bruce, J., Balch, T., Veloso, M.: Fast and inexpensive color image segmentation for interactive robots. In: Proc. 2000 IEEE/RSJ Intl. Conf. on Intelligent Robots and Systems, vol. 3, pp. 2061–2066 (2000)
3. Cameron, D., Barnes, N.: Knowledge-Based Autonomous Dynamic Colour Calibration. In: Polani, D., Browning, B., Bonarini, A., Yoshida, K. (eds.) RoboCup 2003. LNCS (LNAI), vol. 3020, pp. 226–237. Springer, Heidelberg (2004)
4. Hanek, R., Schmitt, T., Buck, S., Beetz, M.: Towards RoboCup without Color Labeling. In: Kaminka, G.A., Lima, P.U., Rojas, R. (eds.) RoboCup 2002. LNCS (LNAI), vol. 2752, pp. 426–434. Springer, Heidelberg (2003)
5. Heinemann, P., Haase, J., Zell, A.: A Novel Approach to Efficient Monte-Carlo Localization in RoboCup. In: RoboCup 2006: Robot Soccer World Cup X. LNCS, vol. 4434, pp. 322–329. Springer, Heidelberg (2007)
6. Heinemann, P., Rückstieß, T., Zell, A.: Fast and Accurate Environment Modelling using Omnidirectional Vision. In: Dynamic Perception 2004. Infix (2004)
7. Heinemann, P., Sehnke, F., Streichert, F., Zell, A.: Automatic Calibration of Camera to World Mapping in RoboCup uing Evolutionary Algorithms. In: Proceedings of the IEEE Congress on Evolutionary Computation (CEC 2006), IEEE Computer Society Press, Los Alamitos (2006)
8. Jüngel, M.: Using Layered Color Precision for a Self-Calibrating Vision System. In: Nardi, D., Riedmiller, M., Sammut, C., Santos-Victor, J. (eds.) RoboCup 2004. LNCS (LNAI), vol. 3276, pp. 209–220. Springer, Heidelberg (2005)
9. Mayer, G., Utz, H., Kraetzschmar, G.: Towards autonomous vision self-calibration for soccer robots. In: Proc. 2002 IEEE/RSJ Intl. Conf. on Intelligent Robots and Systems (2002)
10. Mayer, G., Utz, H., Kraetzschmar, G.: Playing Robot Soccer under Natural Light: A Case Study. In: Polani, D., Browning, B., Bonarini, A., Yoshida, K. (eds.) RoboCup 2003. LNCS (LNAI), vol. 3020, pp. 238–249. Springer, Heidelberg (2004)
11. Sridharan, M., Stone, P.: Towards Illumination Invariance in the Legged League. In: Nardi, D., Riedmiller, M., Sammut, C., Santos-Victor, J. (eds.) RoboCup 2004. LNCS (LNAI), vol. 3276, pp. 196–208. Springer, Heidelberg (2005)
12. Treptow, A., Masselli, A., Zell, A.: Real-Time Object Tracking for Soccer-Robots without Color Information. In: European Conference on Mobile Robotics (ECMR 2003), pp. 33–38 (2003)

Learning to Shoot Goals Analysing the Learning Process and the Resulting Policies

Markus Geipel and Michael Beetz

Department of Computer Science, Technische Universität München

Abstract. Reinforcement learning is a very general unsupervised learning mechanism. Due to its generality reinforcement learning does not scale very well for tasks that involve inferring subtasks. In particular when the subtasks are dynamically changing and the environment is adversarial. One of the most challenging reinforcement learning tasks so far has been the 3 to 2 keepaway task in the RoboCup simulation league. In this paper we apply reinforcement learning to a even more challenging task: attacking the opponents goal. The main contribution of this paper is the empirical analysis of a portfolio of mechanisms for scaling reinforcement learning towards learning attack policies in simulated robot soccer.

1 Introduction

Reinforcement learning is a popular method for solving complex control tasks. It has been applied to a variety of problem domains in various forms. Our contribution will be to provide an in depth analysis of reinforcement learning for a problem of high complexity. For our work we focused on learning a complex control task in the context of simulated robot soccer: the *goalshooting* scenario. In our experiments we investigated the differences between hand-coded policies and learned ones. Our next focus was the process of learning. We monitored not only the success rate but also the action usage in the team as well as the change rate in the policy of each agent. Further more we developed a method to visualize learned policies and used it to compile a video clip of the changing policy during training. Finally we developed a concise method to bring a learned policy into human readable form.

We will now briefly introduce simulated robot soccer, and then specify the goalshooting task and its properties. The perceptions in robot soccer fall into three categories: visual, acoustic and self perception. Visual percepts include the relative distance and angle to all objects within the field of view. They are tainted by random noise according to the distance of the perceived object. The standard agent disposes of three physical actions: `turn`(*angle*), `dash`(*power*) and `kick`(*power*, *angle*) as well as an action for inter-agent communication. An in depth explanation of simulated robot soccer can be found in the server manual [1]. The *goalshooting* scenario is a sub problem of RoboCup simulated soccer. It is

G. Lakemeyer et al. (Eds.): RoboCup 2006, LNAI 4434, pp. 371–378, 2007.

designed to mimic the last phase of an attack on the opponent's goal. A team of three attackers tries to score a goal. The goal is guarded by two defenders and one goalie. In the initial formation the three attackers are positioned equidistantly on the 26 meter line. The ball is positioned one meter in front of a randomly chosen attacker. The two defenders are positioned on the 10 meter line, blocking the direct shooting lanes from the wing attackers to the goal. The shooting lane of the centre attacker is blocked by the goalie. An episode is counted as failure if a time limit is exceeded, the ball gets behind the 30 meter line or leaves the field. We will call the number of successfully completed episodes divided by the number of played episodes the success rate of the attackers. To maximize this success rate is the objective in the *goalshooting* scenario.

One of the most challenging reinforcement learning tasks so far has been the 3 to 2 keepaway task in the RoboCup simulation league, introduced by Stone and Sutton [2]. *Goalshooting* however, exhibits a higher complexity than *keepaway* for the following reasons: Shooting goals demands more coordination and rewards are more scarce. Reward is given only on completing the task either with failure or success. A series of right actions by more than one agent is needed to succeed. This gives also rise to the inter agent credit assignment problem. In *goalshooting* we have chosen to use complex opponents: the "UvA trilearn players 2003" from the binary distribution serve as defenders of the goal.

This paper will be structured as follows. We will describe how the reinforcement learning algorithm SARSA(λ) the *goalshooting* task. In the next section we will describe all the techniques we developed to comprehend and interpret the policies produced by learning agents as well as the process leading to them. Finally we will point out related work and last but not least draw a conclusion.

2 Scaling Reinforcement Learning to *Goalshooting*

We did not implement our soccer agents from scratch but set the learning mechanism on top of the well known UvA Trilearn Open Source Distribution. It is available at [3] for free. Henceforth we will refer to the UvA Trilearn player of the Open Source Distribution as basic player. Figure 1 shows the architecture of the learning agents. The grey parts are already addressed by the basic player, the white ones depict the learning layer that was added. The architecture takes its cues from the one proposed by Russell and Norvig [4] for learning agents. Rectangles represent data structures while boxes with round edges represent algorithmic parts. The critic calculates a feature set based on the belief state. It also estimates the utility of the current state and provides a reward signal. The performance element stores the current policy and is manipulated by the learning element, which implements the actual learning algorithm. The problem generator suggests actions for exploration.

The attackers are the ones to decide what to do when in possession of the ball. They choose an action from the set of parameterless high level actions. `shoot`: shoot the ball towards a randomly chosen corner of the goal. `dribble_goal`: dribble towards the goal. `dribble_free`: dribble towards an open spot. An open

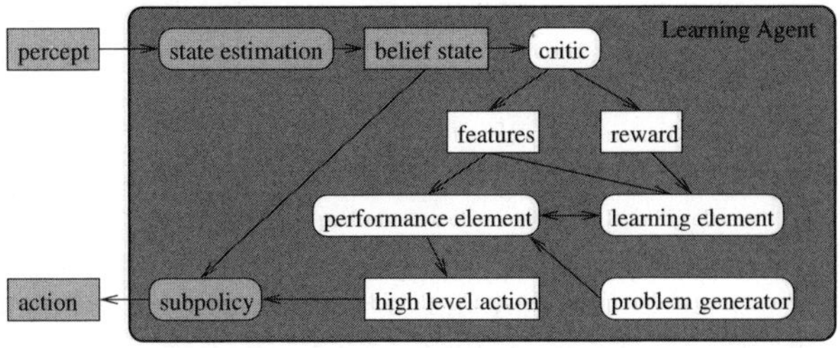

Fig. 1. The agent's architecture

spot is chosen with the SPAR[1] algorithm. `pass_near`: pass to the nearest team mate. `pass_far`: pass to the second nearest team mate. `hold`: wait and see. They were built on top of the basic player's ready to use sub-policies or skills. Henceforth we will call these actions "high level actions" to distinguish them from the atomic actions understood by the soccer server.

The perceptions in simulated robot soccer can not directly be used as input for the learning algorithm. As can be seen in figure 1, state estimation brings the belief state in sync with the current perceptions. This is accomplished by the basic player. Still the state space is too big for learning to be feasible. The critic condenses it to a set of numerical values, called features. The feature set used by Stone and Sutton [2] to built soccer agents that master the *keepaway* task has been the role model for ours. The feature calculation is based on: the set of attackers A and the set of defenders D were the members are ordered by increasing distance from the agent, which is thus defined as a_1. g_r represents the center of the right goal. The distance between x and y is given by $\texttt{dist}(x, y)$ and the angle between x and y with vertex at z by $\texttt{ang}(x, y, z)$. The feature set consists of 7 numerical values. Four distances: $\texttt{dist}(a_1,g_r)$, $\texttt{dist}(a_1,d_1)$, $\texttt{dist}(a_1,a_2)$, $\texttt{dist}(a_1,a_3)$. Three angles: $\texttt{min}(\texttt{ang}(a_1,g_r, d \in D))$, $\texttt{min}(\texttt{ang}(a_1,a_2, d \in D))$, $\texttt{min}(\texttt{ang}(a_1,a_3, d \in D))$. The rationale is, that these angles are a good indicator for the width of an opponent free shooting lane to the goal or the team mates. Not only does the critic calculate an appropriate feature set but it also provides the reward signal.

The learning relies on the performance element, the learning element and the problem generator. The learning element depicted in figure 1 will tune the performance element based on the feature vector and the reward. It uses the SARSA(λ) algorithm as it is presented in "Reinforcement Learning: An Introduction" by Sutton and Barto [6]. The problem generators task is to suggest actions for exploration to the learning element. In our case the problem generator just returns a random action. The performance element holds the learned policy in form of two components: A value-function for each possible action and an arbitration mechanism that returns the action that has the highest value for

---

[1] SPAR stands for Strategic Positioning with Attraction and Repulsion and was introduced by Veloso, Stone and Bowling [5].

the current feature vector. The value-functions themselves are approximated by a CMAC-function-approximator[2].

3 Analyzing Learning and Learned Policies

Now that our test bed is properly defined we will ask the following questions: How well does a learned policy do compared to a hand-coded one and will it be more robust to changes in the environment? Will the policy itself be continuously changed or will there be bursts of changes? Is there a way to visualize the learned policies? Is there a way to bring a learned policy in to a human readable form?

3.1 Robustness of Hand-Coded Policies and Learned Ones

Question: "Are learned policies more robust in respect to changes in the environment than hand-coded ones?". In order to find an answer we used the setup just described to conduct the following experiment: First we will compare the performance of the learning algorithm with the performance of a hand-coded policy. In the next step we will slightly change the environment and look again. In order to make this comparison we need a hand-coded policy. It is essential to make a fair comparison. Thus the hand-coded policy should have exactly the same feature vector available as the learning algorithm. It took several days of testing to find and fine tune a hand-coded policy: The basic idea is to shoot if near enough to the goal. To pass as far as possible or shoot if the enemy is too near. Finally the agent will dribble if no other rule fits. The top plot (90 degrees field of view) in figure 2 shows the result: We plotted the success of the hand-coded policy as well a the success of the learned one. For the next experiment we slightly changed the environment. In the standard configuration, the agents have a view cone of 90°. We changed it to 180°. The bottom plot (180 degrees field of view) in figure 2 shows the result: While the learning from scratch is working just as fine as before (A), the hand-coded policy fails horribly. Furthermore the policy learned in the 90° scenario seems to be quite robust. We see that learning agents starting with it, succeed with an average of 27 percent in the beginning and adapt pretty soon (B). The failure of the hand-coded policy is amazing because the hand-coded policy seemed to be well suited for both scenarios. This example shows that learning is the more flexible and more robust approach.

3.2 Visualization of the Learning Process

The agents basically learn a action-value-function which is represented by the function approximator. But this multidimensional function is not accessible for humans. The objective of this experiment is to visualize policies and the learning process. The key idea is to find a different policy representation. The policy can also be represented by a large number of typically visited points in the feature space. The original policy is used to tag each point with the action it

[2] For a introduction to CMAC consult the work of J. S. Albus [7].

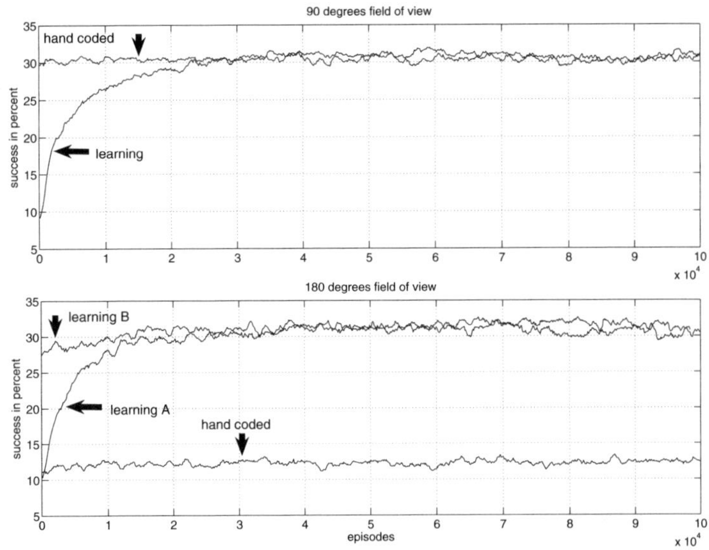

Fig. 2. Each curve is an average of 24 independent runs. See text for an explanation.

would choose in that case. What we get is a set of classified samples. Principal Component Analysis was chosen to project the points in the seven dimensional space down to three dimensional one. The points are colored according to their classification. The policy after every episode was visualized as a picture. All the pictures were compiled to a video. So each frame corresponds to one episode and we can watch the learning taking place in fast motion. The video can be downloaded at `http://home.in.tum.de/geipel/da`.

3.3 The Learning Process

To see what happens to the policy during the learning process we will monitor the following aspects: success rate, action usage by the policies and change rate of the policy of each agent. The results for the training of one attacker team will be presented and interpreted. How do we assess these three aspects? Success rate is the number of successfull episodes divided by the overall number of episodes. The action usage can be extracted from the classified samples by counting all samples that were tagged with one specific action. Please note that not all samples are yet classified at the beginning of the training. For the change rate we count the samples for which the classification changed.

Figure 3 shows the results for one team of attackers. The first plot shows the success of the team. The second the usage of the different actions used by all three players. Finally the third one shows the amount of change in each agent's policy. It can be seen that the success curve consists of three phases of increase with intermediate phases of stagnation. The beginning of each increase phase is

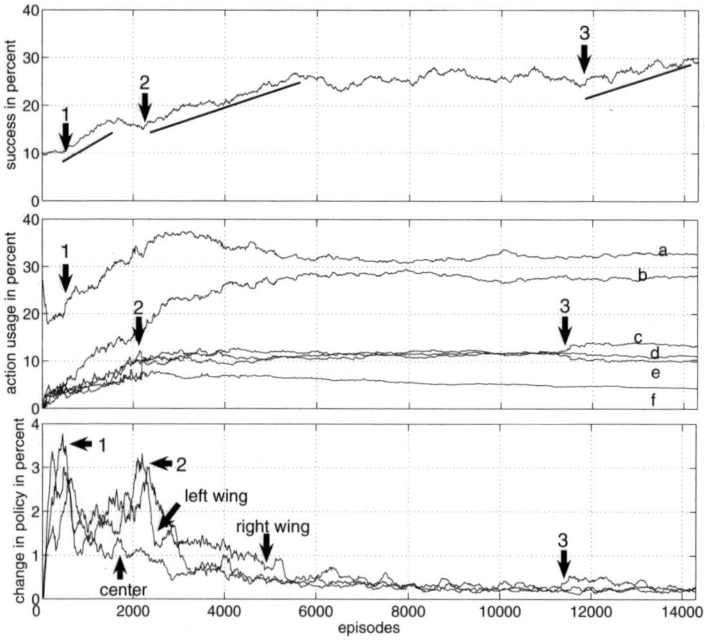

Fig. 3. Top: The success rate of the agents. Center: The usage of the actions. a a for shoot, b for dribble, c for dribble open, d for pass near, e for pass far and f for hold. Bottom: The changes in the policy for each agent.

marked with an arrow. The phase itself is underlined. Each phase of boosting success is launched by a sudden change in the policy of one or more agents. Phase one is most probably caused by steep increase in the usage of the action shoot. The policy of all agents undergoes heavy changes. Phase two goes along with a sudden steep increase of the usage of the action `pass far`. It is marked with an arrow and the number two. The "policy change" plot also shows that this change is only effecting agent one and three, the wing attackers. The third phase seems to be due to a change in the policy of agent one. The usage of hold is slightly increased while the usage of `dribble open` decreases.

3.4 Identifying Situations

In this subsection we describe a method to interpret the learned policy. We would like to express what the agent learned in simple rules like for example: "The agent shoots if the goal is near and free of opponents." We just saw that there are ways to visualize the policies. But this is not yet enough to properly interpret them. Again we will use the before mentioned policy representation based on a set of classified samples. There are nearly single colored regions in the policy, as can be seen in the video. Such a region can be interpreted as a situation in which the policy chooses one action. First we will split the data by actions. This means all samples that were classified with action 0 form a new set, same is done for action 1, and so on. Now a clustering algorithm is applied to each. The clusters that are found are just these "situations". So we get a list of clusters for each action that describes in which situation these actions are applied. Figure 4 shows such a clustering for the shoot action. The Expectation-Maximization-Clusterer which

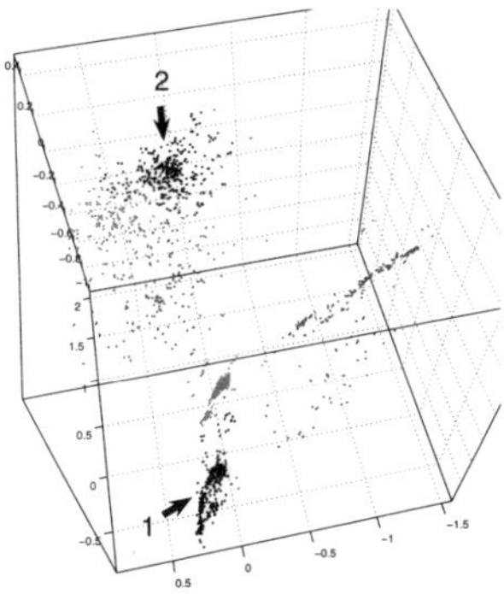

feature	mean	variance
DistanceGoal	**19.75**	7.78
DistanceNearOpponent	3.29	3.18
MinAngleTeammateNear	0.00	40.87
MinAngleTeammateFar	-0.02	0.48
MinAngleGoal	**73.81**	17.52
DistanceNearTeammate	-9.97	0.66
DistanceFarTeammate	-10.00	15.59

Fig. 4. Clustering for the `shoot` action

was used, describes clusters by the mean and variance of every dimension or respectively feature in our case. As an example we will pick one cluster and take a look at the cluster description. The description of the cluster marked 1 with the interesting facts highlighted can also be seen in figure 4. Informally speaking it shows a situation where the agent is relatively close to the goal. The goal itself is quite clear of opponents, as the minimal angle between agent, opponents and goal (`MinAngleGoal`) is comparatively wide. Further more, information about the teammates is inconsistent, which is expressed by negative values for the features `DistanceNearTeammate` and `DistanceFarTeammate`. The action shoot makes perfect sense. However not all clusters make sense. The cluster marked 2 is an example of a rather bad decision: The agent does know the position of its teammates and is far away from the goal. Passing would be the right action, not shooting. This fact that suboptimal clusters prevail, suggests that the learning algorithm does not find a global optimum but gets stuck in a local one.

4 Related Work

There is no directly related work, concerning the analysis tools we presented. There are however several interesting approaches to solve subproblems of simulated robot soccer with reinforcement learning. Riedmiller and Merke [8] employed reinforcement learning to tune the positioning behavior of agents attacking the goal. The scenario included seven attackers and seven defenders. The rest of the attackers behavior is hard wired. Stone and Sutton [2] used reinforcement learning successfully to train soccer agents in the *keepaway* task.

The keepers try to stay in possession of the ball as long as possible while hard wired takers aim to get the ball. The learning keepers where able to outperform hand-coded ones.

5 Conclusion

In this work we concentrated on analyzing the learning process in a challenging domain, namely the *goalshooting* task. Our key findings are the following: The SARSA(λ) algorithm in combination with a CMAC function approximator is able to achieve the same success rate as a tediously tuned hand-coded policy. Even more, for small changes in the environment, the learning shows stable results while the hand-coded policy may suddenly fail. Even though the policy learned by the agents is cryptic for humans there are ways to visualize and interpret it. The fact that we find suboptimal decisions in policies where the learning already leveled off suggests that the learning agent do not find a globally optimal policy. This is backed by the fact that the learned policies are sometimes very different although their success rate is the same. Learning in the *goalshooting* scenario with the agents described in this work, takes place in phases of increasing success with intermediate phases of stagnation. The boost in performance are due to sudden changes in the agent's policy.

References

1. Cheny, M., Foroughi, E., Heintz, F., Huangy, Z., Kapetanakis, S., Kostiadis, K., Kummeneje, J., Noda, I., Obst, O., Riley, P., ens, T.S., Wangy, Y., Yiny, X.: Users manual robocup soccer server for soccer server version 7.07 and later (2002), `http://sserver.sourceforge.net/docs/manual.pdf`
2. Stone, P., Sutton, R.S.: Scaling reinforcement learning toward RoboCup soccer. In: Proceedings of the Eighteenth International Conference on Machine Learning, pp. 537–544. Morgan Kaufmann, San Francisco, CA (2001)
3. Kok, J.: Uva trilearn website (2004), `http://staff.science.uva.nl/~jellekok/robocup/2004/index_en.html`
4. Russell, S., Norvig, P.: Artificial Intelligence: A Modern Approach, 1st edn. Prentice-Hall Inc, Upper Saddle River, New Jersey (1995)
5. Veloso, M., Stone, P., Bowling, M.: Anticipation as a key for collaboration in a team of agents: A case study in robotic soccer (1999)
6. Sutton, R.S., Barto, A.G.: Reinforcement Learning: An Introduction, 4th edn. MIT Press, Cambridge, MA (2002)
7. Albus, J.S.: A theory of cerebellar functions. Mathematical Biosciences ???, 25–61 (1971)
8. Riedmiller, M., Merke, A.: Using machine learning techniques in complex multi-agent domains. In: Stamatescu, I., Menzel, W., Richter, M., Ratsch, U. (eds.) Perspectives on Adaptivity and Learning. LNCS, Springer, Heidelberg (2002)

Cognitive Robotics: Command, Interrogation and Teaching in Robot Coaching

Alfredo Weitzenfeld[1] and Peter Ford Dominey[2]

[1] ITAM, San Angel Tizapán, México DF, CP 0100
alfredo@itam.mx
http://www.cannes.itam.mx/Alfredo/English/Alfredo.htm
[2] Institut des Sciences Cognitives, CNRS, 67 Blvd. Pinel, 69675 Bron Cedex, France
dominey@ isc.cnrs.fr
http://www.isc.cnrs.fr/dom/dommenu-en.htm

Abstract. The objective of the current research is to develop a generalized approach for human-robot interaction via spoken language that exploits recent developments in cognitive science, particularly notions of grammatical constructions as form-meaning mappings in language, and notions of shared intentions as distributed plans for interaction and collaboration. We demonstrate this approach distinguishing among three levels of human-robot interaction. The first level is that of commanding or directing the behavior of the robot. The second level is that of interrogating or requesting an explanation from the robot. The third and most advanced level is that of teaching the robot a new form of behavior. Within this context, we exploit social interaction by structuring communication around shared intentions that guide the interactions between human and robot. We explore these aspects of communication on distinct robotic platforms, the Event Perceiver and the Sony AIBO robot in the context of four-legged RoboCup soccer league. We provide a discussion on the state of advancement of this work.

1 Introduction

Ideally, research in Human-Robot Interaction will allow natural, ergonomic, and optimal communication and cooperation between humans and robotic systems. In order to make progress in this direction, we have identified two major requirements: First, we must work in real robotics environments in which technologists and researchers have already developed an extensive experience and set of needs with respect to HRI. Second, we must develop a domain independent language processing system that can be applied to arbitrary domains and that has psychological validity based on knowledge from social cognitive science. In response to the first requirement regarding the robotic context, we have studied two distinct robotic platforms. The first, the *Event Perceiver* is a system that can perceive human events acted out with objects, and can thus generate descriptions of these actions. The second is the *Sony AIBO* robot having local visual processing capabilities in addition to autonomous mobility. In the latter, we explore human-robot interaction in the context of four-legged RoboCup soccer league. From the psychologically valid language context, we base the interactions on a model of language and meaning correspondence

G. Lakemeyer et al. (Eds.): RoboCup 2006, LNAI 4434, pp. 379–386, 2007.

developed by Dominey et al. [1] having described both neurological and behavioral aspects of human language, and having been deployed in robotic contexts, and second, on the notion of shared intentions or plans by Tomasello et al. [2, 3] that will be used to guide the collaborative interaction between human and robot. The following sections describe the platforms, the spoken language interface for command, control and teaching these systems, and current experimental results with the Sony AIBO platform.

2 Cognitive Robotics: A Spoken Language Approach

In Dominey & Boucher [4, 5, 6] we describe the **Event Perceiver System** that could adaptively acquire a limited grammar based on training with human narrated video events. An image processing algorithm extracts the meaning of the narrated events translating them into *action(agent, object, recipient)* descriptors. The event extraction algorithm detects physical contacts between objects (see [7]), and then uses the temporal profile of contact sequences in order to categorize the events. The visual scene processing system is similar to related event extraction systems that rely on the characterization of complex physical events (e.g. give, take, stack) in terms of composition of physical primitives such as contact (e.g. [8, 9]). Together with the event extraction system, a speech to text system was used to perform translations sentence to meaning using different languages [10].

2.1 Processing Sentences with Grammatical Constructions

Each narrated event generates a well formed <*sentence*, *meaning*> pair that is used as input to a model that learns the sentence-to-meaning mappings as a form of template in which nouns and verbs can be replaced by new arguments in order to generate the corresponding new meanings. These templates or grammatical constructions (see [11]) are identified by the configuration of grammatical markers or function words within the sentences [12].

Table 1. Sentences and corresponding constructions

	Sentence	Construction <sentence, meaning>
1	The robot kicked the ball	<*Agent event object, event(agent, object*>
2	The ball was kicked by the robot	<*Object* was *event* by *agent, event(agent, object*>
3	The red robot gave the ball to the blue robot	<*Agent event object* to *recipient, event(agent, object, recipient)*>
4	The ball was given to the blue robot by the red robot	<*Object* was *event* to *recipient* by *agent, event(agent, object, recipient)*>
5	The blue robot was given the ball by the red robot	<*Recipient* was *event object* by *agent, event(agent, object, recipient)*>

Each grammatical construction corresponds to a mapping from sentence to meaning. This information is also used to perform the inverse transformation from meaning to sentence. For the initial sentence generation studies we concentrated on the 5 grammatical constructions shown in Table 1. These correspond to constructions with one verb and two or three arguments in which each of the different arguments can take the focus position at the head of the sentence. On the left example sentences are presented, and on the right, the corresponding generic construction is shown. In the representation of the construction, the element that will be at the pragmatic focus is underlined.

This construction set provides sufficient linguistic flexibility, for example, when the system is interrogated about the red robot, the blue robot or the ball. After describing the event *give(red robot, blue robot, ball)*, the system can respond appropriately with sentences of type 3, 4 or 5, respectively. Note that sentences 1-5 are specific sentences that exemplify the 5 constructions in question, and that these constructions each generalize to an open set of corresponding sentences.

We have used the CSLU Speech Tools Rapid application Development (RAD) [13] to integrate these pieces, including (a) scene processing for event recognition, (b) sentence generation from scene description and response to questions, (c) speech recognition for posing questions, and (d) speech synthesis for responding.

2.2 Shared Intentions for Learning

Perhaps the most interesting aspect of the three part "command, interrogate, teach" scenario involves learning. Our goal is to provide a generalized platform independent learning capability that acquires new *<percept, response>* constructions. That is, we will use existing perceptual capabilities, and existing behavioral capabilities of the given system in order to bind these together into new, learned *<percept, response>* behaviors.

The idea is to create new *<percept, response>* pairs that can be permanently archived and used in future interactions. Ad-hoc analysis of human-human interaction during teaching-learning reveals the existence of a general intentional plan that is shared between teachers and learners, which consists of three components. The first component involves specifying the percept that will be involved in the *<percept, response>* construction. This percept can be either a verbal command, or an internal state of the system that can originate from vision or from another sensor. The second component involves specifying what should be done in response to this percept. Again, the response can be either a verbal response or a motor response from the existing behavioral repertoire. The third component involves the binding together of the *<percept, response>* construction, and validation that it was learned correctly. This requires the storage of this new construction in a construction database so that it can be accessed in the future. This will permit an open-ended capability for a variety of new types of communicative behavior.

In the following section this capability is used to teach a robot to respond with physical actions or other behavioral responses to perceived objects or changes in internal states. The user enters into a dialog context, and tells the robot that we are going to learn a new behavior. The robot asks *what is the perceptual trigger of the behavior* and the human responds. The robot then asks *what is the response behavior*, and the human responds again. The robot links the *<percept, response>* pair together so that it can be used in the future.

Having human users control and interrogate robots using spoken language results in the ability to ergonomically teach robots. Additionally, it is also useful to execute components of these action sequences conditional on perceptual values. For example the user might want to tell the robot to walk forward until it comes close to an obstacle, using a "command *X* until *Y*" construction, where *X* corresponds to a continuous action (e.g. walk, turn left) and *Y* corresponds to a perceptual condition (e.g. collision detected, ball seen, etc.).

3 Human-Robot Coaching in RoboCup Soccer

In order to demonstrate the generalization of the spoken language human-robot interaction approach we have begun a series of experiments in the domain of RoboCup Soccer [14], a well documented and standardized robot environment thus provides a quantitative domain for evaluation of success. For this project we have chosen as testing platform the Four-Legged league where ITAM's Eagle Knights team regularly competes [15, 16]. In this league two teams of four robots play soccer on a small-carpeted soccer field using Sony's Four-Legged AIBO robots. While no human intervention is allowed during a game, in the future humans could play a decisive role analogous to real soccer coaches adjusting in real-time their team playing characteristics according to the state of the game, individual or group performance. While no such human interaction is possible in the Four-Legged league, RoboCup incorporates a simulated coaching league where coaching agents can learn during a game and then advice virtual soccer agents on how to optimize their behavior accordingly (see [17, 18]).

3.1 Human-Robot Architecture

The human-robot interaction architecture is illustrated in Figure 2. The spoken language interface is provided by the CSLU-RAD framework while communication to the Sony AIBO robots is done in a wireless fashion via the CMU Tekkotsu platform [19] and URBI [21]. The CMU Tekkotsu and URBI systems provide a high level interface for remotely controlling the AIBO. Via this interface, the AIBO can be commanded to perform different actions as well as be interrogated with respect to various internal state variables. Additionally, Tekkotsu provides a vision and motion library where higher level perceptions and movements can be specified. The AIBO architecture shown at the right hand side of Figure 1 describes the robot processing modules. To play soccer robots are programmed with a set of behaviors that are activated depending on information read from sensors and state information that includes ball position, game state, localization, number of robots in the field, team strategies, etc.

3.2 Command, Interrogate and Teach Dialogs

In order to demonstrate the human coaching model we have developed and experimented with simple dialogs that let the user: (1) *command* the robot to perform certain actions; (2) *interrogate* the robot specific questions about its state; and (3) *teach* the robot to link a sequence of lower level behaviors into a higher level command such as "*Go get the ball and walk it into the goal*". Videos for these dialogs can be found in [20]. A sample command and interrogate dialog is shown in Table 2.

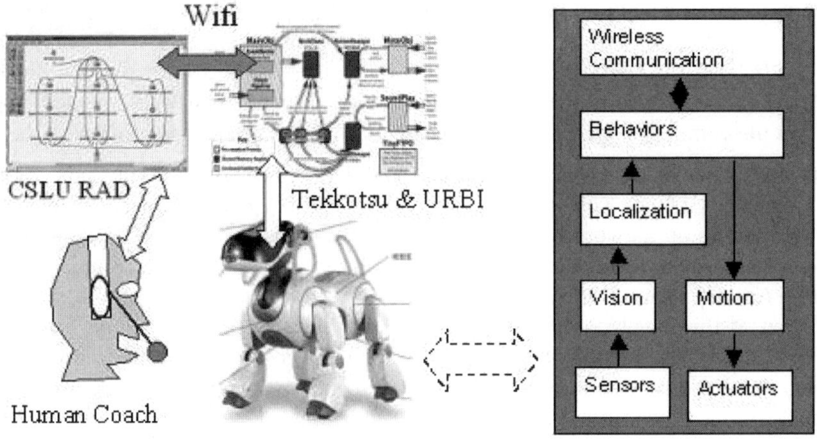

Fig. 1. A schematic view of the CSLU-Tekkotsu/URBI-AIBO system architecture. The diagram to the right shows the internal AIBO processing modules: Sensors, Actuators, Vision, Motion, Localization, Behaviors and Wireless Communication.

Table 2. Example dialog for commanding and interrogating the robot

Robot: *Welcome to the LAFMI Human-Robot Interaction Demonstration. I am the AIBO…* **Human**: *Do you see the ball?* **Robot**: *Looking for the ball…I don't see anything.* **Human**: *Turn right.* **Robot**: *Turning right. (Turns right)* **Human**: *Do you see the ball?* **Robot**: *Looking for the ball… I see the ball.* **Human**: *Get the ball.* **Robot**: *Going for the ball. (Starts walking to the ball).* **Human**: *Stop. Turn right.* **Robot**: *Turning right. (Turns right but misses the ball)* **Human**: *Back up.* **Robot**: *Back up. (Backs up)* **Human**: *Turn left.* **Robot**: *Turning left. (Turns left but misses the ball)* **Human**: *Turn right.* **Robot**: *Turning right. (Turns right and catches the ball)* **Human**: *Walk.* **Robot**: *Walking. (Starts walking, pushing the ball towards the goal)* **Human**: *Walk.* **Robot**: *Walking. (Keeps walking, pushing the ball and "scoring")*

3.3 Human-Robot Coaching

In pursuing coaching capabilities we utilize the three previous levels of human-robot interaction having been defined in the context of soccer playing robots: command, interrogate and teach. We have defined a set of basic commands, action-only and action-perception behaviors that can be instructed to the robot. Additionally the robot may be interrogated with state and perception related queries. Finally, these commands form the basis for teaching new behaviors in the soccer playing domain. While different levels of these commands have already been implemented in the AIBO in the context of soccer playing, we are at this point experimenting with them.

Command. We define a set of action-only and action perception commands. Action-only commands i.e. no perception, include: *Stop, Move, Turn*, *Turn Head*, and *Kick Ball.* Depending on the commands, these may include arguments such as magnitude of rotation, and movement in degrees or steps, etc. For example a rotation command would be *Turn 180 degrees* and a movement command would be *Move 4 steps*. It should be noted that at this level commands such as *Kick Ball* would not use any perceptual information, i.e. the resulting kick will be similar (hopefully) to the current robot orientation. We also define a set of action-perception commands requiring the full perception-action cycle, i.e. the action to be performed depends on the current robot perceptions. These commands include: *Kick Ball* with a specified direction; *Reach Ball* moving to a position behind the ball pointing towards the goal; *Initial Position* during game initialization requiring localization in the field; *Pass the Ball* to gently kick the ball to another team robot; *Move to Location* specifying a position in the field where to move; *Search Ball* resulting in robot looking for a ball nearby; *Explore Field* resulting in a more extensive search for the ball; *Defend Goal* resulting in all robots moving close to the goal requiring knowledge of the robot location in the field; *Defend Kick* in trying to block a kick from the other team, requiring knowledge of ball location, and *Attack Goal* similar although opposite in behavior to defending goal.

Interrogation. We define state and perception interrogation commands returning information on current actions or behaviors. State interrogations include for example: *What was your last action,* e.g. kicked the ball; *Why did you take the last action,* e.g., I saw the ball, so I moved towards it; *What is your current behavior*, e.g. I'm searching for the ball; *What is your current role in the game*, e.g. I am the goalie. Perception interrogations include for example: *Do you see the ball* returning e.g. *I do, I don't*; *What is your distance to the ball*, returning e.g. *30 centimeters*; *What is your current orientation,* returning e.g. *45 degrees* (in relation to field coordinate system); *What is your current position,* returning e.g. *I am in region 9*; *What is the position of object X* returning an estimate of its position.

Teach. The ultimate goal in human-robot coaching in the context of soccer is being able to positively affect the team performance during a game. While part of this interaction can eventually be carried out by agent coaches inside the robot, it is our goal to define the basic capabilities and communication interactions that human coaches should have. For example, being able to transmit strategy knowledge in the form "*if blocked pass the ball to player behind*". Such a command will modify an internal robot database with "*if possess(ball) and goal(blocked) then pass(ball)*".

4 Conclusions and Discussion

The stated objective of the current research is to develop a generalized approach for human-machine interaction via spoken language that exploits recent developments in cognitive science - particularly notions of grammatical constructions as form-meaning mappings in language, and notions of shared intentions as distributed plans for interaction and collaboration. In order to do this, we tested human-robot interaction initially with the Event Perceiver system and later on with the Sony AIBOs under soccer related behaviors.

With respect to social cognition, shared intentions represent distributed plans in which two or more collaborators have a common representation of an action plan in which each plays specific roles with specific responsibilities with the aim of achieving some common goal. In the current study, the common goals were well defined in advance (e.g. teaching the robots new relations or new behaviors), and so the shared intentions could be built into the dialog management system.

An initial evaluation period revealed that while technically we had demonstrated command, interrogation and teaching, the user interface ergonomics was somewhat clumsy. In particular the dialog pathways were somewhat constrained, with several levels of hierarchical structure in which the user had to navigate the control structure with several single word commands in order to teach the robot a new relation, and then to demonstrate the knowledge, rather than being able to do these operations in more natural single sentences. In order to address this issue, we reorganized the dialog management where context changes are made in a single step. Also, in order to focus the interactions, we worked around scenarios in which the human and robot collaborate around the shared goal of finding the ball and moving it towards a landmark so that the robot can see both at the same time.

Acknowledgements

Supported by the French-Mexican LAFMI, the ACI TTT Projects in France and the UC-MEXUS CONACYT, CONACYT grant #42440, and "Asociación Mexicana de Cultura" in Mexico.

References

1. Dominey, P.F., Hoen, M., Lelekov, T., Blanc, J.M.: Neurological basis of language in sequential cognition: Evidence from simulation, aphasia and ERP studies. Brain and Language 86(2), 207–225 (2003)
2. Tomasello, M.: Constructing a language: A usage-based theory of language acquisition. Harvard University Press, Cambridge (2003)
3. Tomasello, M., Carpenter, M., Call, J., Behne, T., Moll, H.: Understanding and sharing intentions: The origins of cultural cognition, Behavioral and Brain Sciences (2006)
4. Dominey, P.F., Boucher, J.D.: Developmental stages of perception and language acquisition in a perceptually grounded robot. Cognitive Systems Research 6(3), 243–259 (2005)
5. Dominey, P.F., Boucher, J.D.: Learning to talk about events from narrated video in a construction grammar framework. Artificial Intelligence 167(1-2), 31–61 (2005)
6. Dominey, P.F., Weitzenfeld, A.: Robot Command, Interrogation and Teaching via Social Interaction. In: IEEE-RAS International Conference on Humanoid Robots, Dec. 6-7, Tsukuba, Japan (2005)

7. Kotovsky, L., Baillargeon, R.: The development of calibration-based reasoning about collision events in young infants. Cognition 67, 311–351 (1998)
8. Siskind, J.M.: Grounding the lexical semantics of verbs in visual perception using force dynamics and event logic. Journal of AI Research 15, 31–90 (2001)
9. Steels, L., Baillie, J.C.: Shared Grounding of Event Descriptions by Autonomous Robots. Robotics and Autonomous Systems 43(2-3), 163–173 (2002)
10. Dominey, P.F., Inui, T.: A Developmental Model of Syntax Acquisition in the Construction Grammar Framework with Cross-Linguistic Validation in English and Japanese. In: Proceedings of the CoLing Workshop on Psycho-Computational Models of Language Acquisition, Geneva, pp. 33–40 (2004)
11. Goldberg, A.: Constructions. U Chicago Press, Chicago and London (1995)
12. Bates, E., McNew, S., MacWhinney, B., Devescovi, A., Smith, S.: Functional constraints on sentence processing: A cross linguistic study. Cognition 11, 245–299 (1982)
13. CSLU Speech Tools Rapid application Development (RAD), http://cslu.cse.ogi.edu/toolkit/index.html
14. Kitano, H., Asada, M., Kuniyoshi, Y., Noda, I., Osawa, E.: Robocup: The robot world cup initiative. In: Proceedings of the IJCAI-95 Workshop on Entertainment and AI/ALife (1995)
15. Martínez-Gómez, J.A., Medrano, A., Chavez, A., Muciño, B., Weitzenfeld, A.: Eagle Knights AIBO Team, Team Description Paper, VII World Robocup 2005, Osaka, Japan, July 13-17 (2005)
16. Martínez-Gómez, J.A., Weitzenfeld, A.: Real Time Localization in Four Legged RoboCup Soccer. In: Martínez-Gómez, J.A, Weitzenfeld, A. (eds.) Proc. 2nd IEEE-RAS Latin American Robotics Symposium, Sao Luis, Maranhao Brasil, Sept 24-25 (2005)
17. Riley, P., Veloso, M., Kaminka, G.: An empirical study of coaching. In: Distributed Autonomous Robotic Systems 6, Springer, Heidelberg (2002)
18. Kaminka, G., Fidanboylu, M., Veloso, M.: Learning the Sequential Coordinated Behavior of Teams from Observations. In: RoboCup-2002 Symposium, Fukuoka, Japan (June 2002)
19. CMU Tekkotsu: http://www-2.cs.cmu.edu/~tekkotsu/
20. Dominey, P.F., Weitzenfeld, A.: Videos for command, interrogate and teach AIBO robots, ftp://ftp.itam.mx/pub/alfredo/COACHING/
21. Universal Real-time Behavior Interface: http://www.urbiforge.com/

Panoramic Localization in the 4-Legged League
Removing the Dependence on Artificial Landmarks

Jürgen Sturm[1], Paul van Rossum[2], and Arnoud Visser[1]

[1] Universiteit van Amsterdam
[2] Technische Universiteit Delft
http://www.dutchaiboteam.nl

Abstract. The abilities of mobile robots depend greatly on the performance of basic skills such as vision and localization. Although great progress has been made in the 4-Legged league in the past years, the performance of many of those approaches completely depends on the artificial environment conditions established on a 4-Legged soccer field. In this article, an algorithm is introduced that can provide localization information based on the natural appearance of the surroundings of the field. The algorithm starts making a scan of the surroundings by turning head and body of the robot on a certain spot. The robot learns the appearance of the surroundings at that spot by storing color transitions at different angles in a panoramic index. The stored panoramic appearance can be used to determine the rotation (including a confidence value) relative to the learned spot for other points on the field. The applicability of this kind of localization for more natural environments is demonstrated in two environments other than the official 4-Legged league field.

1 Introduction

1.1 Context

Mobile robots need to know where they are. Robot localization is therefore an important basic skill of mobile robots, e.g. when playing robot soccer. Many other processes of the robot's cognition - like world modeling and action planning - strongly depend on fast, accurate and robust position estimates.

In the 4-Legged league[1] of the RoboCup, teams consisting of four Sony Aibo robot dogs play soccer fully autonomously against each other on a field of 6x4m. Colored flags, goals and various field lines can be used to achieve localization accuracies below six centimeters [1,2].

The price that these approaches pay is their total dependency on artificial landmarks of known shape, positions and color. Most algorithms even require manual calibration of the actual colors and lighting conditions used on a field and still are quite susceptible for disturbances around the field, as for instance produced by brightly colored clothes in the audience.

[1] RoboCup Four Legged League homepage, last accessed in April 2006, http://tzi.de/4legged

G. Lakemeyer et al. (Eds.): RoboCup 2006, LNAI 4434, pp. 387–394, 2007.

The interest in more general solutions has been (and still is) growing over the past few years. The almost-SLAM challenge [2] of the 4-Legged league and the upcoming challenges in natural environments of the newly founded RoboCup @ home league [3] reflect this growing interest within the robotics community.

1.2 Related Work

As many teams of the 4-Legged league use similar approaches for vision and localization (from the raw camera image to the robot's pose), we briefly want to point out typical main processing steps and their corresponding considerations.

1. In the **image processing** step, the image pixels are **divided** into color classes. Coarse scanning through the image then yields object candidates that need to be post-processed by **object specialists** for object recognition. Finally the complete set of detected objects is filtered by a second-order **sanity checker** to remove invalid percepts. Representative and well-working implementations can be found in [1,3].
2. In the **localization** step, the perceived landmarks are filtered over time and merged with the data from the robot's body odometry. Common approaches include Kalman Filters, Monte-Carlo approaches or combinations thereof. For a good comparison see [4].

Other interesting approaches can be found in the Mid-Size league [4] : as the hardware of a Mid-Size robot is only limited to what it can carry, both different types of sensors as well as different classes of algorithms become feasible. Most Midsize robots use omni-directional high-quality cameras whereof each camera image carries enough information to localize the robot almost perfectly [5]. These approaches are not applicable on an Aibo with its limited camera angle; therefore panorama images have to be constructed of several images.

Also inspiring, but also only partially transferable to the 4-Legged league is the work that has been done on the SLAM problem in general, for instance on panoramic pictures [6,7,8,9]. One of these approaches [10] divides the panoramic image in multiple sectors, but uses as characteristic feature the average color of the sector. Our approach combines sectors with as characteristic feature the frequency of colortransitions.

2 Approach

The main idea is quite intuitive: we would like the robot to generate and store a 360° panorama image of its environment while it is in the learning phase. After

[2] Details about the Simultaneous Localization and Mapping challenge can be found at http://www.tzi.de/4legged/pub/Website/Downloads/Challenges2005.pdf

[3] RoboCup @ Home League homepage, last accessed in March 2006, http://www.ai.rug.nl/robocupathome/

[4] RoboCup Mid-Size league homepage, last accessed in April 2006, http://www.idt.mdh.se/rc/Mid-Size/

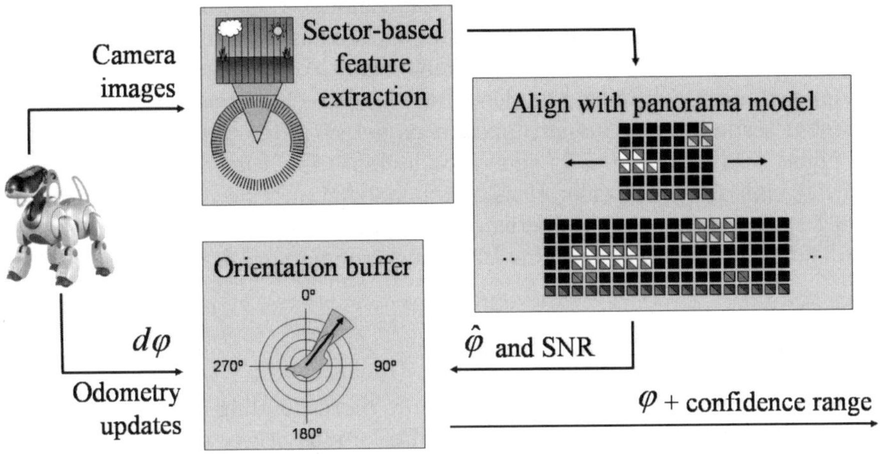

Fig. 1. Architecture of our algorithm

(a) Unsupervised learned color segmentation.

(b) Sectors and strong color transitions visualized.

Fig. 2. Image processing: from the raw image to sector representation. This conversion consumes approximately 6ms/frame on an Sony Aibo ERS7.

that, it should align each new image with the stored panorama, and from that the robot should be able to derive its relative position (in the localization phase). This alignment is not trivial because the new image can be translated, rotated, stretched and perspectively distorted when the robot does not stand anymore at the point where the panorama was originally learned.

Of course, the Aibo is not able (at least not in real-time) to compute this alignment on full-resolution images. Therefore a reduced feature space is designed so that the computations become tractable [5] on an Aibo. Figure 1 gives a quick overview of the algorithm's main components.

The Aibo performs a **calibration** phase before the actual learning can start. In this phase the Aibo first decides on a suitable camera setting (i.e. camera

[5] Our algorithm consumes per image frame approximately 16ms, therefore we can easily process images at the full Aibo frame rate (30fps).

gain and the shutter setting) based on the dynamic range of brightness in the **autoshutter** step. Then it collects color pixels by turning its head for a while and finally clusters these into 10 most important color classes in the **color clustering** step using a standard implementation of the Expectation-Maximization algorithm assuming a Gaussian mixture model [11]. The result of the calibration phase is an automatically generated lookup-table that maps every YCbCr color onto one of the 10 color classes and can therefore be used to segment incoming images into its characteristic color patches (see figure 2(a)). As these initialization steps are not in the focus of this article we kept the explanation short.

2.1 Sector Signature Correlation

Every incoming image is now divided into its corresponding sectors[6]. Using the lookup table from the unsupervised learned color clustering, we can compute the sector features by counting per sector the transition frequencies between each two color classes in vertical direction. This yields 10x10 transition frequencies per sector, which we subsequently discretize into 5 logarithmically scaled bins. In figure 2(b) we displayed the strongest color transitions (bin 5) for each sector.

In the **learning phase** we estimate these 80x(10x10) distributions[7]. We define a single distribution for a currently perceived sector by

$$P_{current}(i,j,bin) = \begin{cases} 1 & discretize(freq(i,j)) = bin \\ 0 & otherwise \end{cases} \tag{1}$$

and the distribution learned from many frequency count samples of a certain sector by

$$P_{learned}(i,j,bin) = \frac{count_{sector}(i,j,bin)}{\sum\limits_{bin \in frequencyBins} count_{sector}(i,j,bin)} \tag{2}$$

Now we can simply multiply the current and the learned distribution to get the correlation between a currently perceived and a learned sector:

$$Corr(P_{current}, P_{learned}) = \prod_{\substack{i,j \in colorClasses, \\ bin \in frequencyBins}} P_{learned}(i,j,bin) \cdot P_{current}(i,j,bin) \tag{3}$$

2.2 Alignment

After all the correlations between the stored panorama and the new image signatures were evaluated, we would like to get an alignment between the stored and

[6] 80 sectors corresponding to 360°; with an opening angle of the Aibo camera of approx. 50°, this yields between 10 and 12 sectors per image (depending on the head pan/tilt).

[7] When we use 16bit integers, a complete panorama model can be described by (80 sectors)x(10 colors x 10 colors)x(5 bins)x(2 byte) = 80 KB of memory.

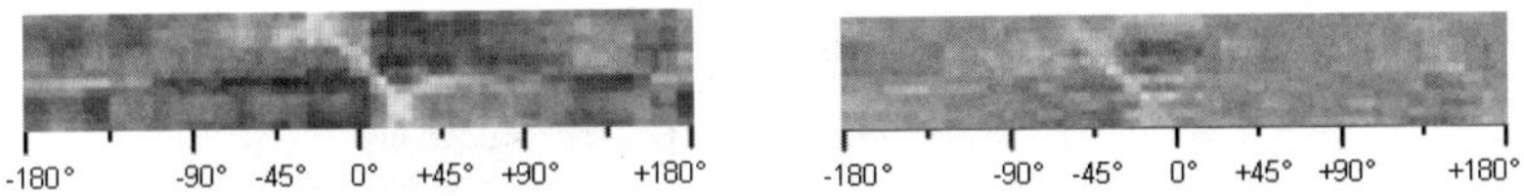

(a) robot looking straight ahead (b) robot looking 45° to the left

Fig. 3. Resulting sector correlation matrix between the 12 seen sectors of the current image and the 80 stored panorama model sectors. The bright diagonal line corresponds to a minimum in the correlation and constitutes the alignment we want to find.

seen sectors so that the overall likelihood of the alignment becomes maximal. In other words, we want to find a diagonal path with the minimal cost through the correlation matrix.

We consider the fitted path to be the true alignment and extract the rotational estimate φ_{robot} from the offset from its center pixel to the diagonal ($\Delta sectors$):

$$\hat{\varphi}_{robot} = \frac{360^\circ}{80} \Delta sectors \tag{4}$$

Further, we try to estimate the noise by fitting again a path through the correlation matrix far away from the best-fitted path.

$$SNR = \frac{\sum_{(x,y)\in minimumPath} Corr(x,y)}{\sum_{(x,y)\in noisePath} Corr(x,y)} \tag{5}$$

The results of eq. 4 and eq. 5 can be found in figure 4.

3 Results

3.1 Environments

We selected five different environments to test our algorithm under a variety of circumstances. The first two experiments were conducted at home on a sunny afternoon and in an office environment[8] to measure performance under real-world circumstances. Furthermore, we conducted exhaustive tests on a (classical) 4-Legged field to test the performance under RoboCup circumstances. Then, we repeated the same measurements on a Midsize field[8],which could become interesting for future 11-against-11 games. Even more challenging, we took an Aibo to a real-world soccer field outdoors. This could be interesting especially for public demonstrations because until now all demonstrations had to be given in closed rooms with artificial lights.

[8] Results omitted due to the lack of space, both located at Delft.

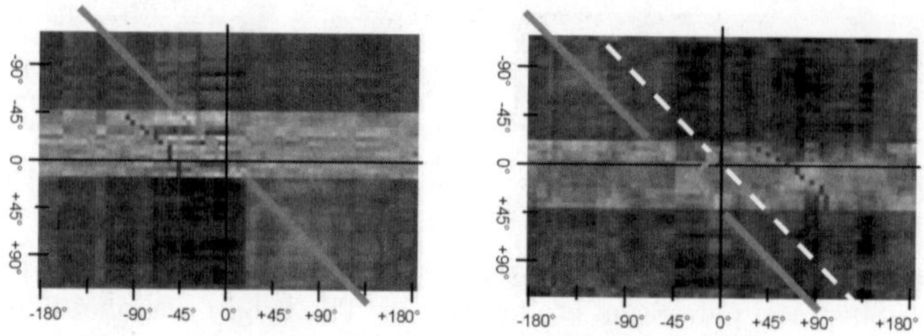

(a) Robot standing on the trained spot (matching line is just the diagonal)

(b) Robot turned right by 45 degrees (matching line displaced to the left)

Fig. 4. Visualization of the alignment step while the robot is scanning with its head. The green line marks the minimum path (assumed true alignment) while the red line marks the second-minimal path (assumed peak noise). The grey line represents the diagonal, while the orange line illustrates the distance between the found alignment and the center diagonal ($\Delta sectors$).

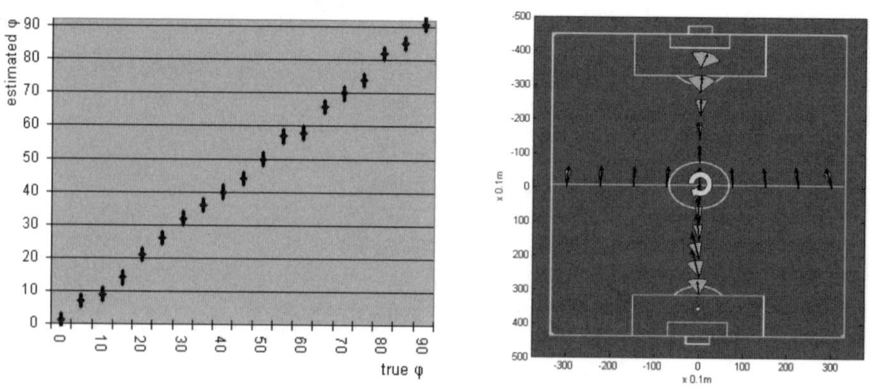

(a) Rotational test in natural environment (living room, sunny afternoon)

(b) Outdoor human soccer field (80x 64m)

Fig. 5. Results of a rotational experiment conducted in a regular living room (left picture) and a translational test conducted on an outdoor soccer field

3.2 Measured Results

Figure 5(a) illustrates the results of a rotational test in a normal living room. As the error in the rotation estimates ranges between -4.5 and +4.5 degrees, we may assume an error in alignment of a single sector; moreover, the size of the confidence interval can be translated into one and two sectors respectively, which corresponds to the maximal angular resolution of our approach.

The next experiments were conducted on different types of soccer fields, exemplarily we present the measurements recorded on a real outdoor soccer field (fig. 5(b)) and a regular 4-Legged field indoor (fig. 6(a)). We developed a

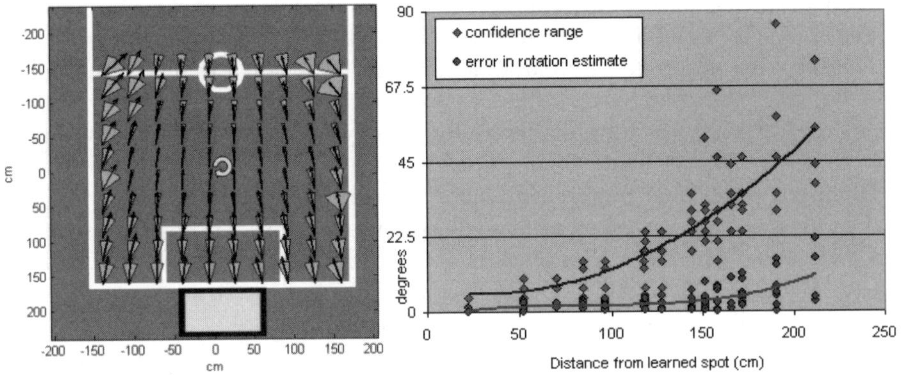

(a) 4-Legged field (6x4m)

(b) Signal degradation as a function of the distance to the learned spot (measured on a 4-Legged field)

Fig. 6. Test results on an official field of the 4-Legged league

suitable visualization that resembles a magnetic field where we can display estimated rotations and confidence ranges in an intuitive way. The direction of an arrow shows the estimated rotation (with respect to the trained spot) and the grey arc shows the 80%-confidence interval around this estimate. Interestingly, it can be seen that the (imaginary) elongations of the arrows all run together in a single spot corresponding to the intersection point with the panoramic index.

In both cases it can be seen intuitively that the rotation estimates are within acceptable range, in the sense that all arrows point in the direction of the original 0-sector. This can also be shown quantitatively (see figure 6(b))): both the rotational error and the width of the confidence interval increase slowly and in a graceful way when the robot is moved away from the training spot.

4 Conclusion

Although at first sight the algorithm seems to rely on specific texture features of the surrounding surfaces, in practice no dependency could be found. This can be explained by two reasons: firstly, as the (vertical) position of a color transition is not used anyway, the algorithm is quite robust against (vertical) scaling. Secondly, as the algorithm aligns on many color transitions in the background (typically more than a hundred in the same sector), the few color transitions produced by objects in the foreground (like beacons and spectators) have a minor impact on the match (because their sizes relative to the background are comparably small).

The lack of absolute position estimates seems to be a clear drawback with respect to the other methods, but bearing information alone can already be very useful for certain applications. For example, an attacking robot can highly benefit from a robust bearing estimation towards the goal. With this bearing estimating the robot can rush into the right direction. After 3 seconds the robot has to shoot, at that moment an additional distance estimation could be advantageous.

Further, this approach of panoramic localization is actually interesting for a broader spectrum than soccer. The requirements for both the robot as well as for its environment are quite moderate (on a Sony Aibo ERS7, the computation time is below 20ms/frame). The robot itself needs only a simple camera and medium computational power, while most natural environments (both indoors and outdoors) carry, as shown, enough panoramic information the algorithm can lock on to. Therefore, this method becomes for example interesting for the newly established RoboCup @ Home league, where fast localization information is needed in natural but completely unknown environments.

As the training on a single spot can be completed in less than one minute on a Sony Aibo in an arbitrary place, small demonstrational games (for example a striker versus a goalkeeper) could be set up more easily especially at non-prepared places. Progress in this domain facilitates the advancement of mobile robots - and thereby robotics research itself - into more natural environments.

References

1. Röfer, T., et al.: GermanTeam RoboCup 2005. Online, 247 pages (2005)
2. Sturm, J., Visser, A., Wijngaards, N.: Dutch aibo team: Technical report robocup 2005. Technical report, Dutch Aibo Team (2005)
3. Quinlan, M.J., et al.: Nubots team report. Technical report, University of Newcastle (2005)
4. Gutmann, J.S., Fox, D.: An experimental comparison of localization methods continued. In: Proceedings of the IEEE/RSJ International Conference on Intelligent Robots and Systems (IROS'02), Lausanne, Switzerland (2002)
5. Terwijn, B., Porta, J., Kröse, B.: A particle filter to estimate non-markovian states. In: Groen, F. (ed.) International Conference on Intelligent Autonomous Systems, IAS'04, pp. 1062–1069. IOS Press, Amsterdam (2004)
6. Jogan, M., Ales, L.: Robust localization using panoramic view-based recognition. In: Proceedings 15th International Conference on Pattern Recognition ICPR'00, Barcelona, Spain, pp. 136–139 (2000)
7. Pajdla, T.: Robot localization using shift invariant representation of panoramic images (research report k335-cmp-1998-170). Technical report, Czech Technical University, Prague (1998)
8. Bunschoten, R.: Mapping and Localization from a Panoramic Vision Sensor. PhD thesis, Universiteit van Amsterdam (2003)
9. Gonzalez, J., Lacroix, S.: Rover localization in natural environments by indexing panoramic images. In: International Conference on Robotics and Automation 2002, Washington, DC (USA) (2002)
10. Gross, H.M., et al.: Omnivision-based probabilistic self-localization for a mobile shopping assistant continued. In: Groen, F. (ed.) IEEE/RSJ Int. Conf. on Intell, pp. 1505–1511. IEEE omnipress, Los Alamitos (2003)
11. Verbeek, J.: Mixture models for clustering and dimension reduction. PhD thesis, Universiteit van Amsterdam (2004)

Orientation Extraction and Identification of the Opponent Robots in RoboCup Small-Size League

Saori Umemura, Kazuhito Murakami, and Tadashi Naruse

Graduate School of Information Science and Technology, Aichi Prefectural University
Kumabari, Nagakute-cho, Aichi, 480-1198 Japan
im051005@cis.aichi-pu.ac.jp,
{murakami,naruse}@ist.aichi-pu.ac.jp

Abstract. In RoboCup small-size league, it is necessary to analyze the opponent robots' behavior in order to make a strategy of the own team. However, it is difficult to prepare image processing methods in advance in order to detect opponent robots' sub-markers used for the orientation detection and identification of the robots, because there is no limitation in the rule in shape, color, arrangement, and the number. This paper proposes a new method to select the most specific sub-marker attached on the top of the robot based on the features such as the size, area, and color values by using the discriminant analysis, and also explains how to extract opponent robots' orientations with some experimental results.

1 Introduction

It is necessary to analyze the opponent robots' behavior in order to make a strategy of the own team in RoboCup. Almost all of the teams utilize one or a set of sub-markers attached on the top of the robot for the orientation extraction and identification[1]. Even though it is not so easy to extract own robots in real time, it becomes more difficult for a team to recognize opponent team's robots, because we have no knowledge about opponent team's sub-markers and also we can't prepare the image processing algorithms for the recognition of them in advance. Figure 1 shows some examples of sub-markers. There is no limitation in the rule in shape, color, arrangement, and the number of the sub-marker in the Small-Size League (SSL)[2]. The freedom of designing sub-markers of own team rises, at the same time the recognition rate of sub-markers of the opponent team falls.

This paper proposes a new method to select the most specific sub-marker attached on the top of the robot based on the features such as the size, area, and color values by using the discriminant analysis, and also presents how to extract opponent robots' orientations. This method realizes a strategic planning of the robots based on the locus and the direction of the opponent robots. First, this paper describes how to utilize the orientation of the robots in the section 2, and then explains the recognition method of the opponent robots and experimental results in the sections 3 and 4, respectively.

G. Lakemeyer et al. (Eds.): RoboCup 2006, LNAI 4434, pp. 395–401, 2007.

(a) team A (b) team B (c) team C

Fig. 1. Examples of sub-markers

2 How to Utilize the Orientations of the Opponent Robots for Planning

SSL's robot has a kicking device and a dribbling device. In this paper, let the orientation of them be the 'front' of the robot. To recognize the opponent robots' orientations realizes a strategic planning as follows.

2.1 Judgment of Kicking or Holding a Ball

If the orientation of the opponent robot nearby a ball is known, the system can judge whether the opponent robot is kicking or holding the ball. More strategic planning as shown Figure 2 could be realized by this information.

Fig. 2. Judgment of having ball

2.2 Judgment of the Shoot Course

In the penalty kick scene, for example, many teams take a strategy to prevent a shoot only by changing the direction of the robot in the same position. In this kind of scene, own robot plays more defensively based on the shoot course in Figure 3 if the orientation of the opponent robot is known.

2.3 Paying Attention to the Opponent Robots

When own robot plays with paying attention to the opponent robots, the orientations of the opponent robots are very important information. Own robot plays more offensively if the robot knows which robot among opponent team's will receive a ball.

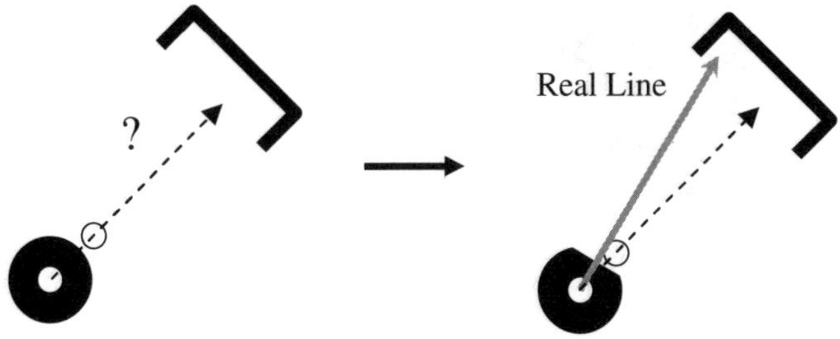

Fig. 3. Judgment of shoot course

3 Orientation Extraction and Identification of Robots

This chapter presents a method of orientation extraction and identification of opponent robots. In general, template matching technique is used to detect unknown patterns, and is a robust method to extract the most certain locus of the template from an input image, it takes much time. RoboCup's image processing system is required to work in real time, for example, 60fps performance. Since there is no information about opponent teams, we can't prepare the image processing algorithms in advance for the recognition of the opponent teams' sub-markers. Hereafter, this section explains a new method to select the most specific sub-marker attached on the top of the robot based on the features such as the size, area, and color values by using the discriminant analysis, and also presents how to extract opponent robots' orientations.

3.1 Selection of the Most Specific Sub-marker

Before the game starts, first input the opponent robots' image (sub-markers' image) and then select the most specific sub-marker, hereafter we call it 'feature marker', among N pieces of sub-markers on a robot. Both of the orientation extraction and identification of each robot is executed based on the locus or arrangement of the feature markers.

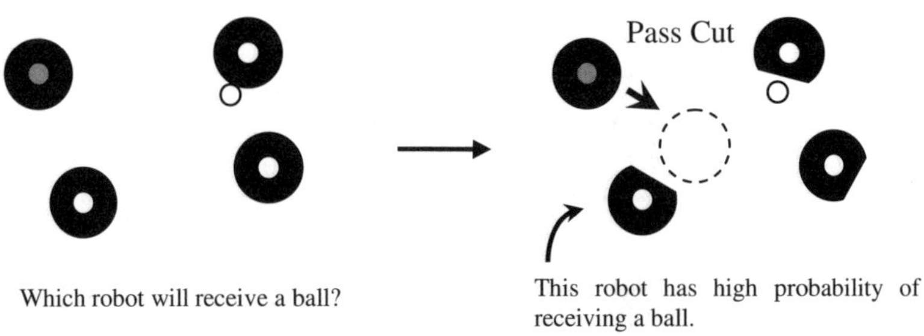

Fig. 4. Paying attention to the opponent robots

The parameter list $L^{(i)}$ of the i-th sub-marker (i=1,...N) is composed of variety of parameters such as color, area, size, the center of gravity, and so on. Let the number of them on a robot be M.

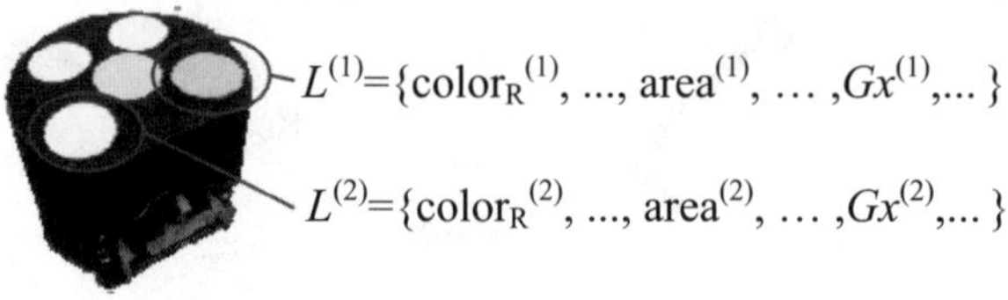

Fig. 5. Example of marker's parameters

First, make a list of each sub-marker $L^{(i)} = \{x_1^{(i)}, x_2^{(i)}, \ldots, x_M^{(i)}\}$($i = 1\ldots N$) as shown in Figure 5 and decide the 'feature marker' which has the most specific parameter value. There are many well known methods to select one of the most specific value among M, but we designed a method whose computation cost is not so high as follows.

Step-1. Make a set $F_j = \{x_j^{(1)}, x_j^{(2)}, \ldots, x_j^{(N)}\}$ of the j-th parameter (j=1,2,...,M) as shown in Figure 6.
Step-2. For the j-th parameter, apply the discriminant analysis method to the set F_j, that is, devide F_j into 2 classes and caluculate the separation ratio S_j between them.
Step-3. If each of devided classes is not composed of one element, then let S_j be 0. Repeat from Step-2 to Step-3 for all parameter j=1,2,...,M.
Step-4. Search the maximum S_j among $\{S_j ; j=1,2,\ldots,M\}$ and let the number be j_{MAX} and the element's number be i_{MAX}.
Step-5. Let the sub-marker which has i_{MAX}-th and j_{MAX}-th element be the 'feature marker' G and terminate the program.

If S_j=0 in Step-4, it means that 'feature marker' could not be decided only one parameter, so we have to combine 2 or more parameters to decide 'feature marker' G.

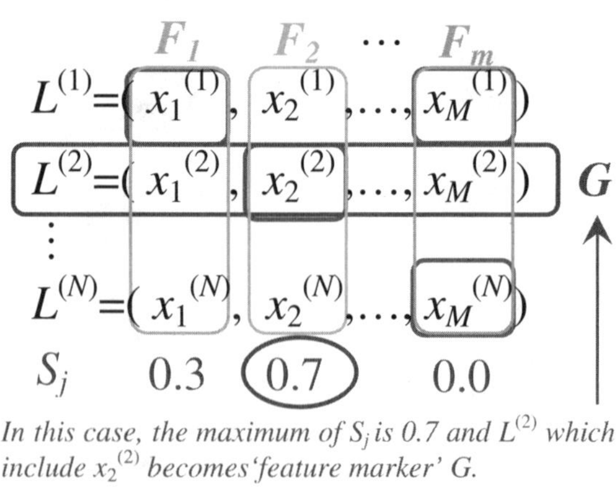

Fig. 6. Example of deciding feature maker

3.2 Identification of the Robot and Orientation Extraction

The robot is identified by matching the parameter lists (*i.e.* $\cos\psi$□0.95, here, ψ is the angle between two vectors corresponding to the lists in M dimensional space. 0.95 is obtained by some experiments).

As shown in Figure 7, the orientation of the robot is calculated by using the 'feature marker' G. If the real front orientation ϕ_0 and the angle θ_0 of the 'feature marker' G are known in advance, the front orientation ϕ during the game is obtained by

$$\phi = (\theta - \theta_0) + \phi_0 \tag{1}$$

where, θ is the angle of the 'feature marker' G. This angle θ is easily calculated by

$$\theta = \tan^{-1} \frac{y_T - y_G}{x_T - x_G} \tag{2}$$

where, (x_T, y_T) and (x_G, y_G) are the centers of gravity of team marker T and the 'feature marker' G, respectively.

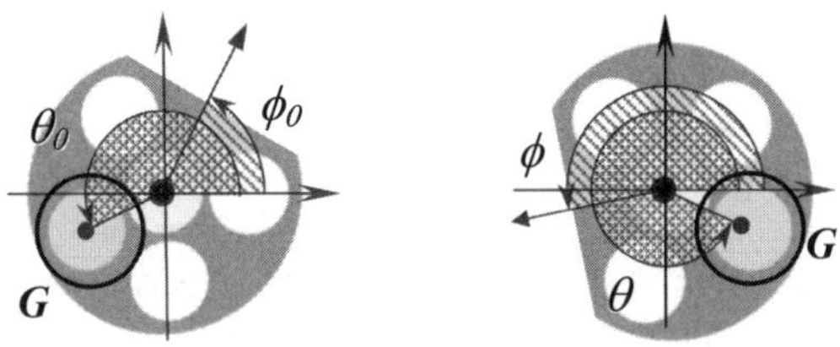

Fig. 7. Feature marker and the definition of the orientation

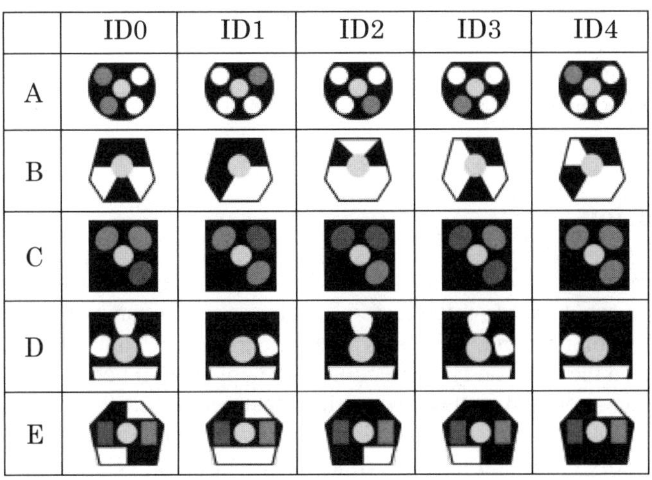

Fig. 8. Marker set of opponent robot's

4 Experiments and Discussions

In order to confirm the effectiveness of the proposed method, we experimented by using several typical sets of sub-markers used in the past games in RoboCup competitions. Figure 8 shows examples of the sub-markers, and Table 1 shows the experimental results. A, B, ...in the figure denotes the team and ID is the identification number of each robot. Orientation error in Table 1 is the difference between the real and measured angles.

Table 1. Identification rate and orientation extraction error

Set of markers	A	B	C	D	E
Identification Rate [%]	100.0	99.44	100.0	100.0	100.0
Maximum of Orientation Extraction Error [deg]	9.818	98.44	7.275	178.01	7.967
Minimum of Orientation Extraction Error [deg]	0.093	0.108	0.040	0.192	0.000
Average of Orientation Extraction Error [deg]	2.956	7.157	2.183	9.729	3.199

From the experiments, there appear some errors in the orientation extraction, especially for B and D team's sub-markers. The differences of size of sub-markers for team B's and the differences of shape, rectangle and circle sub-markers, for team D's are the main causes of the errors.

We compared calculation time with several conventional methods, template matching etc. Here, 'calculation time' is measured only foe the orientation extraction and the identification processes just after the main-marker extraction process. The results are shown in Table 2. As a result of the experiment, the accuracy of the individual identification rate was about 99.8%, and error for the orientation extraction is about 4.95 degrees.

Table 2. Comparison of calculation time

	Calculation Time [msec]	Orientation Extraction Error[deg]
proposed method	0.15	4.95
nearest neighbor method	207	2.0
bi-linear method	214	
bi-cubic method	417	

5 Conclusion

This paper described a new method to select the most specific sub-marker attached on the top of the robot based on the features such as the size, area, and color values by using the discriminant analysis. This method realized a strategic planning of the robots based on the locus and the direction of the opponent robots.

Although the proposed method shows the effectiveness, there remain some subjects to be solved. It is necessary to add and increase the menu of the shape measures such as the complexity of the sub-markers. Since the robot doesn't always kick a ball to the front direction, it is also important to introduce the learning mechanism from image sequences of the game in order to recognize real front orientation of the robot. These are future works.

References

1. RoboCup Official Site: http://www.robocup.org/
2. RoboCup F180 Rules Repository:
3. http://www.itee.uq.edu.au/%7Ewyeth/F180%20Rules/index.htm
4. RoboCup International Symposium (2005)
5. Otsu, N.: A Threshold Selection Method from Gray-Level Histograms. IEEE Transactions on Systems, Man, and Cybernetics SMC-9(1), 62–66 (1979-01)
6. 149th CVIM, pp. 149–22
7. 11th SSII: J-39, pp. 505–506

Rolling Shutter Image Compensation

Steven P. Nicklin, Robin D. Fisher, and Richard H. Middleton

School of Electrical Engineering and Computer Science, The University of Newcastle, Callaghan 2308, Australia

Abstract. This paper describes corrections to image distortion found on the Sony AIBO ERS-7 robots. When obtaining an image the camera captures each pixel in series, that is there is effectively a 'rolling shutter'. This results in a delay between the capture of the first and last pixel. When combined with movement of the camera the image produced will be distorted. The sensor values from the robot, coupled with knowledge of the camera's timing, are used to calculate the effect of the robots movement on the image. This information can then be used to remove much of the distortion from the image. The correction improves the effectiveness of shape recognition and bearing-to-object accuracy.

1 Introduction

Rolling shutters are commonly found in low-cost, low-power CMOS cameras. These cameras are being commonly used in many non-stationary and robotic applications. Cameras that contain rolling shutters do not expose the entire image at one instance, as is done with a global shutter. Instead rolling shutters have pixels that have been exposed at different times and merged together to form a single image. This causes problems when the scene changes in a time which is less than that taken to expose the entire image. This causes some pixels to have newer information than others. The combination between new and old information causes distortions in the image when viewing an object with movement relative to the camera.

These distortions are evident on the CMOS cameras found in the Sony AIBO ERS-7 robots used in the RoboCup Four-Legged League. However these errors will also be found whenever similar camera technology is used in non-stationary cameras. A major contribution to this distortion is the desire to constantly move the camera to gather as much information about the surrounding environment as possible.

These distortions can cause differences in the co-ordinates, as well as the shape of the objects that the robot has seen. Since the distortion stretches or compresses the image of objects, the co-ordinates derived from that image are also altered. The changed co-ordinates, in particular the bearing to an object, have the potential to cause problems when attempting to determine the location of the object. While the distortion causes problems when the shape of the object is used for its identification or measurement. An example of which is the circle fitting on a ball image in the Four-Legged robotic League. If the ball is no longer circular in shape, circle fitting loses some of its effectiveness.

G. Lakemeyer et al. (Eds.): RoboCup 2006, LNAI 4434, pp. 402–409, 2007.

There are two main sources for this relative velocity that causes distortions in the image. The movement may be that of the object, or the movement of the camera itself. In many cases there is a mixture of both. The velocity from the camera can in most cases be measured, or at least estimated, however the velocity of the object is far less easily determined from a single frame.

This paper covers the technique used to correct the distortion caused by the rolling shutter. The correction was designed for the Sony AIBO ERS-7 robot, the hardware used for the RoboCup Four-Legged League. In this case the image is corrected for the movement of the robot's camera caused by the panning motor on the robot. This has been observed to be a major cause of this type of distortion, particularly when calculating the bearing to objects or attempting to identify shapes. A correction method used to improve similar problems with bearings was briefly mentioned in [1] and [3].

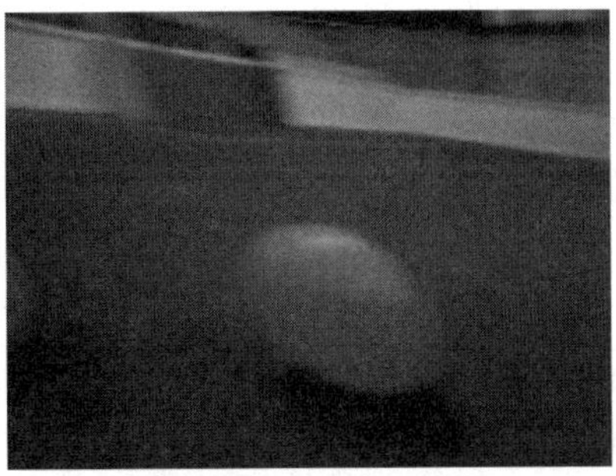

Fig. 1. Effect of image distortion can be seen on both the round ball, and the rectangular goals

2 De-Skewing Approach

The following section describes the principles and equations used to calculate the required constants for image de-skewing. Followed by the equations required to implement the correction using these values.

2.1 Delay for an Entire Image

Before the image can be de-skewed, the timing of the camera must be known. The magnitude of the delay determines the magnitude of the distortion on the image. To find this value an image was taken of a fluorescent light that flickers at a know frequency (100Hz). When viewed, the fluorescent light displayed three light and dark bands upon the image. This shows that the light went through three dark/light cycles in the time one image was taken. Given the period of the cycles (0.01s), the time taken to capture the frame was calculated to be approximately 0.03s (1). This value is very close to the frame rate of the camera, which operates at 30 frames per second. Therefore it was assumed that the

Fig. 2. Image of the fluorescent light shows three distinct light/dark cycles

entire period between frames is used to capture the next frame. This gives an approximate period of 33.3ms.

$$T_{camera} \approx 3 \times T_{FL} = 30ms \approx 33.3ms = T_{Image} \tag{1}$$

2.2 Conversion Between Pan Angle and Pixel Position

All pixels in an image have a equivalent angle offset from the center of the image. The center pixels always have an offset of zero, however the pixels near the edge of the image have an offset determined by the camera's field of view. This offset is required when converting from a pixel to a relative pan location. The relationship between image pixel and offset angle was calculated using the following trigonometry;

First an effective camera distance is found using the field of view of the robot and the resolution of the image in pixels. (figure 3). The effective camera distance is used to convert the pixel position to an offset angle and back again (figure 3).

$$CameraDist = \frac{1/2 \times ImageWidth}{\tan(1/2 \times FOV)} = \frac{104}{\tan(28.45)} = 191.9\, pixels \tag{2}$$

$$x = CameraDist \times \tan(\theta) \tag{3}$$

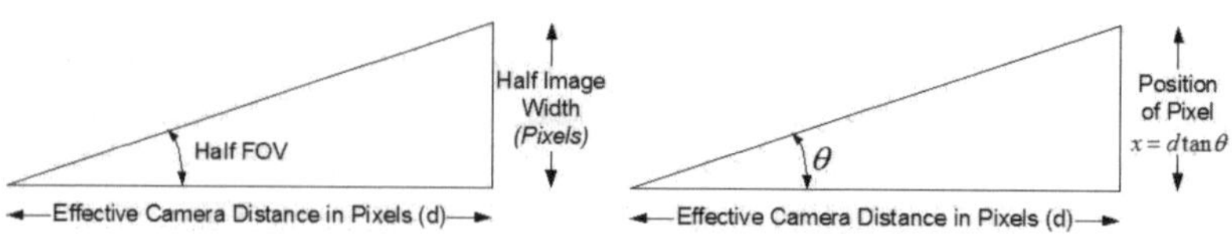

Fig. 3. Calculation of effective camera distance.$(left)$, Pixel position calculation using angle (θ).$(right)$

2.3 Finding the Angle of Distortion

The position of each pixel at the time of capture is required to correct the image. The total pan angle change during the image is proportionally applied over the pixels within the image. Since the first pixel was captured at the camera's original position it has the maximum distortion. The distortion subsequently drops for each pixel captured until the last is assumed to be captured at the camera's current position. This maximum distortion is calculated using the difference between the camera's previous position and the current position. This calculation is simplified since the camera was found to take the entire time between frames to capture a new image.

$$\theta_{MaxDistortion} = \theta_{prev} - \theta_{curr} \tag{4}$$

2.4 Applying Correction

To apply the correction the panning movement of the robot is assumed to be constant between frames. While the velocity of the camera may change within a frame, the inclusion of these changes makes de-skewing much more processor intensive. Also such changes tend to be relatively small and insignificant. For these reasons the total velocity of the camera is used to calculate a linear approximation of the camera's position at the time of pixel capture.

On the AIBO robots the pan sensor reports angles to the right to be negative, while angles to the left are positive. The pixels within the picture are described in (x, y) coordinates with the upper left hand corner $(0, 0)$. For this reason the equations assume these systems, converting between the two coordinate systems as necessary.

First, the angle of the current pixel is found using equation (5). This equation also shifts the coordinates so that the center of the image is at 0 radians. Angles and pixels to the right of this are negative, while angles and pixels to the left are positive.

$$\theta_{original} = \arctan\left(\frac{\frac{ImageWidth}{2} - x}{CameraDist}\right) \tag{5}$$

A linear approximation is then made for the pan angle for this particular pixel. Finding the amount of distortion the pixel has in relation to the bottom right pixel.

$$\theta_{distorted} = \theta_{MaxDistortion} \times \left(\frac{y \times ImageWidth + x}{TotalPixels} - 1.0\right) \tag{6}$$

This calculated distortion is then subtracted from the current pixel's angle, giving the corrected angle for that particular pixel.

$$\theta_{corrected} = \theta_{original} - \theta_{distorted} \tag{7}$$

Using this new corrected angle, the x location of this pixel can be found using equation (3). This is then shifted back to the image's (x, y) coordinate system (8).

$$\left(\frac{ImageWidth}{2} - CamDist \times \tan(\theta_{Corrected}), y\right) \tag{8}$$

Using an equation to correct individual pixels allows the correction to be performed only on the interesting pixels in the image. This means that only the required information, such as an object's X and Y values, have to be corrected, rather then the entire image. This reduction in processing allows de-skewing to be applied to more items with a less noticeable impact on CPU time.

3 Applications of Correction

The following section describes some of the uses for the correction.

3.1 Applying Correction to Shape Fitting

The rolling shutter can cause errors in both the perceived location and shapes of an object. This may cause problems when attempting to identify an object by shape and subsequently determine its position. To over come this, object candidates are first located in the distorted image. These candidates have their edge pixels corrected for skew. From these corrected points the object can then be identified by shape. If further details are needed on the object for position or distance data, the properties of the shape can be used without needing to re-correct the pixels. The use of de-skewing in this manner to verify probable objects reduces the processor load when compared to correcting the entire image.

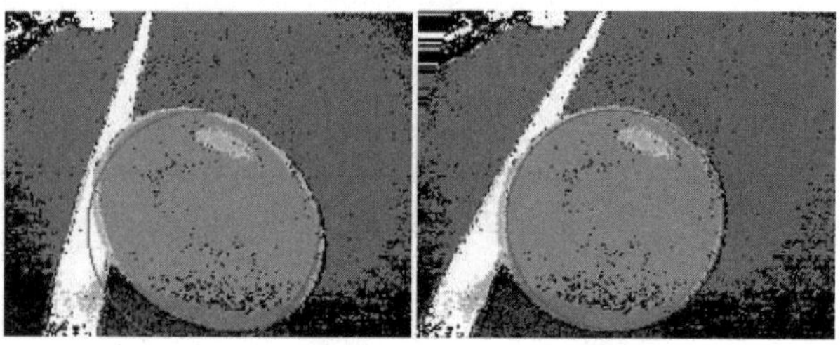

Fig. 4. Visual correction of classified image of ball. Circle fitting shown in blue. *Left* : Original Image, *Right* : Corrected image.

3.2 Line Detection

The use of field lines for localisation provides great benefits, particularly with recent reductions in the number of beacons on the roboCup field. Lines offer a good source of extra information for localisation, since the they are visible from almost anywhere on the field. Although field lines generally provide only ambiguous information for localisation, they may still allow a robot to remain localised without seeing a distinct marker. In particular, the orientation of a field line in the image may be important information.

Field lines are more susceptible to distortion due to camera skew. To use a line from an image in localisation two bearings to that line are needed. This allows

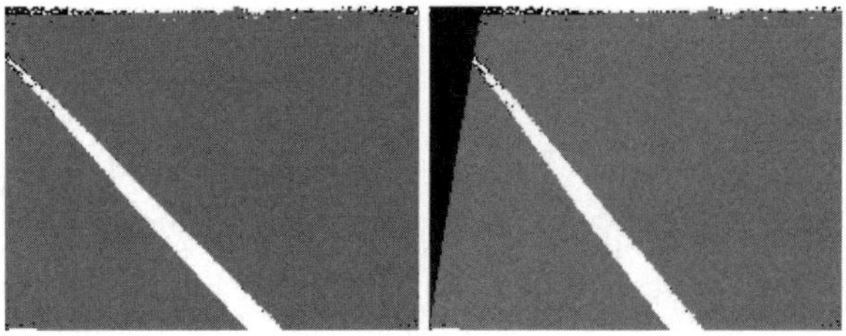

Fig. 5. Visual correction of classified image of line. *Left* : Original Image, *Right* : Corrected image.

the robot to calculate the line's angle and position relative to the camera. Since lines often stretch the length of the image, distortions have a more pronounced effect. Such distortions are then accentuated when the line is translated into real world co-ordinates.

4 Experimental Results

To determine the potential advantages from image correction, a simple experiment was performed. The experiment involved a robot viewing a ball in a known stationary position while quickly panning its head. The data that is most distorted by this kind of action is the bearing to an object. To measure the expected improvement produced by a corrected image the bearing was calculated from both the original and corrected points. Comparing both the values allows the performance of the correction to be evaluated.

The bearing to the ball was first measured by having the robot view a stationary ball with no relative movement. The bearing was calculated to be approximately -1 degree. This value contained very little noise, and was therefore assumed to be the correct bearing.

The robot was then set to pan back and forth. These images were captured and both the corrected and uncorrected bearings recorded. The sensor data from the robot was also recorded for later analysis and verification.

The results show that the correction improved the accuracy of the calculated bearing. While both the corrected and uncorrected values gave an average bearing close to the non panning value, the spread of the corrected data was lower in comparison. However there is still significant noise present in the corrected bearing values. This may be a result of acceleration during the exposure of the frame, which is not taken into account by the linear approximation.

Table 1. Overall results of panning test

	Uncorrected	*Corrected*
Average	-1.434	-1.272
Std.Deviation	6.186	2.961

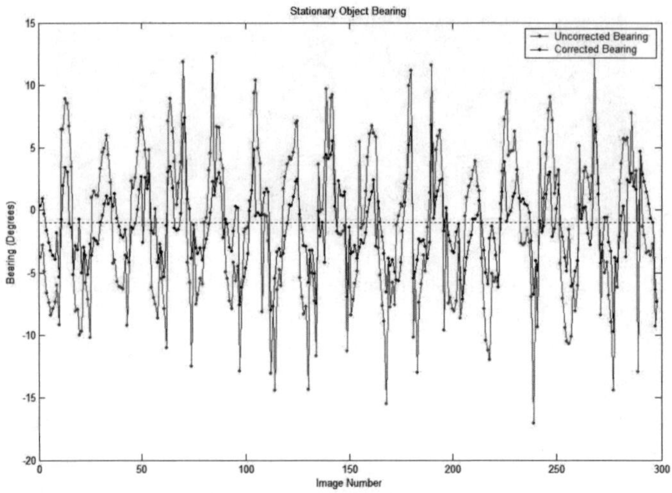

Fig. 6. Resulting bearings to stationary object test

Visual testing was also done to verify that the corrections performed were suitable. In this case the full image was visually corrected (figure 4).

5 Discussion

The correction accounts for the movement of the robot's camera. This however leaves the possibility that once the robot's distortion is corrected, then the perpendicular velocity of the object may be found from the remaining distortion. The distortion left being that created by the movement of the object. If the distance to the object is known and the relative angle that it has moved can be inferred from the distortion, then the perpendicular velocity can be calculated. This would allow a robot to calculate the object's velocity from a single image frame, instead of the normal two or three.

The correction works only on the panning of the head. Vertical movements are not accounted for as they are reliant on many more parameters on the ERS-7 robot. These including two vertical head joints and the body position, and therefore all 12 of the ERS-7s leg joints. Because of a reliance on so many different joints the calculated movement is very noisy. This noise makes it difficult to accurately determine the movements of the robots body.

6 Conclusion

The usage of a rolling shutter causes a distortion in the image when the camera is moving. These distortions skew the objects being viewed resulting in bad shapes and inaccurate bearings. By using the described de-skewing technique these distortions can be reduced or removed. Once the distortion has been removed from the image the data obtained on an object is improved, with improvements in the bearing to an object of better than 50%.

References

1. Nisticò, W., Röfer, T.: Improving Percept Reliability in the Sony Four-Legged Robot League. In: Proceedings of the RoboCup 2005 International Symposium, Osaka, Japan (July 18-19, 2005)
2. Quinlan, M.J., Nicklin, S.P., Hong, K., Henderson, N., Young, S.R., Moore, T.G., Fisher, R., Douangboupha, P., Chalup, S.K., Middleton, R.H., King, R.: The 2005 NUbots Team Report. Technical report (2005) Available online: `http://robots.newcastle.edu.au/publications/NUbotFinalReport2005.pdf`
3. Röfer, T., Laue, T., Weber, M., Stryk, O. v. Brunn, R., Dassler, M., Kunz, M., Oberlies, T., Risler, M., Burkhard, H.-D., Jüngel, M., Göhring, D., Hoffmann, J., Altmeyer, B., Krause, T., Spranger, M., Schwiegelshohn, U., Hebbel, M., Nisticò, W., Czarnetzki, S., Kerkhof, T., Meyer, M., Rohde, C., Schmitz, B., Wachter, M., Wegner, T., Zarges, C.: GermanTeam RoboCup 2005. Technical report (2005) Available online: `http://www.germanteam.org/GT2005.pdf`
4. M. Meingast, C. Geyer, S. Sastry: Geometric Models of Rolling-Shutter Cameras (2005) Available online: `http://arxiv.org/abs/cs/0503076`

Evaluating Learning Automata as a Model for Cooperation in Complex Multi-agent Domains

Mohammad Reza Khojasteh[1] and Mohammad Reza Meybodi[2]

[1] AI & Robotics Laboratory, Computer Engineering Department
Shiraz Islamic Azad University, Shiraz, Iran
mrkhojasteh@persianrobotics.net
[2] Soft Computing Laboratory, Computer Engineering Department
Amirkabir University of Technology (Tehran Polytechnic), Tehran, Iran
meybodi@ce.aut.ac.ir

Abstract. Learning automata act in a stochastic environment and are able to update their action probabilities considering the inputs from their environment, so optimizing their functionality as a result. In this paper, the goal is to investigate and evaluate the application of learning automata to cooperation in multi-agent systems, using soccer simulation server as a test bed. We have also evaluated our learning method in hard situations such as malfunctioning of some of the agents in the team and in situations that agents' sense/act abilities have a lot of noise involved. Our experiment results show that learning automata adapt well with these situations.

1 Introduction

As a model for learning, learning automata act in a stochastic environment and are able to update their action probabilities considering the inputs from their environment, so optimizing their functionality as a result.

Also, as a test-bed, we have used the simulated robotic soccer, "SoccerServer2D" [3] in this paper. Robotic soccer is an example of a complex environment that some agents should cooperate with each other, in order to achieve the team's goal [2][3]. In fact, in this paper we have focused on the systems composed of some autonomous agents that can act in real-time, noisy, collaborative and adversarial environments [1].

To do so, we implemented teams composed of 11 agents that learn using learning automata and compared them to similar teams that have no learning capability or use other learning methods such as Q-learning.

In this paper, we have used Learning Automata [4] as our machine learning method. In the coming sections of this paper, we first present our use of learning automata in a complex multi agent domain with presenting some results of our simulations.

Then, we evaluate our learning method by running some simulated plays with minor and major changes in environment parameters. Also, we evaluate our learning method in hard situations such as malfunctioning of some of the agents in the team and in situations that agents' sense/act abilities have a lot of noise involved.

The goal is to investigate our learning method's adaptation with these changes.

G. Lakemeyer et al. (Eds.): RoboCup 2006, LNAI 4434, pp. 410–417, 2007.

2 Cooperation in a Team Using Learning Automata

Our goal in this section is to use learning automata for cooperation among the members of a simulated soccer team with 11 players in order to achieve the team's goal. By now, various machine learning methods such as Q-learning, genetic algorithms, decision trees, behavioral learning, to mention a few, have been used for training the soccer player agents [1]. To our knowledge, this research is the first attempt to use learning automata in cooperation in multi-agent systems.

Because of the large state space in such a complex multi agent domain, it is vital to have a method for environmental states' generalization. In this paper we have used the technique called the "Best Corner in State Square" for generalizing the vast number of states in agent's domain environment to a few number of states by building a virtual grid in that environment [5]. Our experiments in [6] show that by using the "Best Corner in State Square" technique, each agent performs well in determining its own state and consequently, in determining the proper action in that state.

We have also used 8 learning automata for each agent (one automata for each corner in the "Best Corner in State Square" [6]). Also, we have defined 8 actions for each learning automata; sending the ball to the center of one of agent's 8 immediate squares as defined in the "Best Corner in State Square" technique [6].

In our simulations we used 4-3-3 formation for each team for organizing the eleven players in the field. We implemented some teams using fixed structure learning automata, some teams using variable structure learning automata, and a team using Q-learning (as a team that uses another method of learning). Each of the above teams played against the "without learning" team. Note that the "without learning" team in our simulations is like the "learning" teams from every aspect (architecture, states, actions, and even team formation), except that it can't learn from its previous experiences.

In our "learning automata" teams (fixed or variable), the agents determine their current state by the "Best Corner in State Square" technique [5]. Then the agent that possesses the ball performs the action that is advised by the corresponding automata in its state. The agent then percepts its action's result, and gives itself a reward or a penalty depending on that result.

In fact, our agent simply gives itself a reward if the ball has gone toward opponent's goal and one of its teammates (or even itself, in case of a dribble) has chased the ball (as the result of its action). Similarly, the agent gives itself a penalty if the ball has gone toward its team's goal and one of the opponent players has chased the ball (again as the result of its action). In all other cases, the agent does not give itself any reward or penalty and leaves its learned values unchanged.

Note that we have simulated our agents to learn from zero (i.e. without any previous knowledge of the environment before starting the simulation). Also, we have used the agent itself for the judgement about its action's results and this let us have what we call "distributed judgement", again a multi-agent approach.

We should point that we have used a memory depth of 3 for our fixed structure learning automata teams ($L_{2N,2}$, $G_{2N,2}$, Krylov, and Krinky). Also, for our variable structure learning automata teams, we have used (a = b = 0.1) for L_{rp}, (a = 0.1, b = 0.0) for L_{ri}, and (a = 0.1, b = 0.01) for L_{rep} and Full_ L_{rep}. For our Q-learning team, we have used the TPOT_RL introduced in [1].

The simulation results show that the learning automata teams could defeat the "without learning" team after a few number of training plays. Figure 1, shows one of the results of our simulations.

Note that we have give our results based on the number of games played and each player in our team has a chance of 5 ball kicking (by average) in each game. More results are presented in [6].

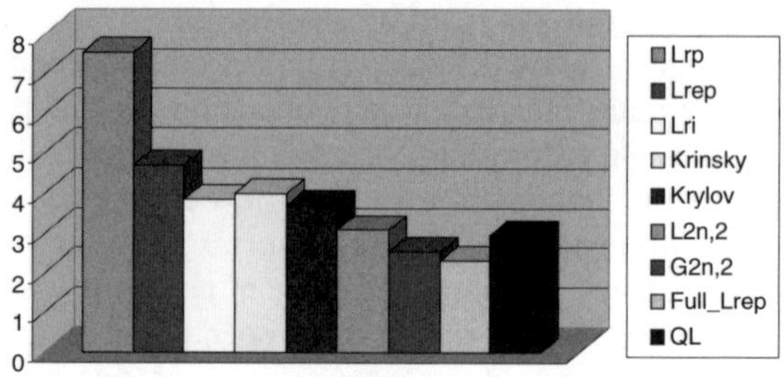

Fig. 1. The ratio of the average goals scored to the average goals received for each team (during 3 test plays), after 15 training plays against the "without learning" team

This figure shows that the variable structure learning automata teams have a better performance in defeating the "without learning" team and so exhibit higher ability of learning in this domain.

We have also discussed the speed of convergence of our learning automata algorithms used by the agents and suggested some techniques for increasing the speed of convergence in [6].

Our simulations [5][6][7] show that learning automata perform well in order to have a cooperative team of agents in a complex multi-agent domain.

3 Evaluation Tests for Teams Using Learning Automata

In previous section, we investigated the efficiency of using learning automata in doing teamwork [5][6][7]. In this section we evaluate our learning method by running some simulated plays with one of the teams that had taken part in previous world RoboCup competitions. Also we investigate the efficiency of learning automata by changing some of the RoboCup SoccerServer2D parameters and observing their effects on our teams' performance. It is necessary to note that our base code is the code of CMUnited98 team [1].

In doing so, we used the team Saloo 2001 [10] that was similar to our team from the agent individual behaviors' (shooting the ball, etc.) point of view. We let our learning automata teams play against this team and observed the results in detail [6].

In this section, we use the teams Full_ L_{rep} and also L_{rep} (both based on L_{rep} automata) for our simulations. Selecting these automata is because of the good results obtained for them in our previous simulations. For more simulations the reader may refer to [5][6][7].

Also in this section, we investigate the performance of our learning method in the environment situations that are more difficult for the agents to adapt with (comparing the situations we have considered so far).

At the end of this section, we have investigated the effects of different formations on the teamwork.

In the first series of simulations, ten consecutive plays between team L_{rep} (learning from zero) and the "without learning" team were simulated. Our goal was to investigate the efficiency of our learning methods and to observe how our team performs against "without learning" team in the presence of noise.

There are several parameters in the RoboCup SoccerServer2D that can be changed [3]. In the first series of simulations conducted, we study the effect of the "rand" parameters (which indicate the amount of noise values) in the RoboCup SoccerServer2D. We changed the parameter "player_rand" from 0.1 to 0.2, the parameter "ball_rand" from 0.05 to 0.1, and at last the parameter "kick_rand" from 0.0 to 0.1. The first parameter mentioned enables us to add noise to the players' movements and the second and the third parameters, add noise to the ball movement and kicking the ball, respectively.

Figure 2 shows the cumulative results of this experiment. As the figure shows, by playing more games, the learning team, adapts itself to the situation more and more and increases its gap with the "without learning" team. This figure shows the efficiency of the proposed learning method when the environment is noisy.

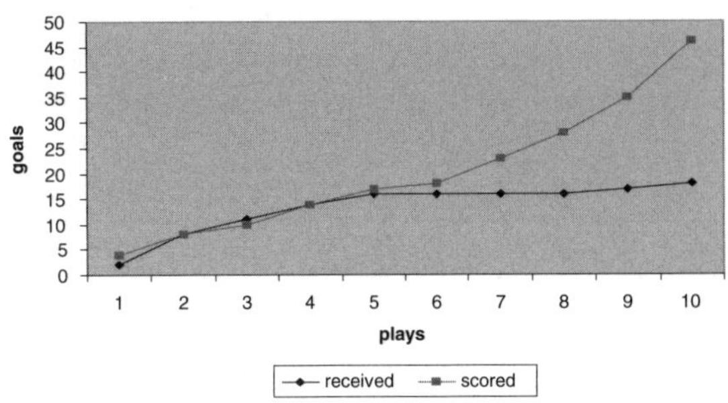

Fig. 2. The number of goals scored versus the number of goals received by the learning team during 10 consecutive training plays versus the team "without learning", as the noise increases

In the second series of simulations, we studied the effects of not using (although single-channel, crowded, and unreliable) communication facility provided by RoboCup SoccerServer2D on the performance of the proposed learning method. We simulated 10 consecutive plays between team L_{rep} (learning from zero) and the "without learning" team. Figure 3 shows the efficiency of our learning method and indicates that by playing more games, the learning team, adapts itself with the situation and increases its gap with the "without learning" team.

In third series of simulations, we eliminated 3 players from the left side of our team. They were player number 2 from the defense line (the left defense), player number 6 from the middle line (the left piston), and player number 10 from the forward line (the left forward) of our team [6]. Our goal was to evaluate the function of our team in the case of failure in some of our agents.

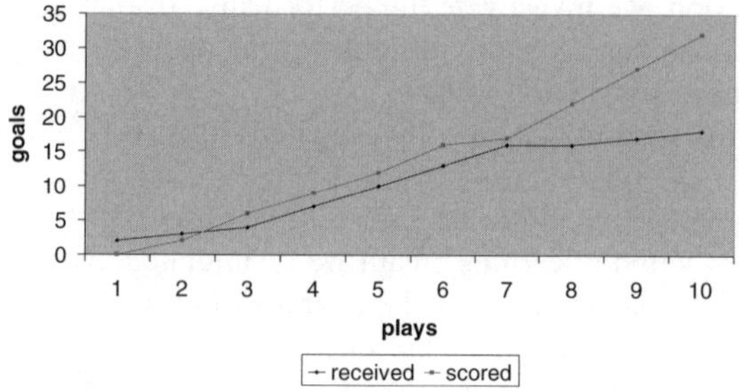

Fig. 3. The number of goals scored versus the number of goals received by the learning team during 10 consecutive training plays versus the team "without learning", without the communication facility between the agents

We simulated 15 consecutive plays between team L_{rep} (learning from zero) and the "without learning" team (with 11 players). Figure 4 shows the results of this experiment. Since each player in our team has a limited freedom around its special post in the field, our simulations show that this elimination causes our team's left side to malfunction. Note that we have used a 4-3-3 formation for our teams.

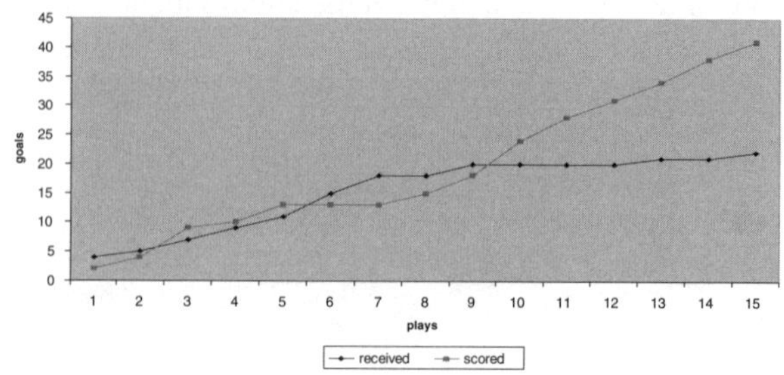

Fig. 4. The number of goals scored versus the number of goals received by the learning team during 15 consecutive training plays versus the team "without learning", with eliminating 3 players from the left side of the learning team

As figure 4 shows, our learning team was able to overcome the absence of its players and adapted itself to fewer number of players and finally defeated the "without learning" team.

In next series of simulations, we made some simulations between team L_{rep} (learning from zero) and team Saloo 2001 and the results of the first seven plays are gathered in table 1. As the table shows, team Saloo 2001 could win all the first games with relatively high goal average (average scored goals of 5.7 versus average received goals of 0.3 in each play), and has an absolute better performance comparing to our team. We conducted more training plays in order to be able to defeat the team Saloo 2001. So, we simulated 150 consecutive plays (equal to 25 hours) between the two teams.

Table 1. The statistics for the first 7 plays between team L_{rep} (learning from zero) and team Saloo 2001

The percentage of possession of the ball for the opponent team (Saloo 2001)	52.6
The percentage of possession of the ball for the our team (L_{rep})	47.4
The percentage of ball movement in opponent's 1/3 of the field	10.5
The percentage of ball movement in the middle 1/3 of the field	47
The percentage of ball movement in our 1/3 of the field	42.5
The maximum continuous time that opponent team has the ball in possession	185
The maximum continuous time that our team has the ball in possession	112
The maximum number of continuous passes between the members of the opponent's team	14
The maximum number of continuous passes between the members of our team	8

Table 2. The statistics The statistics for the last 7 plays (after 25 hours training) between team L_{rep} (learning from zero) and team Saloo 2001

The percentage of possession of the ball for the opponent team (Saloo 2001)	45
The percentage of possession of the ball for the our team (L_{rep})	55
The percentage of ball movement in opponent's 1/3 of the field	24.5
The percentage of ball movement in the middle 1/3 of the field	42
The percentage of ball movement in our 1/3 of the field	33.5
The maximum continuous time that opponent team has the ball in possession	112.7
The maximum continuous time that our team has the ball in possession	134.2
The maximum number of continuous passes between the members of the opponent's team	8.8
The maximum number of continuous passes between the members of our team	12

Table 3. Average wrong actions' percentage for each agent of the team L_{rep} in the first 7 plays versus the last 7 plays (after 25 hours of training) when played against team Saloo 2001

Wrong actions' percentage	
40.1	In the first 7 plays
24.6	In the last 7 plays

During these 150 simulated plays (that their overall results are shown in tables 1-3), our team could improve its performance and gradually move toward "not losing" and finally to continuously "win".

The statistics of the first 7 plays are given in table 1. Also, the statistics of the last 7 plays are given in table 2. In these simulations, an average scored goals of 3.6 and an average received goals of 0.1 is obtained. In table 3, we give the average percentage of wrong actions done [6] by the learning team's players in the first 7 plays and in the last 7 plays, for the sake of comparison.

It is necessary to remind that our agents try to send the ball toward one of their 8 directions, whichever seems to be better for achieving the team's goal [5][6][7]. This action might (relative to position) seem as a pass, a dribble, a shoot, etc. In team CMUnited98, there are two layers for multi-agent behavior (pass evaluation that is trained offline using decision trees) and for team behavior (pass selection that is trained online using a method based on Q-learning which uses the output of the previous mentioned layer as the input) [1].

Our method for learning has combined the above two layers into one layer. We've done some offline training before our team plays against another team. We haven't separated "pass evaluation" from "pass selection". In fact we are dealing with actions that a player chooses and whether or not the action chosen is a right action. We aren't involved with "pass" as a separate problem to solve. Instead, we have looked at the problem of "cooperation between our agents".

As an end to this research, we investigated the effects of different team formations on the agents' cooperation. Note that for all the simulation presented so far we have used 4-3-3 team formation. We created similar teams (L_{rep}) but with different formations 4-4-2, 3-6-1, 4-3-3, 3-5-2, and 3-4-3 and then simulated a series of plays as a tournament between them. By tournament, we mean that we let each of the above mentioned teams play against all other teams and gathered the results. Figure 5 shows the results of these simulations [6].

As the results show it is very important to have a proper team formation in order to achieve a good team performance. Figure 5 shows that the highest number of goals scored was by the team with the formation 3-5-2 and the least number of goals received was by the team with the formation 4-4-2.

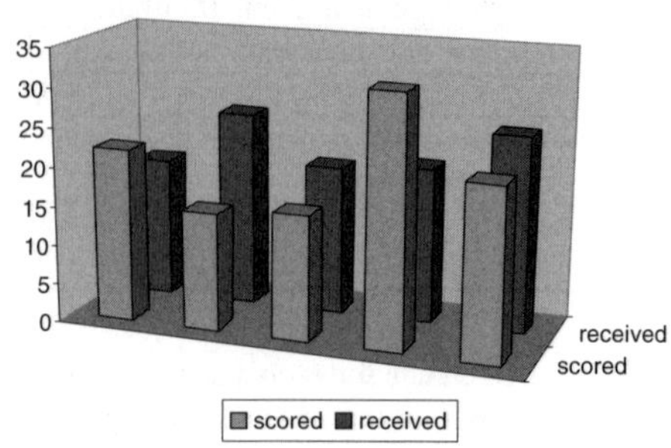

Fig. 5. Scored and received goals by the learning automata teams (L_{rep}) for different team formations playing against each other in a tournament (from left to right 4-4-2, 4-3-3, 3-6-1, 3-5-2, and 3-4-3)

For simulations in which other learning automata such as "Estimator algorithm" [8] and "Discretized Pursuit Learning Automata" [9] are used, the reader may refer to [6]. For a discussion about the speed of convergence of the proposed method and also methods to improve the speed of convergence the reader may again refer to [6].

4 Conclusion

We used learning automata for successful production of a series of actions for agents that were members of a team, such that the resulting team could act well in multi-agent, adversarial, noisy, real-time, and most important collaborative environments.

The methods introduced are general methods that can be implemented, applied, and used in other domains or other test-beds with minor changes.

At last, we evaluated the efficiency of learning automata in cooperation among agents that are seeking a common team goal by implementing some evaluation tests and observing the results.

Our experiments, showed that learning automata adapts itself well with major and minor changes in the environment parameters and also in hard situations such as malfunctioning of some of the agents in the team and in situations that agents' sense/act abilities have a lot of noise involved.

References

1. Stone, P.: Layered Learning in Multi_Agent Systems, PhD thesis, School of Computer Science. Carnegie Mellon University (December 1998)
2. Kitano, H. (ed.): RoboCup-97: Robot Soccer World Cup I. Springer, Heidelberg (1998)
3. Andre, D., Corten, E., Dorer, K., Gugenberger, P., Joldos, M., Kummenje, J., Navaratil, P.A., Noda, I., Riley, P., Stone, P., Takahashi, R., Yeap, T.: Soccer server manual, version 4.0, Technical Report RoboCup –1998-2001, RoboCup (1998)
4. Narendra, K.S., Thathachar, M.A.L.: Learning Automata: An Introduction. Prentice Hall, Inc., Englewood Cliffs (1989)
5. Khojasteh, M.R., Meybodi, M.R.: The Technique "Best Corner in State Square" for Generalization of Environmental States in a Cooperative Multi-agent Domain. In: Proceedings of the 8th annual CSI computer conference (CSICC' 2003), pp. 446–455, Mashhad, Iran (February 25–27, 2003)
6. Khojasteh, M.R.: Cooperation in Multi-agent Systems using Learning Automata, M.Sc. thesis, Computer Engineering Faculty, Amirkabir University of Technology (Tehran Polytechnic), Tehran, Iran (May 2002)
7. Khojasteh, M.R., Meybodi, M.R.: Using Learning Automata in Cooperation among Agents in a Team. In: Proceedings of the 12th Portuguese Conference on Artificial Intelligence, IEEE Conference Publication Program with ISBN 0-7803-9365-1 and IEEE Catalog Number 05EX1157, University of Beira Interior, pp. 306–312, Covilhã, Portugal (December 5th-8th, 2005)
8. Thathachar, M.A.L., Sastry, P.S.: A New Approach to the Design of Reinforcement Schemes for Learning Automata, IEEE Transactions on Systems, Man, and Cybernetics, Vol. SMC-15(1) (Janaury/February 1985)
9. Oomen, B.J., Lanctot, J.K.: Discretized Pursuit Learning Automata. IEEE Transactions on Systems, Man, and Cybernetics SMC-20(4) (July/August 1990)
10. Noda, I.: Team Description: Saloo, AIST & PREST, Japan (2001)

Cooperative 3-Robot Passing and Shooting in the RoboCup Small Size League

Ryota Nakanishi[1], James Bruce[2], Kazuhito Murakami[1], Tadashi Naruse[1], and Manuela Veloso[2]

[1] Aichi Prefectural University, Nagakute-cho, Aichi, 480-1198 Japan
[2] Carnegie Mellon University, Pittsburgh PA 15213, USA
is031032@cis.aichi-pu.ac.jp, bruce@andrew.cmu.edu

Abstract. This paper describes a method for cooperative play among 3 robots in order to score a goal in the RoboCup Small Size League. In RoboCup 2005 Osaka, our team introduced a new attacking play, where one robot kicks a ball and the other receives and immediately shoots the ball on goal. However, due to the relatively slow kicking speed of the robot, top opponent teams could prevent successful passing between robots. This motivates the need for more complex play, such as passing around to several robots to avoid the opponents' passing defense. In this paper we propose a method to realize such a play, i.e. a combination play among 3 robots. We discuss the technical issues to achieve this combination play, especially for a pass-and-shoot combination play. Experimental results on real robots are provided. They indicate that the success rate of the play depends strongly on the arrangement of the robots, and ranges from 20 % to 90 % in tests with an opponent goalkeeper which stands still.

1 Introduction

Year by year, the skill of the robot systems in the Small Size League is growing higher and higher. Cooperation between robots has become a necessary technology in the Small Size League. In these years, passing between two robots has become a stable technology[1]. Last year, in RoboCup 2005 Osaka, our team[2] performed a new attacking play, that is, one robot kicks a ball and the other receives and shoots the ball with no delay. It is an efficient play and is an interesting technique, but due to the rather slow kicking speed by our robot, top opponent teams could prevent the ball from passing between our robots. Kicking the ball faster makes it possible for a successful passing play, but the vision processing and physical robot limits bound the maximal speed of the ball for a reception to work. In order to prevent the opponent robots from stopping the play, a bit more complex play such as passing around between the robots is needed.

One method to achieve such a play is a combination play among 3 robots. In the other words, if a robot A tries to pass the ball to a robot B and an opponent robot intervenes on the pass line, then the robot A should pass it to

G. Lakemeyer et al. (Eds.): RoboCup 2006, LNAI 4434, pp. 418–425, 2007.

a robot C, the robot C just kicks it to the robot B without holding it and the robot B shoots on the goal (or some other target). We call this a **1-2-3 shoot**. Cooperation between 3 robots has several technical issues, mainly the following:

- where should three robots position on the field?
- how is the second robot controlled?

In this paper, we discuss the above issues to achieve a successful 1-2-3 shoot play, and show with experimental results how the play can be carried out effectively with real robots.

2 Cooperation of Multiple Robots

2.1 Why Is Cooperation Necessary?

Humans can easily adapt themselves to suit the environment where they live, while a robot is typically vulnerable to changes in the environment due to its static policies. If a robot could gain the adaptation ability of a human, the robot could be used more often in situations of the human-robot cooperation. We feel that multiple robots cooperating in response to an external opponent is a useful step in that direction. So, in this paper, we discuss the cooperation among robots employed in the RoboCup Small Size League. More specifically, we discuss the methodology for achieving the goal of tight cooperation of three robots.

This situation occurs when a robot holding the ball has its direct shot on the goal blocked by opponent robots, making it difficult to achieve a goal on its own. The possible actions to do next are either moving somewhere else while dribbling the ball, or passing the ball to a teammate. According to the rules of the RoboCup Small Size League, it is not possible to move a long distance while dribbling the ball. Therefore, it is advantageous to achieve the goal with a combination play between robots that uses pass plays aggressively.

2.2 What Kind of Cooperation Do We Achieve?

A primary form of cooperation used in this paper is what we call a **direct play**, where the robot changes the direction of the ball's velocity without holding or dribbling the ball, but by kicking the ball in a new direction as soon as it arrives.

The direct play makes it possible to achieve continuous passing among teammate robots without stopping the ball's motion. This results in a situation where the opponent robots have a difficult time intercepting the ball and consistently defending the goal. Moreover, in a real game, since the opponent robots are likely to move in the direction of preventing the teammate robot holding the ball from shooting on the goal, the second teammate robot (receiving robot) can get the ball from the first robot and achieve the goal relatively easily.

The next kind of cooperation is a **1-2-3 shoot**, which we define as a play where three robots (A, B, C) cooperate and the robot A who is holding the ball kicks the ball toward robot B, and robot B kicks the ball toward robot C by a direct play and finally the robot C shoots on the goal. If successfully carried out, this play makes it possible to achieve a goal with high probability since the fast handling of the ball and direction changes make it difficult for the opponent robots to follow the ball.

3 Achieving a Direct Play

The basis of the 1-2-3 shoot play is a direct play. This section describes an algorithm to achieve the direct play. The direct play is played by two robots (A, B); the robot A holding the ball kicks the ball toward the robot B, and the robot B kicks the coming ball toward in a different direction without holding it. An algorithm to achieve the direct play is as follows:

[**Direct play algorithm**]
Let A be a robot holding the ball and B be a cooperating robot.

Step 1. The robot B moves to an open location that has an open shooting line to the goal (or a pass line to another robot). That is, in the shoot line case, the robot B moves to a position where there are no opponent robots on the line that connects the robot B and the goal (see position B_{t1} in figure 1). If such a position doesn't exist, it looks for the next chance.

Step 2. The robot A kicks the ball toward the robot B. No opponents are assumed to be blocking the direction that the ball follows.

Step 3. Measuring the ball speed using the vision system, calculate the position and the time that the ball meets the robot B.

Step 4. The robot B moves to B_{t2} meeting at exactly time $t2$. (Figure 1)

Step 5. The proximity sensor of the robot B detects the moment that the ball touches to the robot B and kicks the ball at the moment.

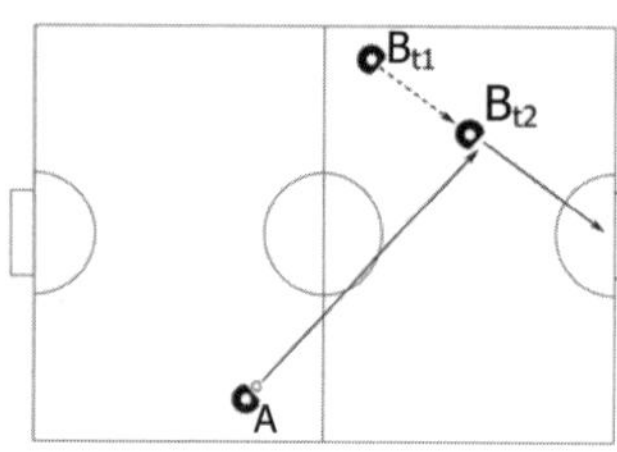

Fig. 1. Direct play

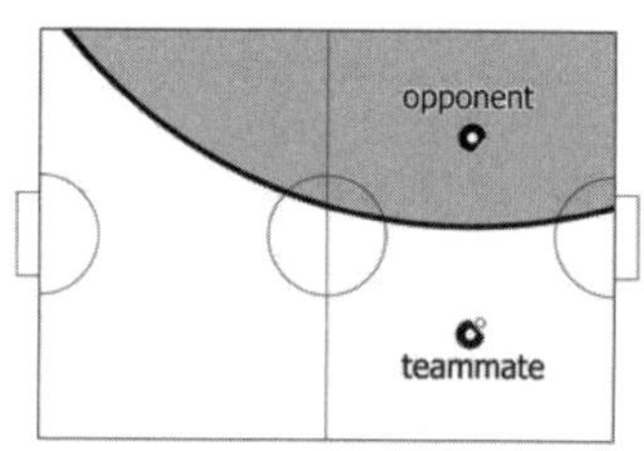

Fig. 2. An example dominant region

4 Achieving a 1-2-3 Shoot

To achieve a 1-2-3 shoot, the teammate robots should have an advantage that a pass is possible between them without being intervened by the opponents. The dominant region method [3] provides a method for calculating a solution to this problem.

4.1 Dominant Region Method

The dominant region method is a kind of Voronoi diagram. It is calculated with respect to two agents, one of which is a teammate and the other which is an opponent. The diagram divides the area of the soccer field into two regions, where one region is the one that the teammate can get to faster than the opponent, while the other is the one that the opponent can get to faster.

4.2 Calculation of the Dominant Region

Figure 2 shows an example of the dominant region. The shaded area is a dominant region for the teammate robot and the other is for the opponent. In the following, we show how to compute the dominant region.

Let v_1 and a_1 be an initial velocity and an acceleration of the teammate robot, and v_2 and a_2 be those of the opponent. Let (x_1, y_1) and (x_2, y_2) be the current positions of the teammate and opponent robot, respectively. Then, for given position (x, y), the distance between each robot and the given position and arrival time are given by the following

$$L_i = v_i t_i + \frac{1}{2} a_i t_i^2 = \sqrt{(x - x_i)^2 + (y - y_i)^2}, \quad i = 1(teammate), \quad i = 2(opponent) \tag{1}$$

Solving Eqs (1) with respect to $t_1$and t_2, respectively, we obtain,

$$t_i = \frac{-v_i + \sqrt{v_i^2 + 2a_i\sqrt{(x - x_i)^2 + (y - y_i)^2}}}{a_i}, \qquad i = 1, 2 \tag{2}$$

In case of zero initial velocity.

$$(a_2^2 - a_1^2)x^2 + (a_2^2 - a_1^2)y^2 + 2(x_2 - x_1)x + 2(y_2 - y_1)y + x_1^2 + y_1^2 - x_2^2 - y_2^2 = 0 \tag{3}$$

Equation (3) expresses the border between the regions.

The dominant region can be generalized to multiple teammates and opponents. It is calculated by considering all pairs of teammates and opponents, and for each location the minimum time value is taken to construct the overall diagram.

4.3 Pass Play and Dominant Region

First, we discuss a direct play based on the dominant region method. Figure 3 is a typical case of such a situation.

The teammate B has an open line from teammate A, as well as an open line to the goal. Teammate A can thus pass the ball to teammate B. The dominant

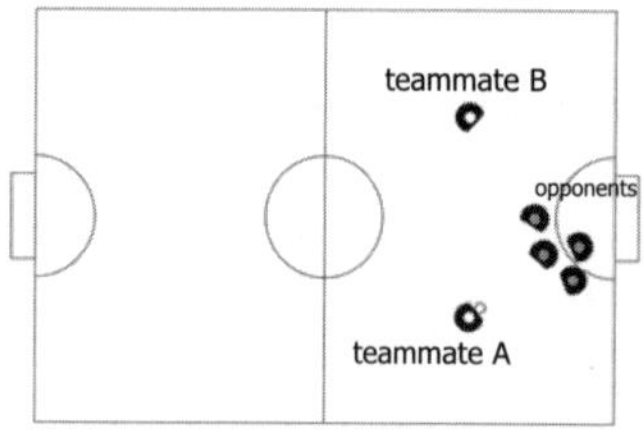

Fig. 3. Typical attack positioning

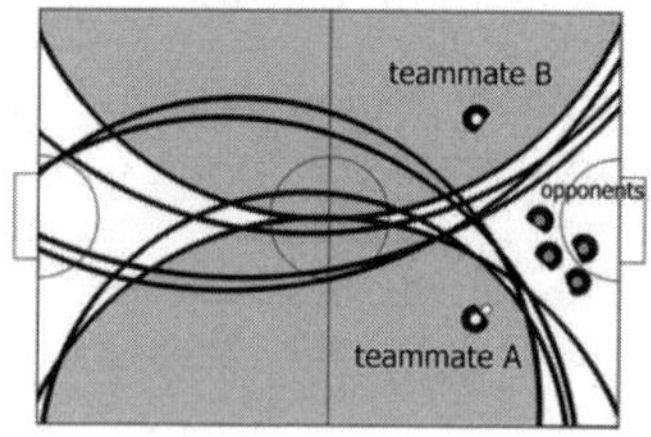

Fig. 4. Dominant region of figure 3

region of the figure 3 is given by the figure 4. In the Figure, each solid curve shows a dominant boundary between one of the teammate robots and one of the opponent robots. Shaded area is the dominant region for the teammates. It is clear from the figure 4 that the opponent can easily intercept the ball if the teammate A kicks the ball to the teammate B too slowly for the direct play.

On the other hand, as shown in figure 5, three teammates make a dominant region which can pass around the opponent robots in as shown in the dominant region.

The dominant region method is useful as a criterion for whether a pass should be done or not. In a real game, we might weaken this criterion slightly. However, we feel it can still act as a good criterion for judging whether to a direct play should be done or not.

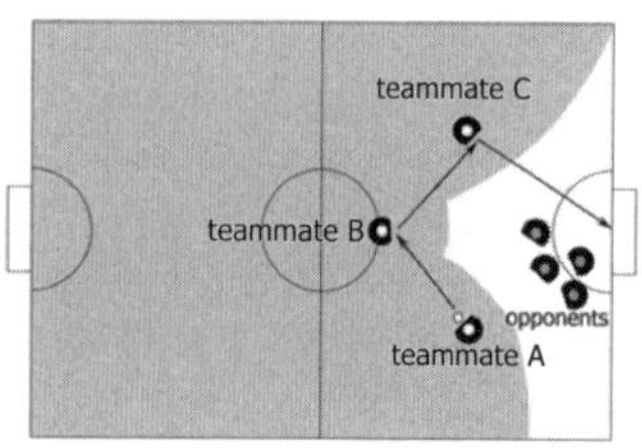

Fig. 5. Example of cooperation among 3 robots

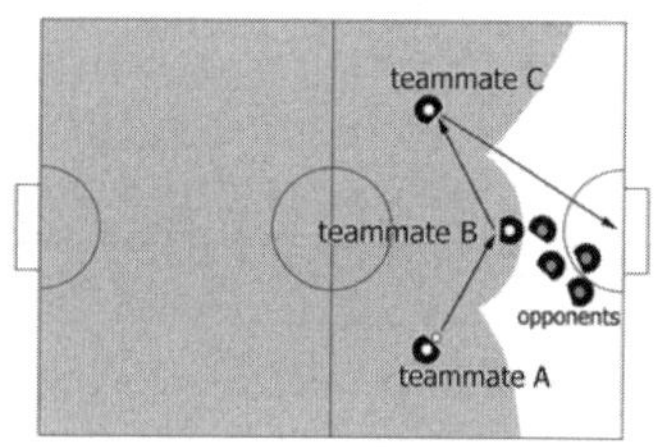

Fig. 6. 1-2-3 shoot position

4.4 Algorithm for the 1-2-3 Shoot

When should we play a 1-2-3 shoot? The following is a basic procedure to select the 1-2-3 shoot play.

[Selection procedure of action]
Let A be a robot holding ball, and let B and C be cooperating robots.

```
if (Robot A have a shoot line)
 { Robot A shoots.} else {
  Search open space which is able to make shoot line.
  Robot C moves to the open space.
  Calculate dominant region.
```

```
if (Pass line is in the dominant region of the teammate)
  {Direct play between robots A and C.} else {
    Robot B moves to the appropriate position.
    1-2-3 shoot play among robots A, B and C.
  }
}
```

When the 1-2-3 shoot play is selected, the following algorithm is executed.

[**1-2-3 Shoot algorithm**]
Let A be a robot holding the ball, and B and C be cooperating robots.
Assume that the positions of the robots are shown in figure 5.

Step 1. The robot C moves to the open space that has a shoot line to the goal. (The same movement as Step 1 in the direct play algorithm)

Step 2. If the pass line crosses the opponent dominant region, the robot B moves to the vertex of the near equilateral triangle as shown in figure 6. As a result, a pass line is made in the teammate dominant region. (If this is not the case, a re-schedule should be done.) The robot B turns to the robot C.

Step 3. The robot A kicks the ball to the robot B, then the robot B kicks it to the robot C according to the direct play algorithm.

Step 4. The robot C kicks the ball toward the goal mouth.

5 Empirical Study

We implemented the 1-2-3 shoot algorithm in our system and measured the success rate of the 1-2-3 shoot under the condition that an opponent goalkeeper stands still, as a first step toward usage in a real game.

5.1 Experimental Environment

Figure 7 shows the robots (with and without cover) we employed in the experiment. The robot consists of 4 omnidirectional wheels with diameter 60mm, 4 motors to drive the wheels, Hitachi's SH2 control processor and its peripheral circuits, dribbling and kicking devices, infrared proximity sensor and radio communication device. The robot moves at the maximal speed of $150cm/sec$. During these experiments the robot is limited to a speed of $100cm/sec$.

The host processor system consists of the Athlon64 3500+ CPU running at 2.2 GHz, 512 MB memory and Debian Linux as the operating system. The host processor system controls the robots by sending commands using the radio communication system. We developed a 1-2-3 shoot program on the system.

5.2 Experimental Results

It is difficult for the robot to kick a moving ball from a direction perpendicular to the initial velocity of the ball while it is rather easy to kick it from the parallel direction to the ball line. So we have tested two cases shown in Figures 8 and 9.

For each case, we ran 20 trials. The success rate of the 1-2-3 shoot is shown in the Table 1.

Fig. 7. Appearance of the robot

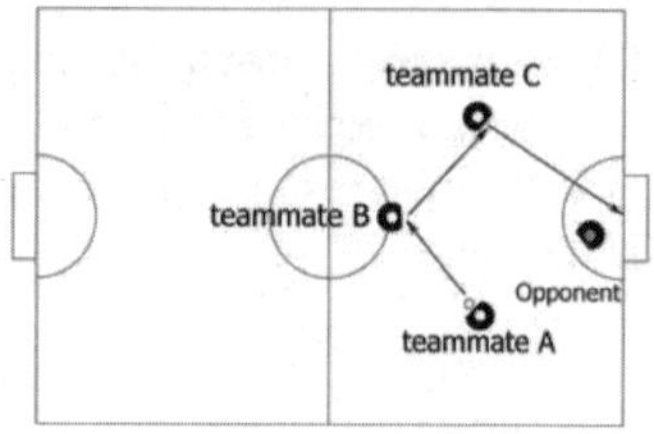

Fig. 8. Experiment 1

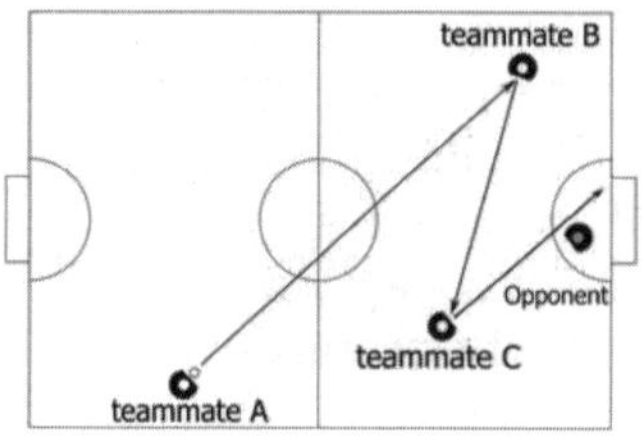

Fig. 9. Experiment 2

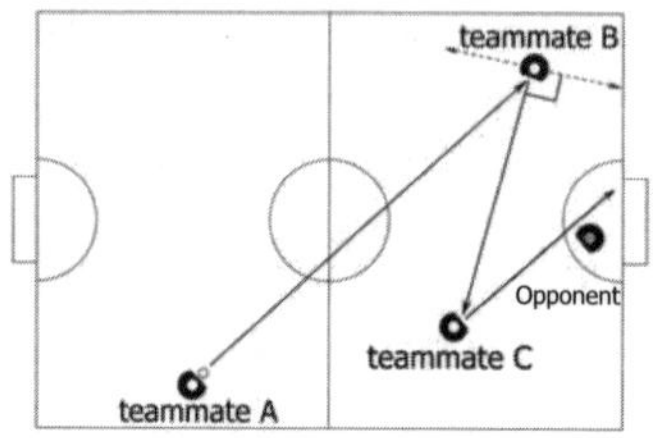

Fig. 10. Improvement of control of robot B

Table 1. Success rate of the 1-2-3 shoot play

	success	fail	rate of success (%)
experiment 1 (figure 8)	4	16	20
experiment 2 (figure 9)	12	8	60
experiment 3 (figure 10)	18	2	90

5.3 Discussion

In addition to the direction between the line on which the ball moves and the kicking direction, the distance between two robots that pass through are important factor to raise the success rate, since the longer the distance the easier the control of the robot becomes. This is because the receiving robot has time to move to the point where the ball comes, compared to a short distance pass. However, the longer distance also gives the opponents a better chance for an interception. Therefore, the positioning shown in figure 6 with long distance should be searched for. Such a strategy is an interesting future research topic.

In the previous implementation, the robot B (in Figures 8 and 9) moves on the pass line back and forth and adjust the kick line to the robot C. However, the success rate did not rise. We improved the control of the robot B as we let the robot move left and right on the line perpendicular to the kick line to the robot C as shown in figure 10. The success rate raised 90 % by this improvement. The reason why the improvement is achieved is that, by the left and right movement, we can keep the distance long enough and form V-shaped kicking lines.

6 Concluding Remarks

We have developed a cooperative skill involving 3 robots, called 1-2-3 shoot, to perform a pair of passes and a goal shot, which is based on the direct play from our previous system and the dominant region for aiding in decisions. This is a highly cooperative play and one of the useful skills for the future of the RoboCup Small Size League, since the goal block skills of opponent robots makes it difficult to achieve a goal by a single robot alone.

Experimental results show that the success rate of the play ranges from 20 % to 90 % depending on the positioning of the robots. Though the success rate varies in wide depending on the positioning, it is important to have shown the continuous cooperation among 3 robots. A successful 1-2-3 shoot play could have a much higher scoring probability than a direct shot, making it worthwhile even if the play is not successfully executed on every attempt. From the experiment, the varying success rate raises the next problem of how best positioning can be achieved. Moreover, a way of effective selection among many skills must not be found by analyzing a game, choosing between plays such as a direct shot, a direct passing play, and the 1-2-3 shoot play.

References

1. Murakami, K., Hibino, S., Kodama, Y., Iida, T., Kato, K., Naruse, T.: Cooperative Soccer Play by Real Small-Size Robot. In: Polani, D., Browning, B., Bonarini, A., Yoshida, K. (eds.) RoboCup 2003. LNCS (LNAI), vol. 3020, pp. 410–421. Springer, Heidelberg (2004)
2. Bruce, J., Hibino, S., Murakami, K., Naruse, T., Veloso, M.: CMRoboDragons 2005 Team Description. In: Bredenfeld, A., Jacoff, A., Noda, I., Takahashi, Y. (eds.) RoboCup 2005. LNCS (LNAI), vol. 4020, Springer, Heidelberg (2006)
3. Taki, T., Hasegawa, J.: Dominant Region: A Basic Feature for Group Motion Analysis and Its Application to Teamwork Evaluation in Soccer Games. In: Proc. SPIE Conference on Videometrics VI, vol. 3641, pp. 48–57 (1999)

Logfile Player and Analyzer for RoboCup 3D Simulation

Steffen Planthaber and Ubbo Visser

Center for Computing Technologies
Universität Bremen, Germany
{steffen,visser}@tzi.de

Abstract. In multi agent environments or systems equipped with artificial intelligence it is often difficult to obtain the function or method which led to a particular behavior that is noticeable from outside. However, this information is crucial if not necessary to optimize the agents behavior. In the RoboCup 3D simulation league this dilemma becomes obvious when replaying logfiles of a game that was simulated before. The 3D soccer simulation league monitor (rcssmonitor-lite) is restricted with regards to replaying logfiles.

This paper describes the concept and the implementation of improvements for the logplaying and analyzing abilities of the monitor. The idea is to provide a tool that is able to assist developers to detect problems of their agents both in single and cooperation mode.

1 Motivation

In the past 10 years the simulation league was two dimensional, all players and even the ball moved on the ground. During this time numerous sophisticated tools were created for analyzing the simulated games such as Logalyzer[1] or Team Assistant[2].

The Logalyzer provides information about detected actions like passes and several visualizations for the collected data about the game. The Team Assistant is able to display information provided via agent logfiles along with statistics about detected actions. The Team Assistant is also mentioned in the 2002 league summary[3] as the winner of the presentation tournament.

In 2003, the 3D simulation was introduced including basic tools to view and replay the simulated game. The tools used in 2D can not be used in 3D simulations because of the lack of one dimension and a different format of the logfiles.

The current monitor is capable of showing the current simulated game (at current time) and of replaying monitor logfiles. The replaying mode can be used to watch previously simulated games again. There is also a "single step mode" which provides slow motion replay.

When trying to develop a behavior for an agent or verifying behaviors acquired by machine learning methods it is hard to determine which methods or functions led to the actions observed in the game. This information is crucial when trying to debug or improve the agents in their behavior and collaboration. Especially

G. Lakemeyer et al. (Eds.): RoboCup 2006, LNAI 4434, pp. 426–433, 2007.

in the case of collaboration it is tedious and time-consuming to check what each agents intention is. This is caused by the fact that every logfile has to be searched for the right record of the actual time displayed in the monitor by hand. Additionally, the agent logfiles are not numbered according to the uniform numbers of the agents which complicates finding the desired logfile.

The aim of this work is to create a logfile player and analyzer for the 3D simulation league with new functions allowing to analyze the agent's behavior and collaboration with other agents. The importance of the evaluation of agent teamwork has been addressed in many papers for 2D simulation [4][5] and [6].

2 Related Work

Analyzing tools in 3D simulation are rare, most of the development efforts that have been done in the past years were made in the conception and development of the 3D simulation server. Even in the 2005 soccer simulation development competition just one tool was introduced, which is the "Persian Robotics Analyst"[1].

The Persian Robotics Analyst (PRA) offers information about ball possession, successful and unsuccessful passes and about good and bad actions.

There is also an unpublished "Studienarbeit"[7] (student project) at the Universität Koblenz which has been worked on in parallel to this work. There are some common functions with this work such as the possibility of agents to draw into the displayed scene or the displaying ability of text messages according to the current scene. That tool also provides the detection of (double-) passes, goal-shots, dribbling and their outcome. Also ball contacts and tackling are detected. Statistics about these detected events are written into a text file. Commentatory text messages can be displayed according to the current detected situation. These comments were the main aspect of this work.

3 Requirements

In consultation with other agent developers, providing an easy and clear way to gather information about what the agent intends and how the world looks like, according to the agent, is essentially needed.

The following features are judged beneficial and essential to debug handwritten behaviors and to verify the decisions of learned behaviors.

3.1 Features

While analyzing a special situation of the game it is obvious that *forward and backward replay in different speeds* is useful. With this possibility the situation can be analyzed again without starting the game from the beginning until the desired situation is reached.

[1] `http://www.persianrobotics.net`

To gain knowledge of the agents intentions in an situation an *output of agents logfile according to current time* is needed.

In addition to the logfile displaying it is valuable to *enable the agents to draw information directly into the displayed scene*, like a line from the agent to the position it intends to move to.

Also *filtering the logfile output* may be helpful to display only those information needed by the developer. This way only those information provided by the current developed behavior could be displayed.

New *camera positions*, like birdview which resides directly over the agent of interest, may be beneficial.

When using the single step mode of the monitor *displaying the ball's and player's movements for some time forward* can be an improvement, the developer could see the next movements without proceeding forward in scene display.

A *Grafical User Interface* would be useful to grant an easy access to all functions of the monitor/ logplayer. This way nobody would have to remember all the keyboard shortcuts.

3.2 Additional Features

Displaying the offside line will become necessary when the 0.4 version of the simulation server becomes official (probably RoboCup 2006, not for qualification), which will include the offside rule.

In some situations it could be difficult to distinguish on which side of the offside line an agent is, so *marking agents that are in an offside position* if the ball would be played to them may be a good feature.

Detecting events and the number of their occurrence may provide essential information about what parts of the agent have to be worked on (i.e. if passes often fail there is some space for improvements). Those events are: pass success/fails, ball dribbling, lost balls, goal shot (success, out, intercepted) and kick out.

By *detecting event sequences* more information can be extracted such as lost balls after dribbling.

Plotting graphs also may provide beneficial knowledge. If the ball is in the own half most of the game, the agents may play too defensive. Or if they move all the time their batteries may run low. Useful graphs could be: goal distance, velocity and ball possession.

4 Implementation

When talking about a logplayer or a monitor this essentially means the same program in two different operation modes, logplayer means that the program replays logfiles of a previously simulated game, montior that a game that is currently simulated is displayed.

4.1 Features

Forward and backward replay in different speeds is achieved by reading and parsing the logfiles at the beginning of the execution of the logplayer into a new data structure with the result that playback is simply done by iterating through the (previously parsed) gamestates. Playback is much faster now, it has to be slowed down artificially leading to the opportunity of different playback speeds.

Output of agent logfiles according to current time can be done if the agent provides time information for its output in its logfile. The times according to the various agent outputs is also parsed at startup. When displaying, the fitting record of the agent logfile is found using the time currently displayed by the monitor.

The rcssmonitor-lite does not really know what the numbers of the agents (0-21) according to their uniform numbers are, this makes it hard to display the right logfile, only if the agent writes its team name and uniform number into its logfile, the right logfile is displayed.

In order to *Enable the agents to draw information directly into the displayed scene* the agent's logfile record for a given time is separated into "draw commands" and plain text. Draw commands are not displayed as text output. They are parsed according to the kind of the command and the desired action is executed (fig. 1).

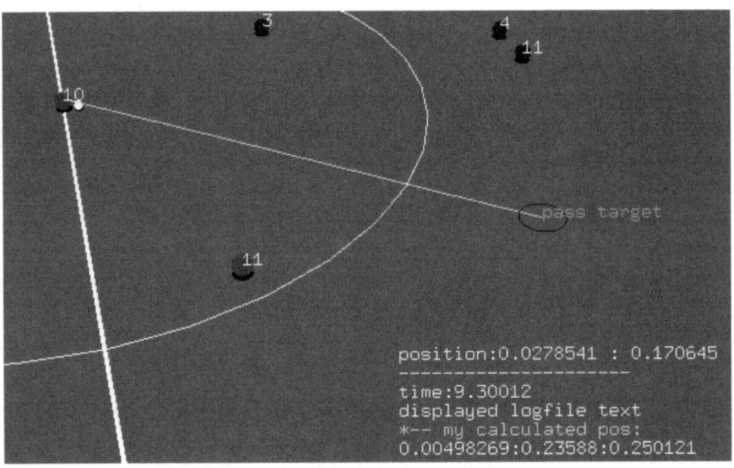

Fig. 1. Output of an agent

Implemented draw commands are: A circle which draws a circle at the given position with a given radius. Lines can be drawn from a starting point to a target point, if a length is given the line starts at the start point into the direction of the target point but stops at the length. Text can be displayed at a given position in the 3D scene. Ground rectangles colorizing the ground in color given though the color command which changes the color of all further items drawn.

In case of *filtering the logfile output* the agent just has to name a filter himself and use it. The filter is detected while parsing the logfile and is made selectable for the developer.

New *camera positions* were implemented, most of them are adjustable in a way (camera height, distance or point to look at).

Displaying the ball's and player's movements for some time forward is solved as lines in the color of the team or white for the ball. The length of the line (how far into the future movements are shown) can be adjusted. The agents track (movements throughout the whole game) can also be displayed, this track can be colored according to the agents movement speed.

The *Grafical User Interface* (GUI) has different layouts according to the current usage. When the program is used as monitor, game control buttons are displayed (drop ball, kickoff, kickoff side). The controls for logfile displaying are shown when used as logplayer.

4.2 Additional Features

For *plotting graphs* the desired informations are piped into the gnuplot[2] program. Information that can be displayed this way are ball position, player position, distance to opponent goal and player speed.

4.3 Other Features Derived from the Implementation

The new data structure of the program allows the user to use single step mode, for- and backward even when watching the game "live". The server is continuing the simulation in background, even if the monitor is in single step mode. Functions that are independent from agent logfiles, such as movement display, can be shown in this mode.

5 Results

The logfile output is useful to understand the agents behavior. This program has replaced the lite monitor that comes with the simulation server within the Virtual Werder team, which started the development of this program. The agent may communicate its decisions to the developer. (i.e. which role it plays, which behavior of the role it has chosen and what action it selected (fig. 1, 2)).

In occurrence of an error in the ball movement prediction the agent was changed to display its data about the ball (its position and movement vector according to the agents world model). This way the error became visible. The line representing the movement vector pointed straight upwards, no matter in which direction the ball was moving. This way the error was resolved very quickly, the x and y component of the movement vector were not written properly into the world model. Without displaying that data, localizing the error would have taken much longer.

When creating a behavior that covers an opponent, it is mostly wanted that only one player covers an opponent at a time. To verify the opponent selection

[2] `http://www.gnuplot.info`

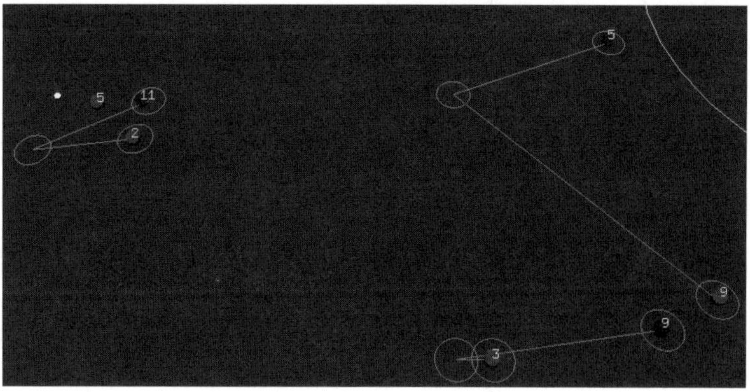

Fig. 2. Covering output

the agent can simply draw a line from himself to the opponent it intends to cover or include the chosen position into its drawings. When displaying the draw commands of all agents it becomes visible if something is wrong with the selection or the position the agent has chosen to move to (lines connecting the agent with the one he has chosen to cover, the circle in middle of the line is the cover position; fig.2). Especially in this case it is useful to display all drawings of the agents to recognize errors in team collaboration.

The new camera features are also useful for verifying behaviors. For example in case of optimizing the ball approaching, a good position for the camera is directly over the agent looking down and moving with it. When checking the goalie behavior a view from behind the goal looking to the ball is a good position for the camera. The view from the side moving left and right with the selected object will be useful to develop offside rule handling.

In case of the ball approach the display of the future movements in addition to the camera view becomes valuable, the developer may directly see possible failures i.e. the agent takes a way that is too long or is moving around the ball too close or far away (fig. 3).

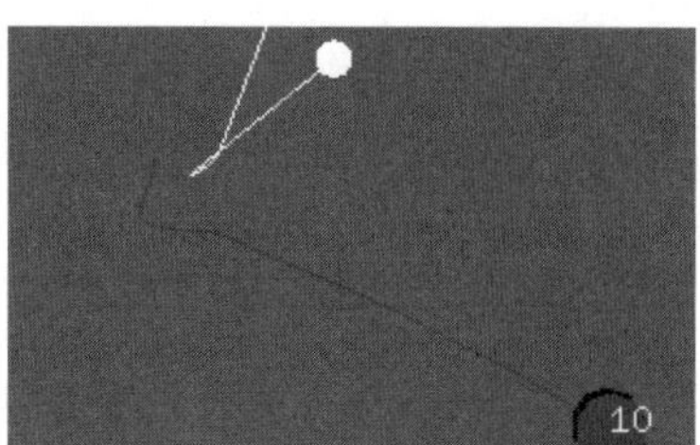

Fig. 3. Approaching the ball

The ability of reverse playback of the game gives the opportunity of analyzing the same scene again without watching the game from the beginning. Also slight transitions in action selection may be analyzed without much interference. This feature in addition to the display of agentlogs (in both ways, text and drawings) gives the opportunity to understand the agents decisions according to its own world model, and not only the real simulated world displayed in the monitor.

The ability of delayed playback and direct analyzing of a currently simulated game provides a real speedup in development in comparison to other tools. Simulating a 3D game may take up to 10 or 20 min which is a long time the developer has to wait in order to in example analyze the previously mentioned ball approach (fig. 3) in slow motion. With the delayed playback that situation may be analyzed while the game is still simulated in background. After the analyzing was finished the game may be watched time shifted in normal speed or the replay may be fastened to reach the most recent scene.

The track display in the 3D scene along with the plotting opportunities were giving the developer essential information about the agents or ball movements throughout the game. This way the developer can find agents which are not moving much or fast, which may point to a suboptimal formation or behavior.

By viewing the ball movement plot (fig. 4) of the 2005 final between Brainstormers3D (left) and Aria2005 (right) it is obvious that the ball was mostly located on the left side of the field and though that Aria2005 dominated the game. It also looks like Aria2005 is able to cross the ball and the Brainstormers3D are not. Aria2005 won this match 2:0.

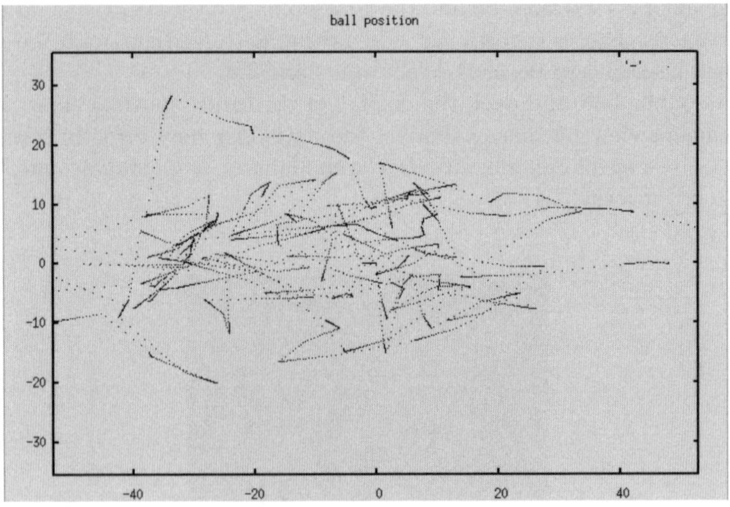

Fig. 4. Plotted ball position

6 Conclusion

The logplayer can be a useful tool when trying to resolve strange behavior of an agent or verifying the proper work of the agent. The Virtual Werder team resolved some problems within their code using the logplayer. These resolved problems are namely the examples given in section 5. The resolving of these problems would have taken longer without the possibility of drawing or text output.

Due to the parsing at the beginning the startup of the logplayer takes longer in comparison to the lite monitor/ logplayer, but the playback is much faster and the monitor does not have to be restarted to watch the game again.

On the other hand the agent logfiles are growing rapidly in size if a lot of draw commands are written to them which also leads to high memory usage after startup.

References

1. Bezek, A.: Modeling multiagent games using action graphs. [Online]. Available: `http://dis.ijs.si/andraz`
2. Nazemi, E., Zareian, A., Samimi, R., Shiva, F.A.: Team assistant team description paper. In: Proc. of Robocup (2002) [Online]. Available `http://www.sbcee.net/files/sbce.pdf`
3. Obst, O.: Simulation league - league summary. In: Kaminka, G.A., Lima, P.U., Rojas, R. (eds.) RoboCup 2002. LNCS (LNAI), vol. 2752, pp. 443–452. Springer, Heidelberg (2003)
4. Raines, T., Tambe, M., Marsella, S.: Automated assistants to aid humans in understanding team behaviors. In: Veloso, M.M., Pagello, E., Kitano, H. (eds.) RoboCup-99: Robot Soccer World Cup III. LNCS (LNAI), vol. 1856, pp. 85–104. Springer, Heidelberg (2000)
5. Takahashi, T.: Logmonitor: From player's action analysis to collaboration analysis and advice on formation. In: Veloso, M.M., Pagello, E., Kitano, H. (eds.) RoboCup-99: Robot Soccer World Cup III. LNCS (LNAI), vol. 1856, pp. 103–113. Springer, Heidelberg (2000)
6. Kaminka, G.A.: The robocup-98 teamwork evaluation session: A preliminary report. In: Veloso, M.M., Pagello, E., Kitano, H. (eds.) RoboCup-99: Robot Soccer World Cup III. LNCS (LNAI), vol. 1856, pp. 345–356. Springer, Heidelberg (2000)
7. Werres, A.: Erweiterung eines 3d-simulators für den robocup (2006) (unpublished)

Local Movement Control with Neural Networks in the Small Size League

Steffen Prüter, Ralf Salomon, and Frank Golatowski

University of Rostock, Faculty of Computer Science and Electrical Engineering, Institute of Applied Microelectronics and Computer Engineering, 18051 Rostock, Germany
{steffen.prueter,ralf.saloman,frank.golatowski}@uni-rostock.de

Abstract. In the RoboCup small-size league, most teams calculate the robots' positions by means of a camera that is mounted above the field as well as different kinds of artificial intelligence methods that run on an additional PC. This processing loop induces various time delays, which require forecasting routines, if more accurate behaviors are desired. This paper shows that by utilizing a combination of a neural network and local sensors, the robot is able to estimates its actual position quite accurately. This paper furthermore shows that the learning procedure is also able to compensate for slip and friction effects that cannot be observed by the local sensors.

1 Introduction

RoboCup small-size league is of particular interest, because it combines engineering tasks, such as building robot hardware and designing electronic components, with computer science applications, such as localization of objects, finding the robots' positions, and calculating the best path through obstacles. Another interesting challenge emerges from the requirement that all team members have to communicate with each other in order to develop a cooperative behavior. Research on artificial intelligence may help find the optimal solution in all of these areas.

In the small-size league, two cameras mounted approximately four meters above the floor observe the field of four by five meters in size on which two teams consisting of five robots play. The processing sequence starting at the camera image and ending at the robots executing their received (action) commands suffers from significant time delays. These time delays have the consequence that when receiving a command, the robot's current position does not correspond to its position in the camera image. Consequently, the actions are either inaccurate or may even lead to improper behavior in the extreme case. For example, the robot may try to kick the ball even though it is not in reach anymore. Section 2 discusses how the position correction can be further improved on the robot itself and how local sensors can alleviate this problem to a large extent.

Fig. 1 shows the omnidirectional drive commonly used by most robots of the small-size league. As can be seen, an omnidirectional drive consists of three wheels, which are twisted to each other by 120 degrees. Other drives with different degrees or four wheeled robots are also present in the small size league. This drives has the

G. Lakemeyer et al. (Eds.): RoboCup 2006, LNAI 4434, pp. 434–441, 2007.

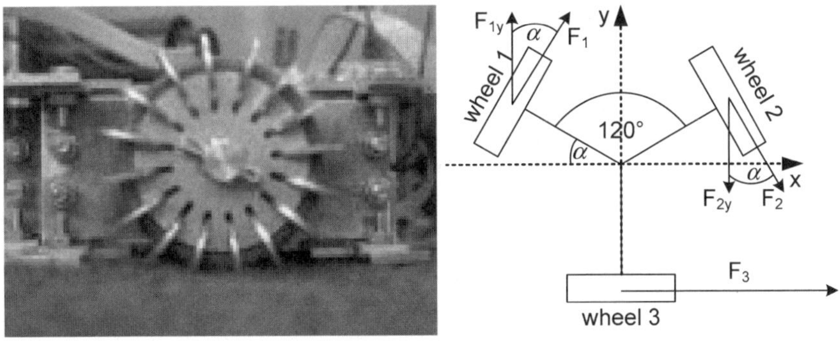

Fig. 1. An omnidirectional drive with its calculation model while the robot is driving in direction of F_3 and wheel 2 and 3 compensate the rotation with F_{1y} equal to $-F_{2y}$.and $F_1+F_2=-F_3$

advantage that a robot can be simultaneously doing both moving forward and spinning around its own central axis. Furthermore, the particular wheels, as shown on the left-hand-side of Fig. 1, yield high grip in the rotation direction, but almost-vanishing friction perpendicular to it. The specific orientation of all three wheels, as illustrated on the right-hand-side of Fig. 1, requires advanced controllers and exhibit higher friction than standard two-wheel drives. The latter requires sophisticated servo loops and (PID[1]) controllers. Depending on the carpet and the resulting wheel-to-carpet friction, one or more wheels may slip. As a consequence, the robot leaves its recalculated moving path. To this end, Section 3 employs a back-propagation network directly on the robot in order to learn the robot's specific slip and friction effects. Section 3 concludes this paper with a brief discussion including possible future research.

2 Local Sensors

As has been outlined in the introduction, the latency caused by the imaging-processing-and-action-generation loop leads to non-matching robot positions. As a measurable effect, the robot starts oscillating, turning around the target position, missing the ball, etc. An approach to solve the latency problem is to do the compensation calculation on the robot itself. The main advantage of this approach is that the robot's wheel encoders can be used to obtain additional information about the robot's actual behavior. However, since the wheel encoders measure only the wheel rotations, they cannot sense any slip or friction effects directly.

2.1 Latency Time

RoboCup robots are real-world vehicles rather than simulated objects. Therefore, all algorithms have to account for physical effects, such as inertia and delays, and have to meet real-time constraints [5]. Because of the real-time constraints, perfectly exact

---

[1] PID is the abbreviation of proportional-integrate-differential. For further detail, the interested reader is referred to [4].

algorithms [1] would usually require too much a calculation time. Therefore, the designer has to find a good compromise between computational demands and the precision of the results [2]. In other words, fast algorithms with just a decent precision are the method of choice here.

For the top-level control software, which is responsible for the coordination of all team members, all time delays appear as a constant-time lag element. The consequences of the latency problem are further illustrated in Fig. 2.

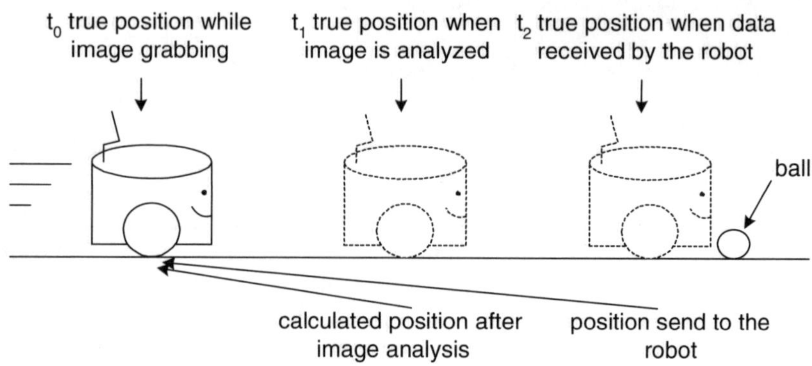

Fig. 2. Due to the latency problem, the robot receives its commands at time t_2 , which actually correspond to the image at time t_0

Fig. 2 illustrates the various process stages and corresponding robot positions. At time $t_{0,}$ the camera takes an image with the robot being on the left-hand-side. At the end of the image analysis (with the robot being at the old position), the robot has already advanced to the middle position. At time t_2, the derived action commands arrive at the robot, which has further advanced to the position t_2 to the right-hand-side. In this example, when being in front of the ball, the robots receive commands which actually belong to a point in time in which the robot was four times its body length away from the ball.

2.2 Experimental Analysis

In order to compensate for the effects discussed above, the knowledge of the exact latency time is very important. The overall latency time was determined by the following experiment: The test software was continuously sending a sinusoidal drive signal to the robot. With this approach, the robot drives 40cm forward and than 40cm backwards. Then, the actual robot position as was seen in the image data was correlated with the control commands. As Fig. 3 shows, the duration of the latency time is seven time slots in length, which totals up to 234ms with 30 frames send by the camera.

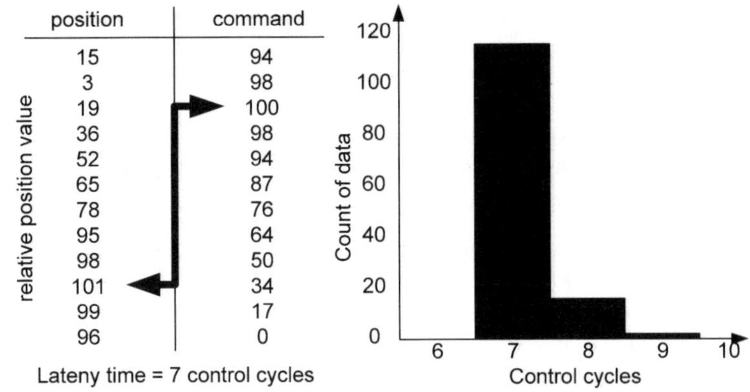

Fig. 3. Detection of the Latency time in the control loop

It might be worthwhile to mention here that for technical reasons, the time delay of the DECT modules is not constant; the jitter is in the order of up to 8ms. The values given above are averages over 100 measurements.

2.3 Increased Position Accuracy by Local Sensors

In the ideal case of slip-free motion, the robot can extrapolate its current position by combining the position delivered by the image processing system, the duration of the entire time delay, and the traveled distance as reported by the wheel encoders. In other words: in case slip does not occur, the robot can compensate for all the delays by storing previous and current wheel tick counts. This calculation is illustrated in Fig. 4.

$$x_{offset} = \sum_{i=1}^{latency} hx_i$$

$$y_{offset} = \sum_{i=1}^{latency} hy_i$$

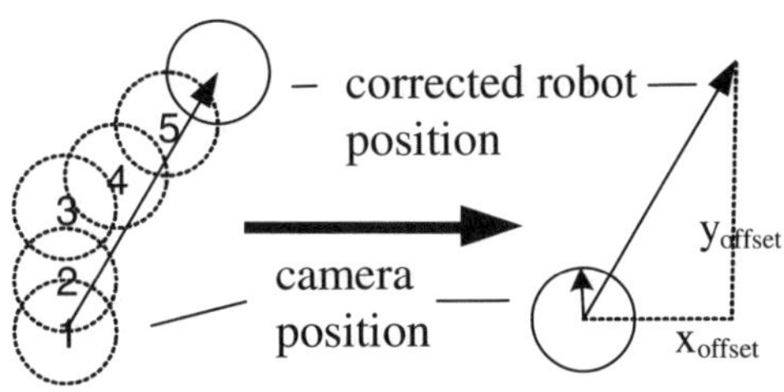

Fig. 4. Extrapolation of the robot's position using the image processing system and the robot's previous tick count

Since the soccer robots are real-world entities, they also have to account for slip and friction, which are among other things, nonlinear and stochastic by nature. The following section employs back-propagation networks to account for those effects.

3 Embedded Back-Propagation Network

Due to the resource limitations of the robot hardware, the number of nodes and connections that the robot can store on its hardware is quite limited. From a hardware point of view, the memory available on the robot itself is *the* major constraint. In addition to the actual learning problem, this section also addresses the challenge of finding a good compromise between the network's complexity and its processing accuracy.

A second constraint to be taken into account concerns the update mechanism of the back-propagation learning algorithm. As is well known, back-propagation temporarily stores the calculated error sums as well as all the weight changes Δw_{ij} [3]. This leads to a doubling of the memory requirements, which would exhaust the robot's onboard memory size even for moderately sized networks. This section stores those values on the central control PC and communicates the weight changes by means of the wireless communication facility.

3.1 Methods

As has been discussed above, the neural network has to estimate the robot position also when slip and/or friction occur. Since the coding of the present problem is but trivial, this section provides a detailed description of it. In order to avoid a combinatorial explosion, the robot is set at the origin of the coordinate system in every iteration. All other values, such as the target position and orientation, are relative to that point.

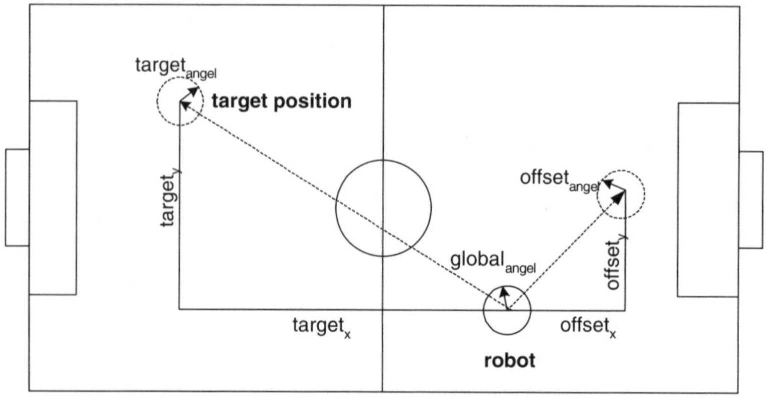

Fig. 5. And example configuration for the slip and friction compensation. For details please see text.

Fig. 5 illustrates an example configuration. This configuration considers three robot positions labeled "global", "offset", and "target". The first robot corresponds to the position as provided by the image processing system. The second position, called "offset", corresponds to the robot's true position, and hence includes the traveled distance during the time delay. The third robot symbolizes the robot's target position. As has already been mentioned above, the neural network estimates the robot's true positions (labeled by "offset") from the target position, the robot's previous position, and its traveled distances. The relative values mentioned above are scaled such that they fit into -40 to 40, and all angels are directly coded between 0 and 359 degrees. With all these values the input layer has to have seven nodes.

Due to the limited calculation capabilities of the microcontroller, all values of the neural network may be stored in integer quantities. In this format, every operation on the microcontroller is done in two processing steps because of the mathematical coprocessor. For this, the feed forward network calculation (FFN) on the robot must be adapted. To this end a simulation of different FFNs on a PC provide important criteria for the implementation on the robot.

Fig. 6 compares the performance of different FFN architectures. All experiments where done with 400 pre-selected training patterns and 800 test patterns. The initial learning rate was set to $\eta = 0.1$. During the course of learning, the learning rate was increased by 2% in case of decreasing error values and decreased by 50% otherwise. It should be noted that in 10% of all experiments, back-propagation got stuck in local optima. These runs were discarded and are not further considered in this paper. Learning was terminated, if no improvement could be achieved over 100 consecutive iterations.

As can be seen, the one and two-hidden layer networks provide a comparable accuracy. Networks with more hidden layers are not considered here, since they would exceed the available computational resources. The variations between the average errors of different node counts are also low. The outcome of this is that the network structure of choice is a one hidden layer network with five nodes. This network is a good compromise between network accuracy and calculation time.

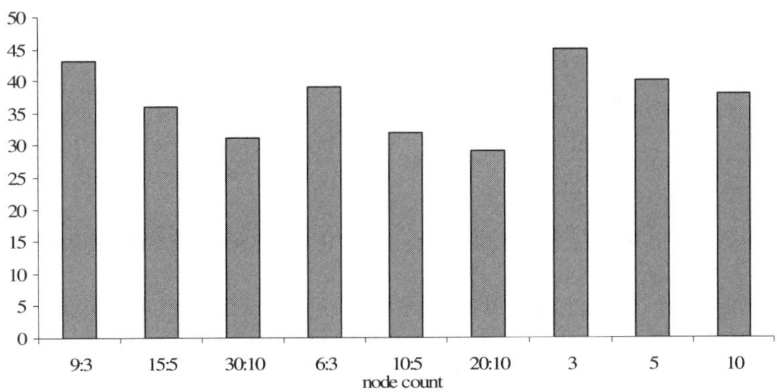

Fig. 6. Average error in mm of two and one hidden layer FFNs

The next step is the adaptation of the selected FFN on the robot. The measurements show that all resulting network weights Δw_{ij} are in the range of -10 to 10. The integer variable on the microcontroller has a range between -32,768 and 32,767. So all weights multiplied by 1,000 to fit into the integer range. The input values, i.e., global, offset, and target position, are multiplied by 100. All nodes calculate their input according to

$$net_i = \sum_i w_{ij} o_j \, .$$

Hence the node input net_i on the microcontroller is by a factor $1{,}000 \cdot 100{=}100{,}000$ larger than on the PC side. As an exception, the node inputs are stored as long integers in consequence of the possible high range. The neural network learning process on the PC side operates with floating-point arithmetic to evaluate the sigmoidal function and the node output, which is also required on the microcontroller. The calculation of the results is difficult to implement, because of the limited calculation time on the robot. The answer to the problem is to store the network's values in a constant predefined array, because no RAM and calculation power is required for the activation function. With this modification, the calculation of the FFN is feasible on the microcontroller.

3.2 Results

The results indicate that the FFN provides a gain of 50% in accuracy. The error caused by the modification of the FFN on the robot is less than 8%. A second measurement evaluates the quality of the correction by driving an 8-shaped figure. In this real world test, the robot is controlled by the camera and the PC outside, as has been suggested by others [5]. This test environment shows how precise and fast the robot can drive. The driven figure has a size of three by one meter and is cut into 64 areas. The PC outside the field checks the robot's position during the measurement and sets the new area as target position when the robot has reached the area before, so the robot cannot derivate from its way. The results shown in Table 1 exemplify that the robot's speed has significantly increased on the field via the employment of FFN.

Table 1. Average time needed to drive the test figure

	Robot	Robot with History	Robot with FFN
Time	8.2s	6.8s	5.9s

This results show that local wheel sensors implemented on the robot advance the accuracy of movement control. But wheel sensors cannot measure slip and friction effects. Back-propagation networks can reduce the positioning errors caused by these effects, but most microcontrollers and embedded devices cannot provide the required calculation power and memory. The adaptation of FFN to accomplish the hardware limitations on autonomous robots was also successful. The measurements show that only marginal variations between the common FFN and the adapted version occur. So, this advancement can be used for further implementations and other developments.

4 Conclusions

The focus of this paper was on the small-size league in which two teams of five robots each play soccer against each other. Since no human control is allowed, the system has to control the robots in an autonomous way. To this end, a control software analyzes images sent by two cameras and derives appropriate control commands for all team members.

Unfortunately, the image processing system exhibits various time delays at different stages, which leads to erroneous robot behavior. Section 0 has shown how local sensors compensate those effects.

The omnidirectional drives used by most research teams exhibit certain inaccuracies due to two physical effects called slip and friction. Section 3 has indicated that neural networks are able to significantly improve the robot's behavior with respect to accuracy, drift, and response.

Furthermore, the architectures presented here still require hand-crafted adjustments to some extent. In addition, the resources available on the mobile robots significantly limit the complexity of the employed networks.

First of all, future research will be addressing the problems discussed above. For this goal, the incorporation of short-cuts into the back-propagation networks seems a promising option. Another important aspect will be the development of complex controllers that would fit into the low computational resources provided by the robot's onboard hardware.

References

1. Gloye, A., Simon, M., Egorova, A., Wiesel, F., Tenchio, O., Schreiber, M., Behnke, S., Rojas, R.: Predicting away robot control latency, Technical Report B-08-03, FU-Berlin (June 2003)
2. Zagal, J., Ruiz-de-Solar, J.: Learning to Kick the Ball Using Back to Reality. In: Nardi, D., Riedmiller, M., Sammut, C., Santos-Victor, J. (eds.) RoboCup 2004. LNCS (LNAI), vol. 3276, Springer, Heidelberg (2005)
3. Rojas, R.: Neural Networks - A Systematic Introduction. Springer, Heidelberg (1996)
4. Astrom, K.J., Hagglund, T.: PID Controllers: Theory, Design, and Tuning, International Society for Measurement and Con, 2nd edn (1995)
5. Gloye, A., Göktekin, C., Egorova, A., Tenchio, O., Rojas, R.: Learning to Drive and Simulate Autonomous Mobile Robots. In: Nardi, D., Riedmiller, M., Sammut, C., Santos-Victor, J. (eds.) RoboCup 2004. LNCS (LNAI), vol. 3276, Springer, Heidelberg (2005)

A Comparative Analysis of Particle Filter Based Localization Methods

Luca Marchetti, Giorgio Grisetti, and Luca Iocchi

Dipartimento di Informatica e Sistemistica
Università di Roma "La Sapienza"
Via Salaria 113 00198 Rome Italy
{firstname.lastname}@dis.uniroma1.it

Abstract. Self-localization is a deeply investigated field in mobile robotics, and many effective solutions have been proposed. In this context, Monte Carlo Localization (MCL) is one of the most popular approaches, and represents a good tradeoff between robustness and accuracy. The basic underlying principle of this family of approaches is using a Particle Filter for tracking a probability distribution of the possible robot poses.

Whereas the general particle filter framework specifies the sequence of operations that should be performed, it leaves open several choices including the observation and the motion model and it does not directly address the problem of robot kidnapping.

The goal of this paper is to provide a systematic analysis of Particle Filter Localization methods, considering the different observation models which can be used in the RoboCup soccer environments. Moreover, we investigate the use of two different particle filtering strategies: the well known Sample Importance Resampling (SIR) filter, and the Auxiliary Variable Particle filter (APF).

1 Introduction

The knowledge of the pose and the orientation of a mobile robot in its operating environment is of utmost importance for an autonomous robot. Therefore self-localization is a well known problem in mobile robotics, and many effective solutions have been proposed. The presence of an initial position guess about robot position determines a distinction between the *position tracking* and the *global localization* problems. The prototype of the algorithms for the position tracking problem is Kalman Filter localization [6], while global positioning encloses common frameworks like Multi Hypotheses Localization, Histogram Filters and Particle Filters [1]. In the last years Particle Filter Localization, also known as Monte Carlo Localization (MCL), became one of the most popular approaches for solving the global localization problem.

Whereas the implementation of a particle filter for localization is straightforward, its performance is strongly affected by the modeling of the process to estimate. Namely the user has to specify the system *motion model*, that is the

G. Lakemeyer et al. (Eds.): RoboCup 2006, LNAI 4434, pp. 442–449, 2007.

probability distribution of successor states conditioned to the odometry readings, and the *observation model* that describes the likelihood of a given observation given the current robot position.

In the RoboCup Four-legged league the localization problem becomes a challenging task, because of the following reasons: i) the only sensor that can be used for acquiring measures of the environment is a low resolution and low quality camera; ii) the robot motion is affected by a considerable amount of noise due to both the presence of opponents in the field of play and to the poor accuracy of the odometry; iii) the computational power available for localization is rather limited.

The dynamic environment strongly violates the Markov assumption which underlies most of the approaches proposed in literature. In order to cope with such violations, several extensions have been proposed to the original Bayes formulation of the localization problem. To this end the most popular technique is known as *sensor resetting* [5]. It consists in bootstrapping the estimator with hypotheses based on the raw observations.

The goal of this paper is to present a systematic analysis performed on the Particle Filter localization, when considering the different observation models which can be used in the RoboCup Four Legged league contexts. Moreover, we investigate the use of two different particle filtering strategies: the well known Sample Importance Resampling (SIR) filter [2], and the Auxiliary Variable Particle filter (APF) [8].

Localization based on APF has been previously proposed by Vlassis *et al.* [9] for solving the vision based localization problem, together with a nonparametric estimate of the likelihood function. The main focus of their work is on how to compute a satisfactory nonparametric estimate of the direct observation model $p(x|z)$, expressing the probability of being in a given location x given the observed panoramic camera image z. Such a distribution is expressed through a Gaussian mixture learned from the data.

In contrast to [9], the contribution of our work is to investigate possible variants of the particle based localization algorithms, for parametric (feature based) observation models, treating APF as an additional degree of freedom.

2 Particle Filtering

One of the most popular algorithms used in localization is the so-called Monte Carlo Localization introduced by Dellaert *et al.* [1]. The core idea of the algorithm is to estimate a robot pose distribution using a particle filter. The system state is represented through a set of samples in the robot pose space $\left\{x^{(i)} = (\mathbf{x}, \mathbf{y}, \theta)^{(i)}\right\} \in \Re^2 \times [0 \ldots 2\pi)$.

$$p(x) \simeq \frac{1}{N} \sum_{i=0}^{N} \delta_{x^{(i)}}(x)$$

here $\delta_{x^{(i)}}(\cdot)$ is the impulse function centered in the sample $x^{(i)}$. The denser are the samples in a state space region the higher the probability that the robot is in that region.

Ideally one wants to sample from the posterior distribution

$$x_t^{(i)} \sim p(x_t | z_{0:t}, u_{0:t})$$

but this is not possible in the general case because such a distribution is not available in closed form. However b.

By representing the distribution through a set of weighted samples $< w^{(i)}, x^{(i)} >$, it is possible to estimate $p(x)$. The samples are drawn from a proposal distribution $q(x_t | z_{0:t}, u_{0:t})$, while the weights for each sample are computed according to the Importance Sampling Principle

$$w_t^{(i)} = \frac{p(x_t^{(i)} | z_{0:t}, u_{0:t})}{q(x_t^{(i)} | z_{0:t}, u_{0:t})} \tag{1}$$

By choosing the motion model $p(x_t | x_{t-1}^{(i)}, u_t)$ as proposal distribution, we can recursively compute the weights as

$$w_t^{(i)} \propto p(z_t | x_t^{(i)}) \, w_{t-1}^{(i)}.$$

Two particle filtering techniques used in self-localization are the Sampling Importance Resampling (SIR) [2] and the Auxiliary Variable Particle Filter (APF) [8,9].

One of the main problems of the SIR algorithm is the *degeneracy problem* [8]. If the ratio between the variance of the proposal distribution and the observation model is high, it can happen that only a few samples generated in the sampling steps have a meaningful weight. The subsequent application of the resampling operation results in the suppression of most of the samples generated, because only those with greater weight are replicated, replacing the low-weighted particles. This fact reduces the chances that the filter achieves to a correct convergence because it impoverishes the set diversity.

The APF has been introduced to lessen the degeneracy problem. The key idea of this algorithm is to select the samples that will be propagated in the subsequent updated estimate. Such a selection is performed by evolving the current filter state using a reduced noise motion model, and evaluating the sample weights according to the Importance Sampling (IS) principle. Then a set of indices is sampled from the weights distribution, and only the surviving particles will be used for computing the updated particle generation, by means of the original motion model. This ensures an increased particle diversity, since the resampling is performed before evolving the filter.

More in detail the APF exploits the following factorization over the joint posterior of particle indices and robot poses

$$p(x_t, i | z_{0:t}, u_{0:t}) \propto p(z_t | \mu_t^{(i)}) p(x_t | x_{t-1}^{(i)}, u_t).$$

Here μ_t is a mean, a mode or some other indicator of the predicted distribution, designed so that

$$p(i | z_{0:t}, u_{0:t}) \propto p(z_t | \mu_t^{(i)}).$$

Under the above hypotheses, we can sample from $p(x_t, i|z_{0:t}, u_{0:t})$ by sampling an index j with probability $\lambda^{(j)} = p(z_t|\mu_t^{(i)})$. We denote with j^i that the value of the jth sampled index is the original index i. Subsequently, we can sample from the motion model $x_t^{(j)} \sim p(x_t|x_{t-1}^{(j^i)})$, according to the value of the state referred to by the drawn index.

According to the IS principle, the resulting weights will be:

$$w^{(j)} = \frac{p(z_t|x_t^{(j)})}{\lambda^{(j)}}$$

3 Algorithm Implementation

Given a filtering method, there are a number of implementation issues to be considered and several parameters to be tuned. The motion model depends on the robot kinematics and on the characteristics of the environment (i.e., the friction with the surface and the presence of collisions). The observation model depends on the characteristics of both the landmarks being observed and the sensor.

Motion Model. When the robot moves, its pose estimate should be updated according to the motion model, incorporating the relative movement u_t, estimated from the odometry. In the Sony AIBO, such a displacement can be obtained by taking into account the joints measures returned from the internal motor encoders. Inaccuracies in the model, as well as environment random phenomena (such as slipping and collisions), affect reliability and precision of measuring such a displacement.

A simple motion model can take into account such a noise by adding an amount of random Gaussian noise to odometry update:

$$x_t \sim x_{t-1} \oplus (u_t + \mathbf{e}_t)$$

where $\mathbf{e}_t$ is the random variable representing the noise affecting the odometry measure u_t, and $\oplus$ is the standard motion composition operator defined as in [7].

A more complex motion model can also consider noise depending on the relative motion of the robot

$$x_t \sim x_{t-1} \oplus (u_t + \alpha(u_t)\mathbf{e}_t)$$

where $\alpha(u_t)$ is a matrix of functions of the odometry motion, and $\mathbf{e}_t$ represents the noise affecting the movement for each time a unity distance is traveled. If α is a constant, the variance of the odometry error grows linearly with the distance traveled by the robot.

Finally, we can extend the previous model considering the presence of random collisions. When a robot hits an object or another robot, it is likely to stack, although the odometry measures a non zero displacement. We can take into account this phenomenon by including a prior of hitting an obstacle and marginalizing it out, as follows.

$$x_t \sim \mathbf{h}x_{t-1} + \neg\mathbf{h}(x_{t-1} \oplus (u_t + \alpha(u_t)\mathbf{e}_t))$$

where $\mathbf{h}$ is a binary random variable that takes into account the probability of hitting an obstacle.

3.1 Observation Model

The environment of RoboCup Four-Legged League includes a set of features that are normally used in localization: unique colored beacons (or markers), unique colored goals, and white lines on the green carpet.

Since the beacons are of limited size they are usually entirely contained in an image. The vision system on the robot can estimate the position of these beacons in the robot reference frame. With these preconditions, developing a localization algorithm should be straightforward, however the noise affecting both the robot sensing and the robot motion, as well as the dynamic environment turns this task into a challenging one.

In particular, the low resolution of the camera, combined with motion blurring effects caused by the robot movements, affects the reliability and precision of feature detection. This is particularly evident when a landmark located quite far away from the robot is perceived. In this case a beacon occupy only a few pixels in the image (as few as 10), and the estimation process is likely to fail.

An adequate observation model for a generic landmarks takes into account these phenomena is

$$p(z|x) = p(z_e) + \sum_i p(z_i|x)$$

here, $p(z|x)$ is the probability of the reading z given the robot pose x. It can be expressed as the conjunction of the following disjoint events:

- the reading is generated by a spurious reading with probability $p(z_e)$ or,
- it is due the landmark γ_i, with probability $p(z_i|x)$.

More in detail, $p(z_i|x)$ can be expressed as

$$p(z_i|x) = p(z|\gamma_i, x)p(\gamma_i|x)$$

here, $p(z|\gamma_i, x)$ is the probability of making the observation z given that it originates from landmark γ_i and the robot is in x, and $p(\gamma_i|x)$ is the prior of perceiving the landmark γ_i from the location x.

The measurement probability is a function of the angular (δ_α) and linear (δ_ρ) distance between the expected and the measured landmark locations. For every type of landmark (beacon b_i, goal post p_i and line l_i) we use these equations:

$$\begin{aligned}
p(z|b_i, x) &\propto e^{(\delta_\alpha^2/\sigma_\alpha^2)} e^{(\delta_\rho^2/\sigma_\rho^2)} \\
p(z|p_i, x) &\propto e^{(\delta_\alpha^2/\sigma_\alpha^2)} \\
p(z|l_i, x) &\propto e^{(\delta_\theta^2/\sigma_\theta^2)} e^{(\delta_\rho^2/\sigma_\rho^2)}.
\end{aligned}$$

In the previous equations we use α denote angular measurements, and ρ indicates the distance. For lines across the fields (and corners, that are intersection of lines), we assume that they are expressed in polar coordinates, so ρ and θ indicates the hough parameters of a single detected line.

For a single beacon observation b_i we use both angle and distance, while for single goal post p_i we only use the measured angle. This is because goal posts are not recognized exactly and they are less reliable. Each σ inside equation is specific to type of observation and are note the same.

The probability for the single observation models are then combined to form definitive probabilities

$$p_{particles} = \prod p(z|b_i, x) \prod p(z|p_i, x) \prod p(z|l_i, x)$$

4 Experiments

In this section we present the results of localization experiments for the methods described in the previous section.

The experiments have been performed using standard Four-Legged soccer field as used in RoboCup 2005 and RoboCup 2006. The scenario is more challenging for localization task with respct to RoboCup 2004 setting (having a smaller field and six beacons instead of four), where experiments in [3] have been performed. We also use two different strategies to track real robot position: external camera to track real position of robot and ground truth made using some measured spot and a smoothing technique. For acquiring a smooth ground truth of robot position we process off-line log data captured from the robot, as follows:

- we take a log containing images and internal information from the robot;
- we manually mark every feature on the images in the log;[1]
- we use the true perceptions to generate a log with reliable and more precise measures about distances and angles for objects (these represent the real ground truth of perception);
- we iterate the localization task several times on same input: first iteration the localization is performed in normal way. The second iteration starts using the last set of particles and use log backward from last to first frame. Running localization many times we obtain a smoothed ground truth.

The same log (with ground truth on perception) is then used to run localization methods (with different parameters) several times and to compare different methods on the same input. Moreover, by using the external camera view and the ground-truth of robot path, we can measure absolute errors in localization.

The subsequent path graphs use these conventions: green is used for field lines, cyan for ground truth, red for SIR and blue for APF.

[1] These data sets have been used also for evaluating color segmentation and image processing and are available from `www.dis.uniroma1.it/~spqr`

Random walk without kidnapping. In the first experiment we consider the behavior of SIR and APF using 100 particles, when the robot is in a random walk, without obstacles. In Fig. 1, the left graph shows the result with sensor resetting disabled, while the right one shows the behavior of sensor resetting. While the behaviour of the two filters is similar in the first case, the SIR particle filter converges faster and recover more precisely the position with sensor resetting enabled, while the APF is less robust and more spikes are present in the calculated path.

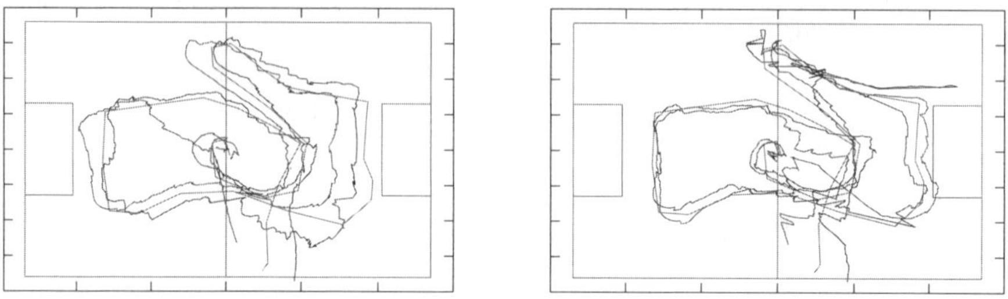

Fig. 1. Random walk with sensor resetting disabled (left) and enabled (right)

Random walk with kidnapping. The second experiment considers another random walk but with a teleporting of robot to simulate the kidnapping problem. The kidnapping was performed from (a) to (b), shown in Fig. 2. Again the SIR filter using sensor resetting recovers from erroneous positions quickly, while the APF changes more slowly.

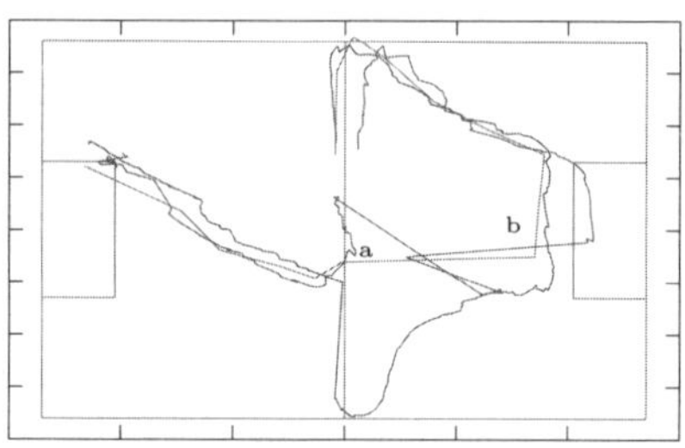

Fig. 2. Second experiment: random walkwith kidnapping from a to b

Playing mode. The third experiment has been done in a typical playing sequence. When robot is in playing state, it interacts with other robot, competing for the ball, occasionally kicking the ball itself. In such situations the odometry give very noisy information to robot. Therefore, sensor resetting needs to be enabled.

Following [3], we performed experiments varying the observation frequency (up to a fraction of 1/256) and adding Gaussian noise (approximately 15% of the real distance and 10 degrees for the angle). As shown in Figure 3 the difference between the two methods are negligible and both degrades its performance in a similar way as in [3].

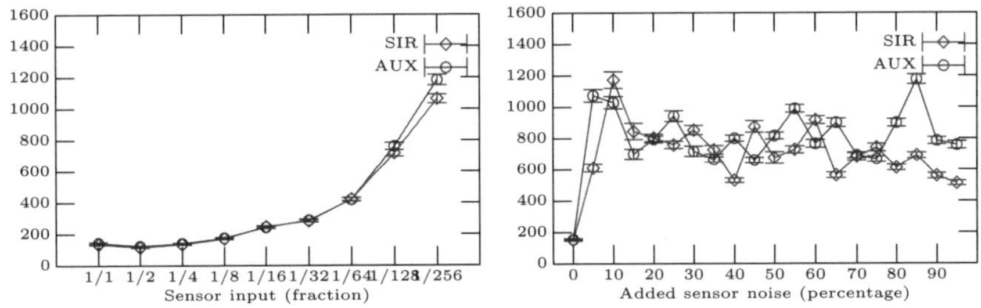

a) b)

Fig. 3. Performance of SIR and AUX compared a) when using only a fraction of observations and b) when additional gaussian noise is present

5 Conclusion

In this paper we have illustrated the performance of SIR-based particle filter and Auxiliary Particle Filter, to accomplish a localization task based on visual landmarks with AIBO robots. The results show that both the method are suitable for this task, and that SIR is slightly better in combination with sensor resetting techniques that are needed in the RoboCup soccer scenario.

Additional results, including the use of datasets used in [3] and [4], are provided in a technical report available from `www.dis.uniroma1.it/~spqr`

References

1. Fox, D., Burgard, W., Dellaert, F., Thrun, S.: Monte carlo localization: Efficient position estimation for mobile robots. In: Proc. of the 16th National Conference on Artificial Intelligence (AAAI99) (1999)
2. Gordon, N., Salmond, D., Ewing, C.: A novel approach to nonlinear nongaussian bayesian estimation. In: Proceedings F., pp. 107–113 (1993)
3. Gutmann, J.S., Fox, D.: An experimental comparison of localization methods continued. In: In Proc. of the IEEE/RSJ International Conference on Intelligent Robots and Systems (IROS) (2002)
4. Jensfelt, P., Kristensen, S.: An experimental comparison of localizationmethod, the mhl sessions. In: IROS (2003)
5. Lenser, S., Veloso, M.: Sensor resetting localization for poorly modelled mobile robots. In: Proceedings of International Conference on Robotics and Automation (ICRA'00) (2000)
6. Leonard, J.J, Durrant-White, H.F.: Mobile robot localization by tracking geometric beacons. Transactions on Robotics and Automation 7, 376–382 (1991)
7. Lu, F., Milios, E.: Robot pose estimation in unknown environments by matching 2D range scans. Journal of Intelligent and Robotic Systems 18, 249–275 (1997)
8. Pitt, M., Shephard, N.: Filtering via simulation: Auxiliary particle filters. Journal of the American Statistical Association 94(446), 590–599 (1999)
9. Vlassis, N., Terwijn, B., Kröse, B.: Auxiliary particle filter robot localization from high-dimensional sensor observations. In: Proc. IEEE Int. Conf. on Robotics and Automation (ICRA) (2002)

A New Mechatronic Component for Adjusting the Footprint of Tracked Rescue Robots

Winai Chonnaparamutt and Andreas Birk

School of Engineering and Science
International University Bremen
Campus Ring 1, D-28759 Bremen, Germany
a.birk@iu-bremen.de

Abstract. There is no ideal footprint for a rescue robot. In some situations, for example when climbing up a rubble pile or stairs, the footprint has to be large to maximize traction and to prevent tilting over. In other situations, for example when negotiating narrow passages or doorways, the footprint has to be small to prevent to get stuck. The common approach is to use flippers, i.e., additional support tracks that can change their posture relative to the main locomotion tracks. Here a novel mechatronic design for flippers is presented that overcomes a significant drawback in the state of the art approaches, namely the large forces in the joint between main locomotion tracks and flippers. Instead of directly driving this joint to change the posture, a link mechanism driven by a ballscrew is used. In this paper, a formal analysis of the new mechanism is presented including a comparison to the state of the art. Furthermore, a concrete implementation and results from practical experiments that support the formal analysis are presented.

1 Introduction

Moving around in an unstructured environment is the principal ability a mobile robot must have to be a rescue robot. Locomotion systems in general can be classified as wheeled, tracked or legged. In the RoboCup Rescue 2005 competition tracked robots were very popular and successful. Figure 1 shows an overview of the different tracked robots in this league in 2005. This type of locomotion is often considered as the most versatile locomotion system as it can handle relatively large obstacles and loose soil [Har97][Won01]. Some versions of tracked vehicles are even used by several teams, namely the Tarantula and RobHaz DT-3. The Tarantula is a typical toy car substantiating the concept that low cost platform can be deployed for rescue application. The Tarantula is R/C vehicle with four tracked articulated arms which can climb over obstacles, steps or stairs. The team Freiburg for example very successfully used this toy [KSD+06]. The RobHaz DT-3 is a sturdy commercial platform. It is based on a passive double track platform. There were three teams using this platform in the competition with impressive performances, namely ROBHAZ-DT3 [LKLP06], CASualty [KKP+06], and the Intelligent Robot Laboratory team [TT06].

G. Lakemeyer et al. (Eds.): RoboCup 2006, LNAI 4434, pp. 450–457, 2007.

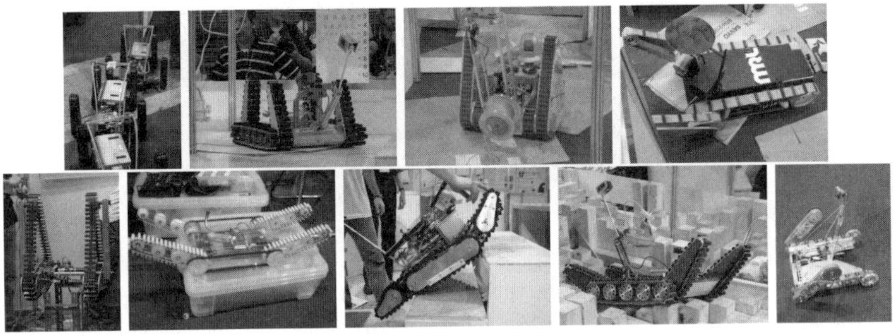

Fig. 1. Flippers as additional tracks that can change their posture relative to the main locomotion tracks are a very common approach to allow for a flexible footprint. The general advantages of this locomotion principle is for example indicated by the many teams that chose this approach for their robots. Some examples from RoboCup 2005 are shown above.

Though a differential drive based on two tracks is simple and in principle already very capable, there is a significant problem especially for rescue robots. A critical aspect is that it is almost impossible to select the right parameters for a single pair of tracks. For some situations, for example when negotiating narrow passages, the footprint of the robot and hence the length of the tracks should be small. When climbing large obstacles, slopes or stairs the footprint should be large. The common solution to this problem is to use additional tracks that can change their posture relative to the main robot body. Note that all robots in figure 1 are equipped with according flippers. The state of the art for changing the posture of the flipper is to directly drive the joint between the additional small track and the main locomotion system. This approach has the tremendous disadvantage that due to the large forces on this active joint it is extremely difficult to construct mechanisms that are sufficiently stable and still within feasible size and weight limits. Broken joints are hence a common phenomena (figure 2). Here a novel design from the IUB rescue robot team [BC06] is presented that circumvents these problems.

2 The Underlying Concept

The standard approach to change the posture of a flipper is to directly drive the joint as shown in figure 3. This can be done by spur or worm gear or a belt or chain drive. But no matter what mechanism is used, it has to take quite some stress. First of all, it has to provide high forces for moving the flipper under load, especially for pushing it down when the full weight of the robot is supported by it. Second, it is subject to shocks and impacts, for example when the robot drives over bumps, stairs, etc. Especially these forces can be very high and they are very hard to predict. It is hence almost impossible to design a fail-safe mechanism

(a) (b)

Fig. 2. Large forces have to be provided when using the standard approach to directly drive the joint of a flipper. In addition, shocks to the tracks are likely to occur in rough locomotion conditions. These can cause large load changes and huge unpredictable dynamic forces directly at the transmission in the joint. Broken transmissions in the flipper-joints are hence a common problem, not only for simple bases like the Tarantula (a) but even for advanced robots like the winner of the RoboCup 2005 rescue competition (b).

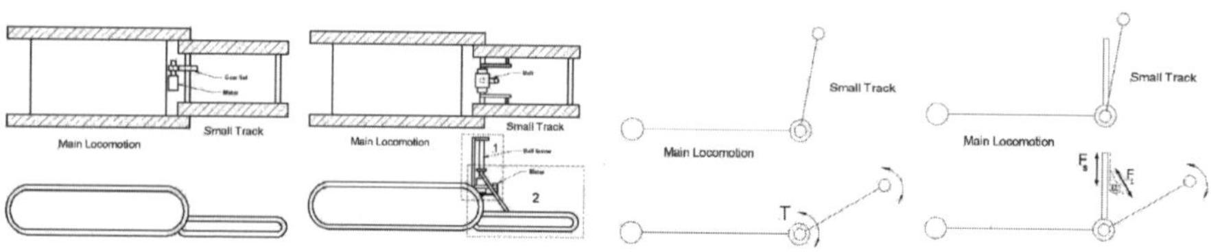

(a) (b) (c) (d)

Fig. 3. A sketch of a classical locomotion system with flipper (a) and its basic free body diagram (c) compared to a sketch (b) and the basic free body diagram (d) of the novel system.

within feasible weight and size limits with this approach. Accordingly, broken flipper joints are a common problem (figure 2).

The novel flipper design presented here consists of a ball screw, a passive link and a motor (figure 3). As shown in the formal analysis later on, the driving force that needs to be provided by the motor is smaller with this set-up. Furthermore, all shocks go against the passive link and the ball screw, which in contrast to spur/worm-gears or belt/chain-drives can be easily laid out to absorb them without any damage. Figure 4 also shows an implementation of the flipper itself.

3 Formal Analysis of the Design

When the robot moves around on the floor, the small track is up to minimize the footprint. Whenever the robot has to move over a big obstacle or up, respectively down a stair, the small track is pushed down to the same level of the big track. The small track is moved up from or down to the floor by a ball screw. The crucial parameters for the ball screw are the thrust force and the stroke of movement. The thrust force of the ball screw determines the force for pulling the small

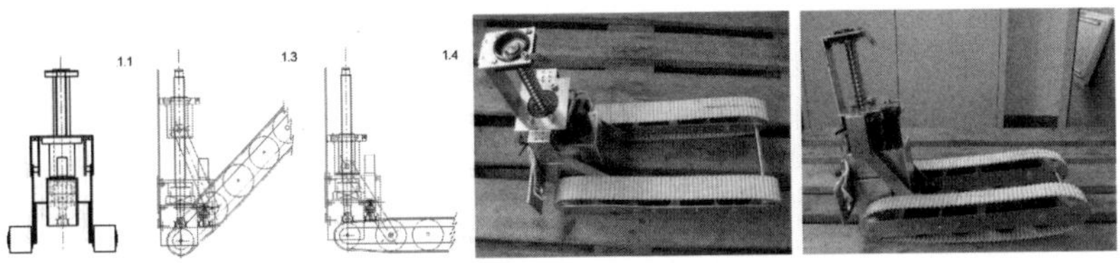

Fig. 4. An implementation of the novel flipper design

track up, respectively pushing the small track down. First, the thrust force is determined that is needed to push the small track down. After that we will find the second parameter, the stroke of movement. First we consider the situation that robot is on a two points support with an angle θ with respect to the floor. To calculate the thrust force of the ball screw, the force in the direction of $\cos\theta$ has to be considered. The maximum value of $\cos\theta$ is one when θ is zero. The maximum thrust force is hence needed in a situation when θ is zero.

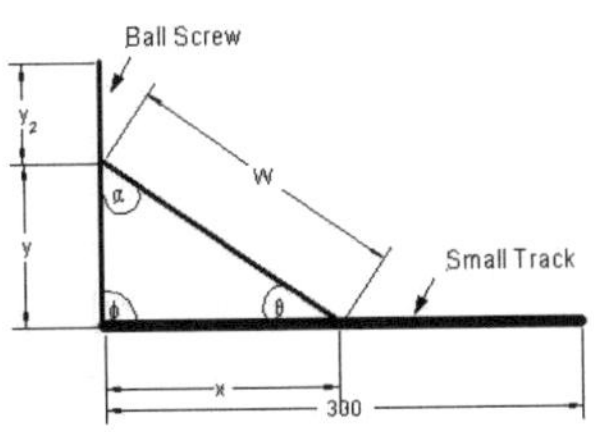

Fig. 5. The core parameters in the free body diagram of the ball screw and the small track

In the following, the maximum thrust force is analyzed following the free body diagrams in figures 5 and 6. First of all, it is assumed that the ball screw is fixed to the robot such that it forms with the robot body and its main locomotion track a single object as shown in figure 6(a). The small track of the flipper is a second object as shown in figure 6(b). As the main locomotion track of Rugbot is 650mm long, a length of 300 mm is chosen for the small track. With this set up, Rugbot is always supported on stairs when the flipper is on the ground. From the stability viewpoint, the stairs can be considered in the worst case like an obstacle on the ground with just two support points at the extrema of the footprint. F and y_2 are most important as they determine the selection of the ball screw and the motor. Also, the length L of the mechanism, i.e., y plus y_2, is of interest. Given the height of a Rugbot that is 550 mm, the mechanism should not extend over it. So, y is the minimum height of the ball screw to which we want to lower down the small track on to the floor. And y_2 is the stroke of the ball screw to move the small track up from the floor. Therefore, $F\cos\alpha$ is a lower constraint on the thrust force. The common condition of the free body diagrams of figure 6(a) and 6(b) is that the robot is stable without any movement in any

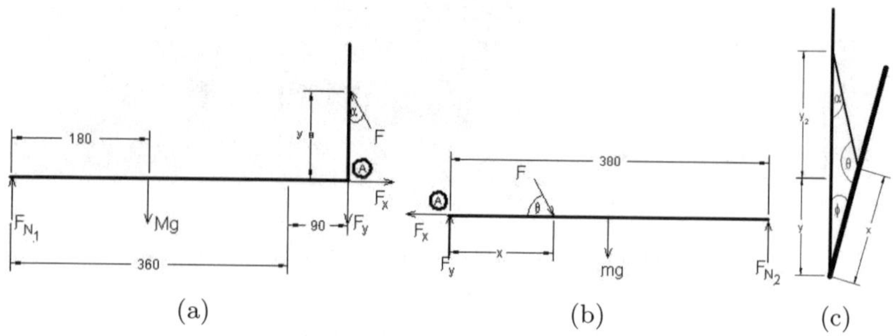

Fig. 6. More detailed free body diagrams and parameters of (a) the main track system with the ball screw and (b) the small track. The free body diagram of the ball screw and the small track when the flipper is moved up (c).

direction. So, the total force in x direction and y direction must be zero. Hence, we get

$$\mathrm{Mg} + \mathrm{F}_y - \mathrm{F}\cos(\alpha) = \mathrm{F}_{N_1} \tag{1}$$
$$\mathrm{F}\sin(\alpha) = \mathrm{F}_x \tag{2}$$
$$\mathrm{mg} - \mathrm{F}_y + \mathrm{F}\sin(\theta) = \mathrm{F}_{N_2} \tag{3}$$
$$\mathrm{F}\cos(\theta) = \mathrm{F}_x \tag{4}$$
$$\mathrm{y} = \mathrm{x}\tan(\theta) \tag{5}$$
$$\theta + \alpha = 90\,\mathrm{deg} \tag{6}$$

Rearrange (1) and (3) with (6)

$$\mathrm{F}_{N_1} + \mathrm{F}_{N_2} = \mathrm{Mg} + \mathrm{mg} + \mathrm{F}(\sin(\theta) - \cos(\alpha)) \tag{7}$$
$$\mathrm{F}_{N_1} + \mathrm{F}_{N_2} = \mathrm{Mg} + \mathrm{mg} \tag{8}$$

The sum of moment about A:

$$\mathrm{Mg} \times 270 + \mathrm{F}\sin(\alpha)\mathrm{y} = \mathrm{F}_{N_1} \times 450 \tag{9}$$
$$\mathrm{F}_{N_1} = \frac{\mathrm{Mg} \times 270 + \mathrm{F}\sin(\alpha)\mathrm{y}}{450} \tag{10}$$
$$\mathrm{mg} \times 150 + \mathrm{F}\sin(\theta)\mathrm{x} = \mathrm{F}_{N_2} \times 300 \tag{11}$$
$$\mathrm{F}_{N_2} = \frac{\mathrm{mg} \times 150 + \mathrm{F}\sin(\theta)\mathrm{x}}{300} \tag{12}$$

From (5), (6), (7) and (8)

$$\mathrm{Mg} + \mathrm{mg} = \frac{\mathrm{Mg} \times 270 + \mathrm{F}\sin(\alpha)\mathrm{y}}{450} + \frac{\mathrm{mg} \times 150 + \mathrm{F}\sin(\theta)\mathrm{x}}{300} \tag{13}$$
$$0.4 \times \mathrm{Mg} + 0.5 \times \mathrm{mg} = \mathrm{F} \times \left(\frac{\sin(\alpha)\mathrm{x}\tan(\theta)}{450} + \frac{\sin(\theta)\mathrm{x}}{300}\right) \tag{14}$$
$$\mathrm{F} = \frac{225}{\frac{\sin(\alpha)\mathrm{x}\tan(\theta)}{450} + \frac{\sin(\theta)\mathrm{x}}{300}} \tag{15}$$

With equation (15), we have the relation between thrust force $F\cos(\alpha)$, the point of push/pull force connect to small track x, and the initial length of the ball screw y. With a numerical analysis, different variations of these parameters can be computed. Based on the size aspect of Rugbot, the size L of the mechanism is the first parameter that should be specified.

Then, the values of x and y are used to calculate the stroke $\mathrm{y_2}$ by the next free body diagram shown in figure 6(c). Given a minimum angle of 10 degrees between the small track and the ball screw, we get the relation between x, y, and $\mathrm{y_2}$ as:

$$\mathrm{W} = \frac{\mathrm{x}}{\cos(\theta)} \tag{16}$$

$$\alpha = \arcsin(\frac{\mathrm{x} \times \sin(10)}{\mathrm{W}}) \tag{17}$$

$$\mathrm{y_2} = \frac{\mathrm{W} \times \sin(170 - \alpha)}{\sin(10)} - \mathrm{y} \tag{18}$$

With the height limit of the robot, we can analyze the set of the data including thrust force, x, θ, y, and $\mathrm{y_2}$. As we specified in the beginning that the first priority in optimization is L, so we analyze the data set with L equal 400 mm, 450 mm, 500 mm, and 550 mm by using the present parameters of the Given the basic parameters of Rugbot with M = 50 kg, m = 5 kg, and g = 10 m/s^2, a numerical analysis can be done with L equals 400 mm, 450 mm, 500 mm, and 550 mm (table 1).

Table 1. Results of the numerical analysis of the parameters for different mechanism lengths L

L: mm	θ: deg	F: N	x: mm	y: mm	$\mathrm{y_2}$: mm
400	50	334.7	158	188.3	211.7
450	52	297.1	173	221.5	228.5
500	52	267.7	192	253.6	253.6
550	51	243.6	214	284.7	284.7

It can be seen that there is an inverse relation between the thrust force and the upper limit. If the upper limit is increased by about 15 cm, there is almost half the thrust force needed than with the shorter upper limit. Based on the parameters of Rugbot and on available ball screws, in the final implementation a ball screw with l=500 mm was chosen leading to a thrust force as 267.7 N.

The benefits of this design concept can also be illustrated by considering the case when the robot faces a ramp as shown in figure 7. Before the robot is going up the ramp, the component has lifted the small track up. When Rugbot is going up the ramp, an inclinometer senses the angle of the ramp φ and the component starts to lay down the track simultaneously. For tele-operated robots this is of course done by the operator. Suppose the robot is moving over the ramp with an approximately constant speed and the same holds for the angle ϕ. So, ϕ is varied

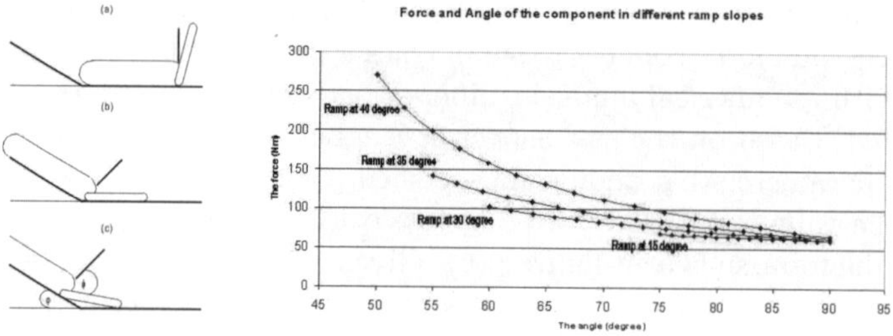

Fig. 7. The new flipper design facing a ramp (left). The forces in dependence of the posture angle for different ramp angles (right).

in each step the robot is moving up the ramp, which can be determined based on φ, robot speed, and the parameters of the flipper. Moreover, the force used to lay down the small track against the floor also can be analyzed to verify the strength of the motor and the ball screw. Figure 7 shows the graphs for several ramp angles with one second period between each data point and a robot speed of 20 mm/sec. As the robot moves up the ramp, the value of ϕ is increasing while the force is decreasing. The analysis also shows that φ not only effects the value of ϕ, but also the force. Larger angles φ need higher forces to put the small track against the floor at the beginning steps to move over the ramp. Note that the overall forces are by far within the allowable range of the ball screw and can be easily provided by the chosen motor.

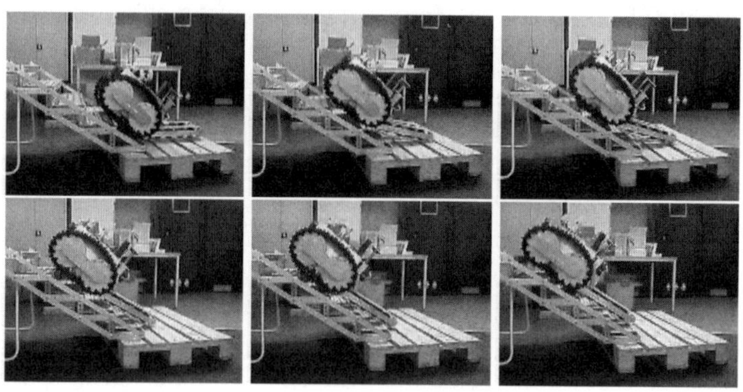

Fig. 8. Rugbot going up stairs

The formal analysis is also supported by all field tests of the robots. The flippers are very sturdy. The can even be used as a handle to pull or lift the whole weight of the robot without the slightest disturbance to the joint between the robot and its flipper. In addition, they support the climbing of obstacles and stairs exactly as they are supposed to do (figure 8).

4 Conclusion

Adjustable support tracks are a common concept for changing the footprint of a robot. Here, a novel mechanism for flipper design was presented that overcomes the flaws of the standard approach to directly drive the joint between the robot body with the main locomotion tracks and the flipper. Instead, a ballscrew and a passive link are used.

References

[BC06] Birk, A., Carpin, S.: Rescue robotics - a crucial milestone on the road to autonomous systems. Advanced Robotics Journal 20(5) (2006)

[Har97] Hardarsson, F.: Locomotion for difficult terrain. Technical report, Mechatronics Lab, Dept. of Machine Design (1997)

[KKP+06] Kadous, M.W., Kodagoda, S., Paxman, J., Ryan, M., Sammut, C., Sheh, R., Miro, J.V., Zaitseff, J.: Robocuprescue - robot league team CASualty (australia). In: Bredenfeld, A., Jacoff, A., Noda, I., Takahashi, Y. (eds.) RoboCup 2005. LNCS (LNAI), vol. 4020, Springer, Heidelberg (2006)

[KSD+06] Kleiner, A., Steder, B., Dornhege, C., Meyer-Delius, D., Prediger, J., Stueckler, J., Glogowski, K., Thurner, M., Luber, M., Schnell, M., Kuemmerle, R., Burk, T., Bräuer, T., Nebel, B.: Robocuprescue - robot league team rescuerobots freiburg (germany). In: Bredenfeld, A., Jacoff, A., Noda, I., Takahashi, Y. (eds.) RoboCup 2005. LNCS (LNAI), vol. 4020, Springer, Heidelberg (2006)

[LKLP06] Lee, W., Kang, S., Lee, S., Park, C.: Robocuprescue - robot league team ROBHAZ-DT3 (south korea). In: Noda, I., Jacoff, A., Bredenfeld, A., Takahashi, Y. (eds.) RoboCup 2005. LNCS (LNAI), vol. 4020, Springer, Heidelberg (2006)

[TT06] Tsubouchi, T., Tanaka, A.: Robocuprescue - robot league team Intelligent Robot Laboratory (japan). In: Noda, I., Jacoff, A., Bredenfeld, A., Takahashi, Y. (eds.) RoboCup 2005: Robot Soccer World Cup IX. Lecture Notes in Artificial Intelligence (LNAI), vol. 4020, Springer, Heidelberg (2006)

[Won01] Wong, J.Y.: Theory of Ground Vehicle.ition, 3rd edn. ch. 4.5, John Wiley and Sons, Chichester (2001)

Vectorization of Grid Maps by an Evolutionary Algorithm

Ivan Delchev and Andreas Birk

School of Engineering and Science
International University Bremen
Campus Ring 1, D-28759 Bremen, Germany
a.birk@iu-bremen.de

Abstract. Mapping is a fundamental topic for robotics in general and in particular for rescue robotics where the provision of information about the location of victims is a core task. Occupancy grids are the standard way of generating and representing maps, i.e., in form of raster data. But vector representations are for many reasons, especially due to their compactness and the possibility to use very efficient computational geometry algorithms, highly desirable for many applications. Here a novel method for vectorization is presented that is intended to work particularly well with maps. It is based on an evolutionary algorithm that generates vector code for a so to say drawing program. The output of the evolving vector code is compared to the input grid map via a special similarity function as fitness. Experiments are presented that indicate that the approach is indeed a successful method to extract vector data out of grid maps.

1 Introduction

Mapping is a very important topic for rescue robotics for two quite different reasons. First of all, it is in general a core problem in robotics [Thr02]. Many fundamental algorithms for mobile robots simply depend on maps without which the system is restricted to be a crude tele-operated device that can hardly be called a robot. Second, maps are a fundamental added value of rescue robots over conventional systems for finding victims [BC06]. While the IUB rescue robot team [Bir05,BCK03,BKR+02] has been the first team to manage mapping in the challenging environment of the RoboCup rescue league in 2003 at the World Championship in Padua, this task meanwhile belongs to the standard challenges in this league [JMW+03,JWM03].

Probabilistic grids, also known as occupancy grids, are a well known approach to represent obstacles and free space by assigning according likelihoods to each cell in the grid [Mor88]. According to [Thr03], occupancy grids are the predominant method for generating, respectively representing maps. Occupancy grids can be easily generated, but they have their disadvantages. They represent maps as raster data, i.e., a collection of values arranged in a rectangular array. Due to its nature, it is difficult to manage and edit, requires a lot of space and contains

G. Lakemeyer et al. (Eds.): RoboCup 2006, LNAI 4434, pp. 458–465, 2007.

Fig. 1. A map (left) generated by an IUB rescue robot (middle) in the entrance hall of a building with a few boxes as obstacles (right)

no intrinsic semantics. Vector data is in contrast a compact format, that describes the geometry of a bar, i.e., a straight line segment with non-zero width, using a small number of attribute values, e.g., two endpoints and line width [DL99].

Many methods, differing in their precision and robustness, have been designed throughout the years for vectorization, i.e., the conversion of raster data into a vector format. The majority share the same structure - group pixels from a raster image into sets, approximate the sets with a set of vectors by some polygonal approximation method and post-process if necessary [TT00,SSTC02]. The classical approaches to vectorization use the following steps [CJ94]:

- Thin the image in order to extract its skeleton.
- Chain-code the thinned image.
- Reduce the chain codes to straight lines.

When the raster data is for example an image based on a line drawing, the standard methods work very well. But they are known to have difficulties with noisy data and "jerks" in the lines, two phenomena that are very common in robot maps. Here a vectorization technique is presented that is based on so-called reproductive perception, which tries to address this task in a non-classical way. Instead of processing the input raster data stepwise to produce its vector representation, exactly the opposite is done: namely vector code is evolved that is used to generate raster data, which is compared to the input data.

2 Overview of the Approach

Some terminology is introduced in this section that will be used in the remainder of the paper. First of all, note that grid maps can be thought of as images. Images, respectively grid maps are rectangular arrays of raster data where each pixel, respectively cell has a value that represents color, respectively the likelihood of free space. As many concepts in this paper are borrowed from image processing, the terms image and grid map are used in an interchangeable way, whatever is more appropriate in respect to its original context.

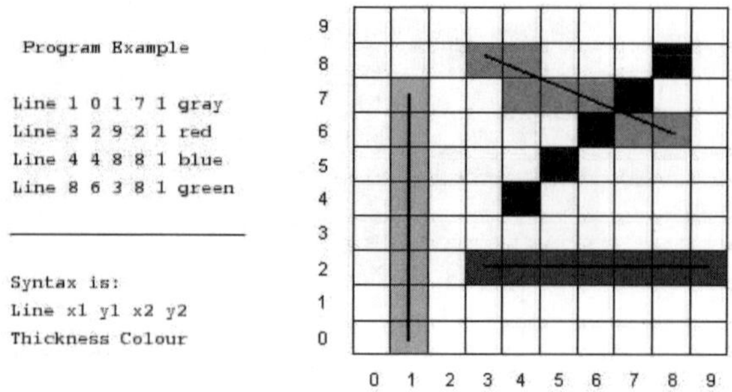

Fig. 2. A simple drawing program and its output

- A *Line* is a data structure that describes a single line - it contains x1, y1, x2, y2 of the endpoints as well as thickness and color.
- A *Program* is a data structure that contains a given number of *Lines*. Each *Program* describes an image containing straight lines and can be easily converted to a PNG or BMP format. The image described by a *Program* will be referred to as the output of the *Program*, shown in figure 2.
- *Population* is a data structure that contains a set of *Programs*.
- *Selection* is a set of *Programs* returned by a selection scheme.

The goal of the Reproductive Perception Vectorization algorithm (RPV) can be summarized in the following way. Given a raster image X like a grid map, create a *Program* Pr such that X', which is the output of Pr satisfies a similarity measure $D_{thresh}(X, X')$, i.e., Pr is the vectorization of X. The set of lines in Pr represents a compact mathematical model of the grid map X. If X' is constructed by using only lines of width 1, it is also the skeleton of X.

3 Fitness Calculation

The *RPV* employs an evolutionary algorithm approach to accomplish its goal. Evolutionary algorithms are a type of heuristic search techniques that incorporate principles of natural selection and "survival of the fittest". They maintain a population P that evolves at each iteration, according to rules called evolutionary operators. A fitness function $F : P \rightarrow \mathbb{R}$ assigns a value, referred to as fitness, to each individual in the population. A selection operator selects the "fittest" individuals from the current population, which are then used as input to the transformation operators. Individuals with a better fitness are more likely to yield superior ones after transformation [S.F93].

The fitness of each individual is calculated after each evolutionary step, therefore the choice of a good fitness function is vital for the performance of the algorithm. In our specific case, where each individual is a *Program*, a way to determine the fitness of an individual is to compare its output X' with the original grid map X, and take into account its length in terms of number of Lines.

We define a measure of difference between two bitmaps called picture distance function $D : Bitmaps \times Bitmaps \rightarrow \mathbb{R}$, where $Bitmaps$ is the population of grid maps, including X. Because P may potentially be a very large population of programs and the algorithm may have to iterate through many generations, it is desirable that D is not computationally expensive. A fast image distance function, linear in the number of pixels in X', is introduced in [BJP00,Bir96]:

$$D(X, X') = \sum_{c} d(X, X', c) + d(X', X, c) \tag{1}$$

$$d(X, X', c) = \frac{\sum_{X[p_1]=c} min\{md(p_1, p_2) | X'[p_2] = c\}}{\#_c(X)} \tag{2}$$

where

- c is a color ($c \in \mathbb{C}$, $\mathbb{C}$ is the set of all colors)
- $X[p]$ is the color at position $p(i, j)$ in image X
- $md(p_1, p_2) = |i_1 - i_2| + |j_1 - j_2|$ is the Manhattan distance between p_1 and p_2.
- $\#_c(X)$ is the number of pixels in X having a color c

A very efficient way to compute $d(X, X', c)$ is described in [BJP00]. Finally, the fitness function F_{fit}: $Programs$ x $Bitmaps \rightarrow \mathbb{R}$, based on the picture distance is defined as $F_{fit}(Pr, X) = D(X, X', c) + 5 * len(Pr)$ where X' is the output of the drawing program Pr. The lower the fitness of Pr, the better the approximation.

4 Evolutionary Operators and Selection Scheme

The algorithm that is used here diverges from the common types of evolutionary algorithms such as genetic programming [RK94,RK92], genetic algorithms [Gol89], evolutionary programming [FOW66] by using a problem-specific set of genetic operators, which are presented below. Each operator creates a new individual, leaving the parents unchanged.

- Random Line Addition: Program → Program. Adds a Line L with randomly generated end point coordinates.
- Random Line Deletion: Program → Program. Randomly picks a Line from the set of Lines in the Program, and removes it.
- Concatenate Programs: Program × Program → Program. Concatenates the two Programs.
- Hill-climbing: Program → Program. Randomly translates the end points of a randomly picked Line from the input Program in a specified range. This procedure is performed k times and the best resultant individual is returned.

As selection operator we are using Roulette Selection - a randomized variant of fitness-proportionate selection. The chance of each individual to be selected is determined by its fitness. This is where the concept of "survival of the fittest" comes into play. However, because in our case a better individual has a smaller fitness (picture distance from the raster image), each individual X_i is assigned a value equal to X_{worst} - X_i. In this way the worst *Program* will have 0% probability of being selected and the best - the largest value compared to all other *Programs* in *Population*.

5 The Algorithm in Pseudo-Code

```
Algorithm Vectorize(Image, Threshold, additionalargs)
Create Initial Population();
while Best Individual Fitness ≥ Threshold
   do Selection ←Roulette Selection();
      Evolve(Selection);
      Add Selection To Population;
      Remove Extra Individuals();
Save Best Individual;
```

```
Algorithm Evolve(Selection)
Random = Random Number(0,1);
Operator ←Determine Evolutionary Operator(Random);
for i ←1 to Selection.Size - 1
   do case: Operator = Random Line Addition
            Add Random Line(Selection[i]);
      case: Operator = Random Line Removal
            Remove Random Line(Selection[i]);
      case: Operator = Hill-climbing
            Hill-Climbing(Selection[i]);
      case: Operator = Concatenate Two Programs
            Concatenate(Selection[i], Selection[i+1]);
            ++i;
```

As a first step an initial *Population* of 50 *Programs* is created. Each *Program* contains a random number of *Lines* (between MIN and MAX specified either as input arguments to the algorithm or as global constants). Each *Line* is also randomly created in the context of a target grid map (i.e the dimensions of the grid map are known so the Line coordinates are on the grid map). At creation time the fitness of each *Program* is evaluated, taking into account both the picture distance and the number of *Lines* present in the *Program*.

The initial *Population* is evolved until a *Program* that is a good enough approximation of the target grid map is produced. At each iteration, a number of individuals called *Selection* is chosen via a selection routine, in our case Roulette Selection. An evolutionary operator that either evolves or mates individuals is applied on *Selection*, the parents remain unchanged. The frequency of each operator is determined by its probability of occurring, specified as an input to the algorithm. As mentioned before the evolutionary operators are - Random Line Addition, Random Line Removal, Concatenation of two Programs and Hill-Climbing on the end points of a Line. The best performance is reached when probabilities of 30%, 25%, 15% and 30% respective to the above-mentioned operators are used.

At the end of each iteration, the same number of *Programs* that were added beforehand is removed via inverse roulette selection, i.e., by favoring worse individuals, in order to keep the population size constant. This approach retains a certain diversity in the population and at the same time quickly converges to the desired target grid map.

6 Experiments and Results

Figure 3 shows the result with an experiment within a typical office environment. The grid map contains about 10 KBytes of data that is reduced to a few hundred bytes to represent 19 vectors. Also, noisy and spurious data in the grid map has more or less disappeared in the vector representation. More important than the compression is that the vectors can serve in contrast to the raster data as basis for efficient computational geometry operations that are very beneficial for many robotic algorithms as discussed in the introduction.

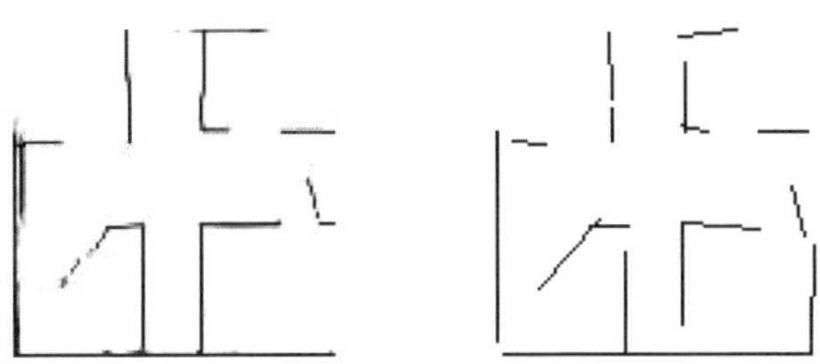

Fig. 3. On the left, a grid map from a typical office environment. Its vectorized representation is shown on the right. The EA has removed some noise and spurious data and generated out of the about 10KBytes of raster data a compact representation with just 19 vectors.

The trend of the average fitness of the whole population is shown in figure figure 4. On a standard PC with a Pentium-4 2.2GHz processor, it takes in the order of 250 msec for computing a whole generation. Analyzing the trends it can be observed that the RPV algorithm behaves like a typical evolutionary algorithm. During the first iterations, the population very quickly improves its fitness, while in the late stages it converges slowly to the target grid map. For many purposes like scan matching in SLAM, the fast rough matches after a few iterations are fully sufficient.

In the following example, a very dense number of lines is used. The purpose of this test case is to test whether the RPV algorithm is able to deal with complicated problems. It managed to evolve a good approximation of the input, although it requires a larger number of iterations. The input and output images are shown in figure 5. The EA performs quite similar to the previous case. The

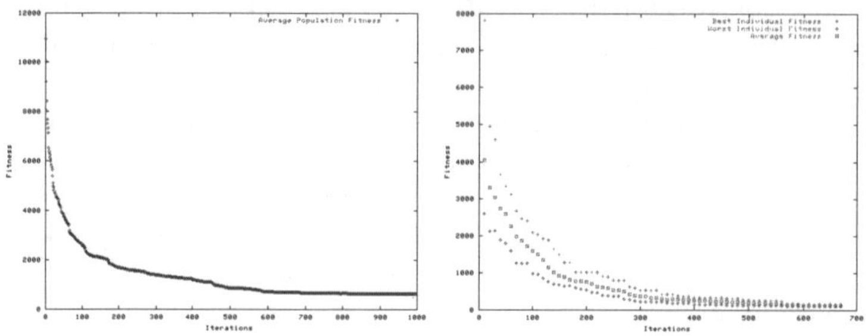

Fig. 4. On the left, the average fitness of the population when generating the vectorization of an office environment. On the right, the best, worst and average Fitness of the population in an experiment with a very crowded set of lines.

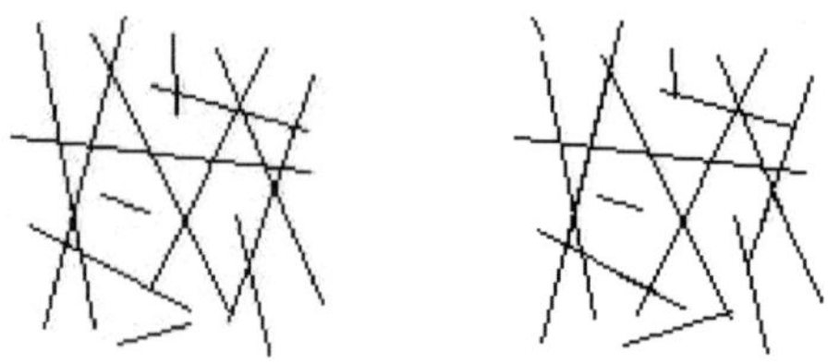

Fig. 5. An example with very "crowded" lines. The original raster data is shown on the left, a result of vectorization on the right.

population does not contain large discrepancies as can be seen from figure 4 - worst to best fitnesses ratio is roughly 2:1, which is a favorable condition, since there are hardly local minima, that could hinder the evolution.

7 Conclusion

Grid maps are widely used to generate a representation of a robot's environment as they are very well studied and as they can be easily generated. But grid maps as a form of raster data have disadvantages, especially when it comes to utilizing the data. Vector data in contrast is very compact and it can be very efficiently processed by computational geometry algorithms, for example for feature extraction or pattern recognition for SLAM or map merging. Here a novel method for vectorization was presented. It is based on an evolutionary algorithm that generates vector code, which represents the input raster data.

References

BC06. Birk, A., Carpin, S.: Rescue robotics - a crucial milestone on the road to autonomous systems. Advanced Robotics Journal 20(5) (2006)

BCK03. Birk, A., Carpin, S., Kenn, H.: The iub 2003 rescue robot team. In: Robocup 2003. Team Description Paper (TDP) (2003)

Bir96. Birk, A.: Learning geometric concepts with an evolutionary algorithm. In: Proc. of The Fifth Annual Conference on Evolutionary Programming, The MIT Press, Cambridge (1996)

Bir05. Birk, A.: The IUB 2004 rescue robot team. In: Nardi, D., Riedmiller, M., Sammut, C., Santos-Victor, J. (eds.) RoboCup 2004. LNCS (LNAI), vol. 3276, Springer, Heidelberg (2005)

BJP00. Birk, A., Paul, W.J.: Schemas and genetic programming. In: Ritter, Cruse, Dean (eds.) Prerational Intelligence, vol. 2, Kluwer, Dordrecht (2000)

BKR+02. Birk, A., Kenn, H., Rooker, M., Akhil, A., Vlad, B., Nina, B., Christoph, B.-S., Vinod, D., Dumitru, E., Ioan, H., Aakash, J., Premvir, J., Benjamin, L., Ge, L.: The IUB 2002 rescue robot team. In: Kaminka, G., Lima, P.U., Rojas, R. (eds.) RoboCup 2002. LNCS (LNAI), Springer, Heidelberg (2002)

CJ94. Jennings, C., Parker, J.R.: Vision knowledge vectorization: Converting raster images into vector form, pp. 311–315 (1994)

DL99. Dori, D., Liu, W.: Sparse pixel vectorization: An algorithm and its performance evaluation. IEEE Transactions on Pattern Analysis and Machine Intelligence 21, 202–215 (1999)

FOW66. Fogel, L.J., Owens, A.J., Walsh, M.J.: Artificial Intelligence through Simulated Evolution. Wiley, New York (1966)

Gol89. Goldberg, D.: Genetic Algorithms in Search Optimization and Machine Learning. Kluwer Academic Publishers, Dordrecht (1989)

JMW+03. Jacoff, A., Messina, E., Weiss, B., Tadokoro, S., Nakagawa, Y.: Test arenas and performance metrics for urban search and rescue robots. In: Proceedings of the Intelligent and Robotic Systems (IROS) Conference (2003)

JWM03. Jacoff, A., Weiss, B., Messina, E.: Evolution of a performance metric for urban search and rescue. In: Performance Metrics for Intelligent Systems (2003)

Mor88. Moravec, H.: Sensor fusion in certainty grid for mobile robots. AI Magazine 9(2), 61–74 (1988)

RK92. Koza, J.R.: Genetic programming. MIT Press, Cambridge (1992)

RK94. Koza, J.R.: Genetic programming II. MIT Press, Cambridge (1994)

S.F93. Forrest, S.: Genetic algorithms - principles of natural selection applied to computation. Science 261, 872–878 (1993)

SSTC02. Song, J., Su, F., Tai, C.-L., Cai, S.: An object-oriented progressive-simplification based vectorisation system for engineering drawings: Model, algorithm and performance. IEEE transactions PAMI 24(8), 1048–1060 (2002)

Thr02. Thrun, S.: Robotic mapping: A survey. In: Lakemeyer, G., Nebel, B. (eds.) Exploring Artificial Intelligence in the New Millenium, Morgan Kaufmann, San Francisco (2002)

Thr03. Thrun, S.: Learning occupancy grids with forward sensor models. Autonomous Robots 15, 111–127 (2003)

TT00. Tombre, K., Tabbone, S.: Vectorization in graphics recognition: To thin or not to thin. In: ICPR, pp. 2091–2096 (2000)

Ego-Motion Estimation and Collision Detection for Omnidirectional Robots

Martin Lauer

University of Osnabrück, Institute of Cognitive Science and Institute of Computer Science, 49069 Osnabrück, Germany
martin.lauer@uos.de

Abstract. We propose an algorithm to estimate the ego-motion of an omnidirectional robot based on a sequence of position estimates. Thereto, we derive a motion model for omnidirectional robots and an estimation procedure to fit the model to observed positions. Additionally, we show how we can benefit from the velocity estimates deriving an algorithm that recognizes situations in which a robot is blocked by an obstacle.

1 Introduction

Autonomous robots which interact with other objects in a dynamically changing world must be able to acquire information about the dynamics in their environment. The faster objects are moving, the more important building dynamic motion models becomes. In RoboCup, the velocities of robots have increased a lot in recent years and therefore the need of dynamic modeling has become crucial to be able to interact with the rolling ball and with teammates and opponent robots. E.g., taking into consideration the internal delays in sensors and motor controllers of about $100 - 200ms$ [1] and robots that drive with $2 - 3\frac{m}{s}$ the robot can cover a distance of at most $60cm$ meanwhile. Furthermore, assuming maximal deceleration of $4\frac{m}{s^2}$ the stopping distance is longer than $1.1m$.

While motion models become more and more important the sensors to determine the robot velocity are simple so far: most teams use wheel encoders without considering problems like wheel slippage and single wheels loosing contact with the ground for short periods of time. Alternatively, desired robot velocities are used instead of the actual velocities ignoring imperfect motor controllers. Both approaches are unreliable and become worse the faster a robot drives. We give an example in Fig. 1.

To avoid these shortcomings we derive a new algorithm to estimate the robot velocity independently of wheel encoders and motor commands. It is based on the idea of fitting a motion model to a sequence of observed robot positions. In contrast to [2,3], our analysis specializes in omnidirectional robots which require the estimation of both the linear velocity as well as the angular velocity.

To illustrate the benefits of velocity estimates we propose in section 3 an algorithm that detects situations of a robot being blocked by an obstacle just by comparing the desired velocity and the estimated velocity without any additional hardware. The algorithms were tested on robots in the Middle Size League.

G. Lakemeyer et al. (Eds.): RoboCup 2006, LNAI 4434, pp. 466–473, 2007.

2 Estimating the Ego-Motion

2.1 Principle Idea

As mentioned before, estimating the robot movement only from the wheel encoders is not reliable. As well, optical flow analysis [4] fails since optical flow is misleading in an environment in which other objects move. Kalman filtering techniques [5] suffer from the assumption of a linear motion model. Even the first order approximation which is made in the extended Kalman filtering approach may be misleading if the robot's angular velocity is large compared to its linear velocity so that the robot is driving on a circle rather than a line.

In recent years, some reliable and accurate self-localization algorithms have been proposed which allow to estimate the robot's pose with an expected error of only a few centimeters [6,7]. These approaches already combine different sources of information like wheel encoders, camera images and laser range scans to eliminate sensor noise. Thus, using these aggregated information makes the estimation of velocity parameters more robust compared to approaches based on the noisy sensory output.

The principle idea of our approach is to use a parameterized motion model of an omnidirectional robot and fitting it to the pose estimates from self-localization. In the remaining part of this section we will first derive the motion model and afterwards show how the model can be fitted to the position estimates.

2.2 Omnidirectional Motion Model

The motion of an omnidirectional robot is determined by its ability to drive in all directions and to turn simultaneously. Here, we want to focus on the trajectories that can be driven by such a robot.

Assuming a robot with differential drive first and assuming constant angular velocity ω and linear velocity v the robot drives on an arc. According to [8] the expected position (x, y) and robot heading ϕ at time t is:

$$\phi(t) = \omega t \quad mod\ 2\pi \tag{1}$$

$$x(t) = \frac{v}{\omega}(\cos(\omega t) - 1) \tag{2}$$

$$y(t) = \frac{v}{\omega}\sin(\omega t) \tag{3}$$

Here, x and y are referring to the robocentric coordinate system at the beginning of the movement.

Things become more complicated in the case of an omnidirectional robot since the linear velocity is not restricted to forward or backward movements but it can be any vector $\boldsymbol{v}$ in the robocentric coordinate system. Hence, we have to generalize (2) and (3) applying first a rotation that turns the coordinate system into the situation discussed in the case of a differential drive, applying (2) and (3) and finally turning the coordinate system back into its original orientation. Additionally, to get coordinates referring to a fixed global coordinate system we have to consider the current position $\boldsymbol{p}$ of the robot and its heading ϕ at the beginning of the movement.

Denoting with $\boldsymbol{u}$ the robot velocity in global coordinates we get after some transformations the completed motion model for an omnidirectional robot:

$$\phi(t) = \phi_0 + \omega \cdot t \quad (mod\ 2\pi) \tag{4}$$

$$\boldsymbol{p}(t) = \begin{cases} \boldsymbol{p}_0 + \boldsymbol{u}(\boldsymbol{t}) \cdot t & \text{if } \omega = 0 \\ \boldsymbol{p}_0 + \frac{1}{\omega} \begin{pmatrix} \sin(\omega t) & \cos(\omega t) - 1 \\ 1 - \cos(\omega t) & \sin(\omega t) \end{pmatrix} \boldsymbol{u}(t) & \text{if } \omega \neq 0 \end{cases} \tag{5}$$

$$\omega(t) = \omega = \omega_0 \tag{6}$$

$$\boldsymbol{u}(t) = \begin{pmatrix} \cos(\omega t) & -\sin(\omega t) \\ \sin(\omega t) & \cos(\omega t) \end{pmatrix} \boldsymbol{u}_0 \tag{7}$$

The parameters ϕ_0, $\boldsymbol{p}_0$, ω_0 and $\boldsymbol{u}_0$ denote the initial heading, position, angular and linear velocity (in global coordinates) at point in time 0, respectively.

Due to our assumption of constant velocities, the angular velocity ω is independent of time (6) but the linear velocity $\boldsymbol{u}_t$ in global coordinates is turning in the same manner as the robot turns (7).

The case-distinction in (5) depends on the angular velocity ω: if it is zero, the robot is driving on a straight line and performs a purely linear movement (first case) while in the case of $\omega \neq 0$ it drives on an arc. Notice that the first case is the limit of the second case for $\omega \to 0$.

2.3 Estimating the Robot Velocity

Once having derived the motion model in (4)-(7) we can fit it to the observed positions of the robot that are calculated by the self-localization approach. Let us denote the pose estimates with tuples $(\boldsymbol{p}_i, \phi_i, t_i)$ where $\boldsymbol{p}_i$ refers to the estimated position of the i-th observation ($i \geq 1$), ϕ_i to the robot heading and t_i to the point in time when this observation has been made.

A direct approach to find the parameters minimizing the discrepancy suffers from three problems:

(a) which error measure do we use to describe the discrepancy between an observed pose and an expected pose of the robot?
(b) how can we deal with the cyclic structure of angles, i.e. the problem that an angle of α is equal to an angle of $\alpha + 2\pi$?
(c) how can we overcome the problem that the motion model contains a case distinction?

The first problem contains the problem of finding a balance between the error that is made in the position of the robot and its heading. Since both parameters are elements of completely different spaces there is no natural common measure of discrepancy. Basically we are faced with the question whether the estimate should be more accurate with respect to the heading or to the position.

The second problem can be solved mapping the angles ϕ_i onto a linear scale of real numbers. Here, we assume that in the time interval $t_{i+1} - t_i$ between two consecutive observations of the robot's pose the robot does not turn by an angle of π or more. Hence, we can "unroll" the measured robot headings on a real axis adding multiples of 2π to the angle in such a way that consecutive values do not differ more than π. To avoid using a new symbol for the unrolled headings we will interpret the symbol ϕ_i as those values instead of the original angles.

To overcome the third problem we propose an algorithm that makes use of the hierarchical structure of the motion model. Our idea is to decompose the whole task into two subtasks:

$$\underset{\phi_0,\omega_0}{minimize}\ \frac{1}{2}\sum_{i=1}^{n}\left(\phi_i - \phi(t_i)\right)^2 \tag{8}$$

$$\underset{\boldsymbol{p}_0,\boldsymbol{u}_0}{minimize}\ \frac{1}{2}\sum_{i=1}^{n}||\boldsymbol{p}_i - \boldsymbol{p}(t_i)||^2 \tag{9}$$

where we use the result of (8) to solve (9).

By estimating the angular velocity first without considering the position of the robot the estimate becomes more sensitive to errors in the heading than errors in the position.

Using the general idea outlined in the previous paragraph we have to derive a solution of (8) firstly. Assuming that the observed values of ϕ_i have been rolled out as described before the task of estimating the angular robot velocity becomes a linear regression task since we have to fit the linear model

$$\phi(t) = \phi_0 + \omega \cdot t \tag{10}$$

to data points of the form (ϕ_i, t_i). The solution can be derived analytically:

$$\hat{\omega} = \frac{n\sum_{i=1}^{n}(\phi_i t_i) - \sum_{i=1}^{n} t_i \sum_{i=1}^{n}\phi_i}{n\sum_{i=1}^{n} t_i^2 - (\sum_{i=1}^{n} t_i)^2} \tag{11}$$

Now, we can tackle (9) using the already estimated $\hat{\omega}$ instead of ω. Depending on the value of $\hat{\omega}$ we have to consider the appropriate case in the motion model (5). If $\hat{\omega} = 0$ we are faced with a linear movement on a straight line. Hence (9) turns out to be a linear regression task. Similarly to (11), the solution can be derived analytically as:

$$\hat{\boldsymbol{u}} = \frac{n\sum_{i=1}^{n}(\boldsymbol{p}_i t_i) - \sum_{i=1}^{n} t_i \sum_{i=1}^{n}\boldsymbol{p}_i}{n\sum_{i=1}^{n} t_i^2 - (\sum_{i=1}^{n} t_i)^2} \tag{12}$$

In the second case of a movement on an arc we get the solution calculating the partial derivatives of $\frac{1}{2}\sum_{i=1}^{n}||\boldsymbol{p}_i - \boldsymbol{p}(t_i)||^2$ and looking for the zeros. After some mathematical transformations we get the system of equations:

$$\begin{pmatrix} n & 0 & \sum s_i & \sum c_i \\ 0 & n & -\sum c_i & \sum s_i \\ \sum s_i & -\sum c_i & \sum(s_i^2 + c_i^2) & 0 \\ \sum c_i & \sum s_i & 0 & \sum(s_i^2 + c_i^2) \end{pmatrix} \begin{pmatrix} p_{0,x} \\ p_{0,y} \\ u_{0,x} \\ u_{0,y} \end{pmatrix} = \begin{pmatrix} \sum p_{i,x} \\ \sum p_{i,y} \\ \sum(s_i p_{i,y} - c_i p_{i,y}) \\ \sum(c_i p_{i,x} + s_i p_{i,y}) \end{pmatrix} \tag{13}$$

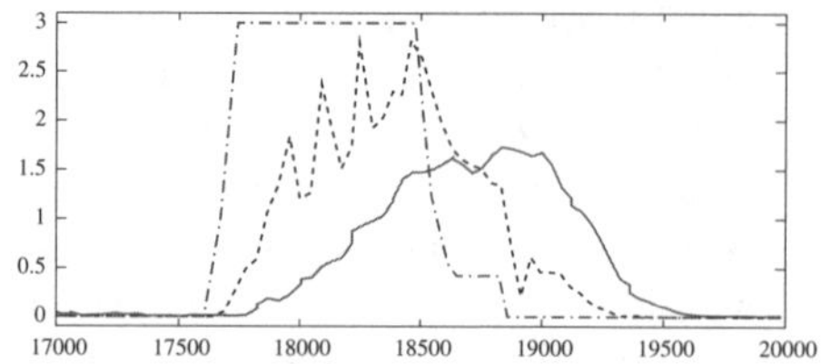

Fig. 1. Acceleration and deceleration of a robot. The dash-dot line shows the desired robot velocity over time, the dotted line shows the odometer values and the solid line shows the estimated velocity using the 10 latest observations.

with $s_i = \frac{\sin(\hat{\omega}t_i)}{\hat{\omega}}$ and $c_i = \frac{\cos(\hat{\omega}t_i)-1}{\hat{\omega}}$. $p_{i,x}$ and $p_{i,y}$ are denoting the first and second coordinate of $\boldsymbol{p}_i$. Resolving (13) with respect to $\boldsymbol{u}_0$ yields:

$$\hat{\boldsymbol{u}}_0 = \frac{1}{d}\begin{pmatrix} \sum s_i \sum p_{i,x} - \sum c_i \sum p_{i,y} - n\sum(s_i p_{i,y} - c_i p_{i,y}) \\ \sum s_i \sum p_{i,y} + \sum c_i \sum p_{i,x} - n\sum(c_i p_{i,x} + s_i p_{i,y}) \end{pmatrix} \tag{14}$$

with $d = (\sum s_i)^2 + (\sum c_i)^2 - n\sum(s_i^2 + c_i^2)$.

The number n of pose estimates that are used for velocity estimation must be at least 2 but should be taken larger than 2 to reduce the noise in the estimates which heavily depends on the number of observations. Beneath the possibility to use a fixed number n like, e.g. 10, it is also possible to use an adaptive strategy that reduces n when the robot accelerates or decelerates and increases n when it drives with constant velocity. This idea is very similar to the approach that was used in [9] to estimate the ball velocity.

Figure 1 shows the velocity estimates for a run on a real robot of the *Brainstormers Tribots* RoboCup team. Obviously, when accelerating the robot, there is some delay until the estimated velocities follow the actual ones but, on the other hand, they do not show the artefacts due to slippage that can be observed considering the odometer values. Furthermore, the slope of the curve is much more realistic than the slope of the odometry or motor command curve. It corresponds to an acceleration of approximately $2\frac{m}{s^2}$.

3 Collision Detection

While we described in section 2 how to estimate the robot velocity we will show in this section an example of how we can benefit from it. In the RoboCup Middle Size League it often happens that robots collide or even push each other. These situations are very undesirable, not only since they are judged as a foul but also since they potentially lead to damaged motors.

There are several ways to recognize pushing situations:

- using elaborated image processing algorithms is a desirable way but needs too much computation time to be applied under hard real time constraints.

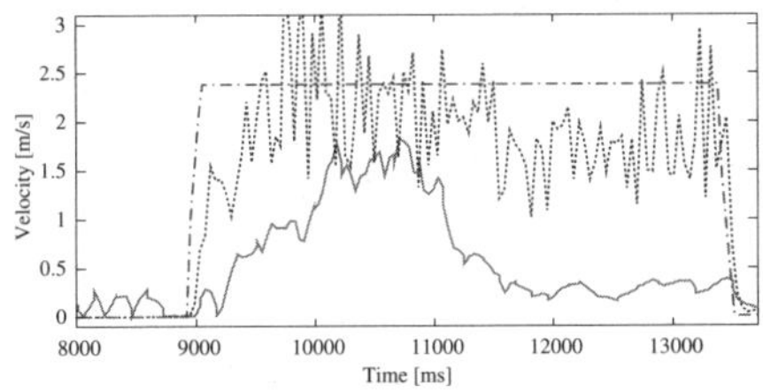

Fig. 2. Acceleration of a robot and collision with an obstacle. After accelerating in the time interval from 9000 until 10000 and driving with constant velocity the robot collided with an obstacle at point in time 11000. The obstacle bared the robot from driving forwardly. The dash-dot line shows the desired robot velocities over time, the dotted line shows the odometer values and the solid line shows the estimated velocity.

- using haptic sensors is a possible way but needs additional hardware that has to be maintained.
- here, we want to propose a simple and at the same time very effective approach that is based on a comparison between the actual robot velocity and the desired velocity. This approach does not need any additional hardware and is also very efficient. It can be seen as a virtual sensor.

Fig. 1 shows what happens if the robot drives without being hindered: the odometer values as well as the estimated velocities follow in principle the driving commands with a certain delay and smoothed over time. Due to slippage the maximal velocity that is actually reached is below the desired velocity.

In contrast, Fig. 2 shows a situation in which the robot was accelerating until it collided with an obstacle at point in time 11000 and after that was blocked by the obstacle. While the motor commands as well as the odometer values indicate a large robot velocity of more than $2\frac{m}{s}$ the estimated velocity decreases to less than $0.5\frac{m}{s}$. Notice that the motor controllers try to turn the wheels with the desired velocity against the resistance of the blocking obstacle. Hence, they need a lot of energy which gets completely lost in slippage.

We can make use of the observed discrepancy between estimated velocity and desired velocity to detect situations of the robot being blocked. Thereto, we calculate the difference between $\hat{\boldsymbol{u}}$ and the desired velocity $\boldsymbol{u}_c$: $e := ||\hat{\boldsymbol{u}} - \boldsymbol{u}_c||$. To be more precise and to take into account the internal delays of the camera, the velocity estimator, and the motor controller [1] we do not compare the current velocity estimate with the most recent motor command but with the driving command that was sent to the motor controller $200ms$ before.

If the discrepancy e is large, say larger than a threshold θ_1, we are faced with a suspicious situation in which the robot potentially is blocked. Unfortunately, large values of e do also occur when the robot is accelerated or decelerated since the motors realize a desired change of velocity only incrementally.

To overcome this problem we propose a combination of two techniques: (a) filtering out times of acceleration and (b) considering intervals of time instead of single points in time. The general idea behind the second technique is that

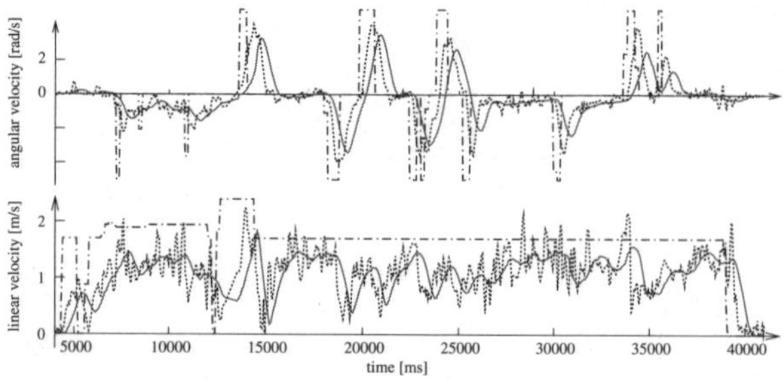

Fig. 3. Motor commands and velocity estimates for a run of the robot of 40 seconds. The dash-dot line shows the desired velocity (motor command), the solid (dotted) line show the estimated velocities using 20 (5) points for estimation.

blocking situations typically hold on for a longer period of time ($> 500ms$) so that for a whole interval in time the values of e are suspicious while the influence of random disturbances is typically shorter.

To ignore periods of acceleration we consider the derivative of the desired velocity with respect to time: $\dot{\boldsymbol{u}}_c$. If it is larger than a threshold θ_2 we do not consider this point in time. The determination of the thresholds θ_1 and θ_2 needs to be done manually.

4 Experimental Results

To evaluate the performance of the velocity estimator we applied it to the robots of the *Brainstormers Tribots* Middle Size League RoboCup team. In several runs where we drove a robot by joystick we observed the velocity estimates. Figure 3 shows the estimates of angular and linear velocity for a run of 40 seconds.

The figure compares velocity estimates with differing number of points used for estimation: while the estimates exhibit a large noise level in the case of only five points used they are very smooth in the case of twenty points. On the other hand, the estimates using only five points react quicker to changes in the velocity than the estimates with twenty points. Hence, an optimal choice depends on the purpose for which the velocity estimates are be used. A good compromise between both extremes is a number of ten observations. Furthermore, Fig. 3 reveals a dependency between angular and linear velocity: as soon as the robot turns the linear velocity decreases.

To test whether the approach to detect situations of a blocked robot works we made experiments driving the robot against a heavy box so that it gets blocked by it. In some cases the robot was able to push the box with reduced velocity or, in the case of a lopsided contact, the robot was turned unintentionally.

The tests contained 20 different blocking situations, in five of it the robot was pushing the box and in three the robot was unintentionally turned. Throughout the experiments which lasted for 320 seconds 17 situations were correctly recognized by the blocking sensor and only one false alarm occurred. The three situations that were missed are pushing and turning situations.

5 Discussion

We proposed two algorithms in this paper, an approach for estimating the ego-motion of the robot and a virtual sensor of blocking situations. As demonstrated the advantage of the ego-motion estimator is its independence of wheel encoders. Hence we get independent information about the robot movement and we are able to reveal kinetic effects of the omnidirectional drive which is a groundwork for controlling fast robot movements accurately.

The algorithm for recognizing blocking situations has also been turned out to be helpful. We used it successfully in tournaments where recognition of blocking situations enables behaviors that free the ball in such a case. Morover, recognizing blocking situations is important to save energy and to prevent the robots from damage. Characteristically, a motor of one of our robots was destroyed when we made experiments with blocking situations.

While we presented in this paper only an approach to estimate the ego-motion and to recognize blocking situations, our research objective is to build a complete dynamic model of the environment incorporating a kinetic model of the robot movements, of its teammates and opponents as well as the movement of the ball. Hence, we will be able to drive faster and more accurate and to interact with other objects.

Acknowledgments. This work was supported by the German Research Foundation (DFG) SPP 1125.

References

1. Behnke, S., Egorova, A., Gloye, A., Rojas, R., Simon, M.: Predicting away robot control latency. In: Polani, D., Browning, B., Bonarini, A., Yoshida, K. (eds.) RoboCup 2003. LNCS (LNAI), vol. 3020, pp. 712–719. Springer, Heidelberg (2004)
2. Shiflett, G.R., Laub, A.J.: The analysis of rigid body motion from measured data. Journal of Dynamics Systems Measurement and Control 117(4), 578–584 (1995)
3. Vithani, A.R., Gupta, K.C.: Estimation of object kinematics from point data. Journal of Mechanical Design 126(1), 16–21 (2004)
4. Stiller, C., Konrad, J.: Estimating motion in image sequences. IEEE Signal Processing Magazine 16, 70–91 (1999)
5. Gelb, A. (ed.): Applied Optimal Estimation. Cambridge (1974)
6. Strack, A., Ferrein, A., Lakemeyer, G.: Laser-based localization with sparse landmarks. In: Proc. RoboCup Symposium 2005 (2005)
7. Lauer, M., Lange, S., Riedmiller, M.: Calculating the perfect match: an efficient and accurate approach for robot self-localization. In: Proc. Robocup 2005 (2005)
8. Borenstein, J., Everett, B., Feng, L.: Navigating Mobile Robots: Systems and Techniques. A. K. Peters, Ltd. (1996)
9. Lauer, M., Lange, S., Riedmiller, M.: Motion estimation of moving objects for autonomous mobile robots. Künstliche Intelligenz 20(1), 11–17 (2006)

Integrating Simple Unreliable Perceptions for Accurate Robot Modeling in the Four-Legged League*

Tim Laue[1] and Thomas Röfer[2]

[1] Faculty 3 - Mathematics / Computer Science, Universität Bremen, Postfach 330 440, 28334 Bremen, Germany
timlaue@informatik.uni-bremen.de
[2] Deutsches Forschungsinstitut für Künstliche Intelligenz GmbH, Sichere Kognitive Systeme, Robert-Hooke-Str. 5, 28359 Bremen, Germany
Thomas.Roefer@dfki.de

Abstract. The perception and modeling of other robots has been a topic of minor regard in the Four-Legged League, because of the limited processing und sensing capabilities of the AIBO platform. Even the current world champion, the GermanTeam, abandoned the usage of a robot recognition. Nevertheless, accurate position estimates of other players will be needed in the future to accomplish tasks such as passing or applying adaptive tactics. This paper describes an approach for localizing other players in a robot's local environment by integrating different unreliable perceptions of robots and obstacles, which may be computed in a reasonable way. The approach is based on Gaussian distributions describing the models of the robots as well as the perceptions. The integration of information is realized by using Kalman filtering.

1 Introduction

The Four-Legged Robot League is one of the official leagues in RoboCup, in which a standardized robot platform is used, namely the Sony AIBO, which has quite limited perceptual capabilities. The main sensor of the Sony AIBO is a camera located in its head. The head can be turned around three axes, and the camera has a field of view of approximately 57° by 42°. As the main sensor of the robot is a camera, all objects on the RoboCup field are color coded. For robots, this leads to two different tricot colors, i. e. red and blue, which are applied to the robots as patches (Fig. 1).

During actual RoboCup games, robots are hard to perceive. Especially the blue tricots are often indistinguishable from black or dark grey. The relatively large distances on the field as well as the limited field of view—compared to robots in other leagues that are allowed using omni-directional sensors—make it

* The Deutsche Forschungsgemeinschaft supports this work through the priority program "Cooperating teams of mobile robots in dynamic environments".

G. Lakemeyer et al. (Eds.): RoboCup 2006, LNAI 4434, pp. 474–482, 2007.

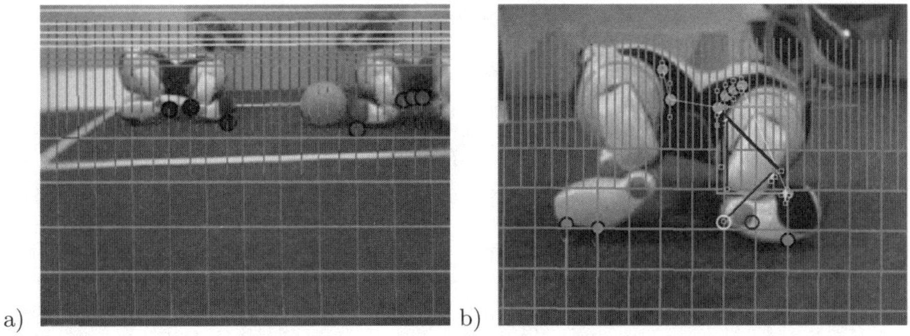

Fig. 1. Detection of robots by using a grid of scan lines

almost impossible for a single robot to compute accurate estimates of all players on a field based on its own perceptions.

Despite of the existence of robot detection algorithms [1,2,3,4], often only simple obstacle information [5,6] is used for navigation. Nevertheless, a localization for robots is needed, if techniques such as passing towards teammates or tactics which are adapting to the opponent's positions should be applied. Because of the limited field of view, and the unreliability of the available robot perceptions, sophisticated techniques for modeling are needed.

The approach presented in this paper aims at computing accurate estimates of player positions in the robot's local environment by using probabilistic modeling techniques. It does not incorporate communication with other robots and depends therefore on visual perceptions. To improve estimates, information different from explicit robot perceptions is additionally integrated, i. e. occupied spaces as well as free spaces on the field.

In the Four-Legged League domain, player position estimation has been a topic of minor regard, so far. Nevertheless, several similar works about modeling position and velocity of the ball using Kalman filters [3] or Rao-blackwellized particle filters [7] have been published. Also the integration of different perceptions for improving estimates of the ball position has been described by [2].

This paper is organized as follows: Section 2 presents the perceptions which are used for computing estimates. The estimation approach is described in Sect. 3. Experimental results are presented in Sect. 4. The paper ends with a conclusion and an outlook on future work in Sect. 5.

2 Perceptions

The work described in this paper is based on the software of the GermanTeam 2005 [4] and therefore uses its vision system. This system processes images of a resolution of 208 × 160 pixels, but actually considers only a horizon-aligned grid of less pixels [8] (see Fig. 1). Each grid line is scanned pixel by pixel. During the scan, each pixel is classified by color. A characteristic series of colors or a pattern of colors is an indication of an object of interest.

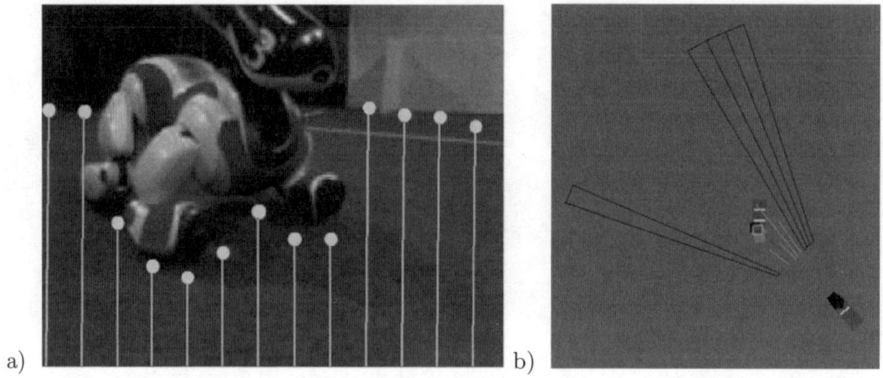

Fig. 2. Detection of obstacles. a) Lines scanning for unoccupied space. The bright dots indicate the end of the free field. b) A similar situation represented in the robot's world model. The short lines indicate the free space among the robot and another robot. The sectors surrounded by black lines are regions which are considered to be unoccupied.

Robot Detection. To find the indications for other robots, the scan lines are searched for the colors of the tricots of the robots. The scan lines are followed until the green of the field appears (cf. Fig. 1a). Thus the foot points of the robot are detected. From these foot points, the distance to the robot can be determined. A refinement for determining the position of the robot is the extraction of the position of a front foot from the image [4] (cf. Fig. 1b). Using this simple approach, a recognition of a robot's rotation is not possible. The only information is the relative position of a robot. The small tricot elements cause a detection of robots at a distance of more than 1.5m to be highly unlikely. The precision of these percepts is shown in Sect. 4, their integration into a robot position estimation is described in Sect. 3.2.

Obstacles and Free Space. A concept different from the recognition of robots is the detection of obstacles [6,5]. Instead of searching robot features in an image, the unoccupied regions, i. e. the green field including the white lines, are determined (cf. Fig. 2a). Thus, areas not classified as free space have to be considered to be obstacles. Though obstacles don't necessarily need to be robots (e. g. beacons, goals, and the feet of a referee would also be classified as obstacles), they can be used for improving the estimation of a robot position (cf. Sect. 3.2). In contrast to this positive information, this perception additionally bears negative information about regions in which no robots are located. The usage of this perception is described in Sect. 3.3.

Collisions. A completely proprioceptive kind of percept is information about the current physical state of the robot, i. e. the correctness of the calculated camera position or the odometry. Both information may be disturbed by collisions with other robots and hence lead to disturbed perceptions. For instance in [9], it has been shown that it is possible to compute reliable information about collisions occurring to a moving AIBO robot. The use of this information is described in Sect. 3.1.

3 Robot Models

Since the number of players that could be observed from a robot's position varies, a set[1] of actual estimations—in the following termed as *hypotheses*—has to be kept and updated. A robot hypothesis H is modeled as a Gaussian distribution. Therefore it is a tuple consisting of a mean μ_h which describes the position of the robot and a covariance Σ_h which models the uncertainty of the position. Since the image processing algorithms used for this work are not capable of recognizing a robot's relative rotation, two-dimensional distributions are used. All hypotheses are kept relative to the observing robot in polar coordinates which consist of a distance d and an angle α.

$$\mu_h = \begin{pmatrix} d \\ \alpha \end{pmatrix}, \quad \Sigma_h = \begin{pmatrix} \mathrm{var}(d) & \mathrm{cov}(\alpha, d) \\ \mathrm{cov}(d, \alpha) & \mathrm{var}(\alpha) \end{pmatrix} \tag{1}$$

New hypotheses may be created from perceptions of robots (cf. Sect. 3.2) whilst existing hypotheses are maintained by a Kalman Filter [10,11] which incorporates the robot's motion (cf. Sect. 3.1) and integrates different perceptions (cf. Sect. 3.2–3.3) to improve the estimation of player's positions. Every hypothesis is considered to be a single robot that is tracked. Nevertheless, it is possible that noisy perceptions lead to different hypotheses describing the same robot. These effects are addressed by the mechanisms described in Sect. 3.4. The general approach—the structure of which is similar to [12]—is depicted in Fig. 3.

3.1 Motion Update

On every execution of the modeling module, all existing hypotheses have to be updated according to the motion $(\Delta x, \Delta y, \Delta\alpha)$ of the observing robot since the last execution. This information is gained from the robot's odometry. The update also includes noise depending on the quantity of the motion. The mean of the hypothesis is updated by

$$\alpha^+ = \mathrm{atan2}(\sin(\alpha^-)d^- - \Delta y, \cos(\alpha^-)d^- - \Delta x) - \Delta\alpha \tag{2}$$

$$d^+ = \sqrt{(\sin(\alpha^-)d^- - \Delta y)^2 + (\cos(\alpha^-)d^- - \Delta x)^2}. \tag{3}$$

The uncertainty caused by the robot's motion is added to the hypothesis' covariance matrix Σ by

$$\Sigma^+ = J_1 \Sigma^- {J_1}^T + J_2(1 + e_c)\Sigma_\Delta {J_2}^T + \Sigma_N \tag{4}$$

where two Jacobian matrices J_1 and J_2 are defined as

$$J_1 = \frac{\partial \begin{pmatrix} \alpha^+ \\ d^+ \end{pmatrix}}{\partial \begin{pmatrix} \alpha^- \\ d^- \end{pmatrix}}, \quad J_2 = \frac{\partial \begin{pmatrix} \alpha^+ \\ d^+ \end{pmatrix}}{\partial \begin{pmatrix} \Delta x \\ \Delta y \\ \Delta\alpha \end{pmatrix}}. \tag{5}$$

[1] Actually, the implementation keeps red and blue robots in two different sets, but this detail is not addressed in this general description of the approach.

```
RobotModeling (Hypotheses, RobotPerceptions, Obstacles, Odometry)
    for each Hypothesis H:
        MotionUpdate (H, Odometry)
    for each RobotPerception P:
        if IntegrationPossible(Hypotheses, P)
            Integrate(P, BestMatch(Hypotheses, P))
        else
            Hypotheses add P
    for each Hypothesis H:
        UpdateByPositiveObstacles (H, Obstacles)
        UpdateByNegativeObstacles (H, Obstacles)
        if LowQuality(H)
            Hypotheses remove H
        else if (H*| MergingPossible(H*, H)) exists
            Merge(H, H*)
    end
```

Fig. 3. The general operation of the robot modeling

The matrix Σ_Δ contains information about the uncertainty of the robot's motion and is provided by the odometry model. Additionally, collisions may be taken into account by multiplying the matrix with a factor e_c. This factor is zero, if no collisions occur. In case of a collision, a positive value reflects the higher uncertainty of odometry. Through the matrix Σ_N, constant white noise is added reflecting the uncertain motion of the observed robots. This causes the variance to grow constantly in absence of any measurements. Adequate values for Σ_Δ, Σ_N and e_c have to be determined empirically.

3.2 Robot Percepts and Positive Obstacle Information

Before adding new hypotheses to the list, all measurements are tried to be integrated with existing estimations. In a first step, a percept is converted to a hypothesis H_m. Its mean μ_m is the position of the measurement. A corresponding covariance matrix Σ_m has to be precomputed from a set of measurements (as those made for Fig. 4a). This can be refined by providing matrices for different distances and angles and using interpolations of these for new measurements.

The new hypothesis has now to be associated to an already existing robot hypothesis H_r. The Mahalanobis distance

$$d_M(H_r, H_m) = (\mu_r - \mu_m)^T(\Sigma_r + \Sigma_m)^{-1}(\mu_r - \mu_m) \tag{6}$$

provides a distance measure that describes the compatibility of two hypotheses, indicating whether both could refer to the same robot. After the closest hypothesis H_r has been found and $d_M(H_r, H_m)$ is below a maximum acceptable

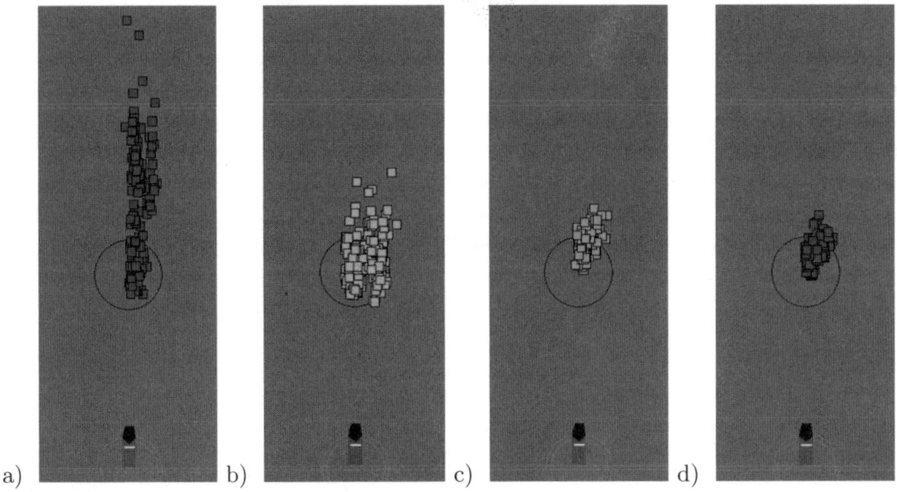

Fig. 4. A robot standing at a distance of 80cm is perceived and its position is estimated. Every dot indicates an estimate, the black circle surrounds the ground truth position. a) The plain perception from the vision system. b) Plain obstacle positions as used by 3.2. c) The modeled position using robot perceptions only. d) The modeled position using robot and obstacle perceptions.

distance, H_m is integrated:

$$\mu_h^+ = \mu_h^- + \Sigma_h^-(\Sigma_h^- + \Sigma_m)^{-1}(\mu_m - \mu_h^-) \tag{7}$$

$$\Sigma_h^+ = \Sigma_h^- - \Sigma_h^-(\Sigma_h^- + \Sigma_m)^{-1}\Sigma_h^- \tag{8}$$

Otherwise, the measurement will be added to the list as a new hypothesis.

In general, perceived obstacles are treated similar to robot percepts, solely the usage of a lower threshold κ_o for hypothesis association is needed and the possibility of adding new hypotheses to the list does not exist. The mean μ_o is computed from a set of adjacent obstacle segments (cf. Fig. 2a). Of course, the values for the covariance matrix Σ_o have also to be determined empirically, since they differ strongly from the robot percept values (cf. Fig. 4b).

3.3 Negative Obstacle Information

In opposite to the previous two perceptions, which denote the presence of robots, the negative obstacle information, i. e. empty regions of the field, denotes absence of any robots. This information is quite useful for the elimination of false positives as well as for a quick update of the world model in case of a robot kidnapping (which have e. g. been picked up by a referee). The incorporation of this information is done via checking the inclusion of every hypothesis' mean μ_h inside every sector recognized as being empty (cf. Fig. 2). In case of such an inclusion, white noise is added to the corresponding covariance matrix.

3.4 Maintenance of Hypotheses

While maintaining a list of hypotheses, it has not only to be taken care of removing elements, e. g. those with an uncertainty above a given threshold. The possibility of having two hypotheses describing the same robot must also be considered. This effect is detected by using a heuristic derived from the limits of the used image processing approaches: Two hypotheses H_1 and H_2 whose means μ_1 and μ_2 are located very close to each other can not be distinguished anymore by robot percepts in a reasonable way. These two hypotheses become merged, i. e. they are viewed as a sum-of-two-Gaussians distribution and replaced by a single Gaussian with the same mean and covariance. This is accomplished by

$$\mu_n = w_1\mu_1 + w_2\mu_2 \tag{9}$$

$$\Sigma_n = w_1\Sigma_1 + w_2\Sigma_2 + w_1 w_2(\mu_1 - \mu_2)(\mu_1 - \mu_2)^T \tag{10}$$

where the relative weight of the two hypotheses is controlled by

$$w_1 = \frac{P_{H_1}(\mu_{H_1})}{P_{H_1}(\mu_{H_1}) + P_{H_2}(\mu_{H_2})}, \quad w_2 = \frac{P_{H_2}(\mu_{H_1})}{P_{H_1}(\mu_{H_1}) + P_{H_2}(\mu_{H_2})} . \tag{11}$$

4 Experimental Results

The approach described in this paper has been implemented using the framework of the GermanTeam. Several experiments using an AIBO on an original Four-Legged League field have been conducted. To demonstrate the improvement of player position estimates by using the proposed modeling techniques, the quality of hypotheses while sensing different robots at different distances has been measured. One example is depicted in Fig. 4.

To demonstrate the capability to model several robots simultaneously as well as assigning measurements to different robots of the same color, different settings including a number of robots have been investigated (cf. Fig. 5). These

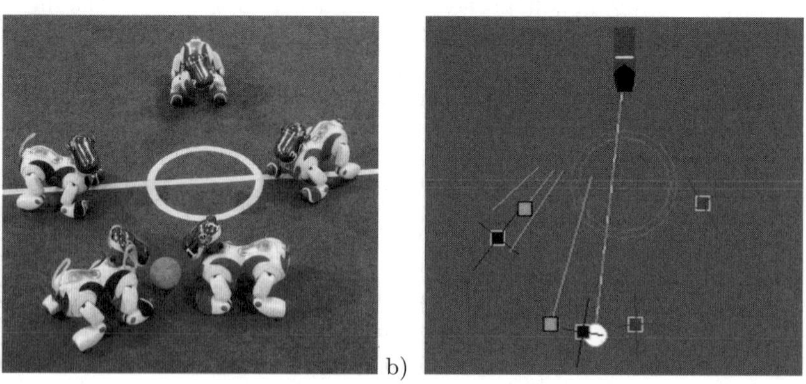

a) b)

Fig. 5. An example with several robots. The large dots indicate the positions of the hypotheses. The lines through the dots illustrate the uncertainty of the estimations.

experiments included only standing robots due to a lack of adequate ground truth data for moving robots.

The implementation of this approach has already been applied to a dynamic scenario by the *Bremen Byters* team which built some tactical behaviors upon the computed robot estimations and used these in a RoboCup competition.

5 Conclusion and Future Works

In this paper, the authors have shown that it is possible to compute accurate position estimations of robots in the Four-Legged League. The low quality of information that is caused by the low perceptual capabilities of the AIBO robot may be compensated by applying sophisticated estimation techniques. The next step will be to create a complete world model that includes the positions of all robots on the field. Due to the limitations of a single robot, this has to be done by communicating information among the robots in a team. The local models described in this paper will be used as a foundation for such a global model.

References

1. Wilking, D., Röfer, T.: Real-time object recognition using decision tree learning. In: Nardi, D., Riedmiller, M., Sammut, C., Santos-Victor, J. (eds.) RoboCup 2004. LNCS (LNAI), vol. 3276, pp. 556–563. Springer, Heidelberg (2005)
2. Veloso, M., Rybski, P.E., Chernova, S., McMillen, C., Fasola, J., von Hundelshausen, F., Vail, D., Trevor, A., Hauert, S., Espinoza, R.R.: CMDash'05: Team Report (2005), `http://www.cs.cmu.edu/~{}coral`
3. Quinlan, M.J., Nicklin, S.P., Hong, K., Henderson, N., Young, S.R., Moore, T.G., Fisher, R., Douangboupha, P., Chalup, S.K., Middleton, R.H., King, R.: The 2005 NUbots Team Report (2005), `http://www.robots.newcastle.edu.au`
4. Röfer, T., Laue, T., Weber, M., Burkhard, H.D., Jüngel, M., Göhring, D., Hoffmann, J., Altmeyer, B., Krause, T., Spranger, M., Stryk, O.v., Brunn, R., Dassler, M., Kunz, M., Oberlies, T., Risler, M., et al.: GermanTeam RoboCup 2005 (2005), `http:// www.germanteam.org/GT2005.pdf`
5. Hoffmann, J., Jüngel, M., Lötzsch, M.: A vision based system for goal-directed obstacle avoidance. In: Nardi, D., Riedmiller, M., Sammut, C., Santos-Victor, J. (eds.) RoboCup 2004. LNCS (LNAI), vol. 3276, Springer, Heidelberg (2005)
6. Lenser, S., Veloso, M.: Visual Sonar: Fast Obstacle Avoidance Using Monocular Vision. In: Proceedings of IROS'03 (2003)
7. Kwok, C., Fox, D.: Map-based multiple model tracking of a moving object. In: Nardi, D., Riedmiller, M., Sammut, C., Santos-Victor, J. (eds.) RoboCup 2004. LNCS (LNAI), vol. 3276, Springer, Heidelberg (2005)
8. Bach, J., Jüngel, M.: Using pattern matching on a flexible, horizon-aligned grid for robotic vision. Concurrency, Specification and Programming - CSP'2002 1 (2002)
9. Hoffmann, J., Göhring, D.: Sensor-Actuator-Comparison as a Basis for Collision Detection for a Quadruped Robot. In: Nardi, D., Riedmiller, M., Sammut, C., Santos-Victor, J. (eds.) RoboCup 2004. LNCS (LNAI), vol. 3276, Springer, Heidelberg (2005)

10. Kalman, R.E.: A new approach to linear filtering and prediction problems. Transactions of the ASME–Journal of Basic Engineering 82, 35–45 (1960)
11. Thrun, S., Burgard, W., Fox, D.: Probabilistic Robotics. MIT Press, Cambridge (2005)
12. Reid, D.B.: An Algorithm for Tracking Multiple Targets. IEEE Transactions on Automatic Control 24, 843–854 (1979)

Distributed, Play-Based Coordination for Robot Teams in Dynamic Environments

Colin McMillen and Manuela Veloso

School of Computer Science, Carnegie Mellon University, Pittsburgh, PA, U.S.A.
{mcmillen,veloso}@cs.cmu.edu

Abstract. We present refinements to our previous work on team communication and multi-robot world modeling in the RoboCup legged league. These refinements put high priority on the communication of task-relevant data. We also build upon past results within the simulation and the small-size leagues and contribute a distributed, play-based role assignment algorithm. This algorithm allows the robots to autonomously adapt their strategy based on the current state of the environment, the game, and the behavior of opponents. The improvements discussed in this paper were used by CMDash in the RoboCup 2005 international competition.

1 Introduction

A common goal of distributed autonomous robotic systems is the development of teamwork and coordination strategies. The benefits of adding multiple robots to a system, such as increased performance and reliability, have been demonstrated in many different situations. However, depending on the domain and the task, different sorts of approaches might be needed. We are interested in multi-robot coordination in domains with high uncertainty and dynamic environments.

In this paper, we present two main contributions: refinements to our previous work in multi-robot world modeling and a novel approach to role assignment. Both contributions are discussed in the context of the RoboCup four-legged league [1], in which two teams of four Sony AIBO robots compete in a robot soccer game. This domain presents many challenges, including: full robot autonomy, distributed robot team control, limited individual robot perception, the presence of robot adversaries, task-dependent temporal constraints, and high communication latency. In this paper, we contribute a new distributed *play-based* system that equips the robots with *plays* – alternative teamwork strategies. This method was developed to overcome limitations of previous approaches. In particular, it assigns roles to robots in a fault-tolerant manner that minimizes role switching and synchronization problems. This paper is explicitly targeted at the very challenging issues posed by the RoboCup four-legged league; however, our approach is designed to be relevant to general multi-robot domains that share some of the challenging features of robot soccer.

In section 2, we discuss our enhancements to communication and multi-robot world modeling. Section 3 introduces our play-based teamwork strategy, and

G. Lakemeyer et al. (Eds.): RoboCup 2006, LNAI 4434, pp. 483–490, 2007.

refers to some positive experimental results. Section 4 presents our conclusions. Related work is discussed throughout the paper as needed.

2 Communication Strategies

Many multi-robot teams make use of communication for world state sharing. Due to the AIBOs' limited perception range and the extensive object occlusion in the RoboCup environment, teams can benefit greatly by building a shared world model. A common approach, used by our team in the past [2], is to have each robot periodically broadcast a packet containing all the shared information, such as the robot's current position, its best estimate of the ball position, and the positions of detected opponent robots. However, some domain information (such as the position of the ball) is inherently more important to the success of the team than other types of information. We have therefore developed a factored communication strategy. In this strategy, there are several different types of message, containing different pieces of information. We can then independently choose the transmission rate for each type of message. This communication strategy allows the robots to respond more quickly to important events (such as a change in the ball's position) without the need to transmit a large message over the communication network. In this section, we present a brief overview of the world-modeling information our robots shared in the RoboCup 2005 competition.

2.1 Ball Messages

These messages are sent by all robots to indicate important information about the status of the ball. Each message contains the following information:

- Ball state. This is a new feature that was added to our world model for RoboCup 2005. This can take on one of the following values:
  - Lost: No reliable estimate of the ball's location is available.
  - Visible: The ball is currently seen.
  - Possession: The ball is believed to be in the possession of the robot (i.e., the robot has grabbed the ball and is lining up for a kick.)
  - NotInFOV: The ball is not currently seen, but is not expected to be seen because it is outside the robot's field of view. This happens (e.g.) when a robot takes its view off the ball to look at a localization marker.
  - InFOVButMissing: The ball is not currently seen, even though the robot believes it is looking at the ball's location.
  - InFOVButOccluded: The ball is not currently seen, but the robot believes that an object (such as another robot) is occluding the ball.

  We now send these symbolic ball states instead of numerical confidence values. These symbolic values allow the team to more accurately characterize the true state of the ball.
- Whether the robot transmitting the message believes that it is lost. This is determined by thresholding the robot's localization uncertainty.

- The global position of the ball. Global ball position estimates are not used from any robot that claims to be lost, since a lost robot is very likely to project its local ball estimate to an incorrect global position.
- The position of the ball relative to the robot. If the ball is very close to the reach of a robot, and that robot intends to kick the ball, the robot's teammates should avoid interfering with the kick, even if the robot believes that it is lost. The transmission of relative ball locations allows robots to back off in this situation, without the need to rely on visually seeing the teammate near the ball.

Since the location of the ball is of utmost importance to the proper functioning of the team, the ball messages are sent frequently. A robot will send a new ball message every 1/8 second if it has a good ball hypothesis and is not lost. If the robot becomes lost or does not have a valid ball hypothesis, it waits a while longer to see if the situation improves. This is done because a valid global ball location provides more valuable information to teammates. After 1/4 second has passed, however, the robot sends a ball message regardless of the circumstances.

2.2 Status Messages and Intentions

Another type of message is the status message. Status messages are sent by each robot at periodic intervals (typically 4 Hz). They include the robot's current position and angle (as reported by localization) as well as the current "intention" of the robot. Intention is a very important concept that we have added to our teamwork this year. When a robot is very close to the ball, its teammates should stay out of the way, to ensure that they do not interfere with the attacker's actions. However, there are specific times when nearby robots might not be intending to go for the ball. In these cases, the teammates should not back away just because another robot is near. The intention of the robot is determined by the robot's top-level behavior, and can take on any of the following values:

- ATTACK: the robot intends to approach the ball and manipulate it.
- WAIT: the robot does not intend to approach the ball. This happens when a robot is returning to position or is searching for the ball.
- YIELD: the robot would intend to approach the ball, except that it is yielding to a teammate instead.

2.3 Periodic Messages

Periodic messages are provided as a form of robustness to failure. The information contained in periodic messages allow the robots to determine when network failures have occurred, when a teammate has crashed, or other anomalous events have occurred. The team can then take appropriate actions to ensure that team play degrades gracefully in the presence of failure. The periodic message is typically sent at a rate of 1 Hz.

3 Distributed Play-Based Role Assignment

It is our experience that it is rather challenging to generate or learn a team control policy in complex, highly dynamic (in particular adversarial), multi-robot domains. Therefore, instead of approaching teamwork in terms of a mapping between state and joint actions [3], we follow a *play-based* approach, as introduced by Bowling *et al.* [4]. A play-based approach allows us to handle the domain challenges introduced in section 2. A play specifies a *plan* for the team; i.e., under some applicability conditions, a play provides a sequence of steps for the team to execute. Multiple plays can capture different teamwork strategies, as explicit responses to different types of opponents. Bowling showed that play selection weights could be adapted to match an opponent. Plays also allow the team to reason about the zero-sum, finite-horizon aspects of a game-playing domain: the team can change plays as a function of the score and time left in the game. Our play-based teamwork approach ensures that robots do not suffer from hesitation nor oscillation, and that team performance is not significantly degraded by possible periods of high network latency. We believe that ours is the first distributed play-based teamwork approach within the context of the RoboCup four-legged league.

3.1 Plays

A *play* is a team plan that provides a set of roles, which are assigned to the robots upon initiation of the play. Bowling [4] introduced a play-based method for team coordination in the RoboCup small-size league. However, the small-size league has centralized control of the robots. One of the significant contributions of our work is the development of a play system that works in a distributed team. The play language described by Bowling assumes that the number of robots is fixed, and therefore always provides exactly four different roles for the robots. In another extension to Bowling's work, our plays also specify which roles are to be used if the team loses some number of robots due to penalties or crashes. This extension to the role-assignment aspects of Bowling's play language allows the team to robustly adapt to the loss or penalization of team members without the need for additional communication.

Our play language itself is also strongly inspired by the work of Bowling. Our language allows us to define *applicability conditions*, which denote when a play is suitable for execution; what *roles* should be assigned when we have a specific number of active robots on the team; and a *weight*, which is used to decide which play to run when multiple plays are applicable.

Applicability. An applicability condition denotes when a play is suitable for execution. Each applicability condition is a conjunction of binary predicates. A play may specify multiple applicability conditions; in this case, the play is executable if any of the separate applicability conditions are satisfied.

Roles. Each play specifies which roles should be assigned to a team with a variable number of robots by defining different `ROLES` directives. A directive applies when a team has k active robots, and specifies the corresponding k roles to be assigned. If a robot team has n members, each play has a maximum of n `ROLES` directives. Since our AIBO teams are composed of four robots, our plays have four `ROLES` directives.

Weight. Weight is used to decide which play to run when multiple plays are applicable. In our current algorithm, the play selector always chooses the applicable play with greatest weight. Future work could include choosing plays probabilistically based on the weight values or updating the weights at execution time to automatically improve team performance. *Playbook adaptation* of this sort was introduced by Bowling for the small-size league [4].

Unlike the work of Bowling, we do not have `DONE` or `TIMEOUT` keywords that specify when a play is complete. Rather, the play selector runs continuously, and each play is considered to be complete as soon as a different play is chosen. This may happen because the current play is no longer applicable or because another play with greater weight has recently become applicable. Each predicate used in an applicability condition is designed with some hysteresis, such that it is not possible for the predicate to rapidly oscillate between true and false. The predicates used in our approach depend on features of the environment – such as the time left in game, the number of goals scored by each team, and the number of robots available to each team – that by their nature cannot rapidly oscillate. This ensures that the play choice also cannot rapidly oscillate.

Figure 1 shows an example of a defensive play. Its applicability conditions specify that this play is applicable 1) when our team is winning and has fewer active players than the opponents or 2) when the game is in the second half and our team is winning by at least two points. If we have only one active robot on our team, we will assign it the Goalkeeper role; if we have two robots, one is assigned the Goalkeeper role and the other is assigned the Defender role; and so on. We have developed a total of sixteen plays, but not all were used in the RoboCup 2005 competition. Figure 2 shows a summary of the seven plays that were used in the competition. (Only the roles used for a 4-robot team are shown.)

```
PLAY Guard
APPLICABLE winning fewerPlayers
APPLICABLE secondHalf winningBy2OrMoreGoals
ROLES 1 Goalkeeper
ROLES 2 Goalkeeper Defender
ROLES 3 Goalkeeper Defender Independent
ROLES 4 Goalkeeper Defender Midfielder Independent
WEIGHT 3
```

Fig. 1. An example play with multiple applicability conditions

Default: Goalkeeper Defender Striker Independent
Defensive: Goalkeeper Defender Midfielder Independent
Guard: Goalkeeper Defender MidfieldDefender Independent
Flankers: Goalkeeper Defender LeftFlanker RightFlanker
Aggressive: Goalkeeper LeftFlanker RightFlanker Independent
PullGoalie: Midfielder LeftFlanker RightFlanker Independent
Kickoff: Goalkeeper Defender Charger KickoffDodger

Fig. 2. Summary of the seven plays used by our team in RoboCup 2005

3.2 Play Selector

The *play selector* runs on one robot that is arbitrarily chosen to be the leader. The play selector chooses which play the team should be running. The leader periodically broadcasts the current play (and role assignments) to its teammates. Distributed play-based coordination is achieved through a predefined agreement among the team members to resort to a *default play* if a robot doesn't hear a play broadcast within a *communication time limit.* A failure of the leader or a network problem may trigger this default coordination plan. A more sophisticated approach could incorporate an algorithm for *leader selection* in the event of failure. However, we did not pursue such an approach for the work presented in this paper. The algorithm used by the play selector is presented in Figure 3.

```
SELECT_PLAY(S: world state, P: playbook, D: default play):
  BEST_PLAY <- D
  BEST_WEIGHT <- WEIGHT(D)
  for each PLAY in P:
    if WEIGHT(PLAY) > BEST_WEIGHT:
      for each CONDITIONS in APPLICABLE(PLAY):
        if all CONDITIONS are satisfied in STATE:
          BEST_PLAY <- PLAY
          BEST_WEIGHT <- WEIGHT(PLAY)
  return BEST_PLAY
```

Fig. 3. Algorithm used by the play selector

3.3 Role Allocator

The selection of a play determines which roles need to be allocated to the robots. However, it does not specify which robots should be assigned to each role. Therefore, a role allocation algorithm is still needed to assign the roles. This algorithm also runs on the leader robot, which broadcasts the assignment along with the selected play. Our role allocator has two features that differentiate it from those used by many other RoboCup teams [5]. First, it only runs when a play is initially selected, as opposed to continuously. Second, it allocates roles in a *role-preserving* manner – minimizing role switching. Formally, if a new play P_t is selected at time t, and P_t specifies n roles $\{R_{1..n}\}$ for the n robots $r_{1..n}$, and r_i was already assigned to R_j in P_{t-1}, r_i is guaranteed to still be assigned to R_j in P_t. (Any remaining roles can be allocated in a greedy fashion.) If two plays share some roles, this strategy guarantees that some of the robots can assume their new roles without any transitional cost. These features provide additional resistance to oscillation in cases in which two plays share common roles.

3.4 Roles

The *role* assigned to each robot determines what behaviors the robot actually runs. Our approach, used in RoboCup 2005, is unique in that it is *region-based*: each robot is assigned to a region of the field. A robot is primarily responsible for going after the ball whenever the ball is in that robot's region. Roles are designed simply by configuring a generic "Player" behavior with appropriate settings for that role. The configurable items include:

- Region: an area of the field that the robot is responsible for covering.
- Ball in Region Policy: the behavior the robot should adopt when it knows that the ball is in its region. Typically, this will involve approaching the ball and trying to clear it downfield or take a shot on goal.
- Ball out of Region Policy: the behavior the robot should adopt when it knows that the ball is not in its region. Some roles specify that a robot is simply to return to a home position, while other roles may have the robot move to block the path between the ball and the goal, or to position for a pass.
- Ball Lost Policy: the behavior the robot should adopt when it believes the ball is lost. This is typically some sort of searching behavior.

Unlike our previous approaches, robots no longer need to negotiate with one another in order to gain the *attacker* role that allows them to approach the ball. In this way, the performance of the team does not degrade significantly under high network latency. We have developed algorithms that prevent the robots from interfering with one another even when they are playing in overlapping regions. To provide robustness against communication failure, these algorithms are designed to operate without the need for communication, using local information such as a robot's vision of its own teammates. If communication is available, our robots use additional features (such as reported teammate positions) that provide added confidence that our robots will not interfere with one another.

3.5 Experimental Results / Discussion

Due to lack of space, we are unable to present detailed experimental results here. Instead, we refer the reader to previous work in which we have presented related results. In [6], we show that using high-level features, such as the presence of opponents, to select a team strategy can improve the goal-scoring performance of a team of two robots. In [7], we explore the problem of *ball advancement* in a team of three robots. These results show that the role-preserving behavior of our role-assignment algorithm allows our team to maintain a consistent level of performance even when the active play is switched at a rapid rate.

The presented role-assignment algorithm and plays have been tested in the RoboCup 2005 competition. Our team came in fourth place in a challenging competition of twenty-four teams. Our team typically rotated through three well-balanced plays in the first minutes of each game, which allowed us to see the performance of each play against the specific opponent. We could manually change the team's strategy at halftime or during a timeout.

Our role assignment system is unique in that it allows role assignments to happen to all robots, including the goalkeeper. If there is not much time left in the game and our team is losing, we have plays that will "pull" the goalkeeper out of the goal box, which provides us with another field player that could score a goal. In fact, in the 3rd-4th place game of the RoboCup 2005 competition, our goalkeeper robot nearly scored a goal in the final seconds of the game.

4 Conclusion

In this paper, we have presented improvements to our team's communication and world modeling strategies. These improvements place high priority on the communication of task-relevant data and ensure that robots communicate some useful information even when lost. We have also presented a distributed, play-based role-assignment algorithm, which aims to solve several important challenges, including the presence of adversaries, task-based temporal constraints, and robustness to network failure. Our future work includes automatic play adaptation within the underlying challenges of a distributed team, and principled reasoning about adversarial temporal constraints, such as changing strategies based on the time left in the game and the current score.

References

1. Committee, R.T.: Sony four legged robot football league rule book (2006)
2. Roth, M., Vail, D., Veloso, M.: A real-time world model for multi-robot teams with high-latency communication. In: Proc. IROS, vol. 3, pp. 2494–2499 (2003)
3. Pynadath, D., Tambe, M.: The communicative Multiagent Team Decision Problem. Journal of Artificial Intelligence Research 16, 389–423 (2002)
4. Bowling, M., Browning, B., Veloso, M.: Plays as team plans for coordination and adaptation. In: Proceedings of ICAPS (2004)
5. Gerkey, B.P., Mataric, M.J.: On role allocation in RoboCup. In: Polani, D., Browning, B., Bonarini, A., Yoshida, K. (eds.) RoboCup 2003. LNCS (LNAI), vol. 3020, pp. 43–53. Springer, Heidelberg (2004)
6. McMillen, C., Rybski, P., Veloso, M.: Levels of multi-robot coordination for dynamic environments. In: Multi-Robot Systems: From Swarms to Intelligent Automata, vol. III, pp. 53–64. Kluwer Academic Publishers, Dordrecht (2005)
7. McMillen, C., Veloso, M.: Distributed, play-based role assignment for robot teams in dynamic environments. In: Proc. Distributed Autonomous Robotic Systems (2006)

Development of an Autonomous Rescue Robot Within the USARSim 3D Virtual Environment

Giuliano Polverari, Daniele Calisi, Alessando Farinelli, and Daniele Nardi

Dipartimento di Informatica e Sistemistica
Università di Roma "La Sapienza"
Via Salaria 113, 00198 Roma, Italy
<lastname>@dis.uniroma1.it

Abstract. The increasing interest towards rescue robotics and the complexity of typical rescue environments make it necessary to use high fidelity 3D simulators during the application development phase. USARSim is an open source high fidelity simulator for rescue environments, based on a commercial game engine. In this paper, we describe the development of an autonomous rescue robot within the USARSim simulation environment. We describe our rescue robotic system and present the extensions we made to USARSim in order to have a satisfying simulation of our robot. Moreover, as a case study, we present an algorithm to avoid obstacles invisible to our laser scanner based mapping process.

1 Introduction

Robotic systems have been proposed in recent years in a variety of settings and frameworks, pursuing different research goals, and successfully applied in many application domains. Technological improvements both in the hardware and in the associated software of robotic platform push their application towards more and more complex scenarios.

Search and Rescue robotics is one of the most challenging and interesting application environments for AI and robotics. Such an application requires the robots to be equipped with several complex sensors and to be able to perform complex manoeuvres in cluttered and unstructured spaces.

When working with an expensive and complex hardware, the presence of a simulator is of significant importance. On the one hand, it enables the evaluation of different alternatives during the robot system design phase leading to better decisions and cost savings. On the other hand, it supports the process of software development by providing a replacement when robots are not available (e.g. broken or used by another person) or unable to endure long running experiments. Furthermore, the simulation offers the possibility to perform an easier and faster debugging phase.

Several robotic simulators for 3D environments have recently been developed, providing a valid alternative to the canonical 2D-oriented ones. A high fidelity 3D environment adds to the simulation the possibility to test extremely realistic interactions, with a superior graphic rendering, extending the range of sensors to be tested.

G. Lakemeyer et al. (Eds.): RoboCup 2006, LNAI 4434, pp. 491–498, 2007.

USARSim is an open source 3D simulator for the urban search and rescue (USAR) environment based on a commercial game engine, currently supported by an international community.

This paper aims to describe the realization of an autonomous robotic system for search and rescue missions using USARSim. The robotic system is based on a Pioneer P3AT[1] commercial platform equipped with a sonar ring. We customized the platform adding a SICK Laser Range Finder, a Stereo Color Camera mounted on a pan-tilt unit, an Infra Red Sensor and a wireless acces point to communicate with a ground station. The purpose of the robotic system is the autonomous exploration of a rescue scenario searching for victims and building the map of the explored area. The autonomous navigation system, which is based on a two level path-planner, is able to guarantee safe navigation in highly cluttered space [8]. The mapping system is based on Laser Range Finder readings and uses a scan matcher based approach so to localize the robot and build the map. Finally, Stereo Vision is used to detect victims.

The first task was to build an interface between USARSim and our robotic development platform to simulate our real robot and its equipment. In particular we both modeled our system with the available built-in features (e.g. Pioneer robotic platform and SICK Laser Range Finder) and extended the simulator, so to correctly represent all our equipment (e.g. the Stereo Color Camera). Moreover, we improved the existing simulation environment, synchronizing sensor readings and correcting the simulation of transparent objects. Interfacing our development platform with USARSim we are able to test the same code on both the real robot and the simulator: as a consequence, we are now able to use USARSim as a powerful debugging environment in the development phase of our robotic applications.

Furthermore, we present a case study concerning path-planning in unknown and cluttered environments. We modeled in USARSim several test scenarios and developed a speed tracking based stall recovery subsystem to deal with invisible obstacles. We tested the algorithm in USARSim, saving time and preserving the robot from dangerous impacts.

The paper is organized as follows: in the next Section we describe the USARSim simulator. Section 3 shows our work with the simulator, the interface we built and the customization we made. In Section 4 we discuss the case study. Section 5 discusses related works and Section 6 concludes the paper.

2 USARSim

USARSim (presented in [1]) is a 3D high fidelity simulator of USAR robots and environments. USARSim can be a valid tool for the study of basic robotic capabilities in 3D environment. USARSim provides a high quality rendering interface and it is able to accurately represent the robotic system behavior.

USARSim development started in the University of Pittsburgh and is currently supported by an international community. It is released as open source

[1] ActiveMedia: Pioneer. http://www.activrobots.com

software[2] and has been adopted as the standard simulation tool for the RoboCup[3] Virtual Robots Competition in the upcoming 2006 edition.

The current version of USARSim consists of: i) standardized environmental sample models; ii) robot models of several commercial and experimental robots; iii) sensor models, like Laser Scanners, Sonars and Cameras; iv) drivers to interface with external control frameworks, like MOAST, Pyro and Player.

USARSim uses Epic Games Unreal Engine 2[4] to provide a high fidelity simulation at low cost. Unreal is one of the leading engines in the first-person shooter genre and is widely used in both the game industry and in the academic community. The use of the Unreal Engine provides several interesting features to USARSim: i) a high-quality, fast 3D scene rendering, supporting mesh, surface (texture) and lighting simulation; ii) a high fidelity rigid body physical simulator, Karma, supporting collision detection, joint, force and torque modeling; iii) a design tool, UnrealEd, that enables developers to build their own 3D robot models and environments; iv) an object-oriented scripting language, UnrealScript, which supports state machine, time based execution, and networking; v) an efficient client-server architecture to support multiple players.

3 Modeling an Autonomous Rescue Robotic System in USARSim

To fully integrate our robotic rescue system within the USARSim virtual environment we performed the following steps: i) we modeled our robotic platform in the USARSim framework and developed a low level interface to the simulator environment; ii) we modified the simulator to improve sensors' realism; iii) we introduced in USARSim a Stereo Vision Camera sensor and a 3D Camera.

In the following, we discuss each phase of the development. Moreover, we show some validation results concerning autonomous exploration in a USARSim simulated environment.

3.1 Modeling Our Robot in USARSim and Building the Interface

The robot we currently use is a Pioneer P3AT. We equipped the virtual chassis (already modeled in USARSim) with a full Sonar ring made of 16 sensors, a SICK Laser Range Finder and Camera mounted on a Pan-Tilt unit. Figure 1 shows a comparison between our real robot and its model in USARSim.

Our development framework, is based on a set of independent modules that interact and communicate among each other using a centralized blackboard-type repository [4]. To interact with the USARSim environment we built specific modules that directly communicate with the USARSim server. Since these modules use the standardized framework interface, they can be directly replaced with those that communicate with real hardware or different simulator environments.

[2] USARSim project page: http://sourceforge.net/projects/usarsim

[3] RoboCup 2006: http://www.robocup2006.org

[4] Epic Games: Unreal Engine. www.epicgames.com

Fig. 1. Our robot and its model in USARSim

This way, we can use all the other modules (e.g. navigation module, mapping module, etc.) without the need of any modification.

In particular, we developed four basic modules: i) the **robot** module, which manages the communication socket, receives and stores odometry and current speed data and sends motion commands to the server; ii) the **laser** module, which stores data gained from the simulated Laser Scanner Sensor, and views/changes its configuration; iii) the **sonar** module, which manages a set of simulated Sonar Sensors; iv) the **camera** module. Camera Sensor simulation is obtained in USARSim using the video feedback of Unreal Client, the Unreal Engine application for 3D scene rendering; in particular, an ImageServer is provided to capture the Unreal Client data and to serve it through TCP/IP. Our camera module holds a dedicated socket to connect the ImageServer and get the virtual Camera data; moreover, the camera module is used to view the Camera configuration and to move the simulated Pan-Tilt unit.

3.2 Improving Sensors' Simulation

USARSim does not provide timestamp information for sensor readings. However, when processing data coming from different sensors, synchronization can be a critical issue. For example, several of our platform subsystems (e.g. the SLAM subsystem) need timestamps for Odometry, Laser and Sonar readings, in order to calculate data confidence and perform coherent state estimation. We added a timestamp information to the Sonar, Laser and Odometry data.

We experienced that the simulated Laser Scanner sensor detected transparent objects as if they were opaque. Every object in USARSim holds a "material" property: we modified the Laser Scanner erroneous behaviour, spreading the laser beam over the transparent objects until it hits another material or it reaches the sensor max range. Thanks to such a modification we have been able to test in USARSim our scanmatcher-based SLAM (simultaneous localization and mapping) and the glass detection subsystem for the identification of transparent materials (which are undetectable by the Laser Scanner) based on the Sonar data.

3.3 Stereo Vision in USARSim

Naturally enough, within a Rescue environment the victim recognition subsystem carries a major weight. Our current approach uses a human detection algorithm driven by a Stereo Vision unit, which is composed of a couple of synchronized cameras with the same orientation.

As seen before, camera sensor simulation is obtained in USARSim through the capture of the video feedback of Unreal Client. Currently, only one running copy of Unreal Client at a time is allowed for each operating system. This limit comes from the single-user nature of the simulation: consequently Unreal-based Stereo Vision seems to be impossible until future versions of the Unreal Engine are released.

Since it is impossible to have multiple camera simulation on the same screen, we extended the robot definition code. Each virtual robot is described in the simulation by an Unreal Script definition code, storing information about its model and instructions to handle input data, to make movements and to draw the camera data.

We modified the function usually used to draw double-exposure images on the screen. Every time a frame is being drawn on the screen, we split vertically the output window overriding the first half with the left camera data and the second with the right camera data, maintaining data synchronization. With this new self-developed Stereo Vision sensor we are now able to have a complete high fidelity simulation of our rescue robot.

3.4 3D Camera Sensor

The Swiss Ranger Camera[5] is a sensor able to add a distance information to every pixel of the image data captured by its internal camera. Such sensor can be extremely useful in the USAR environment, both for navigation and for victim detection.

We added a Swiss Ranger Camera simulation in USARSim introducing a new IRC (Infra-Red Range Camera) sensor providing, for each pixel, the distance from the objects in the scene. By using the IRC sensor together with an ordinary Camera with the same position, orientation and resolution, we add the distance information to every pixel of the camera image, obtaining a simulation of the Swiss Range Camera.

Figure 2 shows, side by side, the Camera feedback (on the left) and the IRC sensor output (on the right, the brightness is proportional to the distance).

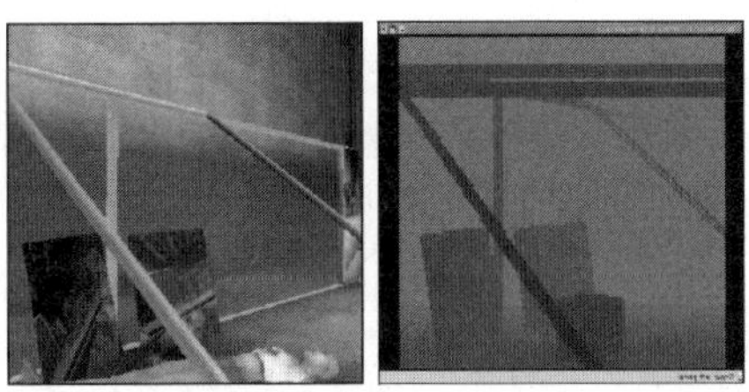

Fig. 2. A Camera image and the corresponding IRC sensor feedback

[5] CSEM: Swiss Ranger Camera. http://www.swissranger.ch/products.php

3.5 Validation Results

We performed several tests to validate the whole system configuration. We placed our robot into different USARSim virtual environments, to perform an autonomous exploration. The system behaviours consistently matched the real robotic system. In particular, we verified that the data gained from the sensors and the motion commands execution were as expected.

Figure 3 shows our rescue robot while autonomously exploring an unknown virtual 3D environment generated by USARSim. In the map on the right the unknown parts are drawn in blue (grey), while walls and obstacles are drawn in black and free space in white. The small table in front of the robot is not drawn by the SLAM module (i.e. in black), because it is invisible to the Laser Range Finder. However, our stall recovery subsystem, described in the following paragraph, identifies the impact surface and draws it on the map.

Fig. 3. Our rescue robot exploring an unknown virtual environment

4 Case Study: Exploration with Invisible Obstacles

During autonomous exploration missions in rescue environments, stall problems ofter arise. Our frontier based autonomous exploration subsystem, presented in [8], uses a two-level approach for navigation. It is based on a global topological path-planner and on a local motion planner, which is an extension of the well-known Randomized Kinodynamic Tree [9]. This kind of algorithms works by building a tree of safe, randomly-generated robot configurations. This local motion planner may be stuck by obstacles that are undetectable by the Laser Scanner, because they do not lie on its scanning plane.

We built a **stall recovery subsystem**, whose development was highly simplified because of the use of USARSim: the simulated environment helped us to save testing time and to preserve the real robot from dangerous impacts with unknown obstacles. We modeled small obstacles such as a tube, a ramp and a small table and observed the reactions of the virtual robot towards these objects.

The main cycle of the subsystem is composed of the following steps:

1. The subsystem first calculates the actual value of linear and angular speed, given the actual and previous robot poses (from the SLAM subsystem).
2. The differences between desired and actual speeds are monitored for several positions around the robot surface, using different stall conditions.

3. To avoid false positives we integrate over time the stall conditions.
4. If a stall condition is verified for several cycles, an obstacle is drawn on the map and an alarm is sent to the navigation subsystem to allow for a fast re-planning.

We tested the stall recovery subsystem in USARSim obtaining valuable results. Figure 3 shows on the left, the robot hitting a small table invisible to the laser; on the right, the obstacle representation in the robot map. Subsequent tests were performed on the real robot, using different obstacles such as chairs and bricks: the subsystem correctly identified stall situations tracking all the objects and allowing complete explorations of the environment.

5 Related Works

Moast[6], is a development framework providing a multi-agent simulation environment, a baseline control system, and a mechanism to migrate algorithms from the virtual world to the real implementation. Moast is intended to provide USARSim users with a customizable control system allowing for a high level interaction with the simulator. Compared to Moast, our system does not need to migrate the developed algorithms to the real implementation; in fact, our system runs indifferently on the real robot and on virtual environments.

Several works related to USARSim focus on validation of sensors such as Laser Scanner [2] or robot mobility [3]. In comparison to these works, we focused more on improving sensor data coherence (e.g. synchronizing sensor readings and testing sensorial fusion tasks) than on validating single sensor simulation.

As for obstacles which are not detectable by Laser Scanner sensors or cameras, different solutions are proposed in literature. Several approaches are based on touch sensors: for example in [5] the authors describe a cylindrical robot with a total coverage bumper, while in [6] an actuated whisker is used to identify objects. Such approaches however require additional sensors. Another way to address the problem is proposed in [7]. In this work the authors describe a mobile robot used as a tour guide, which is able to deal with invisible objects given the known map of the environment, lowering the speed when the localization error is higher. Unfortunately such, a technique is useless in a USAR environment, where the environment map is not known in advance.

To the best of our knowledge, the rescue system presented in this paper is one of the first complete autonomous rescue systems both working on real robots and integrated in the USARSim simulator.

6 Conclusions

Our experience in simulation before USARSim was limited to two dimensions. Several features of our robotic system such as the glass detection subsystem or the victim recognition subsystem were impossible to test during a simulated mission. Using USARSim, we had the widely acknowledged advantages of a high

[6] MOAST Project page. http://moast.sourceforge.net

fidelity 3D simulation, such as an accurate model of robot mechanics, different materials available on 3D surfaces etc.

In this paper we presented the development of an autonomous working system within USARSim. We modeled our robotic system within USARSim, significantly extending the simulation environment. In particular, we added the possibility to use Stereo Vision for our victim recognition subsystem, and synchronized all sensor readings in order to have a coherent map building process. Moreover, we addressed the problem of safe navigation in presence of obstacles which are invisible to the 2D Laser based mapping process. We proposed a solution to this problem and tested our system in the USARSim virtual environment.

The performed tests within the USARSim virtual environment of our robotic system confirm that such a framework is suitable for preliminary validation during the robotic application development phase. In fact, using our virtual robotic system we have been able to conduct experiments involving invisible obstacles preserving the real robot's integrity. Moreover, we can now perform a high fidelity experimental analysis of different rescue system configurations without the need to modify the actual robotic platform.

As a future work we plan to deeply investigate the interactions between the invisible obstacle detection process and the navigation and mapping process. In particular, it would be interesting to represent invisible obstacles as dangerous or forbidden configurations inside the navigation world model, and to study how this different obstacle representation would impact on the system performance.

References

1. Wang, J., Lewis, M., Gennari, J.: USAR: A Game-Based Simulation for Teleoperation. In: Proc. 47th Ann. Meeting Human Factors and Ergonomics Soc. (2003)
2. Carpin, S., Birk, A., Lewis, M., Jacoff, A.: High fidelity tools for rescue robotics: results and perspectives. In: RoboCup International Symposium 2005 (2005)
3. Wang, J., Lewis, M., Koes, M., Carpin, S.: Validating USARsim for use in HRI Research. In: Proc. of the Human Factors And Ergonomics Society 49th Annual Meeting, pp. 457–461 (2005)
4. Farinelli, A., Grisetti, G., Iocchi, L.: SPQR-RDK: a modular framework for programming mobile robots. In: Proc. of Int. RoboCup Symposium (2004) pp. 653–660 (2004)
5. Jones, J.L., Flynn, A.M.: Mobile Robots - Inspiration to Implementation, A K Peters Ltd. Wellesley, Massachusetts (1993)
6. Scholz, G.R., Rahn, C.D.: Profile Sensing with an Actuated Whisker. IEEE Transactions on Robotics and Automation 20(1), 124–127 (2004)
7. Fox, D., Burgard, W., Thrun, S., Cremers, A.: A hybrid collision avoidance method for mobile robots. In: Proc. IEEE Int'l Conf. on Robotics and Automation (1998)
8. Calisi, D., Farinelli, A., Iocchi, L., Nardi, D.: Autonomous Navigation and Exploration in a Rescue Environment. In: RoboCup International Symposium 2004 (2004)
9. LaValle, S.M., Kuffner, J.J.: Randomized Kinodynamic Planning. In: Proc. of IEEE Int'l Conf. on Robotics and Automation (1999)

Appearance-Based Robot Discrimination Using Eigenimages

Sascha Lange and Martin Riedmiller

Neuroinformatics Group,
Institute for Computer Science and Institute for Cognitive Science,
University of Osnabrück, 49069 Osnabrück, Germany
{sascha.lange,martin.riedmiller}@uos.de

Abstract. Transformation of high-dimensional images to a low-dimensional feature space using Eigenimages is a well known technique in the field of face recognition. In this paper, we investigate the applicability of this method to the task of discriminating several types of robots by their appearance only. After calculating suitable Eigenimages for Middle Size robots and selecting the most useful ones, a Support Vector Machine is trained on the feature vectors to reliably recognize several types of robots. The computational demands and the integration into a real-time vision system have an important role throughout the discussion.

1 Introduction

In the Middle Size League of the RoboCup two teams of four to seven fully autonomous mobile robots compete to win a game of soccer. The detection of obstacles (robots or referees) obviously is important for successfully planning motions, dribbling over the field and scoring goals. Whereas the pure detection of objects is simple— according to the rules, robots have to be mostly black— discriminating teammates and opponents is more difficult and has been implemented in the robots seldom. But now, after reaching the needed robustness in the more basic abilities like self localization, ball detection and motion planning, the teams more and more concentrate on implementing cooperative behaviors. Hence, reliably discriminating teammates and opponents becomes important.

The visual discrimination between the robots could be facilitated by the help of the prescribed colored team markers, each robot has to be equipped with (cyan and magenta). But considering the unstructured background, relying on color information only is error-prone. Hence, a robust classifier should consider other features of the robots appearance, too. Recapitulating, a vision system solving this task has to meet the following constraints:

- The system should not only recognize robots but also robustly discriminate between teammates and opponets.
- It should consider color information as well as shape information.
- It must be computational very efficient considering the hard real-time constraints under the availability of only limited computing power.

G. Lakemeyer et al. (Eds.): RoboCup 2006, LNAI 4434, pp. 499–506, 2007.

In this paper, we propose the application of the *Eigenimage* method in combination with the Support Vector Machine (SVM) to the robot discrimination problem. While reaching a generalization accuracy of about 95% in several settings an important strength of the constructed classifiers is their computational efficiency and their real-time capability; thus fulfilling the above constraints.

2 Previous Work

The recognition of obstacles in the Middle Size League of RoboCup has been studied previously [4,9]. The authors proposed to use multi-layer perceptrons to classify regions of interest (ROI). This pioneering method gives a good accuracy for recognizing robots but its real-time capability, due to the computational complexity of the (hand-)chosen features, is at least questionable.

Using the Eigenimages method is well known in the context of face recognition [5,13,12] and has been applied— in less extent— to more general object recognition [11]. In this paper, we give a new application of this method, namely the recognition and discrimination of robots in the RoboCup domain under hard real-time constraints.

3 Visual Learning of Robots

The general scheme applied in this work is to build and train a classifier on a set of labeled training images *offline* that afterwards can be utilized *online* in a real-time vision system.

The first step is to construct a suitable feature space from a number of prototype images. As the dimension of the feature space still may be too big, selection of the most valuable dimensions is the next step. The resulting small feature vectors are used to train a SVM on the robot classification task.

Finally, the resulting classifiers can be utilizied in the real-time vision systems of the robots implementing three sequential processing steps (following [4]): 1. Detecting and preprocessing Regions of Interest (ROI) (see sec. 4), 2. Extracting features from each ROI and 3. Classification of feature vectors using the trained SVMs. The results of the last step can then be integrated into a sensor fusion process. If object tracking is implemented, imperfect classification results can be easily enhanced by considering subsequent frames [10,8].

3.1 Constructing the Feature Space

We consider a set of n $(w \times h)$ images I_i that we could encode each in a $p = w \cdot h$ dimensional vector $\tilde{\mathbf{x}}_i$. $\tilde{X}$ is the $(n \times p)$ matrix containing the image $\tilde{\mathbf{x}_i}$ in its ith row. The Principal Component Analysis (PCA) can be used to construct a (lower-dimensional) feature space, called the Eigenspace that preservers in the projected data $\tilde{Z}$ as much as possible of the variation present in $\tilde{X}$. Before utilizing the PCA it is advisable to center the original data $\tilde{X}$ by subtracting the average image $\bar{\mathbf{x}}$ with entries $\bar{x}_j = \sum_{i=0}^{n} \tilde{X}_{ij} \quad , j = 1, 2, ..., p$ from each original image.

PCA and SVD. In general, the Principal Component Analysis looks for a few derived variables

$$z_k = \sum_{j=1}^{p} a_{kj} x_j = \mathbf{a}_k^T \mathbf{x} \tag{1}$$

that explain most of the variation present in *all* p original variables x_i of a random vector $\mathbf{x}$. Whereas in the original data we expect to find some interrelation between the variables x_i, the derived variables z_k are choosen to be uncorrelated to each other. More formally, it comes down to maximizing $var[\mathbf{a}_k^T \mathbf{x}] = \mathbf{a}_k^T \Sigma \mathbf{a}_k$ subject to

$$\mathbf{a}_k^T \mathbf{a}_l = \begin{cases} 1 \text{ if } l = k \\ 0 \text{ otherwise} \end{cases}$$

where Σ is the covariance matrix of $\mathbf{x}$.

Solving this constrained maximization problem e.g. via applying Lagrange multipliers it turns out that for k = 1,2,...,p the kth $\mathbf{a}_k$ is identical to the eigenvector corresponding to the kth largest eigenvalue λ_k of the covariance matrix Σ [3]. Further, if $\mathbf{a}_k$ is choosen to have unit length, as we did here, then the variance $var[\mathbf{z}_k]$ equals the eigenvalue λ_k.

In practice the underlying covariance matrix Σ of the image sampling process is unkown. Therefore, Σ is replaced by the sample covariance matrix S that for centered data is given by $S = \frac{1}{n} X^T X$ where X is the $(n \times p)$ matrix containing the images $\mathbf{x}_i$ of size p measured about their mean.

We usually have only a few hundreds of observations compared to the thousands of variables, thus $rank(S) = rank(X^T X) = rank(X) \leq n << p$. In this case, the PCA can be calculated computationally efficient by doing a compact Singular Value Decompositin (SVD). The compact SVD factorizes X into three matrices $X = U_r L_r A_r^T$ [3]. Let $r = rank(S)$, A_r is defined as the $(p \times r)$ matrix with its kth column $\mathbf{a}_k$ being the eigenvector corresponding to the kth largest eigenvalue of the sample covariance matrix S. L_r is a $(r \times r)$ diagonal matrix with its kth diagonal element the singular value $l_k^{\frac{1}{2}}$. L contains all non-zero singular values of X in a non-increasing order $l_1 \geq l_2 \geq ... \geq l_k > 0$. It can be shown, that l_k is also the kth largest eigenvalue of $X^T X$ and with (3.1) $\lambda_k = \frac{1}{n} l_k = \frac{1}{n} (l_k^{\frac{1}{2}})^2$. U_r is defined as the $(n \times r)$ matrix with the kth column $\mathbf{u}_k = l_k^{-\frac{1}{2}} X \mathbf{a}_k$. A proof and a more extensive derivation can be found in [3].

With these results we are now prepared to calculate the z_k and thus map images to their representation in the feature space. The mapping of i centered p-dimensional images $\hat{X}$ is given by (in matrix notation) $\hat{Z} = \hat{X} A_r$ where the entry $\hat{z}_{ik}$ of $\hat{Z}$ is the value of the ith image $\hat{\mathbf{x}}_i$ on the kth PC. The eigenvectors $\mathbf{a}_k$ of the sample covariance matrix S span an r-dimensinal orthonormal feature space, where the most variation will be observed along the first dimensions. Since the eigenvectors $\mathbf{a}_k$ have the same dimension as the original data and they can be visualized as sometimes familar and othertimes strange looking coefficient images (see fig. 1), they are called "Eigenimages".

Fig. 1. Top row: the first six Eigenimages of the second experiment of section 5. Gray pixels correspond to coefficients being zero. Lower row: original images of several objects and their reconstruction using only the scores on the first four PCs.

3.2 Selecting PCs

If X is of rank r the compact SVD helped us to construct a r-dimensional feature space. If a further reduction of the dimensionality is desired, a promising strategy is to select the n-first PCs, since the PCs resulting from the PCA/SVD are ordered according to the variance they explain (example in figure 1). An in-depth discussion of several strategies of automatically selecting good PCs in the case of regression analysis can be found in chapter 8 of [3].

3.3 Classification

After projecting a set $\hat{X}$ of images centered about the average image $\bar{\mathbf{x}}$ to the constructed feature space we can now train and test a classifier. We use a standard implementation [1] of a C-Support Vector Machine with Radial Basis Function (RBF) kernels as classifier [2,14]. Good parameters γ and C are chosen doing a grid-search in the parameter space [1]. To handle more than two classes we adopt the "one-against-one"-strategy and train an ensemble of $k(k-1)/2$ classifiers [6].

4 Integrating the Classifiers into the Middle Size Robots

In order to successfully utilize the Eigenimages method in official games, we have to consider several important practical aspects:

Different viewing angles. From the different possibilities of representing objects photographed from different viewing angles [11,12] we have chosen to follow [11] using a single feature space.

Background. The robots should be placed in front of the same background when taking the images that are used to calculate the feature space.

Translation and Scaling. It is necessary to center and scale the objects to always have the same size (see sec. 4).

Fig. 2. Images used in the first experiment. The first three images are from the set that have been used to calculate the PCs. The other images have been used during training and validating the SVM.

Lighting conditions. By constantly adapting the camera parameters according to a software implemented white-balance and auto exposure routine [15], color constancy and a constant brightness are assured up to a certain degree.

Detecting and Preprocessing Regions of Interest (ROI). Here we briefly describe an algorithm to find, center and scale Regions of Interest in the images produced by the directed camera of our robots. This method heavily depends on prior, domain specific knowledge.

We search about 180 carefully placed scanlines for the nearest non-isolated black pixels (according to the rules, robots have to be mostly black). Those pixels are clustered using a distance-threshold. The most central point of each cluster marks the initial position of a Region of Interest. Due to the placement of the scanlines, the initial position is at the bottom of the object. Using a calibrated camera and knowledge about the maximal size of the robots (specified in the rules) it is possible to determine the necessary size of a bounding box for every image location. An estimate of the real horizontal center of the black region is found by a heuristic looking at the black pixels in the lower 15 lines of the slightly enlarged (to compensate bad first guesses of the ROIs center position) bounding box. The initial guess of the position of the ROI is then improved by this estimate. Finally the part of the image in the re-placed bounding box— now using the correct size— is scaled, the particular scaling factor determined by the distance of the object.

Although this method is rather simple, the error in the centering typically is well below the amount that is tolerated by the classifiers (see fig. 4). For an object at the distance of about 1.5m the centering and scaling (using nearest pixels) could be done in less than 2ms on our robots.

5 Results

5.1 Discriminating Similar Looking Robots

The only difference between the otherwise similar looking two types of robots used in this experiment is a grating that extends the chassis of the goalie robot. The set contains 76 images of the robot that is used as fieldplayer and 80 images of the goalie (see fig. 2). Half of the images was taken in front of a non-varying background.

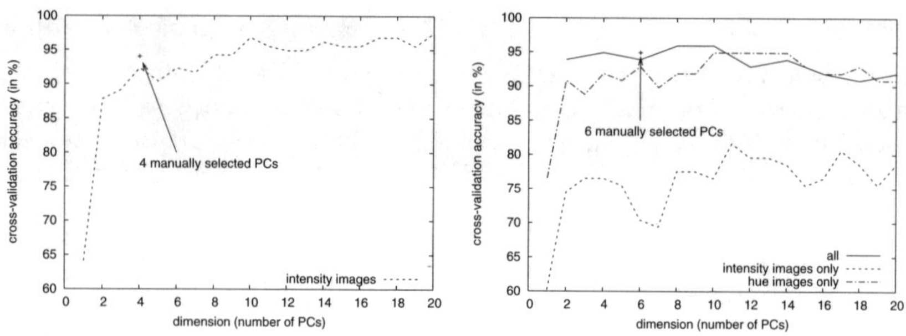

Fig. 3. Results of the 10-fold cross-validation on the first n PCs. Left: Results of the first experiment. Right: Results of the second experiment.

Finding Principal Components. To calculate the PCs we used only the images with the non-varying background. To get the best possible results we verified the correctness of the region detected automatically by the ROI-detecting algorithm and— in the hand full of cases it was necessary— corrected the automatically selected bounding box by hand for the images used to calculate the PCs. The selected image region was scaled to a size of (131×163) pixels.

Classification. All 156 images have been processed by the ROI-detector and were used for the training and testing of the SVM. We did extensive cross-validation experiments to find the best number of PCs to be used for the classification. We formed separate pattern sets for the first $1, 2, 3, ...$ PCs and did a search for good parameters γ and C (see section 3.3) on each of it. Using these parameters, the classifier's generalization accuracy was validated doing 10-fold cross-validations. Typical results of the grid-search for good paramters are $\gamma \leq 0.5$ and $C = 32$.

The classifier works quite accurately for a small number of included features. After including 10 PCs the result is not improved significantly by adding more dimensions. It should be noted that in this case the SVM turned out to be very robust against adding to many dimensions (small risk of overfitting). If reducing the computational demands to a minimum is a must, looking at the class distributions and selecting PCs by hand may be a useful strategy. We found a combination of 4 PCs on that the SVM reaches an accuracy of about 94% clearly outperforming all other classifiers using 4 or less of the first 40 PCs.

5.2 Teammates, Opponents, Boxes and Other Obstacles

In regular competition games we are interested in discriminating between teammates, opponents and other non-robot obstacles. We formed a collection of 98 images of our robots (class 1 and 2), of opposing "bin-bots" (class 4), of black cardboards (class 3) and of other objects (class 5), our ROI-detector may identify erroneously as obstacle (humand legs, dark goal corners, etc.). Our robots are split into two classes, one class of robots equipped with a cyan team marker

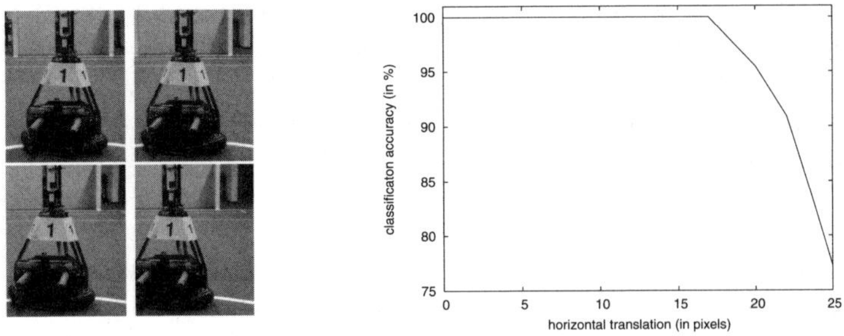

Fig. 4. Left: A bounding box with a correctly centered robot and the same object deliberately shifted to the left for 10, 15 and 25 pixels in order to simulate centering errors. Right: The classification accuracy on the shifted images.

and the other class wearing the magenta team marker. The opposing robots have been photographed wearing both of their team markers, too. Using the proposed assignment of class labels, both types of information, color and shape have to be considered for a correct classification result. All classes contained about the same number of images.

We applied the same procedure as used in the first experiment, but now using two Eigenspaces: one for the intensity- and one for the hue-channel of the color images. The results of using the first n PCs is given in the right diagram of figure 3. Using the PCs of the monochromatic images only results in a poor classification accuracy; the PCs of the color channel are clearly needed. To conclude the results we tested the classifier's vulnerability to incorrectly centered objects (fig. 4).

5.3 Consideration of Computational Demands

Considering the computational demands, the most costly step is to calculate the compact SVD on the image data when constructing the classifier (offline). Using the compact SVD this can be done within less than 10s.

The more critical part that has to be done in real-time, is the transformation of the ROIs to the feature space. We have to calculate equation (1) for each score z_k on each feature k. This needs $(2p)-1$ operations (additions + multiplications). For a classifier using 8-PCs, the complete mapping of a single ROI to the feature space takes about $0.4ms$ in average.

6 Conclusion

We have described and evaluated a module of a vision system capable of finding *and* discriminating robots in real-time. Each possible location of a robot is evaluated— even in worst case— in less than 3ms. Using the Eigenimages method it is possible to automatically construct useful features for discriminating robots in the RoboCup domain. No intuition or prior knowledge about the appearance of the robots is needed. Further dimensionality-reduction can be achieved by selecting only the n-first Principal Components to be presented to the classifier or by using the distribution of the class instances in the Eigenspace to manually identify and select useful features. The resulting classifiers have proven to be very reliable while being computationally efficient at the same time. In direct comparison to [4,9] the extraction of features using the Eigenimags is computational less demanding and solves the more complex task of discriminating robots.

We plan to investigating the possibility of using MCMC-training of GMMs [7] in order to construct a one-class classifier rejecting all non-teammates, thus rendering the training procedure of previously unseen opponents obsolete.

References

1. Chang, C.-C., Lin, C.-J.: LIBSVM: a library for support vector machines (2001), Software available at `http://www.csie.ntu.edu.tw/cjlin/libsvm`
2. Cortes, C., Vapnik, V.: Support-vector network. Machine Learning 20, 273–297 (1995)
3. Jolliffe, I.T.: Principal Component Analysis, 2nd edn. Springer, Heidelberg (2002)
4. Kaufmann, U., Mayer, G., Kraetzschmar, G., Palm, G.: Visual robot detection in robocup using neural networks. In: RoboCup 2004 International Symposium (2004)
5. Kirby, M., Sirovich, L.: Application of the Karhunen-Loeve procedure for the characterization of human faces. IEEE Trans. Pattern Analysis and Machine Intelligence 12(1) (1990)
6. Kressel, U.: Pairwise classification and support vector machines, pp. 255–268. MIT Press, Cambridge (1999)
7. Lauer, M.: Entwicklung eines Monte-Carlo-Verfahrens zum selbständigen Lernen von Gauss-Mischverteilungen. PhD thesis, University of Osnabrück (2004)
8. Lauer, M., Lange, S., Riedmiller, M.: Motion estimation of moving objects for autonomous mobile robots. Künstliche Intelligenz, 1/2006 pp. 11–17 (2006)
9. Mayer, G., Kaufmann, U., Kraetzschmar, G., Palm, G.: Neural robot detection in robocup. In: Biundo, S., Frühwirth, T., Palm, G. (eds.) KI 2004. LNCS (LNAI), vol. 3238, Springer, Heidelberg (2004)
10. Mayer, G., Kaufmann, U., Kraetzschmar, G., Palm, G.: Information integration in a multi-stage object classifier. In: 19. Fachgesprach Autonome Mobile Systeme AMS (2005)
11. Murase, H., Nayar, S.K.: Visual learning and recognition of 3-D objects from appearance. International Journal of Computer Vision 14(1), 5–24 (1995)
12. Pentland, A., Moghaddam, B., Starner, T.: View-based and modular eigenspaces for face recognition. In: IEEE Conference on Computer Vision & Pattern Recognition (1994)
13. Turk, M., Pentland, A.: Eigenfaces for recognition. Journal of Cognitive Neuroscience 3(1) (1991)
14. Vapnik, V.: Statistical Learning Theory. Wiley, Chichester (1998)
15. Wilhelm, T., Kludas, J., Böhme, H.-J., Gross, H.-M.: Automatischer Weissabgleich für eine omnidirektionale Kamera. In: Proc. 9 Workshop Farbbildverarbeitung, Schriftenreihe ZBS, pp. 43–50 (2003)

Fuzzy Naive Bayesian Classification in RoboSoccer 3D: A Hybrid Approach to Decision Making

Carlos Bustamante, Leonardo Garrido, and Rogelio Soto

Center for Intelligent Systems
Monterrey Institute of Technology
Monterrey NL 64849, Mexico
{cfbh,leonardo.garrido,rsoto}@itesm.mx

Abstract. We propose the use of a Fuzzy Naive Bayes classifier with a MAP rule as a decision making module for the RoboCup Soccer Simulation 3D domain. The Naive Bayes classifier has proven to be effective in a wide range of applications, in spite of the fact that the conditional independence assumption is not met in most cases. In the Naive Bayes classifier, each variable has a finite number of values, but in the RoboCup domain, we must deal with continuous variables. To overcome this issue, we use a fuzzy extension known as the Fuzzy Naive Bayes classifier that generalizes the meaning of an attribute so it does not have exactly one value, but a set of values to a certain degree of truth. We implemented this classifier in a 3D team so an agent could obtain the probabilities of success of the possible action courses given a situation in the field and decide the best action to execute. Specifically, we use the pass evaluation skill as a test bed. The classifier is trained in a scenario where there is one passer, one teammate and one opponent that tries to intercept the ball. We show the performance of the classifier in a test scenario with four opponents and three teammates. After a brief introduction, we present the specific characteristics of our training and test scenarios. Finally, results of our experiments are shown.

1 Introduction

Classification is a statistical operation in which certain objects are put into groups or classes according to their characteristics, sometimes called attributes, found on a training set. There are many approaches to classification in literature, like Decision trees, Neural networks, Support vector machines and Bayesian networks, among others. From the aforementioned classifing methods, the bayesian approach is the most commonly used to deal with uncertainty, because it is based on the probability theory.

A well known classifier is the Naive Bayes classifier [1], a simple type of bayesian network [2] that explodes the conditional independence assumption among attributes given the class. In real life, this assumption does not hold most of the time. However, Naive Bayes classifiers have proven to be successful

G. Lakemeyer et al. (Eds.): RoboCup 2006, LNAI 4434, pp. 507–515, 2007.

and more or equally effective than other classification methods in certain domains, like anti-spam filtering [3], information retrieval [4], speech recognition [5], emotion recognition [6] and medicine [7]. An analysis of the reasons why Naive Bayes works well is done in [8].

Generally, in a Naive Bayes classifier the attributes are discrete, but in most real-life situations, attributes are continuous. Crisp partitioning the domain can clearly cause some loss of information [9]. There have been some approaches to overcome this matter with the use of fuzzy variables [10,11].

The aim of this paper is to use the Fuzzy Naive Bayes classifier proposed in [10] as a decision module for a RoboCup [12] Soccer Simulation 3D team. Although the RoboCup simulation 3D is based on SPADES [13] multi-agent discrete event simulator, the data handled by the soccer agents is continuous, i.e. is defined in the range of real numbers. Also, the environment has too many features and the sensed data is influenced by a random noise. These characteristics make the RoboCup simulation 3D domain an excelent platform to test the Fuzzy Naive Bayes classifier effectiveness.

2 Fuzzy Naive Bayes Classifier

The Naive Bayes probabilistic model is one of the simplier Bayesian Network models used in Artificial Intelligence and Machine Learning nowadays. Let C be a class label with k possible values, and $X_1...X_n$ be a set of attributes or features of the world with a finite domain $D(X_i)$ where $i = 1..n$. The objective is to obtain the conditional model $P(C|X_1, ..., X_n)$. We can represent this model using the bayes' rule as follows

$$P(C|X_1, ..., X_n) = \frac{P(X_1, ..., X_n|C)P(C)}{P(X_1, ..., X_n)} \tag{1}$$

where $P(C)$ and $P(X_1, ..., X_n)$ are a priori probabilities, and $P(X_1, ..., X_n|C)$ is the likelihood of event $X_1, ..., X_n$ conditioned on the class C. Notice that the denominator remains constant for every value of C, thus it serves as a normalization constant and will be ommited from now on.

For computing the conditional probabilities of the model, the full joint probability table is needed. When the number of attributes is very large or the domain of such attributes consists of a large set of values, the use of the full joint becomes unfeasible. To overcome this difficulty, the conditional independence assumption is exploded. If we assume that every attribute is independent of each other given the class C, the model can be reformulated again as

$$P(C|X_1, ..., X_n) = P(C)\prod_{i=1}^{n} P(X_i|C) \tag{2}$$

that is known as the *Naive Bayes probabilistic model.* because the assumption of conditional independence does not hold in most scenarios. Besides of that, the Naive Bayes model has proven to be effective in a whole range of applications.

The *Naive Bayes Classifier* combines the Naive Bayes probabilistic model with a so called decision rule (or discriminant function in [8]). Generally, a *maximum a posteriori (MAP)* decision rule is used and we get the definition of a *Naive Bayes Classifier*

$$NBclassify(a) = \arg\max_{c \in C} P(c) \prod_{i=1}^{n} P(x_i|c) \quad (3)$$

where x_i means $X_i = x_i$, c means $C = c$ and a is a complete assignation of attributes, i.e. $a = \{X_1 = x_1, ..., X_n = x_n\}$ and so will be hereafter. In this context, a represents a new example not classified yet. The classifier is used to select a class C given the new example a, based on the previously calculated values of all the probabilities needed by the model. The probabilities are estimated using relative frequencies from data. Sometimes, Laplace-correction is applied to smooth calculations avoiding extreme values obtained with small training sets.

The classical Naive Bayes classifier considers attributes and classes with discrete domains. When dealing with continuous domains, the classifier needs a modification. One way is discretizing or crisp partitioning the domain of attributes into a finite number of classical sets. But that could cause a loss of information [9].

A better method is proposed in [10], consisting of a hybrid classifier bringing together Fuzzy Set Theory and a Naive Bayes classifier, named the *Fuzzy Naive Bayes classifier*

$$FNBclassify(a) = \arg\max_{c \in C} P(c) \sum_{x_1 \in X_1} P(x_1|c)\mu_{x_1} \cdots \sum_{x_n \in X_n} P(x_n|c)\mu_{x_n} \quad (4)$$

where $\mu_{x_i} \in [0,1]$ denotes a membership fuction or degree of truth of attribute $x_i \in X_i$ in a new example a. To be conservative, it is required that all degrees of truth are normalized in the current variable assignation, in this case $\sum_{x_i \in X_i} \mu_{x_i} = 1$. The probabilities for equation (4) can be calculated as below

$$P(C = c) = \frac{(\sum_{e \in L} \mu_c^e) + 1}{|L| + |D(C)|} \quad (5)$$

$$P(X_i = x_i) = \frac{(\sum_{e \in L} \mu_{x_i}^e) + 1}{|L| + |D(X_i)|} \quad (6)$$

$$P(X_i = x_i|C = c) = \frac{(\sum_{e \in L} \mu_{x_i}^e \mu_c^e) + 1}{(\sum_{e \in L} \mu_c^e) + |D(X_i)|} \quad (7)$$

where L is the training set consisting of all examples $e = \{X_1 = x_1, ..., X_n = x_n, C = c\}$, $\mu_c^e \in [0,1]$ denotes the degree of truth of $c \in C$ in a example $e \in L$, and $\mu_{x_i}^e \in [0,1]$ is the membership of attribute $x_i \in X_i$ in such example. As mentioned before, all degrees of truth must be normalized such that $\sum_{c \in C} \mu_c^e = 1$ and $\sum_{x_i \in X_i} \mu_{x_i}^e = 1$. Notice that Laplace-correction is applied to compute the probabilities.

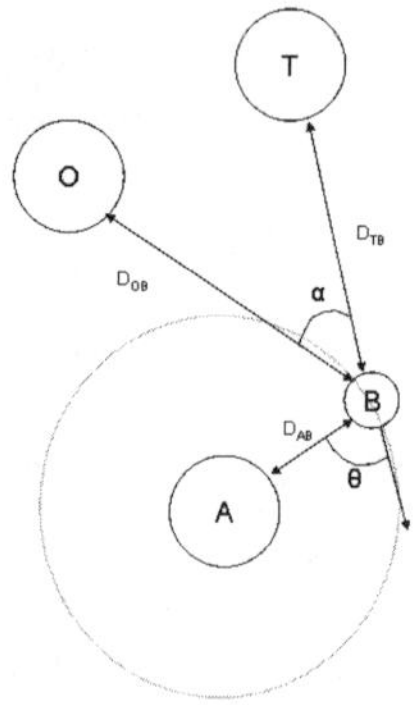

Fig. 1. Training scenario used to learn the probabilities for the Fuzzy Naive Bayes classifier, where we can see a passer agent (A), a receiver teammate (T), an opponent (O) and the ball (B). The features taken into account are: distance to ball d_{AB}, distance to teammate d_{TB}, distance to opponent d_{OB}, alignment angle (θ) and angle bewteen teammate and opponent from the ball's view point (α).

3 Empirical Scenarios

There are many opportunities to apply and test the classifier in the RoboCup domain. We chose the pass skill as a platform because it is a classical test bed for RoboCup. Specifically, we focus on pass evaluation, i.e., the capacity of an agent to evaluate the probability of success of a pass.

We created a training scenario for learning the probabilities for the Fuzzy Naive Bayes model. It is similar to the scenario proposed in [14] for the 2D league. But here we have to consider the alignment angle and ball distance, because in 3D soccer the agents are spheres and sometimes have to surround the ball for kicking in the right direction. The training scenario is shown in figure 1.

The scenario is mounted as follows

1. A passer agent is placed in the center of the field.
2. The ball is randomly placed next to the passer.
3. A teammate is randomly placed at a distance $d_{TB} \in [2, 20]$ from the ball.
4. An opponent is randomly placed at a distance $d_{OB} \in [2, 20]$ from the ball, such that the angle between the teammate and the opponent from the ball's view point is $\alpha \in [0, \frac{\pi}{6}]$.

Before an episode begins, the agent records five variables: distance to the ball d_{AB}, distance to teammate d_{AT}, distance to opponent d_{AO}, alignment angle $\theta \in [0, \pi]$ and the angle between teammate and opponent α. During each episode, the agent aligns with the ball to pass it to the teammate. Then both the teammate and the opponent try to intercept the pass. If the teammate gets the ball the episode is labeled as *SUCCESS*. If the opponent gets the ball first the episode is labeled as *MISS*.

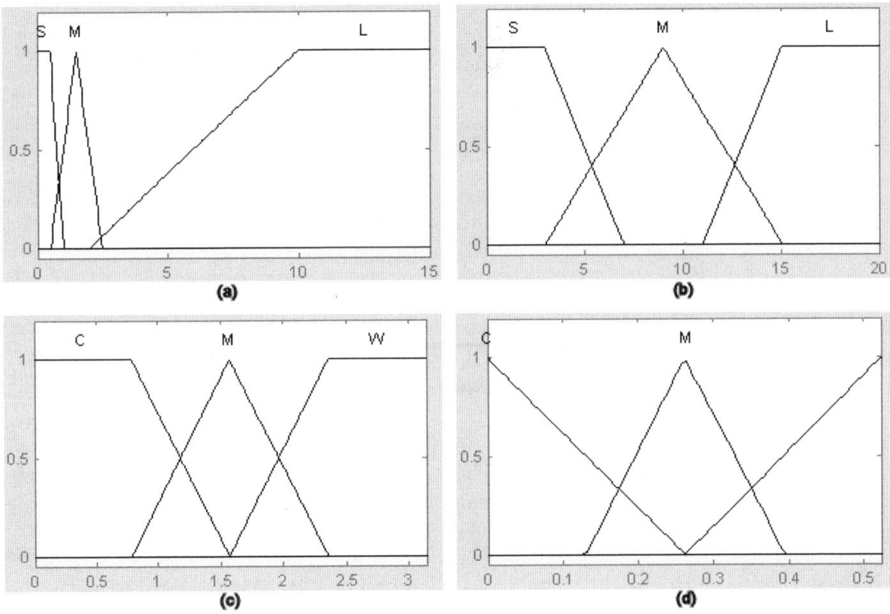

Fig. 2. Fuzzy Sets for each Fuzzy Variable. (a) Distance to the ball d_{AB}, (b) Distance to teammate d_{AT} and distance to opponent d_{AO}, (c) Alignment Angle θ and (d) Angle between teammate and opponent α.

Each variable mentioned above is a fuzzy variable and it is defined by several fuzzy sets. The fuzzy sets for distance to the ball d_{AB}, distance to teammate d_{AT} and distance to opponent d_{AO} variables are {*short, medium, long*}. The fuzzy sets for θ and α variables are {*closed, medium, wide*}. A graphical representation for each fuzzy variable is shown in figure 2.

4 Experimental Results

We obtained 1000 training examples, which we used to calculate the probabilities for the Fuzzy Naive Bayes classifier represented by equations (5), (6) and (7).

We ran 250 episodes to create a test set so we can measure the performance of the classifier. A representative graph is shown in figure 3. As can be seen in the figure, the graph stabilizes quickly, approximately at 120 traning examples. The maximum proportion of correclty classified examples was 0.806 which occured approximately at 350 examples.

To test the performance of the classifier for the pass evaluation skill, we created a test scenario which is shown in figure 4. In this scenario, the ball is in $(x = -20, y = 0)$ and the agent is placed at $(x \in [-22, -18], y \in [kickrange, 2])$. Four opponents and three teammates appear randomly in the area defined by $(x \in [-30, -10], y \in [10, 30])$.

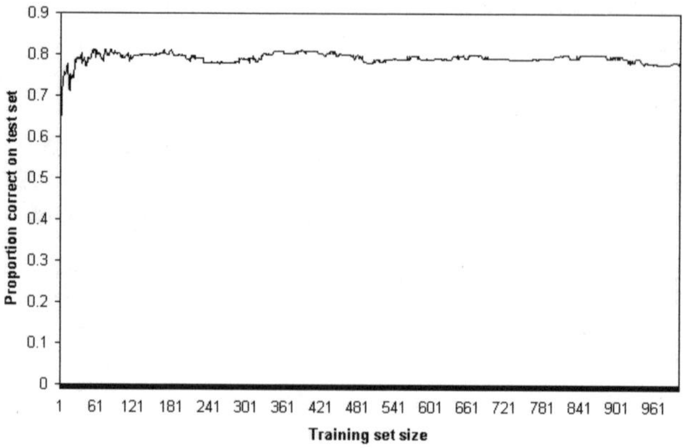

Fig. 3. Performance of the Fuzzy Naive Bayes classifier. The x-axis shows the number of examples used to train the classifier. The y-axis shows the proportion of the correcly classified examples in the test set. The size of the test set is 250.

When the scenario begins, the passer agent chooses the teammate with the best chances to intercept the ball using the Fuzzy Naive Bayes classifier. This is achieved in the following way: the passer uses the classifier to evaluate all 1 vs. 1 competitions between each teammate and each opponent. The lowest probability of success is stored for each teammate given all its 1 vs. 1 competitions and finally the teammate with the maximum probability of success is chosen. Formally,

$$Receiver = argmax_{t \in T}\ argmin_{o \in O}\ P(SUCCESS_{to}) \tag{8}$$

where T is the set of all teammates, O is the set of all opponents and $P(SUCCESS_{to})$ is the probability of success given the 1 vs. 1 competition between teammate $t \in T$ and opponent $o \in O$.

An episode is considered *SUCCESS* if a teammate is able to intercept the ball before any opponent does. If an opponent gets the ball before a teammate does, the episode is classified as *MISS*.

We obtained 300 examples using the test scenario described above. The percentage of SUCCESS and MISS classified examples is shown in table 1.

Table 1. Percentage of SUCCESS and MISS for a total of 300 classified examples using the test scenario of figure 4

Class	Percentage	Number of examples
SUCCESS	76	228
MISS	24	72

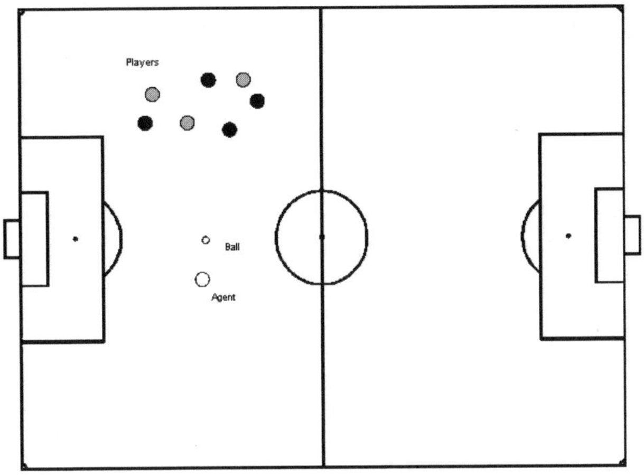

Fig. 4. Test scenario for the pass evaluation skill. Four opponent agents (black circles) and three teammates (gray-circles) appear randomly in a certain area of the field. The passer (white circle) and the ball (little circle) are placed a few meters away from them. The passer chooses the teammate with the best chances to intercept the ball using the classifier.

5 Conclusions

In this paper, we researched the application of a Fuzzy Naive Bayes classification algorithm to the decision making process of a RoboCup 3D team, specifically in the pass evaluation skill. The Naive Bayes method has proved to be effective in a wide range of situations although conditional independence assumtion is not met. As in RoboCup simulation we can consider variables as being continuous, we suggested seeing them as fuzzy variables so we could apply the Fuzzy extension to Naive Bayes proposed in [10].

We trained the classifier under a specific scenario where one agent passes the ball to a teammate and an opponent tries to intercept such pass. This 1 vs. 1 competition for pass evaluation can be easily extended to be used for the pass selection skill as in equation (8). As shown in figure 3 the classifier (trained with 1000 training examples) correclty classified a proportion of 0.806 examples of a test set of 250. Although the performance is not so good as in other implementations of Fuzzy Bayes in similar domains as in [15], we think it is not a bad performance at all. In table 1 we can see that 76% of the passes in our test scenario where successful passes. This is relatively better than the results shown by Stone [16], who used a Decision Tree for the pass evaluation procedure and a similar test scenario, and just 65% of all passes where successful.

We plan to extend our implementation in a near future to other skills like dribble and shoot. We think the performance may increase if the fuzzy sets are constructed more carefully. Perhaps it would be possible to extend this classifier to use fuzzy k-means clustering to obtain the fuzzy sets directly from data. We would like to implement other classifiers and compare their performance against the Fuzzy Naive Bayes classifier. Another possibility is to use the classifier to decide when to execute an air kick and take advantage of the characteristics of the 3D environment. Our final goal is to have a 3D team fully based on the fuzzy-bayes hybrid approach for the world cup competitions.

Acknowledgements

This work was supported in part by the research grant CAT011 and the Center for Intelligent Systems at Monterrey Institute of Technology (ITESM), Mexico.

References

1. Langley, P., Iba, W., Thompson, K.: An Analysis of Bayesian Classifiers. In: Proc. 10th Nat. Conf. on Artificial Intelligence, pp. 223–228. AAAI Press and MIT Press, Cambridge, USA (1992)
2. Heckerman, D.: A tutorial on learning with bayesian networks. Technical Report MSR-TR-95-06, Microsoft Research, Redmond, Washington (1995)
3. Androutsopoulos, I., Koutsias, J., Chandrinos, K.V., Paliouras, G., Spyropoulos, C.D.: An Evaluation of Naive Bayesian Anti-Spam Filtering. In: Proceedings of the workshop on Machine Learning in the New Information Age (2000)
4. Lewis, D.: Naive Bayes at forty: The independence assumption in information retrieval. In: Proceedings of European Conference on Machine Learning, pp. 4–15 (1998)
5. Tóth, L., Kocsor, A., Csirik, J.: On Naive Bayes in Speech Recognition. Int. J. Appl. Math. Comput. Sci. 15(2), 287–294 (2005)
6. Sebe, N., Cohen, I., Garg, A., Lew, M.S., Huang, T.S.: Emotion recognition using a Cauchy naive Bayes classifier. In: Proceedings of 16th International Conference on Pattern Recognition, pp. 17–20 (2002)
7. Demsar, J., Zupan, B., Kattan, M.W., Beck, J.R., Bratko, I.: Naive Bayesian-based nomogram for prediction of prostate cancer recurrence. Medical Informatics Europe '99, Studies in health technology and informatics 68, 436–441 (1999)
8. Rish, I.: An empirical study of the naive bayes classifier. In: Proceedings of IJCAI-01 workshop on Empirical Methods in AI, pp. 41–46 (2001)
9. Friedman, N., Goldszmidt, M.: Discretization of continuous attributes while learning Bayesian networks. In: Saitta, L. (ed.) Proceedings of 13-th International Conference on Machine Learning, pp. 157–165 (1996)
10. Störr, H.-P.: A compact fuzzy extension of the Naive Bayesian classification algorithm. In: Proceedings InTech/VJFuzzy, pp. 172–177 (2002)
11. Tang, Y., Pan, W., Li, H., Xu, Y.: Fuzzy Naive Bayes classifier based on Fuzzy Clustering. In: IEEE International Conference on Systems, Man and Cybernetics (2002)
12. Kitano, H., Asada, M., Kuniyoshi, Y., Noda, I., Osawa, E.: RoboCup: The robot world cup initiative. In: Proceedings of the First International Conference on Autonomous Agents, pp. 340–347 (1997)
13. Riley, P., Riley, G.: SPADES: - A Distributed Agent Simulation Environment with Software-in-the-Loop Execution. In: Winter Simulation Conference Proceedings, pp. 817–825 (2003)

14. Buck, S., Riedmiller, M.: Learning Situation Dependent Success Rates Of Action In A RoboCup Scenario. In: Pacific Rim International Conference on Artificial Intelligence, p. 809 (2000)
15. Mostafa, M.G.-H., Perkins, T.C., Farag, A.A.: A Two-Step Fuzzy-Bayesian Classification for High Dimensional Data. In: 15th International Conference on Pattern Recognition (ICPR'00), vol. 3, pp. 3421–3424 (2000)
16. Stone, P.: Layered Learning in Multiagent Systems: A Winning Approach to Robotic Soccer. MIT Press, Cambridge (2000)

A Novel Omnidirectional Wheel Based on Reuleaux-Triangles

Jochen Brunhorn, Oliver Tenchio, and Raúl Rojas

Institut für Informatik
Freie Universität Berlin
Takustr. 9, 14195 Berlin, Germany

Abstract. This paper discusses the mechanical design and simulation of a novel omnidirectional wheel based on Reuleaux-triangles. The main feature of our omniwheel is that the point of contact of the wheel with the floor is always kept at the same distance from the center of rotation by mechanical means. This produces smooth translational movement on a flat surface, even when the profile of the complete wheel assembly has gaps between the passive rollers. The grip of the wheel with the floor is also improved. The design described in this paper is ideal for hard surfaces, and can be scaled to fit small or large vehicles. This is the first design for an omnidirectional wheel without circular profile, yet capable of rolling smoothly on a hard surface.

1 Motivation and Reuleaux Triangles

It has been thought for many years, that the only way of providing a smooth rolling effect when using omnidirectional wheels[1] with gaps between rollers is: a) by stacking two wheels on the same axis, producing a combined circular profile; b) by using several synchronized wheels which combine in order to support the vehicle keeping the distance to the floor constant; or c) by using spheres or quasi-spheres as wheels. There is a fourth alternative, which is to design omniwheels with a circular profile, in which the gaps are almost closed by using two kinds of rollers alternatively, as in [3]. Long and short rollers alternate on the periphery of the wheel. In this paper we show for the first time that it is possible to build an omnidirectional wheel *with gaps* between the transversal rollers, that is without circular profile, which is nevertheless able to drive smoothly.

A so-called Reuleaux triangle (named after the German engineer Franz Reuleaux, who was a professor of mechanical engineering at the Technical University of Berlin) is a geometric shape whose width remains constant during rotation. This means that two parallel lines in contact with the shape's boundary stay at the same distance independently of the shape's orientation. The simplest shape of constant width is a circle. Other geometric figures can be modified to have constant width. Fig. 1 shows how a Reuleaux triangle is constructed. Starting from an equilateral triangle of side length l, constant width is

[1] The very first omnidirectional wheel was patented in 1919 by J. Grabowiecki in the US [1]. Bengt Ilon patented another in 1973 [2].

G. Lakemeyer et al. (Eds.): RoboCup 2006, LNAI 4434, pp. 516–522, 2007.

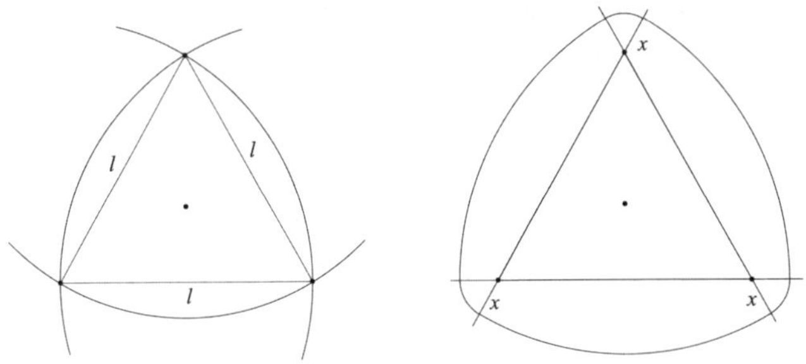

Fig. 1. Construction of the Reuleaux triangle

achieved adding circular arcs with radius l to each of the triangle's sides. The center of each arc is placed at the corner opposite the respective side. Reuleaux first mentioned these triangles in 1876 [4].

The shape's area can be increased by extending the triangle's sides beyond the corners by a distance x (Fig 1, right). The arcs' radii become then $l + x$. The gaps outside the original triangle and between two crossing sides are closed by circular arcs of radius x. In the following, the smaller arcs of radius x will be referred to as l_x-sections, and the larger arcs as l_r-sections.

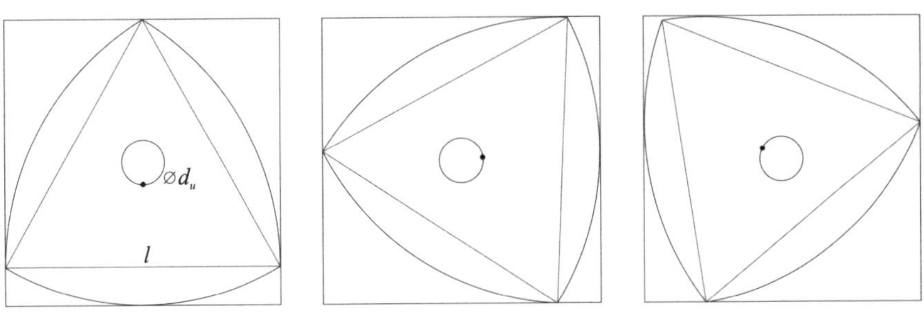

Fig. 2. The Reuleaux triangle rotating inside a square

The characteristics of a Reuleaux triangle allows such shape to rotate inside a square, as a circle also can (Fig. 2, see [5] for an explanation). Notice that the triangle's center *does not* remain in the same place while rotating the triangle, but moves along a curve consisting of four elliptical arcs. For the wheel design presented in this paper, this curve is approximated by a circle[2]. The circle's diameter d_u is obtained using the following formula:

$$d_u = l \cdot (\frac{4}{3} \cdot \cos(\pi/6) - 1)$$

[2] The curve is nearly the superellipse $x^{2.36} + y^{2.36} = c$, where c is a constant [5].

obtained from elementary geometrical considerations. This diameter *remains the same* for a given l, independent of the chosen extension of length x. Therefore, the principal movement can be obtained from considering simple Reuleaux-triangles (that is, triangles for which $x = 0$) instead of extended triangles. This fact is important since the wheel design presented here makes use of extended Reuleaux-triangles.

2 Wheel Design

The omnidirectional wheel proposed in this paper consists basically of the following parts:

1. Two discs based on Reuleaux-triangles, each carrying three passive wheels.
2. A gear connecting these discs, which allows the transmission of rotation between both.
3. An excenter which holds both discs and the gear.

In our wheel, a Reuleaux-triangle of enhanced area is used for the shape of the component which carries each group of three passive wheels (Fig. 3). The shape of the passive wheels' profile is determined by the l_x-sections described above. The l_r-sections remain empty, except for small supporting structures.

The structures do not allow the passive wheels to lose contact with the ground as long as the next passive wheel has not reached the ground yet. The path described by such passive wheels is explained below.

The passive wheel is built by using the l_x-section shape as the profile of a roller. Note that the position of the center axis of this roller (which is actually the passive wheel's axis) may be chosen arbitrarily. Thus it is possible to construct passive wheels of different sizes.

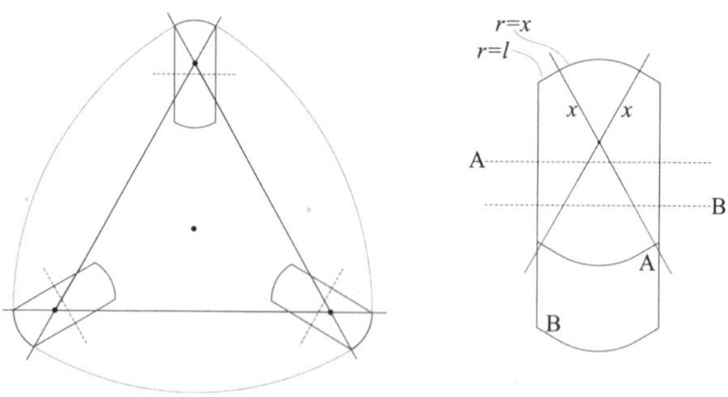

Fig. 3. Profile of the Reuleaux wheel (left). The wheel has three passive rollers. The rollers can have a smaller radius (centerline A), or a larger radius (centerline B). In each case, the effect is the same.

Before describing the remaining parts, it is necessary that the reader understands the principle behind the functionality of this omnidirectional wheel.

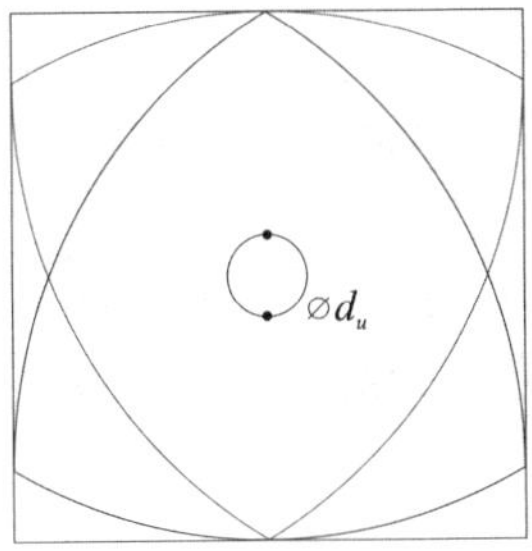

Fig. 4. Two Reuleaux discs shifted in phase

As described above, a Reuleaux-triangle can be rotated inside a square [5]. The square's sides are touched alternately by the triangle's corners and the circular arcs (in the extendend triangle by the l_x-section and the l_r-section). Imagine one of the square's sides to be the ground over which the Reuleaux discs roll, with the passive wheels touching the ground (remember that the passive wheels' profile matches the curve that describes the l_x-section). Each disc has contact with the ground only part of the time, because when the l_x-section loses contact with one side, the l_r-section reaches that side. Remember that the l_r-section-surface is *empty* in the Reuleaux discs.

To guarantee ground contact the entire time, a second Reuleaux disc is added to the wheel design, describing the exact same path of movement, but with a phase shift. The phase shift corresponds to exactly one half of a rotation of the Reuleaux-triangles' center around the displacement path of diameter d_u, as illustrated in Fig. 4 (one triangle looks as "flipped" by vertical reflection).

This ensures that the Reuleaux discs' passive wheels always touch the ground, alternating from one disc to the other. A simulation of a working omnidirectional wheel following this principle is shown in Fig. 5. The advantage of this wheel design becomes clear immediately: The passive wheels describe a purely linear path along the ground, preventing any up- and down-movement. On the contrary, since conventional omnidirectional wheels are shaped as n-side polygons (see Fig. **??**, left diagram), this leads to an up-and-down movement of the robot, which is lifted up every time a passive wheel touches the ground . At high velocities, the entire wheel might lose ground contact, making it difficult to drive accurately (as often happens during RoboCup competitions.

3 Design of Excenter and Gears

In order to achieve the circular movement of the center of each Reuleaux disc an excenter is needed. This excenter is held in place by the motor axis to which the torque is applied in order to drive the omnidirectional wheel. The first Reuleaux disc center axis is placed at a distance $\frac{1}{2}d_u$ of the motor axis (black rod in Fig. 6). When the excenter is rotated, the Reuleaux disc's center describes the desired curve (a circle of diameter d_u).

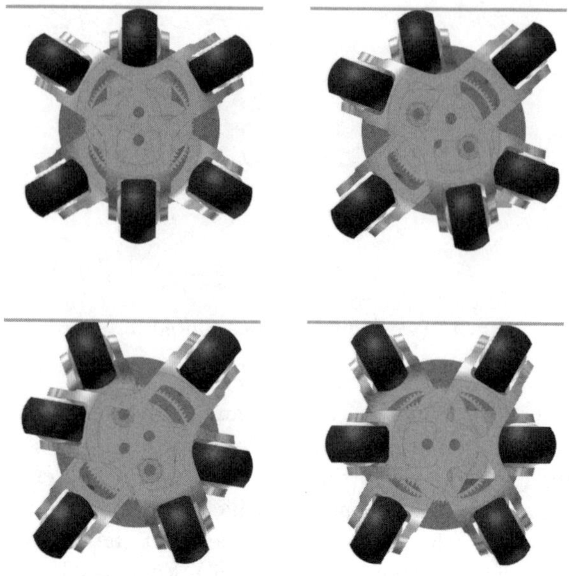

Fig. 5. Rotation of the wheel and relative displacement of the rollers. The wheel starts on the upper left and rolls counterclockwise. The passive roller on the ground loses contact only when the next roller has reached the ground-contact position. The combined movement looks as if the next roller "stretches a leg" before reaching ground.

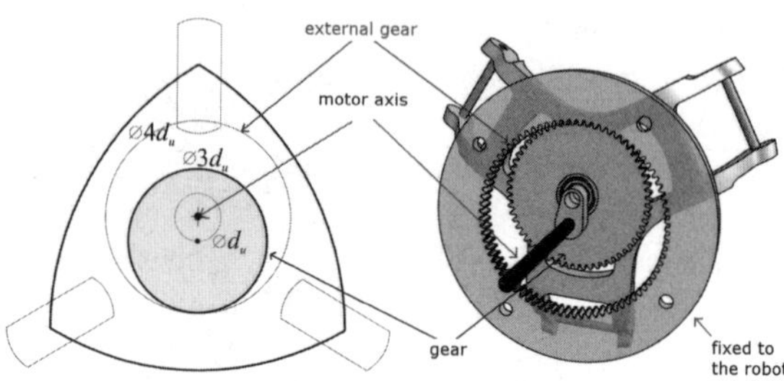

Fig. 6. Gear wheels and their diameter. When rotating around the gear fixed to the robot, the center of the Reuleaux disc describes a circle of radius d_u.

Additionally, for each complete rotation around the excenter, the Reuleaux disc itself needs to rotate by $-\frac{2}{3}\pi$ around its center. This combined movement is realized by two connected gear wheels. The smaller gear wheel, with the sprockets on the outside, is fixed to the Reuleaux disc. Its diameter must be $3 \cdot d_u$. Since this gear wheel rotates around the motor axis excentrically, a counterweight may be attached to the excenter to compensate the gear wheel's centrifugal force. The second larger gear wheel, with its sprockets on the inside, is fixed to the robot. Its diameter is $4 \cdot d_u$. Fig. 6 shows the complete configuration.

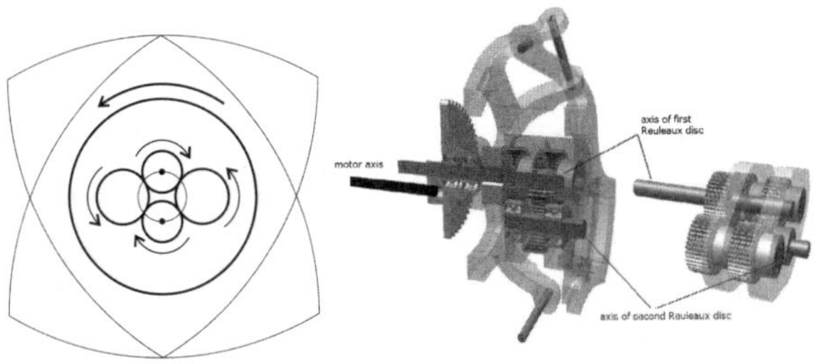

Fig. 7. Assembly and movement of the two Reuleaux discs. To the left a diagram, in the middle a cross section of the assembly, and to the right the transmission gear.

Now that the first Reuleaux disc describes the desired movement, the second disc needs to be connected to imitate the exact same movement with a phase shift. To obtain a phase shift of $\frac{1}{3}\pi$, we just flip the Reuleaux structure by 180 degrees, as shown in Fig. 7. Building a gear for this task is not difficult: the second disc's center axis is placed at a distance $\frac{1}{2}d_u$ from the motor axis, as before, but in the opposite direction of the first disc's axis. In order to rotate the second disc, a transmission gear is constructed to provide a 1:1 transmission of rotation from the first to the second disc (Fig. 7).

4 Conclusions

This paper has presented a new design for an omnidirectional wheel based on Reuleaux triangles. We have shown that even when the complete profile of the wheel mount has "gaps" between the passive rollers, it is indeed possible to achieve smooth rotational movement. For this to occur, the wheel center rotates around an excenter. A conventional wheel with profile gaps resembles a polygon and a rotating polygon moves the wheel's center up and down – if nothing is done against that movement. The wheels used by almost all teams in the small-size league resemble such polygons. Mid-size teams have used also wheels with perpendicular passive rollers and gaps in the wheel profile. The wheels presented in this paper are an alternative solution.

We are aware that the wheel design described in this paper is more complex than other omnidirectional wheels and requires more mechanical parts. Nevertheless, the wheel presented here has some interesting theoretical properties. Firstly, its profile is not circular, correcting the long held misconception that only circular-profile omniwheels can roll smoothly. Secondly, this design allows a robot (or other vehicle) highly precise and controllable omnidirectional driving, even at high speeds. Wheel grip is optimal because driving is vibration free, and the individual passive wheels have contact with the ground over a long time. Modelling Mecano wheels is extremely complex (because of the angle at which the passive rollers are placed) [6]. Kinematic modeling of the Reuleaux wheels is much simpler.

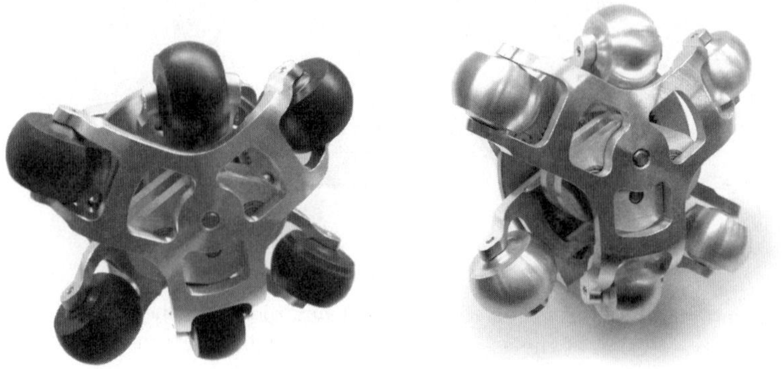

Fig. 8. Our first two Reuleaux omnidirectional wheel prototypes

Fig. 8 shows a ready-to-build sample design, including all the components described above and a photograph of the first prototype. The smooth rolling movement was validated in the computer before the prototype was built. Two Reuleaux omnidirectional wheels will be on display at the RoboCup 2006 competition in Bremen. It is a new attempt at reinventing the wheel, right, but the omnidirectional wheel.

References

1. Grabowiecki, J.: Vehicle Wheel. US Patent 1,303,535 (June 3, 1919)
2. Ilon, B.E.: Wheels for a course stable self-propelling vehicle movable in any desired direction on the ground or some other base, US Patent 3,876,255 (1975)
3. Byun, K-S., Song, J-B.: Design and Construction of Continuous Alternate Wheels for an Omnidirectional Mobile Robot. Journal of Robotic Systems 9(9), 569–579
4. Reuleaux, F.: The Kinematics of Machinery: Outlines of a Theory of Machines. MacMillan, London (1876)
5. Weisstein, E.W.: Reuleaux Traingle from MathWorld - A Wolfram Web Resource, `mathword.wolfram.com/ReleauxTriangle.html`
6. Muir, P.F., Neuman, C.P.: Kinematic Modeling for Feedback Control of an Omnidirectional Wheeled Mobile Robot. In: Cox, I.J., Wilfong, G.T. (eds.) Autonomous Robot Vehicles, Springer, Heidelberg (1990)

Development of Three Dimensional Dynamics Simulator with Omnidirectional Vision Model

Fumitaka Otsuka[1], Hikari Fujii[2], and Kazuo Yoshida[2]

[1] School of Science for Open Environmental Systems, Keio University,
3-14-1, Hiyoshi, Kohoku-ku, Yokohama, Japan
otsuka@yoshida.sd.keio.ac.jp
http://www.yoshida.sd.keio.ac.jp/~robocup/
[2] Department of System Design Engineering, Keio University,
3-14-1, Hiyoshi, Kohoku-ku, Yokohama, Japan

Abstract. In the field of robotics, simulators are important tools to verify algorithms. They are required to generate movements and physical interactions of objects based on the dynamics and to simulate outputs of sensors. However, few simulators of robots simulate dynamics of objects and outputs of sensors, in particular omnidirectional cameras, which are effective sensors because they can acquire an omnidirectional field of view at one time. In this study, omnidirectional vision simulators are developed based on the ray casting method. The proposed simulators enables to render accurate omnidirectional images and an intersection test is developed for speeding up. These are able to be used in "Gazebo" which is open source 3D dynamics simulator. The proposed methods are verified through comparison of the rendered images with the images which are obtained by real omnidirectional vision sensor.

1 Introduction

Omnidirectional vision sensors, which can acquire an omnidirectional field of view at video rate, have been applied in a variety of fields such as autonomous mobile robot, telepresence, virtual reality and remote monitoring [1]-[2]. In robotics, many kinds of robots such as soccer playing robots, communication robots, and monitoring robots equip this kind of sensors. The omnidirectional vision sensors are widely used. Therefore, the omnidirectional vision simulators are required. Some omnidirectional vision simulators have been developed [3][4], but they render only simplified images. More accurate one is required to verify processing algorithms using omnidirectional vision sensor. In addition, it is necessary for accurate simulations to generate movements and physical interactions of objects based on the dynamics and the kinematics of objects in 3D worlds. Some 3D dynamics simulators are developed; "Webots" [5]-[7], "Gazebo" [8][9], and some simulators for RoboCup Legged League [10]-[13]. However, most simulators of robots are for 2D worlds, and images from an omnidirectional vision sensor are rarely simulated.

This paper aims at developing a 3D dynamics simulation system for autonomous mobile robots with an omnidirectional vision sensor which is albe to render accurate images from sensors. For realization of them, an omnidirectional vision simulator using

G. Lakemeyer et al. (Eds.): RoboCup 2006, LNAI 4434, pp. 523–531, 2007.

omnidirectional vision model is developed based on 3D dynamics simulator "Gazebo". The omnidirectional vision simulator in which ray casting method is applied can render accurate image based on the model, and can easily be used as robot sensor because it is developed as "sensor" model of Gazebo. Moreover, an intersection test algorithm is developed for the purpose of speeding up rendering. The effectiveness of the omnidirectional vision simulator is verified through comparison of the rendered images with the images which are obtained by a real omnidirectional vision sensor.

2 Gazebo

Gazebo is a 3D dynamics simulator and can simulate some sensors such as odometer, ray proximity sensor, and camera. Some robot models, actuator models and sensor models are prepared in Gazebo, and the addition of new model is easy. Therefore, Gazebo is suitable as a basic system for developing sensor simulator. The architecture of Gazebo is shown in Fig.1. Client programs receive simulated outputs of sensor models and return commands data to actuator models through a shared memory interface so that simulator and client programs can be developed independently. Gazebo utilizes Open Dynamics Engine [14] as a physics engine and OpenGL [15] as a graphics interface. These are briefly described below.

Open Dynamics Engine (ODE). The ODE is a widely used physics engine which was developed by Russel Smith under open source licenses. It is designed to simulate rigid body dynamics. This engine includes many features such as numerous joints, collision detection, many geometries and so on. Gazebo utilizes these features by providing a layer of abstraction situated between ODE and Gazebo models.

OpenGL and GLUT(OpenGL Utility Toolkit). OpenGL is a standard library for the creation of 2D and 3D interactive applications. It is platform independent, reliable, portable, scalable, stable, and continually being developed. GLUT is a toolkit independent of window system for writing OpenGL programs. It implements a simple windowing API for OpenGL. Gazebo uses these for visualization and providing user interface with standard input devices such as keyboards and mice.

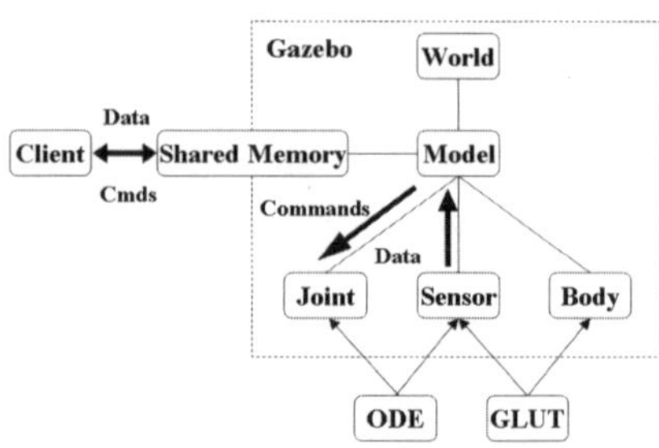

Fig. 1. Architecture of Gazebo

3 Omnidirectional Vision

3.1 Omnidirectional Vision Model

In this study, a model which consists of a camera and a hyperboloidal mirror is used as an omnidirectional vision model. The optical axis of the camera is aligned with the hyperboloidal mirror's one, as shown in Fig.2. This model is proposed by Yamazawa et all[16]. In this model, a two-sheeted circular hyperboloid is chosen as a design of the mirror surface. The hyperboloidal mirror is defined as follows.

$$\frac{x^2 + y^2}{a^2} - \frac{z^2}{b^2} = -1 \tag{1}$$

where a and b are parameters for determining the shape of hyperboloid. The hyperboloid has two focal points. The focal point of mirror is fixed at one of focal points O_M $(0, 0, +c)$ and the center of camera is fixed at the other point O_C $(0, 0, -c)$. The parameter c is defined as follows.

$$c = \sqrt{a^2 + b^2} \tag{2}$$

The three dimensional coordinate system O-XYZ, aligned with the image coordinate system o-xy and the Z-axis pointed toward the vertex of the hyperboloidal mirror is used. The point P at (X, Y, Z) is projected onto the image point p at (x, y). The azimuth is described as follows.

$$\tan\theta = Y/X = y/x \tag{3}$$

The following equation of the hyperboloidal projection is derived from simple geometrical analysis of the vertical section through P and the Z-axis as shown in Fig.3.

$$Z = \sqrt{X^2 + Y^2}\,\tan\alpha + c \tag{4}$$

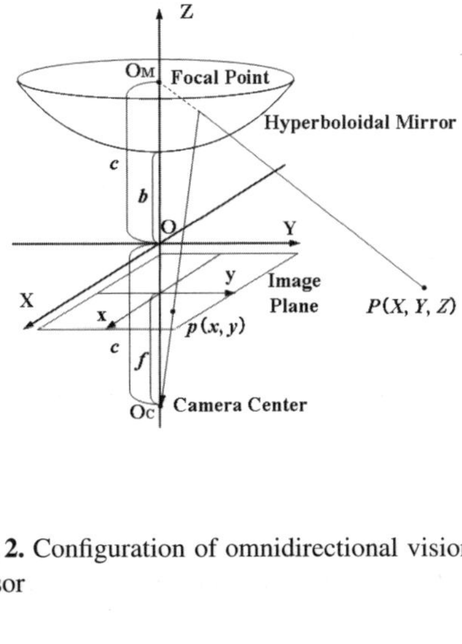

Fig. 2. Configuration of omnidirectional vision sensor

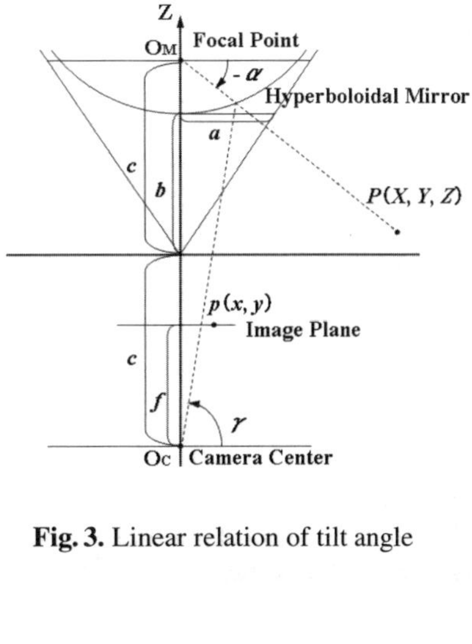

Fig. 3. Linear relation of tilt angle

$$\alpha = \tan^{-1} \frac{(b^2 + c^2)\sin\gamma - 2bc}{(b^2 - c^2)\cos\gamma} \tag{5}$$

$$\gamma = \tan^{-1} \left(\frac{\sqrt{x^2 + y^2}}{f} \right) \tag{6}$$

where f is the focal distance of the camera lens, and α is the tilt angle from the horizontal plane. These equations mean that the azimuth θ and the tilt angle α are given uniquely by fixing the center of the camera lens at the focal point O_C. Therefore, the image on the image points $p\,(x, y)$ can be simulated by rendering the image from the focal point of the hyperboloidal mirror O_M.

3.2 Omnidirectional Vision Simulator

An omnidirectional simulator is constructed according to the omnidirectional vision model described in section 3.1. First, a simulator based on the camera model of Gazebo which utilizes OpenGL camera is constructed. The procedure is as following.

1. Set resolution
2. Set field of view
3. Set view point
4. Set sight line
5. Construct all light models
6. Construct all object models
7. Capture image

In order to render an omnidirectional image, the resolution of OpenGL camera is set up to 1×1 pixel, the view angle is set up to $0°$, and the view point is fixed at the focal point of the mirror O_M. The sight line is determined based on θ and α. Thus, an omnidirectional image can be rendered by the implementation of the process from 1. to 7. on all image points respectively, and the resolution of it is equal to the number of the image points. However, the calculation cost of the process 6. becomes enormous according to increasing complexity of the environment, so that an intersection test algorithm is developed to speed up rendering. By this method, the nearest object in the field of view is detected, and the reduction of the calculation cost is achieved by constructing only this object on the process 6. An intersection of a sight line and a object is equivalent to the collision between the ray and the object. From this concept, two kinds of omnidirectional vision simulators using collision detection function in ODE are constructed.

Omnidirectional vision simulator based on intersection test with light model. In this method, the sight line model of OpenGL and the ray model of ODE are used for rendering an omnidirectional vision with shade. First, the ray model from view point is constructed by θ and α of image point p. Next, intersection test between the ray model of ODE and objects is carried out by using collision detection function in ODE, and the distance from the view point is calculated. If the intersection point is the nearest

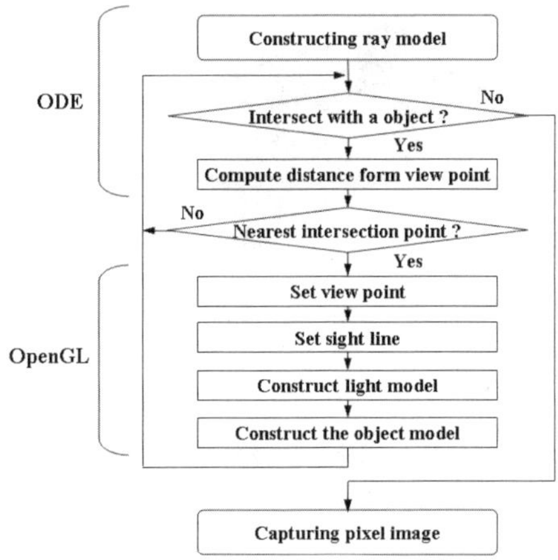

Fig. 4. Flow chart with light model

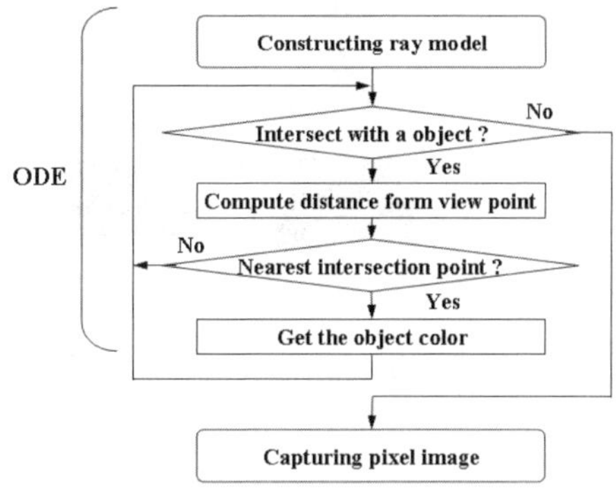

Fig. 5. Flow chart without light model

from the view point, sight line of OpenGL camera is set to get the image of the point. The model of the light and the object which collides with the ray is constructed to create accurate image of the intersection point. By iterating this process on all p, an omnidirectional vision image with shade is rendered. In this method, the number of constructed model is reduced, so that great reduction of the processing time is brought about. The flow chart of this method is shown in Fig.4.

Omnidirectional vision simulator based on intersection test without light model. In the method with light model, the rendering time is reduced. However, it still takes much time for reconstruction by OpenGL to simulate shade. To reduce the processing time, the method not using OpenGL camera is proposed. The flow chart of this method is shown in Fig.5. In this method, the intersection test is carried out as well as the method described above, and the color of object is directly obtained when the nearest intersection is detected. By iterating this process on all image point p, an omnidirectional vision is rendered. In this method, the shade of the objects can not be simulated, but the processing time is much shorter than the method with light model. The basic idea of this method is similar to the ray casting method. The ray casting is carried by using ODE to reduce the proseccing time.

4 Construction of the Model of RoboCup Middle Size League

To verify the omnidirectional vision sensor model proposed in chapter 3, a robot model which equips the proposed sensor and an environment model are constructed. This robot equips an omni-drive system with four roller omni-wheels and kicking device as actuators, and it also equips odometers and omnidirectional vision camera as sensors. The mass of this robot is 17.5 kg. A soccer field is constructed as simulation environment according to the regulation of RoboCup MSL. Robot models and environment models are

Fig. 6. Soccer robot model

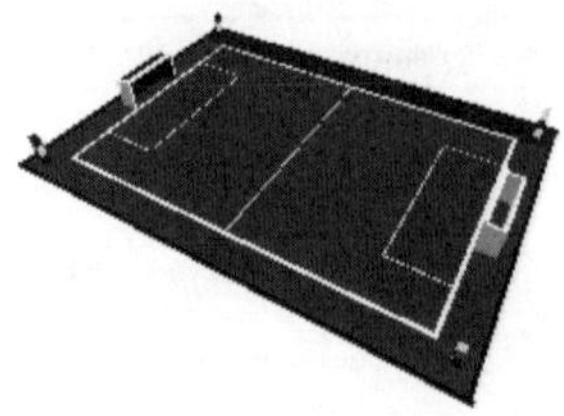

Fig. 7. Soccer field model

easily constructed.The constructed robot model and the environment model are shown in Figs.6 and 7.

5 Evaluation of the Developed Simulator

The omnidirectional vision simulator is verified by using the robot model and the environment model described in chapter 4. First, an image from a real omnidirectional vision sensor and an image rendered by the simulator are compared. The environment for verification is shown in Fig.8. The field size of real and simulation environments are same; 6.2m $\times$ 4.5m. The marked robot shown in Fig.8 equips an omnidirectional vision sensor. The image from a real sensor is shown in Fig.9, and the image rendered by the fastest method without light model is shown in Fig.10. The resolutions of the real image and the simulated image are 320 $\times$ 240 pixels and 240 $\times$ 240 pixels respectively. The simulated image is very similar to the real image. Next, the three kinds of the simulation methods described below are compared to verify the methods and the intersection test algorithm.

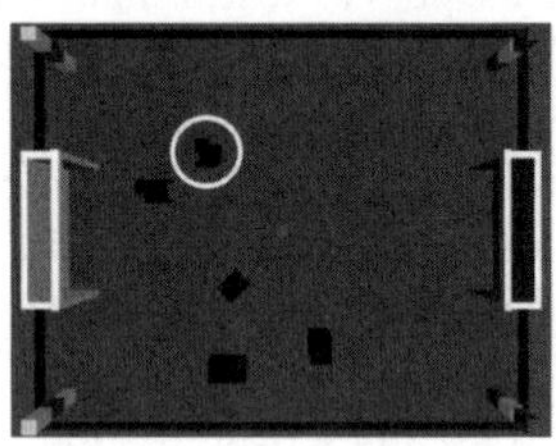

Fig. 8. Simulation environment

a) the method without intersection test
b) the method with intersection test and the light model of OpenGL
c) the method with intersection test and without the light model

The environment for verification is shown in Fig.11 where the omni directional simulator is marked, and the rendered image and the processing time are investigated. The results are shown in Fig.12. In this verification, the resolution of the image is 50 $\times$ 50 pixels and 100 $\times$ 100 pixels. These images show that these methods are able to simulate

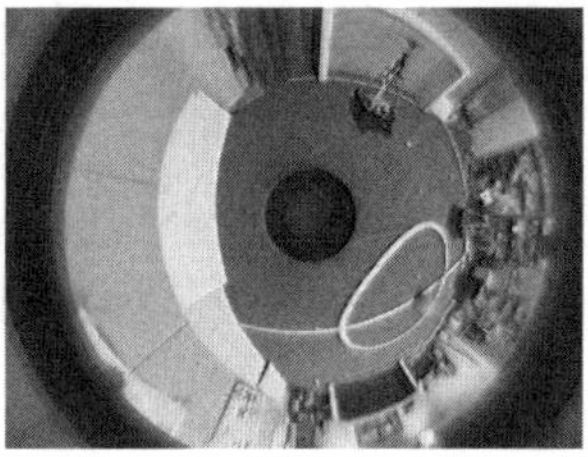

Fig. 9. Real image

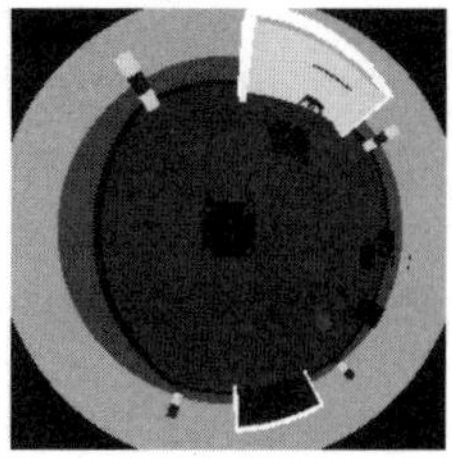

Fig. 10. Simulated image

the omnidirectional vision accurately, and the mirror reversed images are simulated correctly. In the methods a) and b), images with shade are rendered, and an image without shade is rendered in the method c). The processing time of each method is shown in Tables.1 and 2. The computer simulation environment is a laptop PC of which CPU is Pentium M 1.8GHz, memory is 1.0GB, and video chip is ATI MOBILITY RADEON (AGP 4X) 16MB. From the results shown in Tables.1 and 2, the processing time is reduced in the methods b) and c). This shows that the intersection test algorithm is effective. The processing speeds depend on the conditions of objects in the environment, but it is considered that the method c) is sufficiently fast for simulation. Figure 13 shows the scenes of the simulation of robot behavior using the methods c) (50 × 50 pixels) as

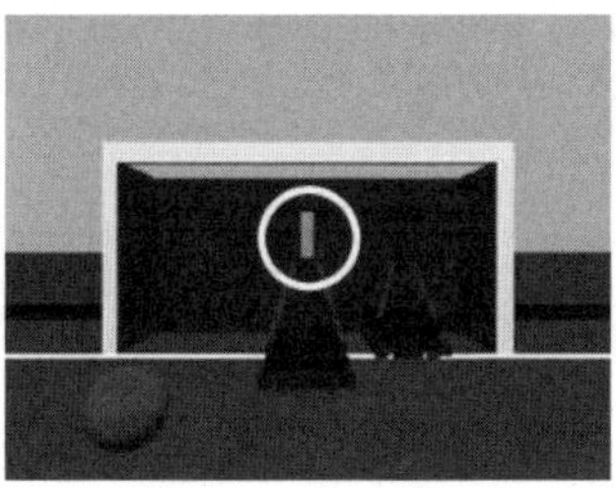

Fig. 11. Simulation environment

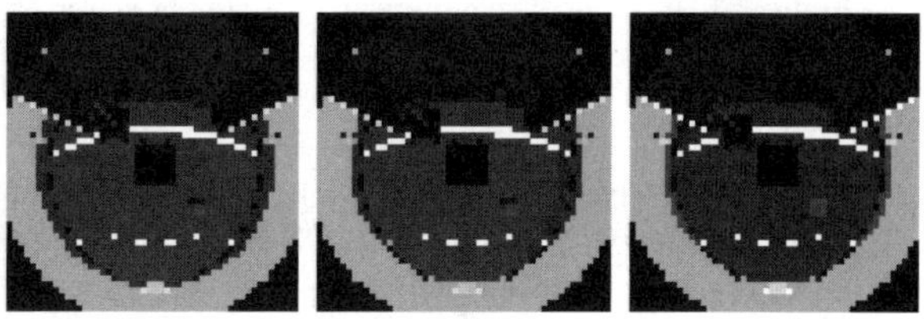

(a) Method a) (b) Method b) (c) Method c)

Fig. 12. Simulated images of the three models

Table 1. Average processing time (50x50)[sec]

method a)	method b)	method c)
12.9	0.300	0.0238

Table 2. Average processing time (100x100)[sec]

method a)	method b)	method c)
50.6	1.36	0.282

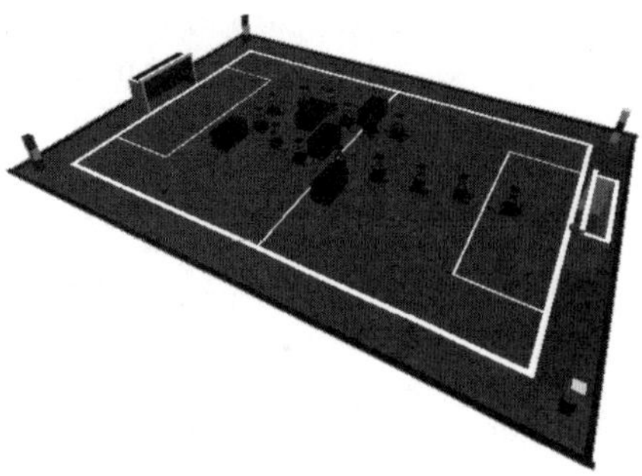

Fig. 13. Simulation result

sensor input. By considering the processing time and the resolution, the method c) with 50×50 pixels is suitable for a sensor for robots. If the accurate simulation with the lighting is required, the method b) is suitable.

6 Conclusion

The 3D dynamics simulation system with omnidirectional vision was developed. An intersection test algorithm was developed for the purpose of speeding up rendering the omnidirectioan vision image. The effectiveness of the omnidirectional vision simulator is verified through comparison of the rendered images with the images which are obtaind by real omnidirectional vision sensor.

Acknowledgement. This research work is supported in part by Grant in Aid for the 21st century center of excellence for "System Design: Paradigm Shift from Intelligence to Life" from Ministry of Education, Culture, Sport, and Technology in Japan.

References

1. Yagi, Y.: Omnidirectional Sensing and Its Applications. IEICE Trans. on Information and Systems E82D(3), 568–579 (1999)
2. Yagi, Y., Yokoya, N.: Omnidirectional Vision: A Survey on Sensors and Applications. Journal of Information Processing Society of Japan. vol. 42(SIG13 (CVIM3)), pp.1–18 (2001) (in Japanese)
3. Fujiwara, H., et al.: Development of a simulator for robotic soccer research. In: the 17th Annual Conference of the Robotics Society of Japan (RSJ1999) pp. 255–256 (1999) (in Japanese)
4. Takahashi, Y., et al.: Simultaneous Learning to Acquire Competitive Behaviors in Multi-Agent System based on a Modular Learning System. In: Proc. of IEEE/RSJ Int. Conf. on Intelligent Robots and Systems (IROS 2005), pp.153–159 (2005)
5. `http://www.cyberbotics.com/`
6. Michel, O.: Webots: Symbiosis between Virtual and Real Mobile Robots. In: Heudin, J.-C. (ed.) VW 1998. LNCS (LNAI), vol. 1434, pp. 254–263. Springer, Heidelberg (1998)
7. de Meneses, Y.L., Michel, O.: Vision Sensors on the Webots Simulator. In: Heudin, J.-C. (ed.) VW 1998. LNCS (LNAI), vol. 1434, pp. 264–273. Springer, Heidelberg (1998)
8. `http://playerstage.sourceforge.net/gazebo/gazebo.html`
9. Koenig, N., Howard, A.: Design and Use Paradigms for Gazebo, An Open-Source Multi-Robot Simulator. In: Proc. of IEEE/RSJ Int. Conf. on Intelligent Robots and Systems (IROS 2004), vol. 3, pp. 2149–2154 (2004)
10. Röfer, T.: An Architecture for a National RoboCup Team. In: Kaminka, G.A., Lima, P.U., Rojas, R. (eds.) RoboCup 2002. LNCS (LNAI), vol. 2752, pp. 417–425. Springer, Heidelberg (2003)
11. Asanuma, K.: Development of a Simulator of Environment and Measurement for Autonomous Mobile Robots Considering Camera Characteristics. In: Polani, D., Browning, B., Bonarini, A., Yoshida, K. (eds.) RoboCup 2003. LNCS (LNAI), vol. 3020, pp. 446–457. Springer, Heidelberg (2004)
12. Umeda, K., et al.: Development of a Simulator of Environment and Mesurement for Multiple sutonomous Mobile Robots Considering Camera Characteristics. Journal of the Robotics Society of Japan 23(7), 112–119 (2005) (in Japanese)

13. Zagal, J.C., Ruiz-del-Solar, J.: UCHILSIM: A Dynamically and Visually Realistic Simulator for the RoboCup Four Legged League. In: Nardi, D., Riedmiller, M., Sammut, C., Santos-Victor, J. (eds.) RoboCup 2004. LNCS (LNAI), vol. 3276, pp. 34–45. Springer, Heidelberg (2005)
14. http://ode.org/
15. http://www.opengl.org/
16. Yamazawa, K., et al.: Visual Navigation with Omnidirectional Image Sensor HyperOmni Vision. Journal of The Institute of Electronics, Information and Communication Engineers(D-II) J79-D-II(5), 698–707 (1996) (in Japanese)

Real-Time Randomized Motion Planning for Multiple Domains*

James Bruce and Manuela Veloso

Computer Science Department
Carnegie Mellon University
Pittsburgh, Pennsylvania, USA
{jbruce,mmv}@cs.cmu.edu

Abstract. Motion planning is a critical component for autonomous mobile robots, requiring a solution which is fast enough to serve as a building block, yet easy enough to extend that it can be adapted to new platforms without starting from scratch. This paper presents an algorithm based on randomized planning approaches, which uses a minimal interface between the platform and planner to aid in implementation reuse. Two domains to which the planner has been applied are described. The first is a 2D domain for small-size robot navigation, where the planner has been used successfully in various versions for five years. The second is a true 3D planner for autonomous fixed-wing aircraft with kinematic constraints. Despite large differences between these two platforms, the core planning code is shared across domains, and this flexibility comes with only a small efficiency penalty.

1 Introduction

Motion planning is problem central to autonomous robotics. As soon as a robot needs to move within a nontrivial environment, the question arises as to *how* it should move to satisfy the constraints posed by its environment. In its general form, the simplest motion planning problem is that of a single query [1]. That is, given some configuration space C that the robot operates within, find a free path from an initial position q_i to a final or goal position q_f. Obstacles pose constraints on valid configurations in C, thus we can subtract them from the full space, leaving the remaining "free" configuration space C_f in which the robot can move without hitting any obstacles. The path planning problem then becomes finding a continuous curve $p(s) \in C_f$ for $s \in [0, 1]$ where $p(0) = q_i$ and $p(1) = q_f$. For some robots, additional constraints are needed due to limitations of the robot itself. These could be kinematic limitations, such as a car-like robot's

* This work was supported by United States Department of the Interior under Grant No. NBCH-1040007, and by Rockwell Scientific Co., LLC under subcontract No. B4U528968 and prime contract No. W911W6-04-C-0058 with the US Army. The views and conclusions contained herein are those of the authors, and do not necessarily reflect the position or policy of the sponsoring institutions, and no official endorsement should be inferred.

G. Lakemeyer et al. (Eds.): RoboCup 2006, LNAI 4434, pp. 532–539, 2007.

steering limitations, or they could be dynamics constraints such as maximum acceleration. These are collectively known as kinodynamic constraints and take the form of additional constraints on $p(s)$.

While the problem of motion planning has been studied extensively, the general trend of research has been concerned with solving successively more difficult problems. While it is important to expand the problems solvable given generous time and computing resources, advances may not translate directly to improvements at the other end of the spectrum. This other end consists of relatively "easy" problems, but with much tighter time bounds and limited available computation. It is these latter problems which abound in mobile robotics, and with which this work is concerned. The vast majority of mobile robots exist on a 2D surface, while those that do move freely in 3D, such as unmanned aerial vehicles (UAVs) typically do not encounter dense obstacles (i.e. C_f occupies a large fraction of C).

In fairly static domains, two-stage "multi-shot" planners such as PRM [2, 3] work well. PRMs separate planning into a learning phase, which builds a finite graph model G of C_f, and the query phase, which maps the problem to graph search on G. In highly dynamic environments however, learned models quickly become obsolete. This encourages the use of "one-shot" planners which only concern themselves with solving a single query given no apriori model of C_f. Among the fastest one-shot planners are the *RRT* family of randomized planners [4, 5, 6]. RRT planners incrementally build a tree in C_f while they search for the solution to a planning problem. A typical RRT search is as follows. First, q_i is added as the root of a tree. Then we iterate the following: Pick a draw a random sample q_r from C, find the closest vertex v in the current tree, then grow the tree toward q_r using an extend operator. The end of the extension is added to the tree with v as its parent. The first few steps of such a tree are shown in Figure 1. As this process is iterated, the RRT grows to fill the free space, tending toward an even distribution. The major variables in the method are in how we draw the random samples and in how the extend operator works. Two adjustments turn the space-filling RRT into a planner. First, we can throw out any extension segment that would hit an obstacle, thus restricting the tree to C_f and guaranteeing that any path from node to node is a valid path in free space. Next, we can alter the random target distribution by picking the goal configuration some fraction of the time, thus biasing the tree to grow toward the goal in a more directed fashion. Once a node is added to the tree that is sufficiently close to the goal configuration, we can trace up the parent pointers in the tree to recreate the path from q_f to q_i, the reverse of which is a plan [4].

1.1 Approach

The ERRT planner developed in previous work [6] builds on RRT and offers a navigation approach for mobile robots using iterated replanning. Each control cycle, a new plan is developed, rather than waiting for an error condition to occur before replanning. This allows the planner to deal with both small and large errors in the same way, and thus is highly tolerant of position jumps and action

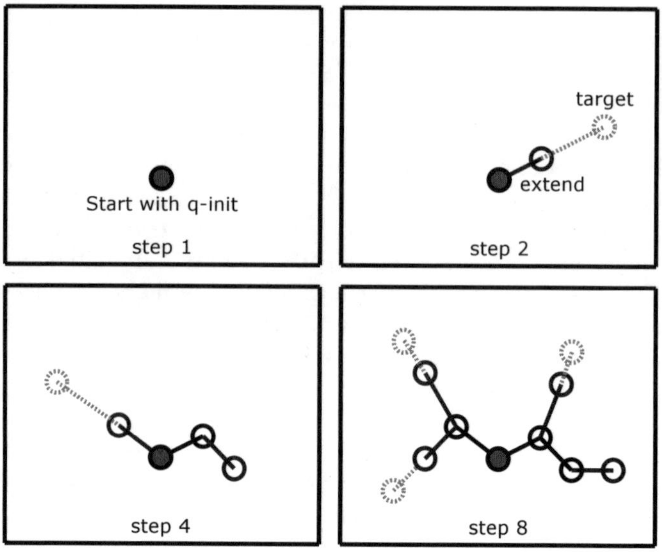

Fig. 1. Example growth of an RRT tree for several steps. At each iteration, a random target is chosen and the closest node in the tree is "extended" toward the target, adding another node to the tree.

error which is invariably present in physical robots. To speed up replanning, and decrease the variance from cycle to cycle in the plans, ERRT also introduces the concept of a waypoint cache. This is a fixed-size bin with random replacement into which states from previous plans are added. During RRT's target point selection, some part of the time waypoints are chosen instead of random configurations or q_f. This biases the planner to search along previously successful plans, both decreasing the running time of the planner and resulting in more stable solutions during successive iterative replans.

In our work, we have built upon ERRT to create a planner for mobile robots which span a wide range of parameters. The first platform are robots built for the RoboCup F180 "small size" league [7]. The field of play is a carpet measuring 4.9m by 3.8m, similar to that shown in Figure 2. Due to its competitive nature, teams have pushed robotic technology to its limits, with the small robots travelling over $2m/s$, accelerations between $3-6m/s^2$, and kicking the golf ball used in the game at up to $10m/s$. These speeds require every module to run in realtime to minimize latency, all while leaving enough computing resources for all the other modules to operate. Since five robots must be controlled at up to 60Hz, this leaves a realistic planning time budget of about 1ms for each robot. The robots themselves are holonomic, and controlled through a local obstacle avoidance mechanism which incorporates dynamics. Thus the planner is free to operate without kinodynamic constraints.

The second platform is an autonomous unmanned air vehicle (UAV), and in particular an autopilot designed for the small UAV shown in Figure 2. The

Fig. 2. Two teams are shown playing soccer in the RoboCup small size league (left), and the RnR RPV-3 unmanned air vehicle (UAV) (right)

UAV and thus the planner must operate in 3D at low to intermediate altitude, avoiding both the terrain and user specified "no-fly" zones as obstacles. The UAV has highly constrained kinodynamics; Despite is small size and $3.5m$ wingspan, the minimum turning radius is $300m$, while climb and descent rates are limited to $5m/s$. The autopilot can accept new sets of waypoints every few seconds, leading to a much less constrained timing schedule compared to the RoboCup robots. However due to the 3D nature of the problem and the constrained kinodynamics, the problem to be solved is much more difficult.

While widely varied, the two domains still offer similarities we may take advantage of. They are both mobile agents operating primarily in ordinary 2D or 3D space, and both can be conservatively but acceptable modelled by a bounding circle or sphere. The remainder of this paper describes the planner we have implemented based off of a generalized extension of the ERRT approach. The next section describes an abstract domain interface that allowed us to share core planning code across the two domains without sacrificing the planners' execution speed. The following section then describes our collision detection approach which takes advantage of bounding spheres to implement exact swept-volume collision checks with high efficiency, while keeping a straighforward implementation to add new types of obstacles.

2 The Domain Interface

The domain interface resulted from an attempt to unify platform interfaces so that common planning code could be developed. In traditional planning work, there are typically three primary modules: Planner, collision detection, and a platform model. While this approach works well for robots of relatively similar type, the communication required between the platform model and collision detection are through the planner, complicating the interface so that the planner needs to know much more about the domain than is really necessary. Thus the current approach was devised, in which platform model and collision detection are wrapped into a single "domain" module that the planner interacts with. Internally, collision detection and the platform model are implemented separately, but importantly the planner doesn't depend on anything involved in the communication between the two. This allows states in the configuration space to be a wholly opaque type to the planner (denoted by S), which needs only to be copied and operated on by the domain's functions. Additionally, to speed up nearest neighbor lookups, the state must provide bounds and accessors to its individual component dimensions. This allows the planner to build a K-D tree of states so that linear scans of the tree are not necessary for finding the nearest state to a randomly drawn target. While the traditional architecture can be made nearly as flexible, it typically does so at a cost in efficiency. The domain interface approach leaves open the opportunities for improved collision detection speed that can come with constraints or symmetry present in the agent model. The domain operations are as follows:

– *RandomWorldTarget():S* - Returns a state uniformly distributed in C
– *RandomGoalTarget():S* - Returns a random target from the set of goal states
– *Extend(s_0:S,s_1:S):S* - Returns a new state incrementally extending from s_0 toward s_1
– *Check(s:S):bool* - Returns true iff $s \in C_f$
– *Check(s_0:S,s_1:S):bool* - Returns true if swept-sphere from s_0 to s_1 is contained in C_f
– *Dist(s_0:S,s_1:S):real* - Returns distance between states s_0 and s_1
– *GoalDist(s:S):real* - Returns distance from s to the goal state set

Using these primitives, an RRT planner can be built which operates across multiple platforms, and does so without sacrificing runtime efficiency. One could say that it moves most of the important code into the domain itself, making the planner itself simplistic. However, the crucial difference is that the code in the domain is relatively straightforward and self contained, while the intricate interactions and practicality driven fallback cases of planning reside in the core planning code. Thus one could implement a domain with little or no knowledge of path planning, reaching a core goal of general modular programming.

3 Fast Collision Checking

Collision detection is a research area in its own right, and has been extensively studied (for a good survey see [8]). However for our planner we can achieve higher performance than general solutions by taking advantage of some simplifications present in our domains. First, since the mobile robots are bounded by circles or spheres, we need not check two complex shapes against one another to test for collision; We merely need to be able to test a sphere against the possibly complex environment. This results in our planner being pessimistic, but allows us one critical advantage: The ability to model continuous time trajectories in the collision check. Many implemented planners using general collision checkers represent time trajectories as fixed steps along the path. Unfortunately this creates an uneasy tradeoff between planning time and safety. We take the conservative approach of bounding the agent in a simple shape, but then use exact collision checking for time trajectories given that shape. The result is a planner that does not sacrifice safety in obtaining its fast execution times.

In our originally developed 2D implementation, checking a line-swept circle against various geometries proved easy enough to implement for various obstacles geometries, although adding a new type of obstacle proved quite tedious. Supporting 3D queries for the UAV would have resulted in a much more complex implementation to solve the trajectory-swept-sphere problem, so a different solution was sought. Our new approach required implementing only a single primitive for each obstacle: distance from the obstacle to a point in space. From this primitive, all the other required queries could be derived numerically. In particular, this included the swept-sphere query required for checking trajectories. The method works as shown in Figure 3. For any particular point, the current distance to obstacles, or clearance, determines how far along a trajectory is safe. The checker can then step forward by that distance, and recursively check the remaining swept area. The figure shows a dark blue trajectory to be checked, the light blue is the current clearance, and the red spheres show the steps that can be safely taken each iteration. Normally, few iterations are required, although the algorithm can take many steps if it is very close to an obstacle. This is handled by failing after a certain number of iterations have been exceeded. Though this is yet another pessimistic approximation, long paths running very close to obstacles are not typically desirable for execution by mobile robots anyway.

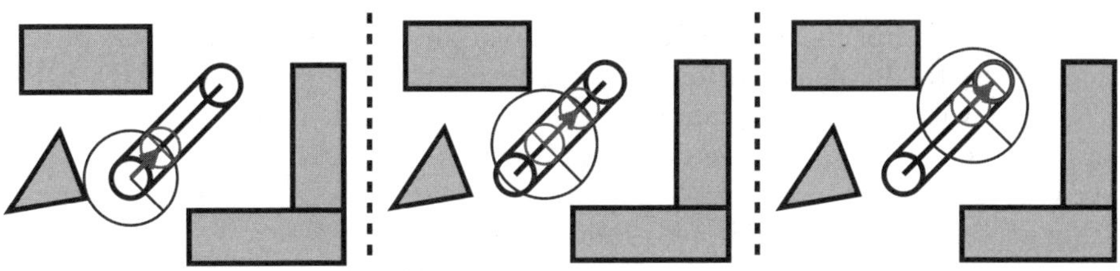

Fig. 3. An example of checking a swept circle using only obstacle distance queries to iteratively step forward along the swept path

While in our current implementation, only line-swept-spheres are supported, the distance query stepping method of checking a swept sphere can be applied to other trajectory functions as well. For some continuous trajectory function $x(t)$ where time $t \in [0, t_f]$, starting from an initial position $x(0)$ with a distance of at least $D(x(0)) = d$ from all obstacles, we need only find the first t such that $\|x(t) - x(0)\|^2 = d^2$, or verify that no such t exists for $t \in [0, t_f]$. For functions with bounded curvature, such as lines and circular arcs, these calculations are straightforward. In particular, for a linear trajectory defined by $x(t) = a + bt$, then the solution is $t_i = \sqrt{d^s - r^2}/\|b\|$. If $t_i > t_f$, then the trajectory is verified to be free, otherwise a recursive check must be made with a new trajectory starting at time t_i. Thanks to the square root, t_i increases quite rapidly from zero even with very small clearances, resulting in few steps needed for a typical check, and making the approach efficient in practice.

The obstacles implemented by our collision checker range from the obvious simple geometric shapes all the way up to 200x200 terrain meshes for the UAV planner. While the geometric shapes are straightforward to develop a distance metric for, the terrain posed the challenge of efficient distance calculation. Since the terrain is a 2D grid projected into 3D as a low-curvature mesh (i.e. a relatively flat manifold), we broke it up using a 2D K-D tree in grid space, and bounded the individual nodes with an axis-aligned bounding box in the 3D space. To query the distance from a point to the terrain, we follow the branches of the K-D tree nearest first, calculating a distance to the actual terrain once a leaf node is reached. After that, the remainder of the traversals and be compared against this known minimum and pruned if the bounding box is further than the current minimum. This aggressive pruning and the relatively flat nature of actual terrain meshes yields near logarithmic access times for the queries.

4 Results and Conclusion

In the RoboCup domain, we found the planner to work well in practice, helping our team consistently place within the top four teams, with comparatively few penalties for aggressive play. In testing, the planner in the RoboCup achieved execution times below 1ms to meet the tight timing requirement of the small-size system. It averaged just 0.5 ms per run, compared to an average of 0.9 ms for a baseline RRT implementation lacking a waypoint cache. In practice this means that the ERRT implementation can expand more nodes than plain RRT while remaining within the 1 ms planning envelope. Next, while reusing the same code and the same generic collision detection framework, a UAV planner was created by writing a new domain implementation. An example is shown in Figure 4, using actual data for the 12km x 12km area surrounding Reno, Nevada, USA. It has been tested driving a vendor-provided UAV simulator and a real hardware autopilot for an existing UAV. Depending on the problem difficulty, it runs from 0.5s to 2.0s per query on a modern computer. Eight ERRT searches are generated per query, and the shortest plan from a successful search is returned. Due to local minima in the distance metric resulting from kinematic constraints, running several independent runs generated more consistent results than running one large plan. In the RoboCup environment, path consistency is achieved by a high waypoint cache bias in ERRT [6].

Development of a unified planner for multiple mobile robot platforms provided many insights that would be difficult to determine if only one platform or similar platforms were considered for an implementation. However there are still many interesting areas of further work. First and foremost, the relationship between distance metric, kinodynamic constraints, and accelerated nearest-neighbor search

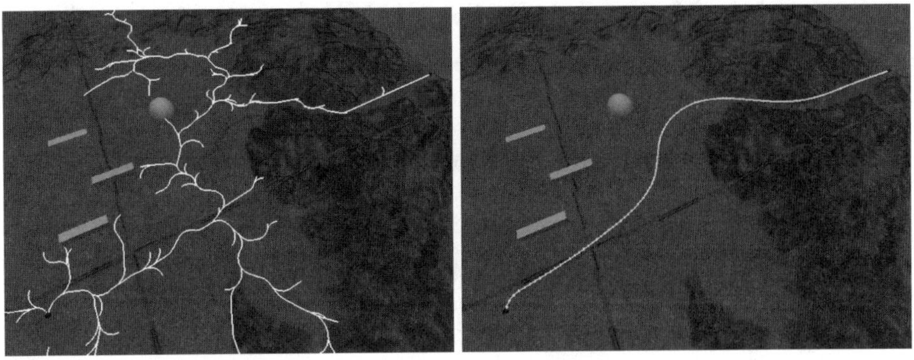

Fig. 4. A kinodynamicly-limited search tree (left), and the corresponding simplified plan (right) for the UAV. The plan length is approximately 13km with waypoints every 100m.

should be explored. Developed good distance metrics for kinematically constrained platforms is possible, but tedious, and more seriously it prevents most forms of accelerating nearest-neighbor search to fail because the triangle inequality is no longer satisfied. Using Euclidean distance worked, but generated local minima that could only be avoided by rerunning the planner several times, which is an inelegant solution. Better approximations which still allow the use of fast geometric data structures most likely exist.

References

1. Latombe, J.-C.: Robot Motion Planning. Kluwer, Dordrecht (1991)
2. Kavraki, L.E., Svestka, P., Latombe, J.-C., Overmars, M.H.: Probabalistic roadmaps for path planning in high-dimensional configuration spaces. IEEE Transactions on Robotics and Automation 12, 566–580 (1996)
3. Kavraki, L.E., Latombe, J.-C.: Randomized preprocessing of configuration space for fast path planning. In: Proceedings of the IEEE International Conference on Robotics and Automation, pp. 2138–2145. IEEE Computer Society Press, Los Alamitos (1994)
4. LaValle, S.M.: Rapidly-exploring random trees: A new tool for path planning. In Technical Report No. 98-11 (October 1998)
5. James, J., Kuffner, J., LaValle, S.M.: Rrt-connect: An efficient approach to single-query path planning. In: Proceedings of the IEEE International Conference on Robotics and Automation, IEEE Computer Society Press, Los Alamitos (2000)
6. Bruce, J., Veloso, M.: Real-time randomized path planning for robot navigation. In: Proceedings of the IEEE Conference on Intelligent Robots and Systems (IROS) (2002)
7. Kitano, H., Asada, M., Kuniyoshi, Y., Noda, I., Osawa, E.: Robocup: The robot world cup initiative. In: Proceedings of the IJCAI-95 Workshop on Entertainment and AI/ALife (1995)
8. Lin, M., Gottschalk, S.: Collision detection between geometric models: A survey. In: Proc. of IMA Conference on Mathematics of Surfaces (1998)

Towards a Methodology for Stabilizing the Gaze of a Quadrupedal Robot

Marek Marcinkiewicz, Mikhail Kunin, Simon Parsons, Elizabeth Sklar, and Theodore Raphan

Department of Computer & Information Science
Brooklyn College, City University of New York
2900 Bedford Avenue, Brooklyn NY 11210 USA
marek@sci.brooklyn.cuny.edu

Abstract. When a quadrupedal robot moves, the body and head pitch, yaw and roll, because of its stepping. This natural effect of body and head motion adversely effects the use of visual sensors embedded in the robot's head. Any object in the visual frame of the robot will, from the perspective of the robot, be subject to considerable unmodeled motion or slip. This problem does not affect mammals, which have vestibulo-collic and vestibulo-ocular reflexes that stabilize their gaze in space and maintain objects of interest approximately fixed on the retina. Our work is aimed towards constructing an artificial vestibular system for quadrupedal robots to maintain accurate gaze. This paper describes the first part of this work, wherein we have mounted an artificial vestibular system in a Sony AIBO robot.

1 Introduction

Robot locomotion has been studied using a wide range of wheeled and legged robots [3]. Although wheeled robots move quickly, they can only move on smooth terrain and lack the versatility of legged robots in handling rough terrain. As a result, there has been a concerted effort within the robot community to understand the motion of legged robots [11]. This is particularly true within the RoboCup community, where there has been considerable work on both the Sony AIBO used by the legged league [11,12] with a long-term aim of developing sophisticated humanoid soccer players and a humanoid league [4].

When playing soccer, using AIBO robots, one aim is to maintain specific landmarks and the ball within the visual range of a camera [17], which is mounted in the robot's head. A number of mechanisms that move the head with the aim of keeping the relevant object in the visual frame have been implemented [6] by obtaining feedback from motor sensors. These implementations are not accurate representations of head position in space, resulting in considerable errors during tasks that make use of vision information, such as self-localization [18], navigation [7] and identification of the ball [21]. This, in turn, leads to non-optimal trajectories in adjusting robot motion towards the ball and in team coordination activities such as passing the ball.

Additional problems arise during actual motion of the robot. The head, in which the camera is mounted, will pitch, yaw, and roll as well as linearly accelerate because of

G. Lakemeyer et al. (Eds.): RoboCup 2006, LNAI 4434, pp. 540–547, 2007.

that motion. The very fact that legged motion generates this kind of disturbance makes it difficult to keep the visual frame stable.

One approach to dealing with this motion of the camera is to accept the motion and use a Kalman filter to track objects in the visual frame [6]. Another approach is to move the head to compensate for the unwanted motion, guided not by the direct feedback from the motor sensors, but instead from a learned response to the motor sensors which indicates what the motion really is. Such an approach could be based either on the model-based method introduced in [19], or on the neural-network method of [13].

In our work, we take a different approach, hypothesizing that the estimation of landmarks and ball position could be significantly improved if we had *a priori* knowledge of the statistics and spectral content of head rotations and linear accelerations when the robot executes particular tasks. As a result, we set out to measure these accurately. In mammals, the vestibular system compensates and orients the head and body as it moves through space. The purpose of this paper is to describe an enhancement that we have made to an AIBO robot using an an artificial vestibular system that allows us to estimate these statistics and spectra during particular motions that are presently used to identify landmarks and the ball. To this end, we have augmented the AIBO with a system that mimics the inertial sensing mechanisms of the vestibular system of mammals — see [15] for review. In this paper, we show the considerable angular head perturbations that exist during robot motions. We derive the signals in terms of the Euler angles of head rotation so that these signals can be utilized in estimating the landmarks and ball in space, relative to the camera during the various head maneuvers and locomotion. Rotational head compensations for linear perturbations are more complicated [16] and beyond the scope of this paper.

2 Related Work

A number of studies have utilized robot enhancement using similar principles associated with the vestibular system [10,13]. One approach attached gyrosensors to a walking robot to reduce the shaking effect on the camera caused by the walk [10]. This was done by using a high resolution camera and cutting out a subimage. The subimage frame moved according to the rotations measured by the gyrosensors. Despite this enhancement, the sensors alone did not provide satisfactory image clarity and additional template matching was utilized to refine the image. Our work to define the statistical and spectral content of the head perturbation in space should help to define the corrections needed to reduce the errors in the visual processing of the landmark and ball images.

3 Robot Modification for Studying Quadrupedal Locomotion

For our investigation, we used a Sony AIBO ERS-210 robot. The ERS-210 is a quadrupedal robot that has has three perpendicular degrees of freedom (DOF) in the head. The three degrees of freedom makes it possible to study the 3D head motion in space.

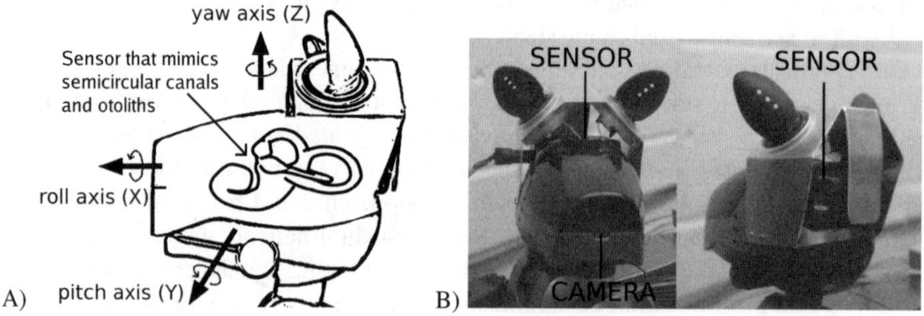

Fig. 1. A) Diagram of the head of an AIBO ERS-210 with embedded sensor mimicking the semicircular canals and otoliths. The heavy arrows show the roll (X), pitch (Y), and yaw (Z) axes of the head. The circular arrows indicate the positive direction of rotation. B) Modified AIBO ERS-210 that was used for these experiments, showing the sensor firmly mounted in the back of the head.

3.1 Sensor for Measuring Head Perturbations

The AIBO has a linear acceleration sensor built into its body that can be used to obtain some useful data about its motion [20]. However, there is no accurate information about how the head moves. To establish information about the head, we used an Xsens MTX sensor[1]. This sensor is factory calibrated and can detect 3D linear acceleration ($\pm 17m/s^2$) and 3D rotational velocity ($\pm 1200deg/s$). This mimicks the peripheral vestibular system in live animals, which contains the semicircular canals and otoliths, structures that are embedded in the inner ear and sense angular and linear acceleration, respectively — see [15] for a review of work on the vestibular systems of humans and monkeys. The system of canals and otoliths is important for compensating and limiting head perturbations via the vestibulo-collic reflex as well as maintaining the stability of the visual world via the vestibulo-ocular reflex [15].

The sensor is $38 \times 53 \times 21mm$ and weights $30g$. Communication with the sensor is over an RS-232 interface to an off-board computer at a rate of $100Hz$ using a baudrate of $57.6kbits/s$. We have implemented an interface in Matlab 7 SP3[2] that runs on a Linux-based PC and can access signals from the sensor via Matlab. We have also implemented the matrix transformations that convert the rotary signals from the sensor into the Euler angular perturbations corresponding to the AIBO's head motor axes. The sensor was firmly embedded in the head of the AIBO at the approximate positions at which the peripheral vestibular system is located in the head of humans and other animals. This location is close to the origin of the axes of rotation of the head[3] (see Figure 1). As a result, the weight and displacement did not significantly alter the head's moment of inertia or its dynamic properties.

---

[1] www.xsens.com

[2] www.mathworks.com

[3] Note that the ears of the AIBO are still attached despite the modification to the head of the robot. This was necessary since if the ears are removed, the AIBO will not boot up.

4 Relationship Between Robot Head Motors and the Sensor

The AIBO has two coordinate frames: the body coordinate frame (BCF) and the head coordinate frame (HCF). The sensor is attached to the head so that its 3 axes correspond to the 3 axes of the HCF. The motors rotate the head with Euler angles defining a Helmholtz gimbal, i.e., the pitch axis fixed relative to the body [8]. The first motor performs a *pitch* of angle θ at the neck and moves the head, which contains the other two motors. The second motor performs a *yaw* of angle ϕ and moves the part of the head that contains the remaining motor. The last motor performs a *roll* of angle ψ and moves the final part of the head containing the sensor. Each rotation can be represented as a rotation around an axis:

$$roll = R_x(\psi) \qquad pitch = R_y(\theta) \qquad yaw = R_z(\phi)$$

The rotations may be combined in the proper order to create one pitch-yaw-roll rotation $R_{PYR} = R_x R_z R_y$ that defines the transformation from the BCF to the HCF.

4.1 Sensor Reading

The sensor provides the 3D rotation velocity in space in terms of the head coordinate frame. Let us call this velocity ω_s. To convert this to Euler angles we need to determine the rotation of the head, add the incremental rotation caused by ω_s at time t, and determine what Euler angles would be the equivalent of such a rotation. Let $P_{cur} = [\psi_{cur}, \theta_{cur}, \phi_{cur}]$, equal to the original motor position. The original rotation matrix R_{cur} is obtained by inserting P_{cur}into R_{PYR} [8].

Any rotation in space may be represented by a single axis and a rotation angle. To obtain the axis of incremental rotation $\hat{n}_{inc}$, we need to normalize the velocity vector ω_s. To obtain the angle of incremental rotation Φ_{inc}, we multiply the magnitude of this vector by Δt. In our case, Δt is the amount of time between sensor readings or $.01s$.

$$\hat{n}_{inc} = \frac{\omega_s}{||\omega_s||} \qquad \Phi_{inc} = ||\omega_s||\Delta t$$

Now to obtain the new rotation, we simply apply the incremental rotation R_{inc} $(\hat{n}_{inc}, \Phi_{inc})$ to the current rotation R_{cur} to generate R_{new}:

$$R_{new} = R_{inc} R_{cur}$$

From this matrix and the definition of the R_{PYR} matrix, we can extract the Euler angles of the new position P_{new}:

$$\psi_{new} = \tan^{-1}\frac{r_{32}}{r_{22}} \qquad \theta_{new} = \tan^{-1}\frac{r_{13}}{r_{11}} \qquad \phi_{new} = -\sin^{-1} r_{12}$$

The change of Euler angles is then $P_{new} - P_{cur}$. (See [14] for a complete derivation of these results.)

5 Experimental Results and Data Analysis

We measured the positions as output by the external sensor and the internal motor sensors while the robot is walking. To generate the motion we utilized the motion module of the Carnegie Mellon University (CMU) team CMPack'04 from the 2004 RoboCup competition [5][4]. The robot gait utilized was the standard trot gait at the maximum forward velocity of $240mm/s$.

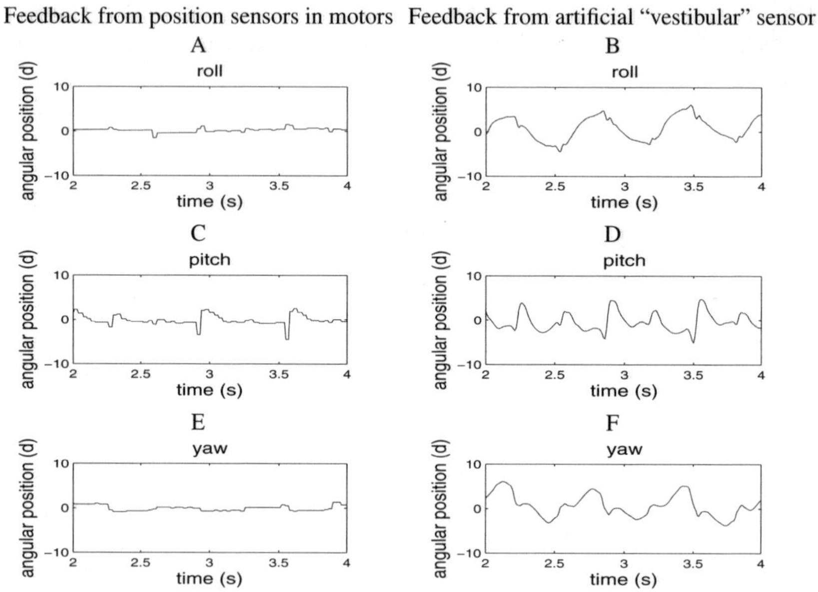

Fig. 2. Comparison of pitch, yaw, roll of the motor position sensors (A, C, E) and the artificial "vestibular" sensor (B, D, F) during walking

The test was run over a period of 8 seconds. The data were filtered by removing linear trends in position to remove the drift appearing in the external sensor readings. The resulting graphs are shown in Figure 2. The position information was then run through the Welch function available in Matlab to obtain power spectrums. The resulting graphs are shown in Figure 4. The averaged cycles were determined and are shown in Figure 3.

During walking at 240 mm/sec with a period of 640ms, the feedback from the external sensor shows considerable motion of the head in all three axis. The roll component of the head, which is approximately 6° peak to peak as reported by the vestibular sensor (Fig. 2 B) appears as less than 1° as reported by the motor sensors(Fig. 2 A). Similar strong discrepancies in rotation angles of the motor and vestibular sensor were found for pitch and yaw (Compare Fig. 2 C, E to Fig. 2 D, F).

We next considered how to best utilize the information in order to make corrections for the head movement. To accomplish this, we determined the average roll, pitch

[4] This gait is a variation of the trot gait used by most RoboCup legged league teams.

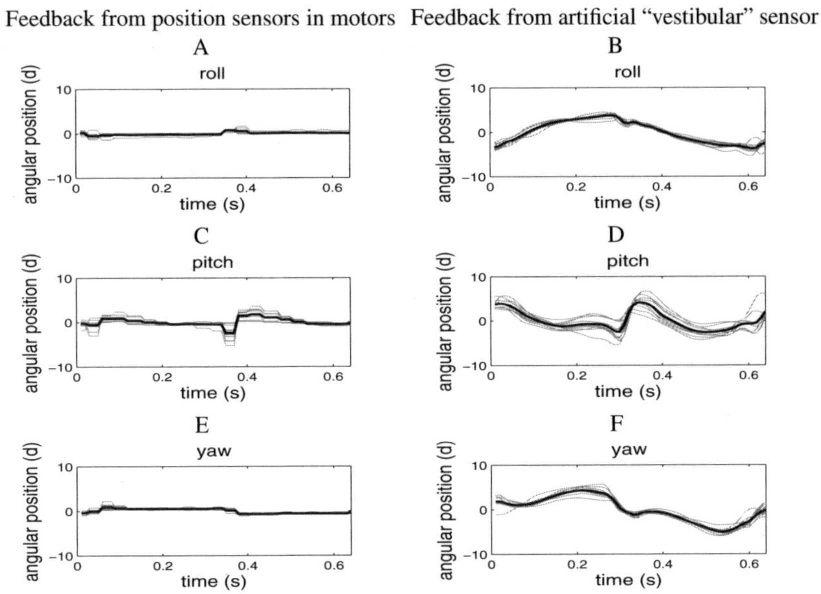

Fig. 3. Comparison of the average cycle of pitch, yaw, roll of the motor position sensors (A, C, E) and the artificial "vestibular" sensor (B, D, F) during walking

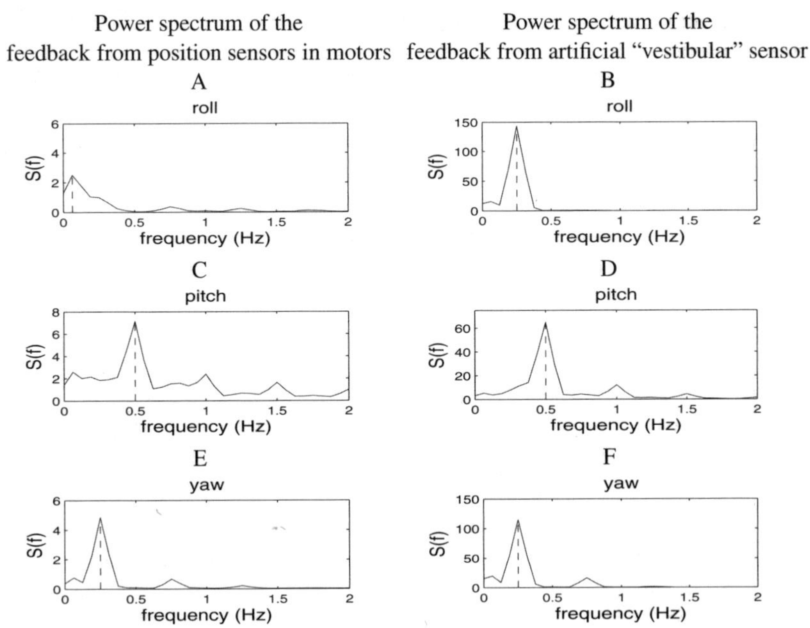

Fig. 4. Comparison of the pitch, yaw, roll power spectral density of the motor position sensors (A, C, E) and the artificial "vestibular" sensor (B, D, F) during walking. The peak frequency is .0625 for A, .25 for B, E, F and .5 for C, D.

and yaw waveform during locomotion. The vestibular sensor determined clear periodic oscillatory patterns with small standard deviations around the mean (Fig. 3 B, D, F) whereas the motor sensors reported a negligible oscillation in all components of head movement.

The spectra of the comparable signals from motor and vestibular sensors were also significantly different (Fig. 4). The peak powers as reported by the external vestibular sensor were several orders of magnitude higher. The spectral content is also much narrower and more concentrated around the dominant harmonics in the vestibular sensor output as compared to the motor sensors. These are important parameters for determining the control that would be needed for optimizing compensatory head movements for maintaining head stability.

6 Conclusions

The results of this study indicate that all components of the head movements of an AIBO robot are periodic during locomotion. The spectra are fairly narrow and the average waveforms have small standard deviation over the period of movement. This indicates that the waveforms, as reported by the vestibular sensor, could form a basis for making corrections to images in camera coordinates and provide a stable platform for identifying objects of interest, including the ball and other robots.

We are currently working on using the readings to stabilize the visual frame of the AIBO. Initially we plan to add sensor feedback into existing approaches to gait development — for example [9] — delivering new gaits that exhibit better head stability. Subsequently, we aim to have the AIBO respond in real-time, adjusting the head motors in response to detected head motion. Eventually, we plan to extend this work to humanoid robots. To our knowledge, there is currently no research on using vestibular feedback in gait development — despite the use of gyroscopic sensors [2] that provide similar information — nor is there any work on having robots dynamically adjust their gait to help stabilize head movement despite much work on gaits [1].

While RoboCup rules prevent us from using the modified AIBO in competition, we anticipate using the "head-stable" gaits we develop in future RoboCup events.

Acknowledgements. This work was supported by NIH grant DC 05222 from the NIDCD (TR), NSF CNS-0520989 (TR and SP) and CUNY Collaborative Grant 80209-09-12 (TR and SP).

References

1. Asada, M., Katoh, Y., Ogino, M., Hosoda, K.: A humanoid approaches the goal — reinforcement learning based on rhythmic walking parameters. In: Polani, D., Browning, B., Bonarini, A., Yoshida, K. (eds.) RoboCup 2003. LNCS (LNAI), vol. 3020, pp. 344–354. Springer, Heidelberg (2004)
2. Baltes, J., McGrath, S., Anderson, J.: The use of gyroscope feedback in the control of the walking gaits for a small humanoid robot. In: Nardi, D., Riedmiller, M., Sammut, C., Santos-Victor, J. (eds.) RoboCup 2004. LNCS (LNAI), vol. 3276, pp. 628–635. Springer, Heidelberg (2005)

3. Bekey, G.: Autonomous Robots. MIT Press, Cambridge, MA (2005)
4. Christaller, T.: Lessons learned from Fukuoka 2002 humanoid league. In: Kaminka, G.A., Lima, P.U., Rojas, R. (eds.) RoboCup 2002. LNCS (LNAI), vol. 2752, pp. 485–488. Springer, Heidelberg (2003)
5. (accessed February 2, 2006), `http://www.cs.cmu.edu/~robosoccer/legged/`
6. del Solar, J.R., Vallejis, P.A.: Motion detection and tracking for an AIBO robot using camera motion compensation and Kalman filtering. In: Nardi, D., Riedmiller, M., Sammut, C., Santos-Victor, J. (eds.) RoboCup 2004. LNCS (LNAI), vol. 3276, pp. 619–627. Springer, Heidelberg (2005)
7. Fukase, T., Yokoi, M., Kobayashi, Y., Ueda, R., Yuasa, H., Arai, T.: Quadruped robot navigation considering the observational cost. In: Birk, A., Coradeschi, S., Tadokoro, S. (eds.) RoboCup 2001. LNCS (LNAI), vol. 2377, pp. 350–355. Springer, Heidelberg (2002)
8. Goldstein, H.: Classical Mechanics. Addison-Wesley, Reading, MA (1980)
9. Kohl, N., Stone, P.: Machine learning for fast quadrupedal locomotion. In: Proceedings of the 19th National Conference on Artificial Intelligence, San Jose, CA (July 2004)
10. Kurazume, R., Hirose, S.: Development of image stabilization system for remote operation of walking robots. In: Proceedings of the IEEE International Conference on Robotics and Automation, vol. 2 (2000)
11. (accessed February 2, 2006), `http://www.tzi.de/4legged/`
12. Lima, P., Custódio, L., Akin, L., Jacoff, A., Kraetzschmar, G., Kiat, N.B., Obst, O., Röfer, T., Takahashi, Y., Zhou, C.: Robocup 2004 competitions and symposium: A small kick for robots, a giant score for science. AI Magazine 26(2), 36–61 Summer (2005)
13. Panerai, F., Metta, G., Sandini, G.: Learning visual stabilization reflexes in robots with moving eyes. Journal of Neurocomputing 48, 323–337 (2002)
14. Raphan, T.: Modeling control of eye orientation in three dimensions. I. Role of muscle pulleys in determining saccadic trajectory. Journal of Neurophysiology 79, 2653–2667 (1998)
15. Raphan, T., Cohen, B.: The vestibulo-ocular reflex in three dimensions. Experimental Brain Research 145, 1–27 (2002)
16. Raphan, T., Imai, T., Moore, S.T., Cohen, B.: Vestibular compensation and orientation during locomotion. Annals of the New York Academy of Sciences 942, 128–138 (2001)
17. Schmitt, T., Hanel, R., Buck, S., Beetz, M.: Probabilistic vision-based opponent tracking in robot soccer. In: Kaminka, G.A., Lima, P.U., Rojas, R. (eds.) RoboCup 2002. LNCS (LNAI), vol. 2752, pp. 426–434. Springer, Heidelberg (2003)
18. Sridharan, M., Kuhlmann, G., Stone, P.: Practical vision-based Monte-Carlo localization on a legged robot. In: Proceedings of the IEEE International Conference on Robotics and Automation, Barcelona (April 2005)
19. Stronger, D., Stone, P.: A model-based approach to robot joint control. In: Nardi, D., Riedmiller, M., Sammut, C., Santos-Victor, J. (eds.) RoboCup 2004. LNCS (LNAI), vol. 3276, pp. 297–309. Springer, Heidelberg (2005)
20. Vail, D., Veloso, M.: Learning from accelerometer data on a legged robot. In: Proceedings of the 5th IFAC Symposium on Intelligent Autonomous Vehicles, Lisbon, Portugal (2004)
21. Zagal, J.C., del Solar, J.R., Guerrero, P., Palma, R.: Evolving visual object recognition for legged robots. In: Polani, D., Browning, B., Bonarini, A., Yoshida, K. (eds.) RoboCup 2003. LNCS (LNAI), vol. 3020, pp. 181–191. Springer, Heidelberg (2004)

Automatic Acquisition of Robot Motion and Sensor Models

A. Tuna Ozgelen, Elizabeth Sklar, and Simon Parsons

Department of Computer & Information Science
Brooklyn College, City University of New York
2900 Bedford Avenue, Brooklyn NY 11210 USA
{tuna,sklar,parsons}@sci.brooklyn.cuny.edu

Abstract. For accurate self-localization using probabilistic techniques, robots require robust models of motion and sensor characteristics. Such models are sensitive to variations in lighting conditions, terrain and other factors like robot battery strength. Each of these factors can introduce variations in the level of noise considered by probabilistic techniques. Manually constructing models of noise is time-consuming, tedious and error-prone. We have been developing techniques for automatically acquiring such models, using the AIBO robot and a modified RoboCup Four-Legged League field with an overhead camera. This paper describes our techniques and presents preliminary results.

1 Introduction

Robots in RoboCup have two main requirements in order to play effective soccer. They have to be able to self-localize with reasonable accuracy [6], and they have to be able to detect and track the ball [13][1]. The current state-of-the-art in localization is to use Bayesian filter models [22, chap. 3–4], and a particularly popular approach is the particle filter [23]. This is especially popular in RoboCup because it allows robots to track multiple position hypotheses, helpful when robots are regularly kidnapped by referees, while running on modest computational hardware. To apply any Bayesian filter model, a robot requires a model of its own motion, which it uses to predict new poses from old ones following motion, and a model of its sensor behavior, which the robot uses to choose between multiple possible poses. The sensor model is clearly also important for detecting and tracking the ball.

Now, it is clear that the sensor and motion models are of importance to obtaining effective behavior from any robot, but they are especially important in vision-based soccer-playing robots. As a number of authors have pointed out, for example [6,15], vision-based robots have much less sensor data to work with than robots equipped with sonar or laser range-finders (at least when the vision is based on landmark detection as it so often is in RoboCup). This comparative paucity of sensor data argues for the importance of making each datum as accurate as possible (though it should be noted

[1] Successful soccer-playing robots clearly need to be able to do a lot of other things as well, but these other things — effective moving of the ball, tactical positioning, and coordinated team play, for example — have good self-localization and ball-detection as pre-requisites.

G. Lakemeyer et al. (Eds.): RoboCup 2006, LNAI 4434, pp. 548–555, 2007.

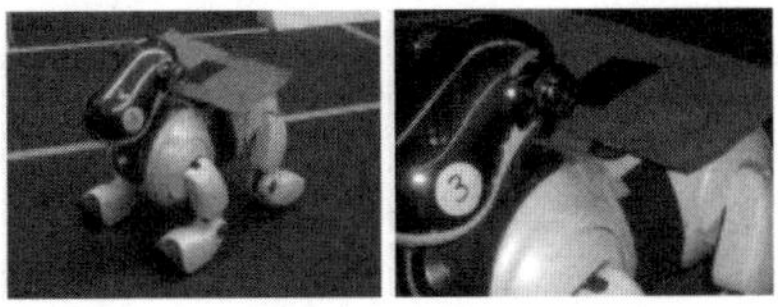

Fig. 1. An AIBO with a color marker

that if sensor data is too accurate, the performance of the particle filter degrades slightly [23]). The paucity of sensor data also argues for making the motion model as accurate as possible — with infrequent sightings of landmarks, robots have to run for several seconds at a time without sensor data [15], and during that time can only update their notion of where they are using motion data. Furthermore, when tracking the ball, the robot may not see a landmark for considerably longer, and so will have to rely on what is effectively dead-reckoning from its last confirmed position.

This requirement on the vision sensor model holds not only for models of the kind that we deal with here, which use information about distance and bearing to landmarks, but also for models that deal only with bearing [12] (and recent work [15] shows that distance information helps to improve the precision of localization provided that the distance information is adequately calibrated).

In this paper, we are concerned with the Sony AIBO ERS-7, the robot used by our Legged League team MetroBots[2]. To construct both motion and sensor models for the AIBO we are usually reduced to taking measurements "by hand and tape measure" [18]—running the robot for a given time and measuring how far it moved, or having the robot estimate how far it is from a landmark and comparing that with the measured distance. This gives relatively few measurements from which to construct and evaluate models, and the work described here is a response to that situation.

In this paper we describe how we have been using a global vision system, a system which uses an overhead camera, and from that image data determines the position of the robot, to automatically acquire motion and sensor data. This approach allows us to collect data sufficiently easily and rapidly — several hundred data points in an hour[3] — that we can use data-intensive machine learning techniques to construct models of motion and sensor error.

2 Experimental Setup

For our experimental work, we have adapted a modified setup derived from the RoboCup E-League [1]. The E-League makes use of a simplified small-size league environment, where global vision data is provided by a common vision server. This data is sent to both teams using UDP broadcast. Teams decide how to move their robots, and

[2] `http://agents.sci.brooklyn.cuny.edu/metrobots`

[3] A limit set, effectively, by the fact that at the moment we have to have the robot write image data to its memory stick, which takes several seconds, and then upload the image by ftp and that we use just a single robot. A group of several robots could collect data faster, as suggested in [11].

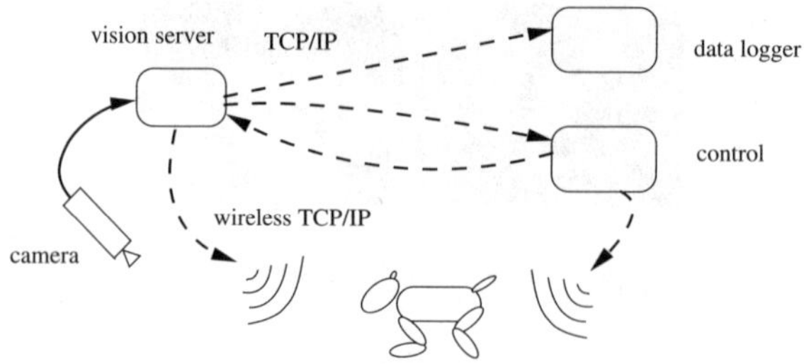

Fig. 2. The experimental setup

package instructions for the robots into a common format. These instructions are then combined into messages by the communication server, and broadcast to the robots via an infra-red transmitter. Each robot on each team unpacks the messages to find out what to do next.

At the heart of our setup is the Mezzanine visual tracking package [9]. Suitably calibrated, this software provides 2D tracking of objects — establishing x, y coordinates and orientation — provided that the objects are color coded for easy recognition from above. Mezzanine provides accurate tracking even with very unsophisticated camera hardware and can handle considerable image distortion. We currently use an XCam2 WideEye from X10, an inexpensive wide-angle security camera[4]. The original vision tracking system used by the E-League was Doraemon [2], which provides robust position estimates even when the camera is mounted at an angle rather than directly overhead. We are using Mezzanine because it more accurately handles the type of fish-eye images obtained from the wide-angle camera that is needed in order to get the whole soccer pitch in a single field-of-view.

As mentioned above, instead of the type of small, wheeled robots that have typically been used in the E-League, we have been working with Sony AIBO ERS-7 robots. To make them visible to Mezzanine, we simply attach a color marker to the back of the robot as in Figure 1. Since the AIBO is equipped with a wireless ethernet card, we can send data between the robot and the computer that is running the control code and the data logger (both are the same machine, though logically distinct), and we can send the position data from Mezzanine directly to the robot as well. The setup is as in Figure 2.

The idea of the experimental setup is to provide a completely automated mechanism for data-collection. The control module polls Mezzanine for location data and simultaneously sends instructions to the AIBO telling it how to move around the pitch, and when to gather data from its internal camera. When the robot is moving, we can continuously collect data about its position, and collate this position data with the motion commands sent to the robot. As we discuss below, this data can be used, amongst other things, to learn a motion model for the robot.

In addition to collecting this motion data, we can collect sensor data from the robot. Of particular interest, given the fact that the data used by the robot for self-localization is visual data, is the collection of camera images. Currently we do this by causing the

[4] `www.x10.com`

robot to pause—thus allowing us to get an accurate idea of where each picture was taken without having to synchronise the clocks on the robot and the machine running Mezzanine—and then take a picture (which takes a few seconds to write to the robot's memory stick) and then upload the picture to the data logger.

3 Results

We used the setup described in the previous section to construct models for the robot's standard trot gait and the error in its perception of the Legged League markers. The robot gait is that from the motion module of the Carnegie Mellon University Legged League team CMPack'04 from the 2004 RoboCup competition [4].

3.1 Motion Model

Data for the motion model was collected by making the AIBO walk forwards and backwards for 10 seconds at a time, while Mezzanine measured the coordinates of the robot at one second intervals. From these measurements, we computed the velocity of the robot over the relevant period in the three coordinate directions of the global frame of

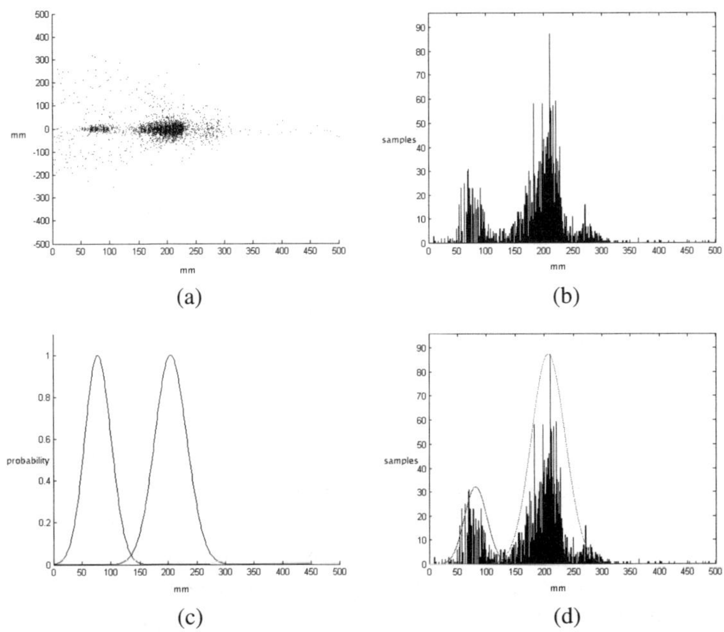

Fig. 3. The motion model for walking forward: (a) scatter plot of rates of motion in the x and y directions (x on the horizontal axis, y on the vertical) when walking forward; (b) histogram of motion in the x direction when walking forward; (c) Gaussian mixture fitted to the forward motion data; and (d) Gaussian mixture scaled and plotted with histogram of forward data

reference used by Mezzanine[5]. Since the robot takes time to accelerate and decelerate, we effectively had two sets of data—measurements for the robot moving continuously, and measurements for the robot when it was speeding up or slowing down.

For both forwards and backwards motion, we then plotted a histogram of around 3600 velocity measurements, obtaining two-peaked distributions — the lower valued peak corresponding to times when the robot was changing veocity, and the larger peak corresponding to constant velocity motion — that were approximately Gaussian. We then learnt the parameters of a two-Gaussian mixture that fitted the data. This learning was carried out using the standard EM algorithm [5]. A sample of this procedure for the x component of forward motion (that is the component in the direction of motion) is provided in Figure 3. Looking at Figure 3 (a) the two sets of measurements are clear, and these emerge as two distinct peaks in the histogram in Figure 3 (b) and (d). As Figure 3 (c) and (d) show, the two-Gaussian mixture closely fits the data.

The two forward motion distributions have means of 77 and 204, and standard deviations of 22 and 28 respectively, while the two backward motion distributions have means of 87 and 174, and standard deviations of 27 and 23, respectively.

3.2 Sensor Model

Our second use of the experimental setup was to measure the error in the robot's estimates of its distance from the Legged League beacons. To do this, we first used the experimental setup to have the robot move around the pitch taking pictures, and recording the robot's position when these pictures were taken[6]. We used these images to build a color map and to calibrate a distance coefficient, based on the number of pixels counted for each beacon shape and the robot's distance measured from the beacon by hand. We then used the experimental setup to have the robot take a much larger set of images, again recording the position at which each picture was taken. For each of this second set of images we had the perception system of the robot calculate the distance to the beacon, and we compared this with the real distance as measured by the global vision—the difference is then the error in the local vision system.

Given this error data, we then carried out exactly the same kind of learning as in the previous section, and the steps in this process are as depicted in Figure 4.

3.3 Discussion

The main thrust of the work described here has been the use of the external camera to measure robot pose and the subsequent use of this information, in conjunction with information computed on board the robot, to develop a motion model for the robot and a sensor error model. This is rather different to most existing work on developing vision models within RoboCup, for example [3,10,16,17,24], which has tended to concentrate on the automated segmentation of images, especially with an eye to handling changing illumination of the playing field, or work such as [21], which has concentrated on automatically identifying landmarks from sensor data.

[5] Taking due account of the orientation of the robot in that frame of reference.

[6] In fact we combined taking pictures with the motion measurements required for the motion model.

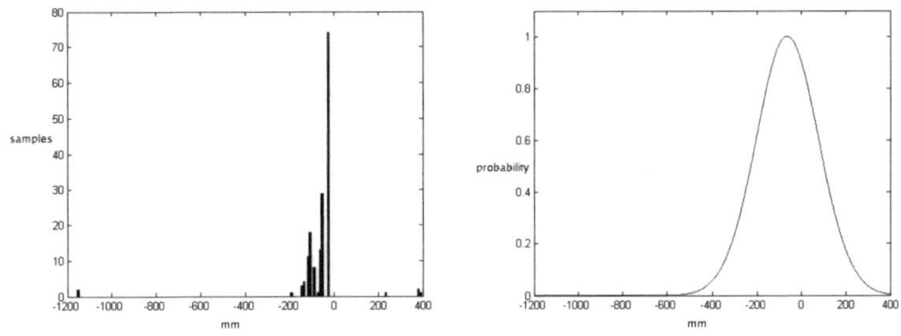

(a) (b)

Fig. 4. The sensor error model (a) histogram of the error established from the global vision; (b) the model learnt from the data (adjusted for a measurement of 1200 mm)

Of course, there are problems with using the overhead camera as a measure of "ground truth", since, as [14] points out, overhead camera-based global vision systems tend to suffer from quantization problems and are adversely affected by noise in the image. However, these problems are much reduced for us in comparison with [14] thanks to the unique beacons used by the Legged League — [14] studies the small size league setup. These beacons greatly simplify the problem of localizing a robot by uniquely anchoring points on the image. Furthermore, while an occasional error in robot localization can have catastropic effects on the way that the robot plays soccer, which is the concern in [14], in our work an error will only introduce a little more noise, and create distributions with slightly more variance.

Our work described here is clearly related to the simultaneous learning of sensor and motion models described by [18,19]. That exciting work promises to supercede what we are doing here, but for now is only capable (at least as reported in the literature) of learning models that work in the same single dimension — in the case of [18,19] that is motion towards and away from a beacon, along with sensing of the distance to the beacon. In contrast our approach, while requiring data external to the robot — which is clearly a limitation in some domains — can aquire multi-dimensional models (and so, for example, can easily acquire models for the y direction and rotation).

4 Future Work

We began this work not just to obtain data from which we could learn motion and sensor models off-line, but in order to be able to learn them on-line. In particular, we wanted to be able to run the robot, have it self-localize, and then adjust the parameters that control its motion and sensor models in order to improve its self-localization in much the way that [11] adjust parameters in order to optimize the robot gait (though clearly in a less autonomous way). This is still our aim, and we are continuing to work towards it. At the moment, as an intermediate between our overall goal and what we have reported here, we are using the experimental setup we have described to evaluate our use of particle filtering to localize the AIBO while it is playing soccer.

There have been many previous evaluations of localization. For example [7] examine a range of different probabilistic algorithms, while [6], and [20] evaluate RoboCup specific approaches, and [8,12] look at the quality of localization on the AIBO in a RoboCup setting. However, all of these use rather contrived scenarios. For example, [6] required the robot to be manually placed around the pitch in order that the true location be known, while [8] controlled the robot with a joystick and obtained measurements by moving the robot over a known location and seeing where the robot thought it was as it passed over that location. [12] comes closest to what we are working on, using a laser range-finder to monitor continuously the real location of the robot, but never carried this out during a game (the addition to the robot to allow the laser to detect the robot presumably prevented this). As a result, we have no data on the extent to which actually playing, and thus, as described above, having to focus on the ball, affects the quality of the localization.

5 Summary

This paper has described the use of a global vision system as a means of automatically acquiring motion and vision sensor data for a legged robot. Despite the fact that these models are essential in order that robots can accurately self-localize, there has been little work to try and acquire them automatically. In addition to describing the process by which we collect the data in order to construct the motion and sensor models, we have demonstrated the kinds of results that it is possible to obtain in this way. In particular, we gave two components of the motion model for an AIBO ERS-7 that we learnt in this way, and the error model for the extraction of the beacons on the Four-Legged League pitch. While the learning process currently involves some human intervention, and is run on an off-board computer, there is no especial reason why the process could not be completely automated and run on-board.

Acknowledgements. This work was partially supported by NSF CNS-0520989 and CUNY Collaborative Grant 80209-09-12. Thanks to Aleksandr Barkan for help with calibrating Mezzanine and setting up the overhead camera.

References

1. Anderson, J., Baltes, J., Livingston, D., Sklar, E., Tower, J.: Toward an undergraduate league for RoboCup. In: Polani, D., Browning, B., Bonarini, A., Yoshida, K. (eds.) RoboCup 2003. LNCS (LNAI), vol. 3020, pp. 670–677. Springer, Heidelberg (2004)
2. Baltes, J.: Yue-Fei: Object orientation and ID without additional markers. In: Birk, A., Coradeschi, S., Tadokoro, S. (eds.) RoboCup 2001. LNCS (LNAI), vol. 2377, pp. 257–262. Springer, Heidelberg (2002)
3. Cameron, D., Barnes, N.: Knowledge-based autonomous dynamic color calibration. In: Polani, D., Browning, B., Bonarini, A., Yoshida, K. (eds.) RoboCup 2003. LNCS (LNAI), vol. 3020, pp. 226–237. Springer, Heidelberg (2004)
4. (accessed February 2, 2006), `http://www.cs.cmu.edu/~robosoccer/legged/`
5. Dempster, A.P., Laird, N.M., Rubin, D.B.: Maximum likelihood from incomplete data via the EM algorithm. Journal of the Royal Statistical Society, Series B 34, 1–38 (1977)
6. Enderle, S., Ritter, M., Fox, D., Sablatnög, S., Kraetzchmar, G., Palm, G.: Soccer-robot localization using sporadic visual features. In: Proceedings of the 6th International Conference on Intelligent Autonomous Systems (2000)

7. Gutmann, J-S., Burgard, W., Fox, D., Konolige, K.: An experimental comparison of localization methods. In: Proceedings of the IEEE/RSJ International Conference on Intelligent Robots and Systems (1998)
8. Gutmann, J-S., Burgard, W., Fox, D., Konolige, K.: An experimental comparison of localization methods. In: Proceedings of the IEEE/RSJ International Conference on Intelligent Robots and Systems (2002)
9. Howard, A.: Mezzanine user manual, version 0.00. Technical Report IRIS-02-416, Institute for Robotics and Intelligent Systems, USC (2002)
10. Jungel, M., Hoffmann, J., Lötzsch, M.: A real-time auto-adjusting vision system for robot soccer. In: Polani, D., Browning, B., Bonarini, A., Yoshida, K. (eds.) RoboCup 2003. LNCS (LNAI), vol. 3020, pp. 214–225. Springer, Heidelberg (2004)
11. Kohl, N., Stone, P.: Machine learning for fast quadrupedal locomotion. In: Proceedings of the 19th National Conference on Artificial Intelligence, San Jose, CA (July 2004)
12. Röfer, T., Jüngel, M.: Fast and robust edge-based localization in the Sony four-legged robot league. In: Polani, D., Browning, B., Bonarini, A., Yoshida, K. (eds.) RoboCup 2003. LNCS (LNAI), vol. 3020, pp. 262–273. Springer, Heidelberg (2004)
13. Schmitt, T., Hanel, R., Buck, S., Beetz, M.: Probabilistic vision-based opponent tracking in robot soccer. In: Kaminka, G.A., Lima, P.U., Rojas, R. (eds.) RoboCup 2002. LNCS (LNAI), vol. 2752, pp. 426–434. Springer, Heidelberg (2003)
14. Sekimori, D., Usui, T., Masutani, Y., Miyasaki, F.: Evaluation of self-localization performance for a local vision robot in the small size league. In: Kaminka, G.A., Lima, P.U., Rojas, R. (eds.) RoboCup 2002. LNCS (LNAI), vol. 2752, pp. 41–52. Springer, Heidelberg (2003)
15. Sridharan, M., Kuhlmann, G., Stone, P.: Practical vision-based Monte-Carlo localization on a legged robot. In: Proceedings of the IEEE International Conference on Robotics and Automation, Barcelona (April 2005)
16. Sridharan, M., Stone, P.: Autonomous color learning on a mobile robot. In: Proceedings of the 20th National Conference on Artificial Intelligence, Pittsburgh, PA (July 2005)
17. Sridharan, M., Stone, P.: Towards eliminating manual color calibration at RoboCup. In: Bredenfeld, A., Jacoff, A., Noda, I., Takahashi, Y. (eds.) RoboCup 2005. LNCS (LNAI), vol. 4020, Springer, Heidelberg (2006)
18. Stronger, D., Stone, P.: Simultaneous calibration of action and sensor models on a mobile robot. In: Proceedings of the IEEE International Conference on Robotics and Automation, Barcelona (April 2005)
19. Stronger, D., Stone, P.: Towards autonomous sensor and actuator model induction on a mobile robot. Connection Science, 18(2) (to appear)
20. Stroupe, A.W., Sikorski, K., Balch, T.: Constraint-based landmark localization. In: Kaminka, G.A., Lima, P.U., Rojas, R. (eds.) RoboCup 2002. LNCS (LNAI), vol. 2752, pp. 8–24. Springer, Heidelberg (2003)
21. Thrun, S.: Bayesian landmark learning for mobile robot localization. Machine Learning 33(1), 41–76 (1998)
22. Thrun, S., Burgard, W., Fox, D.: Probabilistic robotics. MIT Press, Cambridge, MA (2006)
23. Thrun, S., Fox, D., Burgard, W., Dellaert, F.: Robust Monte Carlo localization for mobile robots. Artificial Intelligence 128(1–2), 99–141 (2001)
24. Zagal, J.C., del Solar, J.R., Guerrero, P., Palma, R.: Evolving visual object recognition for legged robots. In: Polani, D., Browning, B., Bonarini, A., Yoshida, K. (eds.) RoboCup 2003. LNCS (LNAI), vol. 3020, pp. 181–191. Springer, Heidelberg (2004)

Ambulance Decision Support Using Evolutionary Reinforcement Learning in Robocup Rescue Simulation League

Ivette C. Martínez, David Ojeda, and Ezequiel A. Zamora

Grupo de Inteligencia Artificial
Universidad Simón Bolívar
Caracas 1080-A, Venezuela
{martinez,david,ezequiel}@gia.usb.ve

Abstract. We present a complete design of agents for the RoboCup Rescue Simulation problem that uses an evolutionary reinforcement learning mechanism called XCS, a version of Holland's Genetic Classifiers Systems, to decide the number of ambulances required to rescue a buried civilian. We also analyze the problems implied by the rescue simulation and present solutions for every identified sub-problem using multi-agent cooperation and coordination built over a subsumption architecture. Our agents' classifier systems were trained in different disaster situations. Trained agents outperformed untrained agents and most participants of the 2004 RoboCup Rescue Simulation League competition. This system managed to extract general rules that could be applied on new disaster situations, with a computational cost of a reactive rule system.

1 Introduction

RoboCup Rescue has become a standard problem for the artificial intelligence, intelligent robotics and multi-agents communities. In particular, the RoboCup Rescue Simulation League problem (RCRSL) has proven to be a excellent environment for AI and Machine Learning software testing.

Tadokoro *et.al.* [9] define RCRSL as a *semi optimal behavior planning problem with extremely complex constraints having widely time-varying multiples objectives*, these constraints include limited time for decision making, limited communication, constantly changing conditions and incomplete and partial information.

In this work we decided to focus on one of the multiple challenges RCRSL offers, the victim rescue problem, which has the strongest impact on the team's performance. This problem depends on various simulation factors. We chose four of them: *world time* and victim *buriedness*, *damage* and *health points*. All this factors have a large feasible domain, which combined generate a very large state-space.

Usually large state-space problems are managed through generalization techniques such as neural networks and other function approximator; *which allow compact storage of learned information and transfer of knowledge between "similar" states and actions* [3].

G. Lakemeyer et al. (Eds.): RoboCup 2006, LNAI 4434, pp. 556–563, 2007.

In order to manage this large state-space problem under RCRSL time restrictions, we use an accuracy-based evolutionary reinforcement learning mechanism called XCS [11]. In particular the XCS decides how many Ambulance Teams are required to make an effective rescue of a victim buried in a building.

Evolutionary reinforcement learning (ERL) is an approach to reinforcement learning that takes advantage of Darwin's theory of evolution. Evolutionary algorithms can find satisfactory solutions in large state-spaces at a low computational cost. *Methods from genetic algorithms, evolutionary programming, genetic programming, and evolutionary strategies could all be used in this framework to form effective decision making agents* [5].

We compared our agents with other successful teams from the 2004 competition and obtained satisfactory results. We believe this technique is able to process the pertinent information from the environment and give the appropriate output to solve this problem. Additionally we give a short description of our RoboCup Rescue decomposition into sub-problems and the solutions we designed and implemented for each one of them.

2 XCS

The XCS classifier systems [10], as well as Holland's Learning Classifier Systems (LCS) [2], are domain independent adaptive learning systems. Its main distinguishing features are the base of classifier fitness on the accuracy of classifier reward prediction instead of the prediction itself, and the use of a niche genetic algorithm, i.e., a GA that operates on a subset of the classifier population.

The structure of *XCS* rules' conditions are the translation of the conditional part of the logical rules. Rules' actions are binary strings that represent motion actions.

A classifier is a compact representation of a complex set of environment states. Rules have the form $\langle condition \rangle \rightarrow \langle action \rangle$. Conditions are strings of length l in the alphabet $\{0, 1, *\}$. A classifier's condition satisfies a message if its condition matches the input message. A condition c matches message m if and only if: $\forall i, (1 \leq i \leq l) \rightarrow \Pi_i(c) = \Pi_i(m) \vee \Pi_i(c) = {'*'}$ [1]. Actions are fixed length strings in the alphabet $\{0, 1\}$.

XCS are composed by three subsystems: A performance system, a learning system and a rule discovery system.

The performance system takes an input from the environment, selects an action and transforms it into an output message.

The learning system takes feedback signals from the environment and updates the values of the four parameters that replaces the traditional fitness of LCS: prediction, prediction error, accuracy, and fitness. This change allows a more complete $State \times Actions \rightarrow Prediction$ mapping than traditional LCS.

The rule discovery system uses a GA in order to create new rules. XCS's rule discovery system has two operations: niche GA and covering. The niche GA acts over the Action Set $[A]$, choosing random parents in proportion to the rules'

[1] Where Π_i represents the character located at the position i of the string.

fitness. Offsprings are copies of the parents, modified by crossover and mutation. Covering is triggered when the matches set is empty or its mean prediction is a small fraction of the population [P] average prediction. Covering creates a new classifier whose condition matches the current input message and its action is generated randomly.

In the Reinforcement Learning (RL) research, two different approaches have been stated. These two approaches are known as: searching in value function space, and searching in policy space. In the first approach, RL algorithms try to find the optimum value function for the problem. Then, to find the optimal policy given the optimal values function, is immediate. The second approach is to search an optimal policy directly over the space of the policies. For this purpose, evolutionary algorithms are frequently used [5].

This RL approach based on evolutionary algorithms is called Evolutionary Reinforcement Learning (ERL). The ERL algorithms vary in terms of the policies representation method and the fitness evaluation for individual policies.

The two methods of policies' representation are: a chromosome representation and distributed rules-based representations. LCS [2] as well as XCS are examples of a rules-based ERL.

The advantage of XCS from the ERL point of view is its generalization capacity. For this reason, the XCS must be able to scale to more complex problems, in contrast with the RL traditional algorithms [11].

3 Design

We present the result of our analysis and decomposition of RoboCup Rescue into sub-problems. We use a hybrid approach for decision making, i.e. some decisions are centralized while others are taken by platoon agents. Therefore, platoon agents can take decision with slight relevance but central agents must decide the most important matters.

3.1 Problems Categories

Sub-problems were divided into four categories: Common Problems, Fire Extinction, Rubble Cleaning and Victim Rescue. Now we describe some of the identified problems and their solutions.

1. **Common Problems**

   Civilian search: Our agents look for civilians in all the buildings in the city. Each platoon must do this job when there is no other higher-priority task to do. All agents have a "world model" which they share in every turn to avoid visiting an already explored site and to provide central agents with the necessary information to make decisions.

   Route planning: This problem was solved by implementing the idea of LongRoads proposed by ResQ Freiburg [4].

Communication: In order to make information available for as many agents as possible, we designed and built a communication protocol which intends to use the messages as much as possible and gives preference to high-priority information.

2. **Victim Rescue**
   We decided to use a centralized approach for this task. The Ambulance Center must decide which victim is going to be rescued next, the number of ambulances that will be sent to the rescue site, which ambulances must go and which one takes the victim to the refuge.
   The *next victim selection* algorithm we are using is a based on the strategy proposed by *Damas Team* [6]. It goal is to minimize future casualties considering the next rescue.
   The number of ambulances for each victim is determined using an XCS classifier system whose structure and parameters are explained in Section 3.3. Once the number of ambulances is fixed, the nearest ambulances are sent to the rescue. If we do not have enough ambulances, all free ambulances are sent and the rest will be sent when they report themselves as freed. The nearest ambulance of all takes the victim to the rescue.

3. **Fire Extinction**
   The Fire Station Agent decides which fires to extinguish. It uses a set of fixed rules that chooses how many fire brigades are going to be sent to each fire and sends the nearest units. We consider a fire as a group of burning buildings relatively close to each other. Each fire is built using clustering techniques. Once the agent gets to the fire, it chooses which building is going to put out. Considering that all agents have a similar "world model", it is highly probable that the fire brigades assigned to this fire will choose the same building.

4. **Rubble Cleaning**
   We implemented several techniques for choosing the next road to be cleaned, the Police Station sorts police agents to each one at the beginning of the simulation.

   - **Road selection techniques:**
     - Select the nearest LongRoad and clean all its roads.
     - Select the roads belonging to the most frequently used LongRoads.
     - Select the roads around a certain spot of the map.
     - Select the nearest road (only when all LongRoads are passable).
   - **Cleaning requests:** If a platoon agent needs to go through a blocked area, it sends a cleaning request to the Police Station, who assigns a police force agent to clean it.

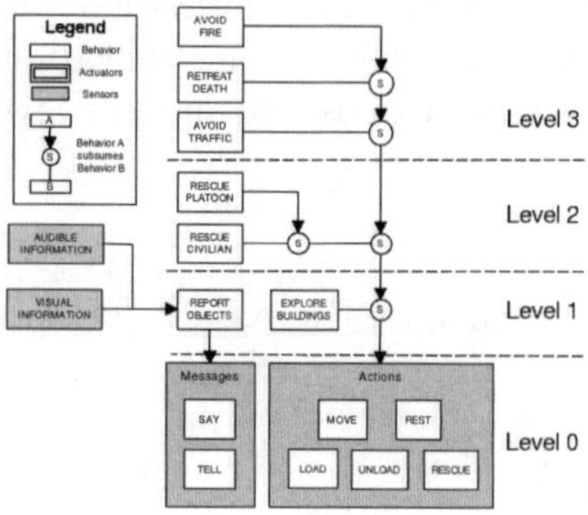

Fig. 1. Subsumption Architecture Diagram - Ambulance Teams

3.2 Subsumption Architecture

Our agents' conduct was modeled using *Brooks' Subsumption Architecture* [1]. Each kind of agent has different behaviors but they all have the same structure. We decided to sort our agent behaviors in four levels, which we describe next:

Level 0. This level contains the most basic behaviors, which are RoboCup Rescue commands and actions.
Level 1. This level contains the default behavior of agents such as building exploration and victim search.
Level 2. This defines those behaviors that are only activated by a central order such as victim rescue and extinction of fires.
Level 3. This is where the highest-priority behavior are located.

Fig. 1 shows the subsumption diagram for ambulance platoon agents. Level 3 behaviors are shared by all platoon agents, Level 2 has the behaviors that entail those agents main tasks, while Level 1 behaviors encode cooperation mechanisms.

3.3 Description of Genetic Classifiers

As we mentioned in Section 2, we use XCS genetic classifiers to support our decision making. In particular we decide how many ambulances are required to rescue a victim using this kind of system.

XCS Design for Victim Rescue. Our classifier system takes into account the following attributes: *health points*, *damage*, *buriedness* and *world time*. Each classifier contains 24 bits as shown in Fig. 2.

<table>
<tr><th colspan="8">HP/Damage</th><th colspan="7">Buriedness</th><th colspan="6">World-Time</th><th colspan="3">Output</th></tr>
<tr><td>0</td><td>1</td><td>1</td><td>1</td><td>0</td><td>0</td><td>1</td><td>0</td><td>0</td><td>1</td><td>1</td><td>0</td><td>0</td><td>0</td><td>1</td><td>0</td><td>1</td><td>0</td><td>0</td><td>0</td><td>1</td><td>1</td><td>0</td><td>1</td></tr>
</table>

Fig. 2. Ambulance Center's Classifier Structure

Bits 0 to 7 contain the ratio of the victim's health points to its damage, bits 7 to 14, its degree of buriedness, and the other 6 bits from the input represent the simulation time in order to inform the system about how much time can be used to rescue the victim. The translation between integer and binary representation is accomplished by creating predefined ranges.

Each classifier has 3 output bits that represent the number of ambulances that will be sent to rescue the victim.

4 Experiments and Results

Two different classifier sets were defined in order to train the XCS system using the parameters shown in Table 1. The rules of the first set, called *Foligno-Rules*, were generated using the classifier system on a learning phase over two maps of the same city: FolignoEasy and Foligno. The second set of rules, *Kobe-Rules*, was generated with the same procedure, only changing the maps to Kobe and KobeEasy.

Table 1. XCS and GA Parameters

XCS Parameter	Value
Genetic algorithm probability	0.2
Reinforcement update rate (α)	0.1
Min error (ε_0)	0.5
Error Penalty (n)	5
Covering Parameter	**Value**
Don't care bit probability	0.2
Initial prediction	0
Initial error	100
Initial fitness	0
Max population size($\|[P]\|$)	100

GA Parameter	Value
Replacement algorithm	Elitist
Selection algorithm	8-tournament
Crossover algorithm	One point
Mutation algorithm	One point
Mutation probability	0.02

Each classifier set was initially empty. All new rules generated by covering or by the evolutive steps of the XCS. After each simulation resulting rules were stored and used in the next simulation.

We used the percentage of alive agents and the score at the end of each simulation to measure the performance of the decision system for the rescue operations. This procedure was repeated 900 times. The analyzed data are shown in Fig. 3.1.

We selected the trained classifier that showed the highest score and alive agents at the end of the simulation. The rules used around the 250th simulation for the *Foligno-Rules* training were selected for our experiments.

In order to observe the performance of the classifier system, we selected three maps from different cities and compared the rescue task of our agents using trained rules and random rules. Each group of agents' results are a mean of 20 simulations, except for the results of ResQ Freiburg [4], DAMAS Team [6] and 5Rings [8], which were extracted from the 2004 competition logs.

Fig. 3.2 shows the results of the experiments on all maps. It can be noticed that on all test cases our agents show better results when using the trained set of rules. These agents achieve higher scores and percentage of alive agents than the ones using randomly generated rules. This demonstrates that the evolutionary reinforcement learning system tends to refine and keep better rules for the XCS.

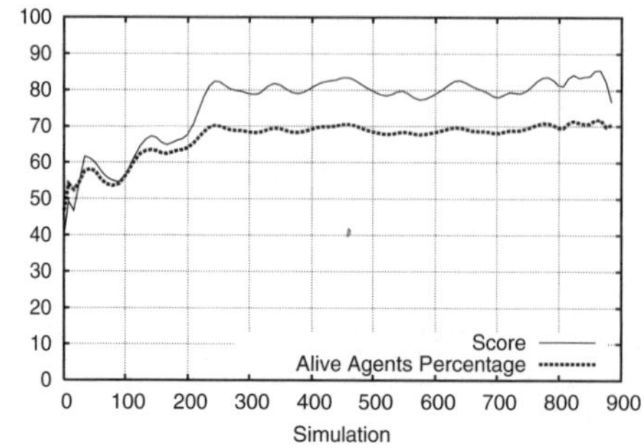

3.1. Score and Alive Agents Evolution over *Foligno-Rules* Training

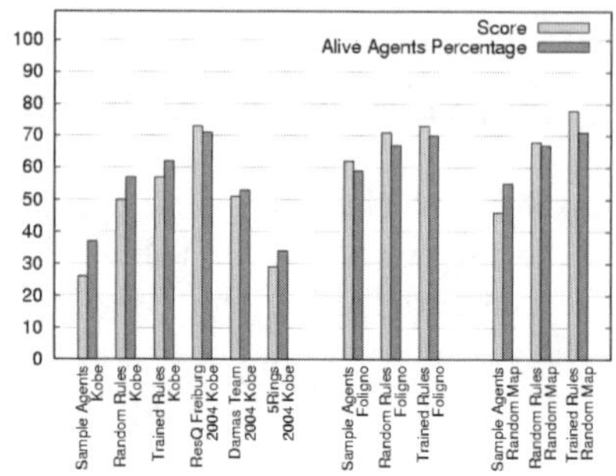

3.2. Teams result over Kobe, Foligno and Random maps

Our trained agents also managed to rescue more agents than DAMAS Team and outperformed the 5Rings agents. However, ResQ Freiburg agents can solve the victim rescue problem with better results by using all available Ambulance Teams to rescue a civilian and choosing the rescue order with a GA [4].

5 Concluding Remarks and Future Work

This paper presents an approximation to the RoboCup Rescue simulation problem that uses an evolutionary reinforcement learning technique, particularly XCS classifier systems, to support the decision making of a central agent that coordinates several platoon agents on the complicated victim rescue task.

The agents can solve numerous coordination problems presented by RoboCup Rescue using a distributive coordinated search for civilians, as well as road cleaning, and a centralized coordination for victim rescue and fire extinction.

Many ideas and approaches used by our agents are based on previous studies and agent teams. These included informed search using LongRoads, victim selection minimizing future casualties, building clustering, token-based communications, distributed civilian search and road unblocking petitions.

The design proposed by this paper proved to be an effective solution to the problem and is competitive with other agent teams. Reinforcement learning techniques proved to be a feasible method to extract general rules that can support decision making on RoboCup Rescue. In particular, the number of ambulances needed to save a victim depends on several non-predictable factors; this study found that a trained classifier system provides a good approximation at a low computational cost. We conclude that evolutionary reinforcement approaches are appropriate for the RoboCup Rescue domain.

Even though this paper presents a successful decision support system and a design for a RoboCup Rescue agents team, further development is needed to present these agents on a competition. The agents must be revised in order to assure compatibility with the current competition rules, since our agents were developed and tested with the rules published in 2004.

Parameters of the GA should be examined in future studies. Determination of which mutation and crossover strategies work best with this problem should be considered.

The current design of the XCS classifier system for rescue task is currently very simple. An extension of the elements taken into account by the rules shall outperform the current system with a longer training procedure tradeoff.

We are currently meticulously studying an appropriate design for a XCS that can determine the number of fire brigades needed to control and extinguish a fire.

References

1. Brooks, R.: How to build complete creatures rather than isolated cognitive simulators. Architectures for Intelligence, 225–240 (1991)
2. Holland, J.H.: Escaping Brittleness: The possibilities of general-purpose learning algorithms applied to parallel rule-based systems. In: Mitchell, Michalski, Carbonell (eds.) Machine Learning: an artificial intelligence approach, Morgan Kaufman, San Francisco (1986)
3. Littman, K.L.P., Moore, A.P: Reinforcement learning: A survey. Journal of Artificial Intelligence Research, 237–285 (1996)
4. Kleiner, A., Brenner, M., Bräuer, T., Dornhege, C., Göbelbecker, M., Luber, M., Prediger, J., Stückler, J.: ResQ Freiburg: Team Description Paper and Evaluation. RoboCupRescue simulation league (2004)
5. Moriarty, D.E., Schultz, A.C., Grefenstette, J.J.: Evolutionary Algorithms for Reinforcement Learning. Journal of Artificial Intelligence Research. 11, 199–229 (1999)
6. Paquet, S., Bernier, N., Chaib-draa, B.: DAMAS-Rescue Description Paper. RoboCupRescue simulation league (2004)
7. RoboCup Rescue Commitee: Robocup 2004 Rescue Simulation League Official Home Page (2004), `http://robot.cmpe.boun.edu.tr/rescue2004`
8. Silva, P., Coelho, H.: The 5Rings Team Report. RoboCupRescue simulation league (2004)
9. Tadokoro, S., Kitano, H., Takahashi, T., Noda, I., Matsubara, H., Shinjoh, A., Koto, T., Takeuchi, I., Takahashi, H., Matsuno, F., Hatayama, M., Nobe, J., Shimada, S.: The RoboCup-Rescue Project: A Robotic Approach to the Disaster Mitigation Problem. ICRA (2000), `http://www.rescuesystem.org/robocuprescue/icra00resc.pdf`
10. Wilson, S.: Classifier Fitness Based on Accuracy. Evolutionary Computation Journal, 149–175 (1995)
11. Wilson, S.: Generalization in the XCS Classifier System Genetic Programming 1998. In: Proceedings of the Third Annual Conference, pp. 665–674. Morgan Kaufmann, San Francisco (1998)

Author Index

Printing: Mercedes-Druck, Berlin
Binding: Stein+Lehmann, Berlin

Lecture Notes in Artificial Intelligence (LNAI)

Vol. 4660: S. Džeroski, J. Todoroski (Eds.), Computational Discovery of Scientific Knowledge. X, 327 pages. 2007.

Vol. 4651: F. Azevedo, P. Barahona, F. Fages, F. Rossi (Eds.), Recent Advances in Constraints. VIII, 185 pages. 2007.

Vol. 4632: R. Alhajj, H. Gao, X. Li, J. Li, O.R. Zaïane (Eds.), Advanced Data Mining and Applications. XV, 634 pages. 2007.

Vol. 4617: V. Torra, Y. Narukawa, Y. Yoshida (Eds.), Modeling Decisions for Artificial Intelligence. XII, 502 pages. 2007.

Vol. 4612: I. Miguel, W. Ruml (Eds.), Abstraction, Reformulation, and Approximation. XI, 418 pages. 2007.

Vol. 4604: U. Priss, S. Polovina, R. Hill (Eds.), Conceptual Structures: Knowledge Architectures for Smart Applications. XII, 514 pages. 2007.

Vol. 4603: F. Pfenning (Ed.), Automated Deduction – CADE-21. XII, 522 pages. 2007.

Vol. 4597: P. Perner (Ed.), Advances in Data Mining. XI, 353 pages. 2007.

Vol. 4594: R. Bellazzi, A. Abu-Hanna, J. Hunter (Eds.), Artificial Intelligence in Medicine. XVI, 509 pages. 2007.

Vol. 4585: M. Kryszkiewicz, J.F. Peters, H. Rybinski, A. Skowron (Eds.), Rough Sets and Intelligent Systems Paradigms. XIX, 836 pages. 2007.

Vol. 4578: F. Masulli, S. Mitra, G. Pasi (Eds.), Applications of Fuzzy Sets Theory. XVIII, 693 pages. 2007.

Vol. 4573: M. Kauers, M. Kerber, R. Miner, W. Windsteiger (Eds.), Towards Mechanized Mathematical Assistants. XIII, 407 pages. 2007.

Vol. 4571: P. Perner (Ed.), Machine Learning and Data Mining in Pattern Recognition. XIV, 913 pages. 2007.

Vol. 4570: H.G. Okuno, M. Ali (Eds.), New Trends in Applied Artificial Intelligence. XXI, 1194 pages. 2007.

Vol. 4565: D.D. Schmorrow, L.M. Reeves (Eds.), Foundations of Augmented Cognition. XIX, 450 pages. 2007.

Vol. 4562: D. Harris (Ed.), Engineering Psychology and Cognitive Ergonomics. XXIII, 879 pages. 2007.

Vol. 4548: N. Olivetti (Ed.), Automated Reasoning with Analytic Tableaux and Related Methods. X, 245 pages. 2007.

Vol. 4539: N.H. Bshouty, C. Gentile (Eds.), Learning Theory. XII, 634 pages. 2007.

Vol. 4529: P. Melin, O. Castillo, L.T. Aguilar, J. Kacprzyk, W. Pedrycz (Eds.), Foundations of Fuzzy Logic and Soft Computing. XIX, 830 pages. 2007.

Vol. 4511: C. Conati, K. McCoy, G. Paliouras (Eds.), User Modeling 2007. XVI, 487 pages. 2007.

Vol. 4509: Z. Kobti, D. Wu (Eds.), Advances in Artificial Intelligence. XII, 552 pages. 2007.

Vol. 4496: N.T. Nguyen, A. Grzech, R.J. Howlett, L.C. Jain (Eds.), Agent and Multi-Agent Systems: Technologies and Applications. XXI, 1046 pages. 2007.

Vol. 4483: C. Baral, G. Brewka, J. Schlipf (Eds.), Logic Programming and Nonmonotonic Reasoning. IX, 327 pages. 2007.

Vol. 4482: A. An, J. Stefanowski, S. Ramanna, C.J. Butz, W. Pedrycz, G. Wang (Eds.), Rough Sets, Fuzzy Sets, Data Mining and Granular Computing. XIV, 585 pages. 2007.

Vol. 4481: J. Yao, P. Lingras, W.-Z. Wu, M. Szczuka, N.J. Cercone, D. Ślęzak (Eds.), Rough Sets and Knowledge Technology. XIV, 576 pages. 2007.

Vol. 4476: V. Gorodetsky, C. Zhang, V.A. Skormin, L. Cao (Eds.), Autonomous Intelligent Systems: Multi-Agents and Data Mining. XIII, 323 pages. 2007.

Vol. 4455: S. Muggleton, R. Otero, A. Tamaddoni-Nezhad (Eds.), Inductive Logic Programming. XII, 456 pages. 2007.

Vol. 4452: M. Fasli, O. Shehory (Eds.), Agent-Mediated Electronic Commerce. VIII, 249 pages. 2007.

Vol. 4451: T.S. Huang, A. Nijholt, M. Pantic, A. Pentland (Eds.), Artifical Intelligence for Human Computing. XVI, 359 pages. 2007.

Vol. 4438: L. Maicher, A. Sigel, L.M. Garshol (Eds.), Leveraging the Semantics of Topic Maps. X, 257 pages. 2007.

Vol. 4434: G. Lakemeyer, E. Sklar, D.G. Sorrenti, T. Takahashi (Eds.), RoboCup 2006: Robot Soccer World Cup X. XIII, 566 pages. 2007.

Vol. 4429: R. Lu, J.H. Siekmann, C. Ullrich (Eds.), Cognitive Systems. X, 161 pages. 2007.

Vol. 4426: Z.-H. Zhou, H. Li, Q. Yang (Eds.), Advances in Knowledge Discovery and Data Mining. XXV, 1161 pages. 2007.

Vol. 4411: R.H. Bordini, M. Dastani, J. Dix, A.E.F. Seghrouchni (Eds.), Programming Multi-Agent Systems. XIV, 249 pages. 2007.

Vol. 4410: A. Branco (Ed.), Anaphora: Analysis, Algorithms and Applications. X, 191 pages. 2007.

Vol. 4399: T. Kovacs, X. Llorà, K. Takadama, P.L. Lanzi, W. Stolzmann, S.W. Wilson (Eds.), Learning Classifier Systems. XII, 345 pages. 2007.

Vol. 4390: S.O. Kuznetsov, S. Schmidt (Eds.), Formal Concept Analysis. X, 329 pages. 2007.

Vol. 4389: D. Weyns, H.V.D. Parunak, F. Michel (Eds.), Environments for Multi-Agent Systems III. X, 273 pages. 2007.

Vol. 4384: T. Washio, K. Satoh, H. Takeda, A. Inokuchi (Eds.), New Frontiers in Artificial Intelligence. IX, 401 pages. 2007.

Vol. 4371: K. Inoue, K. Satoh, F. Toni (Eds.), Computational Logic in Multi-Agent Systems. X, 315 pages. 2007.

Vol. 4369: M. Umeda, A. Wolf, O. Bartenstein, U. Geske, D. Seipel, O. Takata (Eds.), Declarative Programming for Knowledge Management. X, 229 pages. 2006.

Vol. 4342: H. de Swart, E. Orłowska, G. Schmidt, M. Roubens (Eds.), Theory and Applications of Relational Structures as Knowledge Instruments II. X, 373 pages. 2006.

Vol. 4335: S.A. Brueckner, S. Hassas, M. Jelasity, D. Yamins (Eds.), Engineering Self-Organising Systems. XII, 212 pages. 2007.

Vol. 4334: B. Beckert, R. Hähnle, P.H. Schmitt (Eds.), Verification of Object-Oriented Software. XXIX, 658 pages. 2007.

Vol. 4333: U. Reimer, D. Karagiannis (Eds.), Practical Aspects of Knowledge Management. XII, 338 pages. 2006.

Vol. 4327: M. Baldoni, U. Endriss (Eds.), Declarative Agent Languages and Technologies IV. VIII, 257 pages. 2006.

Vol. 4314: C. Freksa, M. Kohlhase, K. Schill (Eds.), KI 2006: Advances in Artificial Intelligence. XII, 458 pages. 2007.

Vol. 4304: A. Sattar, B.-h. Kang (Eds.), AI 2006: Advances in Artificial Intelligence. XXVII, 1303 pages. 2006.

Vol. 4303: A. Hoffmann, B.-h. Kang, D. Richards, S. Tsumoto (Eds.), Advances in Knowledge Acquisition and Management. XI, 259 pages. 2006.

Vol. 4293: A. Gelbukh, C.A. Reyes-Garcia (Eds.), MICAI 2006: Advances in Artificial Intelligence. XXVIII, 1232 pages. 2006.

Vol. 4289: M. Ackermann, B. Berendt, M. Grobelnik, A. Hotho, D. Mladenič, G. Semeraro, M. Spiliopoulou, G. Stumme, V. Svátek, M. van Someren (Eds.), Semantics, Web and Mining. X, 197 pages. 2006.

Vol. 4285: Y. Matsumoto, R.W. Sproat, K.-F. Wong, M. Zhang (Eds.), Computer Processing of Oriental Languages. XVII, 544 pages. 2006.

Vol. 4274: Q. Huo, B. Ma, E.-S. Chng, H. Li (Eds.), Chinese Spoken Language Processing. XXIV, 805 pages. 2006.

Vol. 4265: L. Todorovski, N. Lavrač, K.P. Jantke (Eds.), Discovery Science. XIV, 384 pages. 2006.

Vol. 4264: J.L. Balcázar, P.M. Long, F. Stephan (Eds.), Algorithmic Learning Theory. XIII, 393 pages. 2006.

Vol. 4259: S. Greco, Y. Hata, S. Hirano, M. Inuiguchi, S. Miyamoto, H.S. Nguyen, R. Słowiński (Eds.), Rough Sets and Current Trends in Computing. XXII, 951 pages. 2006.

Vol. 4253: B. Gabrys, R.J. Howlett, L.C. Jain (Eds.), Knowledge-Based Intelligent Information and Engineering Systems, Part III. XXXII, 1301 pages. 2006.

Vol. 4252: B. Gabrys, R.J. Howlett, L.C. Jain (Eds.), Knowledge-Based Intelligent Information and Engineering Systems, Part II. XXXIII, 1335 pages. 2006.

Vol. 4251: B. Gabrys, R.J. Howlett, L.C. Jain (Eds.), Knowledge-Based Intelligent Information and Engineering Systems, Part I. LXVI, 1297 pages. 2006.

Vol. 4248: S. Staab, V. Svátek (Eds.), Managing Knowledge in a World of Networks. XIV, 400 pages. 2006.

Vol. 4246: M. Hermann, A. Voronkov (Eds.), Logic for Programming, Artificial Intelligence, and Reasoning. XIII, 588 pages. 2006.

Vol. 4223: L. Wang, L. Jiao, G. Shi, X. Li, J. Liu (Eds.), Fuzzy Systems and Knowledge Discovery. XXVIII, 1335 pages. 2006.

Vol. 4213: J. Fürnkranz, T. Scheffer, M. Spiliopoulou (Eds.), Knowledge Discovery in Databases: PKDD 2006. XXII, 660 pages. 2006.

Vol. 4212: J. Fürnkranz, T. Scheffer, M. Spiliopoulou (Eds.), Machine Learning: ECML 2006. XXIII, 851 pages. 2006.

Vol. 4211: P. Vogt, Y. Sugita, E. Tuci, C.L. Nehaniv (Eds.), Symbol Grounding and Beyond. VIII, 237 pages. 2006.

Vol. 4203: F. Esposito, Z.W. Raś, D. Malerba, G. Semeraro (Eds.), Foundations of Intelligent Systems. XVIII, 767 pages. 2006.

Vol. 4201: Y. Sakakibara, S. Kobayashi, K. Sato, T. Nishino, E. Tomita (Eds.), Grammatical Inference: Algorithms and Applications. XII, 359 pages. 2006.

Vol. 4200: I.F.C. Smith (Ed.), Intelligent Computing in Engineering and Architecture. XIII, 692 pages. 2006.

Vol. 4198: O. Nasraoui, O. Zaïane, M. Spiliopoulou, B. Mobasher, B. Masand, P.S. Yu (Eds.), Advances in Web Mining and Web Usage Analysis. IX, 177 pages. 2006.

Vol. 4196: K. Fischer, I.J. Timm, E. André, N. Zhong (Eds.), Multiagent System Technologies. X, 185 pages. 2006.

Vol. 4188: P. Sojka, I. Kopeček, K. Pala (Eds.), Text, Speech and Dialogue. XV, 721 pages. 2006.

Vol. 4183: J. Euzenat, J. Domingue (Eds.), Artificial Intelligence: Methodology, Systems, and Applications. XIII, 291 pages. 2006.

Vol. 4180: M. Kohlhase, OMDoc – An Open Markup Format for Mathematical Documents [version 1.2]. XIX, 428 pages. 2006.

Vol. 4177: R. Marín, E. Onaindía, A. Bugarín, J. Santos (Eds.), Current Topics in Artificial Intelligence. XV, 482 pages. 2006.

Vol. 4160: M. Fisher, W. van der Hoek, B. Konev, A. Lisitsa (Eds.), Logics in Artificial Intelligence. XII, 516 pages. 2006.

Vol. 4155: O. Stock, M. Schaerf (Eds.), Reasoning, Action and Interaction in AI Theories and Systems. XVIII, 343 pages. 2006.